FLORA OF GUNTUR DISTRICT
Andhra Pradesh, India

T. Pullaiah
V. Ramakrishnaiah
S. Sandhya Rani
Department of Botany
Sri Krishnadevaraya University
Anantapur 515003

and

P.N. Rao
Department of Botany
Nagarjuna University
Nagarjuna Nagar 522 510
Guntur

Regency Publications
New Delhi

ISBN 81-87498-16-1

Published by Regency Publications, 20/36-G, Old Market, West Patel Nagar, New Delhi 110 008 and printed at Radiant Printers, New Delhi 110 008, Phones: 571 2539, 574 0038; Telefax: 578 3571.

Website: *www.regencypublications.com*
Email: *regency@satyam.net.in*

PREFACE

A district flora of a state is a macro-unit in the assessment of our regional biodiversity. The flora of Guntur is an addition to the data base which ever needs to be enriched. Nearly 70 years back Mr. Lemon Uhl, in his Guntur flora contributed to an annotated notes on plants of Guntur town. Subsequently there is no authentic flora embracing Guntur city not to speak of Guntur District except for a few scattered reports.

The Guntur district is located in the heart of coastal Andhra Pradesh. It has a diversity of soils, ground water levels, erratic rainfall, records of high summer temperatures, minor altitudinal variation, an attractive coastline and an extensively rich and fertile plains which are centers of cultivation of variety of cereals, pulses, vegetables, fruits, sugar cane, chilies, and tobacco. The topography further shows low lying, deciduous, hillock-scrub jungles along the bank of river Krishna on the Eastern side of the gigantic Nagarjunasagar dam and at the same time there are barren waste lands with very deep ground water levels. The heterogenous physiography is bound to decide the floristic content. Further, industrialisation, urbanisation, laying of new railway tract, highways, construction of a fishing harbour at Nizampatnam, spread of modern information net work far and wide and other anthropogenic factors influence the evolution of the Guntur District Flora. Afforestation with plants as sources of timber and NTFP, conservation of exiting resources and increased action by plant introduction, ornamentals and weeds, cultivation of High Density Energy plantations to harness waste lands all provide great scope to assess floristic wealth of Yesterday, Today and Tomorrow. Herbal renaissance in medicine bloomed all over and influence this District as well. Educationally, intellectually and culturally, the district is famous. There is considerable degree of tourist interest also.

Against this back drop, an up to date and authentic work on the Flora of Guntur District is sure to cater to the needs of students, teachers, personnel of the Department of Agriculture, Forestry, NG Organizations, Vana Samrarakshana Samithis (VSS), herbopathic physicians, tourists and other public. There is a long felt, pressing need for a regional flora of Guntur District.

The content of the book adhered to the basic conventional format of a floristic work and is self explanatory. Suggestions for improvement are welcome.

The authors are thankful to the personnel of Forest Department for their help in field trips. They also thank the authorities of Botanical Survey of India, Southern Circle, Coimbatore, Northern Circle, Dehradun and Central National Herbarium for permitting to consult their herbaria and libraries. Authors express their heartfelt thanks to Dr. Vatsavaya S. Raju, Dr. B.V. Raghavaswamy, Dr. D.L. Reddy, Dr. K.K. Neelai, Dr. K. Sri Ramamurthy and Mr. D. Muralidhara Rao for their help in various ways.

Authors

CONTENTS

Chapter 1

INTRODUCTION

Most of the 1700 million hectares of tropical forests, rich in biodiversity are located in poor countries. While such forests covered barely 7 per cent of land surface, they harbour half of the species of the world's flora and fauna. These forests are being cut down at an alarming rate. According to WWF estimate, 17 million hectares of forests are plundered every year world wide. This rich resource is lost even before a basic assessment is satisfactorily completed. Hence, there is an urgent necessity to explore, assess and document this wealth.

India is a vast country endowed with rich floristic biodiversity. It is of paramount importance for a developing country like ours to particularly harness and conserve these plant resources for the welfare of mankind. Though India is very rich in its plant wealth, the exploration of plant resources in India has not been taken up fully intensively or extensively but exposed limitless gaps in our data base.

The importance of district floras has been emphasized by several taxonomists. Jain (1978) pointed out that the district floras serve as local identification manuals. Micro level floristic survey of a smaller region opens up immense possibilities in enriching the knowledge of plant wealth from grass root level. Such floras of smaller limited geographical regional areas are often referred to as 'Florulas'.

Basic data on plant resources of Guntur district is fragmentary. Foundation data for researchers, teachers, students, visitors, foresters and district administration is highly needed. The district has active Botanical Research Centre at Nagarjuna University established in 1976. There are more than 80 colleges offering Plant Science course with relevance to floristics and taxonomy. A course on Medical Botany is offered at Under-Graduate level as a special paper in many colleges. Eco-floristics is one of the thrust areas in Department of Botany, Nagarjuna University. There is a Pulse and Chilly Research Centre at Lam Farm area of Andhra Pradesh Agricultural University and Weed Science section at Bapatla, APAU. Guntur city is the

Headquarters of Indian Leaf Tobacco Company Limited (ILTD) with a strong Research Wing and Cotton Corporation of India. Clean and Green is the recent motto. Hence awareness of Green Wealth, its conservation and consolidation are imperative. To achieve this, preparation of an authentic and up to date plant inventory is essential.

Though work on Flora of Andhra Pradesh (Pullaiah, 1997; Pullaiah and Ali Moulali, 1997, Pullaiah and Chennaiah, 1997; Pullaiah and Surya Prakash babu, 1998) has been published embracing the plant wealth of the vast state, the present work has been undertaken to investigate the plant wealth of Guntur district, keeping in view the objectives cited. It is also advisable to consult published or unpublished floras of the neighbouring districts to obtain a further picture of the continuity of floristic wealth. Reference to Flora of Nalgonda district (P.N. Rao *et al.* in press), Krishna district (Lakshminarayana *et al.*, 1997), Prakasam district (Krishnamohan, 1986; Bhiravamurthy and Krishnamohan, 1991), Mahabubnagar (Raghava Rao, 1987) helps in a better assessment of occurrence of a species of the 'District' a segment which is only an administrative geographical unit earmarked for operational convenience.

Each species is likely to show enormous infraspecific biodiversity. It is in fact a fundamental evolutionary unit. Information on prevailing variation pattern within a species, if met with, may be communicated to the authors by various individual researchers for future incorporation in later editions of this book.

The district is named after its major city Guntur, which is elevated now to a Corporation status. The name Guntur is pronounced to have been derived from the word 'gunta' meaning 'pond'. The original village having been situated on what used to be a 'gunta' (pond) and hence it was called Gunta uru (pond village) or Guntur. Another etymological explanation is that the village was on a 'Kunta'-a local unit for measuring land and it came to be called 'Kunta uru or Guntur'.

The Guntur district was first formed in 1794 A.D. with 14 taluks. It was restructured in 1904 with the following 9 taluks—Guntur, Bapatla, Repalle, Ongole, Vinukonda, Sattenapalle, Narasaraopet and Palnadu. In 1970, the taluk of Ongole, part of Bapatla (Chirala) and Narasaraopet were transferred to newly formed Prakasam district. The taluk of Palnadu was divided into Macherla, Gurajala and Piduguralla. Now the district comprises three Revenue Divisions and 57 Revenue Mandals as follows:

1. **Guntur Revenue Division** (19 Mandals).

1) Amaravathi, 2) Tadikonda, 3) Tulluru, 4) Tadepalli, 5) Mangalagiri, 6) Vatticherukum, 7) Prathipadu, 8) Pedakakani, 9) Guntur, 19) Medikonduru, 11) Phirangipuram, 12) Sattenapalli, 13) Pedakurapadu, 14) Achampeta, 15) Muppalla, 16) Krosuru, 17) Rajupalem, 18) Bellamkonda, 19) Pedanandipadu.

2. Tenali Revenue Division (18 Mandals)

1) Duggirala, 2) Tenali, 3) Vemuru. 4) Amruthalur, 5) Kollipara, 6) Chunduru, 7) Chebrolu, 8) Kolluru, 9) Ponnuru, 10) Pittalavani palem, 11) Kakumanu, 12) Bapatla, 13) Karlapalem, 14) Nagaram, 15) Nizampatnam, 16) Cherukupalli, 17) Battiprolu, 18) Repalle.

3. Narasaraopet Revenue Division (20 Mandals)

1) Veldurthi, 2) Macherla, 3) Durgi, 4) Rentachintala 5) Karempudi, 6) Gurajala, 7) Dachepalle, 8) Machavaram, 9) Piduguralla, 10) Ipuru, 11) Bollapalli, 12) Vinukonda, 13) Nadendla, 14) Savalyapuram, 15) Rompicharla, 16, Nakrekal, 17) Narasaraopet, 18) Chilkaluripet,

The publication of the Flora of Guntur District gains greater significance in view of the valuable aspect of tourism, a thrust area of high priority of the State Government of Andhra Pradesh. The following are important places of tourism and pilgrimage in the heart of plant wealth.

1. Undavilli Caves: Near the western part of the district beyond which the Krishna river bank occurs. The caves are on a hillock with impressive greenary during monsoon.
2. Mangalagiri: Thousands of pilgrims throng the town throughout the year but especially in December. A temple is located on a hillock—a repository of valuable medicinal species.
3. Amaravati: (a) Abode of Lord Shiva—considered as one of the twelve Jyothirlingas by some—is located on Guntur District side bank of river Krishna. (b) There is a famous archaeological museum also. Mahayana .buddhism flourished here.
4. Nizampatnam, Adavula Divi, Hamsala Divi—Mangrove forest areas on the interior eastern coast of the district—the landscape of the unique mangrove ecosystem is a treat to watch.
5. Right bank of Nagarjunasagar Dam with its green cover—named after the great saint and scientific philosopher Nagarjuna—also a promoter of herbal medicine.
6. The little known Ethipothala water falls are rich in biodiversity of species—30–40 Kms from Macherla.

A floristic work, thus may find additional use for tourists interested in Plant diversity of the area.

Chapter 2

TOPOGRAPHY AND GENERAL FEATURES

Geographical Position

Guntur, one of the 9 coastal districts of Andhra Pradesh, lies between 15° 18' and 16° 50' of the Northern latitudes and 79° 10' and 80° 55' of the Eastern longitudes (Figs. 1, 2). The district which is in the middle of coastal Andhra Pradesh, is bounded on the north by Krishna and Nalgonda districts, west by Mahaboob Nagar and Kurnool districts, south by Prakasam district and east by Bay of Bengal. The district is over arched by Krishna river on west, and north. Guntur district with an area of 11,400 Sq. Km occupies 13th position in the State and 3rd position in coastal districts. The shape of the district resembles roughly the picture of an open umbrella spread out with the handle pointing out southwards.

The district can broadly be dividied into three regions:

1. The delta region comprising the eastern coastal part
2. The stony uplands and hills consisting of the southern and western parts of the district.
3. The open plains covering the rest of district abounding in black cotton soils.

Hills and Elevation

The principal hill ranges of the district are the Nallamalais, the Venkatayapalem range and the Kondavedu hills.

1. The Nallamalai off-shoots

The lofty Nallamalais of the Kurnool district with their 'U' shaped terminal off-shoots skirt the Palnadu area. The highest point on the range is known as 'Swamikonda' or 'Vami konda' (605 m), a flat-topped hill on the Macherla-

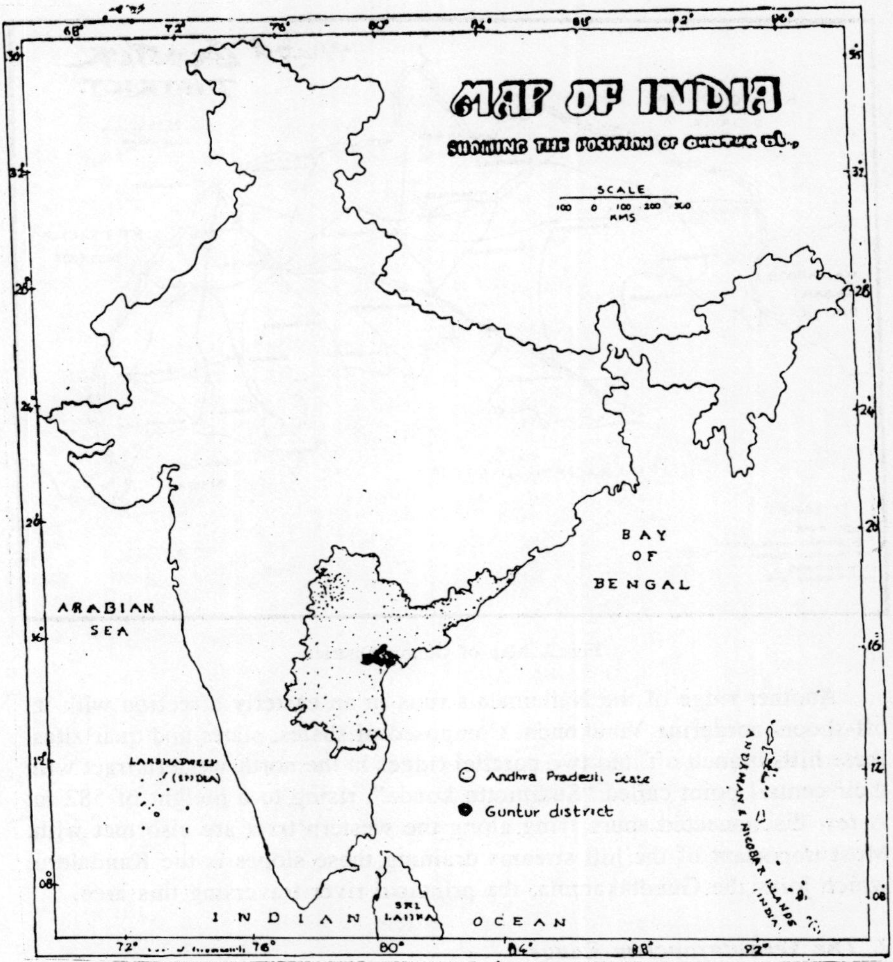

Fig. 1. Map of India showing the Position of Guntur District

Yerragondapalem range. The other peak of appreciable height in this stretch is Kairalakonda (590 m). A north-western edge of the Nallamalai off-shoot runs in an easterly direction along the Krishna river until it reaches Mallavaram. The hills in this portion are mostly composed of slates and quartzites.

An important water fall in this part of the country is "Ethipothala" situated on a range of hills (Nallamalais) on the rivulet Chandravanka about 10 Km from Macherla. The water falls registering a height of about 21 metres, are very pretty to look at.

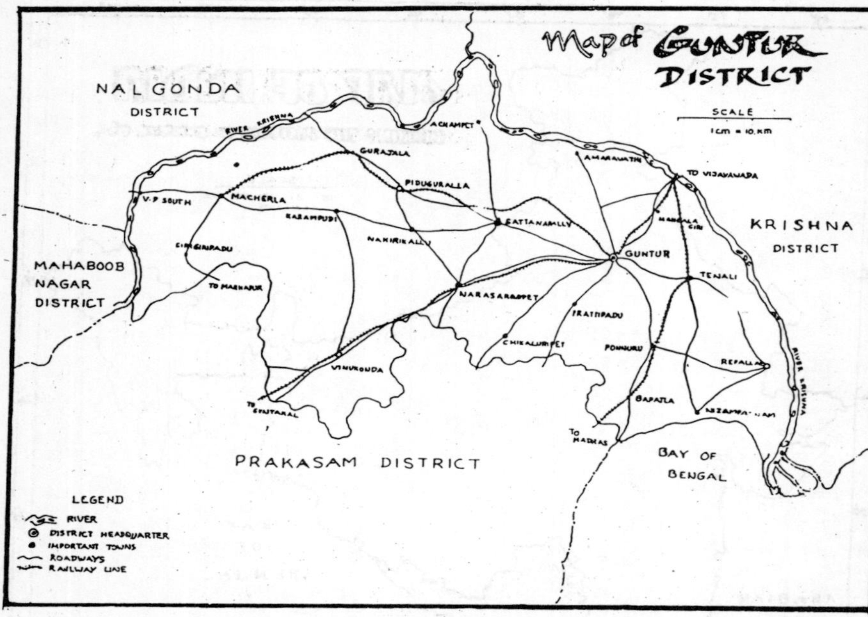

Fig. 2. Map of Guntur District

Another ridge of the Nallamalais runs in an easterly direction with its off-shoots bordering Vinukonda. Composed of shales, slates and quartzites, these hills branch off into two parallel ridges in the north-western tract with their central point called "Sudimottu konda", rising to a hieight of 582 m. A few disconnected spurs lying along the western tract are also met with. Most important of the hill streams draining these slopes is the Kandaleru, which joins the Gundlakamma, the principal river traversing this area.

2. The Venkatayapalem Range

The next important hill range is Venkatayapalem named after a village in Sattenapally range and is composed of slates and quartzites. It runs for a length of about 40 Km in north-east to south-west direction. It has long been known for the ancient working of diamonds. Its highest point is 'Maidarsal' (447 m). At Narasaraopet, where the borders of Palnadu, Vinukonda and Sattenapalle converge, a continuous chain of hills tapering gradually towards the Krishna river is in evidence.

3. The Kondavedu Range

The Kondavedu range near Narasaraopet is composed of granite rock and extends about 19 Km, registering a maximum height of 523 m. A few kilometers to its west is the isolated hill of Yellamanda (489 m), otherwise known as 'Kotappakonda'. To its south are detached hills spreading themselves towards Addanki. Some of the isolated peaks projecting themselves into the area lying within the Krishna basin between Mangalagiri and Vijayawada can also be seen especially towards the north-west. The other hills of some historic antiquity and religious association are Guttikonda, Mangalagiri and Undavilli.

Rivers and Drainage System

The important rivers that traverse the district are the Krishna, the Chandravanka and the Naguleru.

1. The Krishna

The Krishna, one of the mighty rivers in Peninsular India rises about 64 Km from the Arabian sea in the Western Ghats, north of the hill station of Mahabaleshwar in Maharashtra State. Taking a south-easterly course through the States of Maharashtra and Karnataka, it forms the border between the districts of Mahaboob Nagar and Kurnool and enters the Guntur district at the south-western portion of Macherla range near the Ganikonda hills at an elevation of roughly 182 m above sea leavel. It runs through a gorge in a broad northerly direction, separating Macherla from Achampet in Mahaboobnagar. It then flows forward collecting in quick succession, the Jarrivagu from its right bank and the Dindi river from its left. Taking loop it winds its way in a north-easterly direction through the spurs of the Nallamalais. It then placidly traverses along the ancient Buddhist shrine of Nagarjunakonda. Widening considerably almost for 1.6 Km, it encompasses a sizeable island to the east of which the mighty Nagarjunasagar dam has ben constructed. Passing by a ridge of low hills, the river receives the Chandravanka, Golivagu, Dandi vagu and Naguleru, which fertilize the adjoinging tracts. It continues its easterly course through Venkatayapalem range and scales the slopes of Gudimettakonda and broadens gradually flowing with a wide and shallow bed in an easterly direction with a slight inclination towards the south, it touches Vijayawada and then runs eastwards and southwards. It ultimately turns sharply south-eastwards and joins the sea at Hamsaladivi and Nachugunta in Krishna district.

2. The Chandravanka

The Chandravanka is an important tributary of the Krishna. It rises in the Mutukuru extensions of the eastern Nallamalais near the south-west corner of Macherla Range. It initially contacts Mutukuru village and flows north until it is joined by its tributary the "Edibogula vagu" at a place where the Atmakur tank project is constructed. Emerging from the tank it takes a north-westerly course until it contacts Macherla and flows to the north. Before it enters the Thumurukota reserve forest, it falls from a height of about 21 metres and is popularly called "Ethipothala" water falls. The stream continues to run northwards and then joins the Krishna river.

3. The Naguleru

Another tributary of the Krishna is the Naguleru, which rises in the Nallamalais off-shoots near the Nayakurali pass in Vinukonda range. It contacts initially Karempudi and then flows northwards across Macherla range over a distance of 32 Km till it joins the Krishna near Ramapuram.

Other Minor Streams

Among the other minor hill streams and rivulets in the district, the most important are the Gundlavagu near Durgi, the Golivagu near Rentachinthala, the Dandivagu near Gurazala.

Drainage System

Apart from the above mentioned rivers and streams, the Romperu drainage basin, the Tungabhadra drain, the Battiprolu drain and the Repalle drain are the drainage facilities available in the district.

Geology and Mineral Resources

The Geology of the district consists of various types of rocks which ranges from the Archaean to the recent. From the point of distribution of various rock types, the district can broadly be divided into three parts from west to east. In the western part are found the rock formations of the upper Cuddapah system and the Palnadu series. The slates, phyllites and the quartzites form a series of parallel hill ranges trending roughly in a north-east, south-west direction. The Palnadu formation constitute a gently undulating terrain without any significant relief. The most ancient rocks, the gneisses, schist and granites are exposed in the central part. The general strike of the gneisses and schist is north-east south-west. These rock types form essentially a monotonous plain country relieved here and there by hills made up to more resistant rocks like charnockite, granite, quartzite and khondalite. The thick alluvial deposits in the eastern part form a plain sloping gently eastwards. The different geological formations exposed in the district are given below.

Age	Description
Recent and sub-recent	Soils, tufa, kankar, beach sands, alluvium and lateritic gravels.
Mio-pliocene	Ippatam conglomerate
Upper Gondwana	Pavalur Sandstones, Vemavaram shales, Budavada Sandstones
Algonkian	a. Palnadu Series; Shales, limestones and quartzites b. Upper Cuddapah system: Srisailam quartzites and Cumbum slates.
Archaeans	Basic dykes, quartze veins, pegmatites, gneisses, charnockites and leptynites. Grey granites and gneisses quartzites and magnetite quartzites, micra-schists and khondalities.

1. Archaeans

The schistone rocks comprising hornblende-schists, quartz-felspar-schists and mica-shists and amphibolites are exposed around Vinukonda, Narasaraopet, Sattenapalle and Amaravathi. The kondalites are the hills around Mangalagiri.

2. Cuddapahs

In the district only the younger rocks of the cuddapah system are exposed near Vinukonda. The granite is now considered responsible for the copper and lead mineralisation in Agnigundala areas. In the Macherla-Nagarjunasagar area the Srisailam quartzite with intercalated shales forms the plateau in which the Krishna river has cut deep gorges.

3. Palnad Series

The rock formations of the Palnad series are correlated with those of the Kurnool system. They consists of Nandhyal shale, Koilkuntla Limestoe, Paniam quartzite, Owk shale, Narjee Limestone and Banganipalle quartzite. The Narjee limestone is referred to as the Palnadu limestone. It has an extensive development in Piduguralla, Dachepalli, Gurazala area. The Owk shales and the younger formation are exposed only in the Veldhurthi-Durgi area. The Palnadu in general are undisturbed and unfossiliferom.

4. Upper Gondwanas

The district remained an elevated land exposed to denudation by several natural agencies during the palaeozoic era for over 300 million years and as such there are no sedimentary rocks of this period. Upper Gondwanas contain intercalated marine sediments, suggesting the transgression of the sea on to the land during upper Gondwana times. The alluvial rocks are found between Bapatla and Tenali. The upper Gondwana beds are very little disturbed and not yielded any fossils.

5. *Mio-pliocene Formation*

Near the village of Ippatam, there are exposures of a very coarse conglomerate bed with a lateritic matrix. This has been tentatively correlated with a Cuddalore sandstones of Mio-pilocene age.

6. *Recent and sub-Recent Formation*

Laterite, lateritic shingle and gravel occur in numerous patches in the Guntur-Bapatla area as superficial deposit over the gneisses etc. Palaeolithic stone implements were found in these deposits.

Mineral Resources

The district contains immense reserves of cement grade limestone, iron ores, copper and lead minerals. The occurrence and deposits of economic mineral are as follows:

1. *Limestones*

Narjee limestone is very extensively present in the Palnad region which is used for the manufacture of cement.

2. *Copper and lead ores*

Ancient workings for copper and lead exist in the vicinity of Agnigundala and Karempudi. The ores of copper and lead are extensively found in these areas and extraction is being carried out.

3. *Diamonds*

Extensive mining was carried out in the past in the vicinity of Kollur village in Sattenapalle range on the banks of Krishna river. It is believed that the famous 'Kohinoor' diamond was found at Kollur. The diamond mines in the district are situated near Madugula, Mallavaram and Sarangapalle hills.

4. *Diatomaceous Earth*

It was reported to occur at Thimmayapalem and Ainavolu near Vinukonda.

5. Iron-ore

Low grade iron-ore the magnetite quartzites occur near Thumurukota near Macherla.

6. Gypsum

Gupsum occurs as spasely distrubuted plates at a depth of one to three metres in the marine silts near Santaravuru.

7. Quartz

There are several reefs of quartz in Palnadu, Sattenapalle, Narasaraopet and Vinukonda which might find use in glass manufacture.

8. Kankar

Kankar or calcareous nodules, used for manufacture of lime, has a wide distribution in the district near Chebrolu, Mangalagiri, Pedakakani, Venkatayapalem and Nadendla.

9. Clays

White clay occurs near Macherla.

10. Granite

Gondwana granite stones are useful in building construction. Palnadu limestone was utilized for the constructin of stupas during the Buddhist period.

Soils

The soils of Guntur district can broadly be divided into 4 types as follows:

1. Red Gravelly Soils

The red gravelly soils are essentially found over the area covered by the Archaean formation which occupy the major portion of the district. Macherla and Vinukonda ranges mostly have red gravelly soils.

2. Black Cotton Soils

These are present generally in the deltaic area on the banks of the Krishna and also in northern parts of Sattenapalle and Macherla range, where these are derived by the weathering of the limestone. These have formed due to the former existence of large and thick forest when a more moist climate prevailed.

3. Sandy Alluvial Soils

These are found along the sea coast and in the area covered by some of the Gondwana rocks. Kankar or calcareous nodules occur in the soils at several places in the district.

4. Saline swampy soils

These are present in the areas where tidal waves penetrate into the coast. These are characterised by mangrove vegetation. Repalle, Kothapalem, Sarlagondi and Nizampatnam have these soils.

Climate

The district suffers from hot climate, the summer especially being extremely trying. The year may be divided into four sessions.

December	—	February -	Dry and Cool Winter season
March	—	May	Summer season
June	—	September -	South-west monsoon season
October	—	November -	Post Monsoon or retreating monsoon season.

Rainfall

The average rainfall in the district is 830 mm. The rainfall generally decreases from east to the west. The rain is experienced mostly by both the south-west monsoon and the retreating monsoon. October in the rainiest month of the year (Tables 1, 2). On the average there are 47 rainy days in a year. The highest rainfall recorded was 386 mm at Sattenapalle on 19th of November, 1879.

Temperature

The day temperature will be higher by 3°C–4°C in summer and the night temperatures lower by 3°C–4°C in winter. Rentachintala is the hottest place, recording a maximum temperature of 49°C on 18th May 1948 and the lowest temperature recorded is 10°C at the same place on 31st December

1936. The annual minimum and maximum temperatures are 15°C and 47°C respectively (Table 3).

Special Weather Phenomena

Storms and depression, originating in the Bay of Bengal, cross the east coast of the district causing wide spread heavy rain and strong winds. Thunderstorms occur during March–October, being more frequent in the post-monsoon season.

Table 2
Normal Rainfall and Actual Rainfall for last three Years

Sl. No.	Name of the Month	Normal Rainfall m.m.	Actual Rainfall 1996-97 Rain fall	1996-97 No. of Rainy days	1997-98 Rain fall	1997-98 No. of Rainy days	1998-99 Rain fall	1998-99 No. of Rainy days
South West Monsoon								
1.	June	94.8	141.1	5	29.4	4	81.9	8
2.	July	154.7	133.8	6	97.3	9	167.2	14
3.	August	150.8	286.1	12	81.4	8	164.7	10
4.	September	159.1	135.5	7	376.3	17	235.3	13
	Total	559.4	696.5	30	584.4	38	649.1	48
North East Monsoon								
1.	October	162.9	228.9	11	98.2	5	290.5	16
2.	November	72.2	35.7	2	75.0	7	60.8	8
3.	December	16.7	7.8	1	44.9	4	-	-
	Total	251.8	272.4	14	218.1	16	351.3	24
Winter Period								
1.	January	2.6	18.1	2	3.0	1	-	-
2.	February	5.3	-	-	3.3	1	-	-
	Total	7.9	18.1	2	6.3	2	-	-
Hot Weather Period								
1.	March	9.1	1.6	Nil	1.9	Nil	Nil	Nil
2.	April	13.9	71.9	3	30.0	1	-	-
3.	May	47.0	11.2	1	12.9	2	31.0	2
	Total	70.0	84.7	4	44.8	3	31.0	2

Chapter 3

FOREST BIODATA AND GENERAL VEGETATION TYPES

Forests

The total area under forests is 1,74,844 hectares which in about 13.7% of the total geographical area of the district. The bulk of the forest area is confined mainly to the hill ranges of Macherla, Vinukonda and Sattenapalle. The entire forest of Guntur district is under one Forest Division with four ranges and 60 beats as detailed given in the next page:

General Vegetation Types

The vegetation in the ditrict can broadly be divided into two types—Forest vegetation and plain vegetation.

Forest Vegetation

The forest vegetation ranges from mangroves in the coast to the deciduous category in uplands. The majority is of mixed dry deciduous type varying considerably in composition and density. It is of poor growth in height and girth, except in places where the soil conditions are favourable. The combination of severe climatic, edaphic and biotic factors acting simultaneously makes the land incapable of producing any high forest except scrub type. The soil, particularly has undergone much deterioration due to heavy grazing incidence. The poverty of growth is due not only to the local factors such as soil, climate, rainfall, but definitely to the influence of man as manifested by fire, theft, and grazing. This is more apparent in the forest adjoining the villages.

Type succession is not well pronounced. Timber forests as such do not exist in the district except to a limited extent and that too in a deteriorated form. The peripheral forest consists of thorny scrubs due to depredation of

Ranges	Sections	Beats
1. Sattenapalle	1. Atchempet	1. Atchempet, 2. Challagariga 3. Chintapalli, 4. Pulichinta
	2. Papayapalem	1. Venkatayapalem 2. Papayapalem, 3. Bodanum, 4. Dodleru
	3. Piduguralla	1. Tipurapuram, 2. Konangi 3. Vemavaram, 4. Madinapadu
	4. Nakkarikal	1. Gutti konda, 2. Gundlapalle, 3. Nakkarikal,
	5. Guntur	1. Mangalagiri, 2. Edroy, 3. Ameenabad, 4. Kondavedu, 5. Enamadala
2. Repalle	1. Adavula Deevi	1. Adavula Deevi, 2. Dindi
	2. Kothapalem	1. Kothapalem, 2. Mollaguntla, 3. L.V. Dibba
	3. Nizampatnam	1. Nizampatnam, 2. Bapatla
	4. Nagayalanka	1. Yellichetti Dibba, 2. Nachagunta, 3. Sorlagandi.
3. Vinukonda	1. Vinukonda	1. Ravvaram, 2. Ipuru 3. Pomgirala, 4. Vaddemgunta, 5. Tangedumalli, 6. Kukutlapalli
	2. Bollapalle	1. Bollapalle, 2. Kandriga, 3. Naidupalem
	3. Remidicherla	1. Lomalogudem, 2. Ravulapuram, 3. Remidicherla, 4. Garikapadu
	4. Mellavagu	1. Mellavagu, 2. Jayanthipuram, 3. Chejarla, 4. Karempudi
4. Macherla	1. Macherla	1. Macherla, 2. Daida, 3. Madugula
	2. Mutukur	1. Veldurthy, 2. Loyapalle, 3. Mutukur, 4. Minchalapadu, 5. Kakirala, 6. Adigoppala
	3. V.P. South	1. Nagalavaram, 2. Kandlagunta, 3. Javuku, 4. Koppanur

man and the evil of grazing. The growth of grass is sparse and stunted. Generally the forest became poorer and poorer as we approach closer to the villages. Better forest do exist on hills and valleys which are far away from human habitation.

Classification

Based on the classification of forest types by Champion and Seth (1968), the forests are broadly classified into two major types basing on physiography. 1. Coastal and 2. Inland.

I. Coast Forests

Again these are divided into two types: (1) Mangrove Forests or Tidal Swamp Forests; (2). Southern dry evergreen Forests.

1) Mangrove Forests

Locally these are called Mada forests. These are typically a closed and impenetrable evergreen forests of moderate height. These are composed of trees specially adopted to survive on tidal mud which is permanently wet with salt water and submerged by every tide. Stilt roots, pneumatophores, vivipary and thick coriaceous leaves are the predominant and specialised features of the forest. These are found on the mud banks of delta streams, the tail of islands and over more extensive areas near the sea face. These are gregarious types-wholly or largely composed of a single species coupled with the gregariousness of the flora.

Common Tree Species

Avicennia officinalis, A. marina, Exoecaria agallocha, Rhizophora mucronata, R. apiculata, Ceriops tagal, Bruguiera cylindrica, B. gymnorhiza, Sonneratia apetala, Aegiceras corniculatus, Lumnitzera racemosa etc.

Shrubs

The common shrubs include *Acanthus ilicifolius, Salicornia brachiata, Clerodendrum inerme* etc. Some of the herbs on the salt marshes include, *Arthrocnemum indicum, Suaeda maritima, S. monoica* and *S. nudiflora*. Climbing shrubs like *Derris trifoliata* are common on trees.

Mangroves are really the builders and guardians of the mainland. They grow sea-wards sending their spreading roots into the shallow water. The mud and sand which is washed down by rivers and creeks, is caught amongst the dense roots and thus the soil erosion into the sea in prevented. They also prevent the high tidal waves during cyclones to some extent.

2. *Southern Dry Evergreen Forest*

The forest spread along the coasts of Repalle range about 9 to 12 Km from sea and consists of small coriaceous leaved evergreen trees and shrubs with short boles and spreading crowns. This forest consists of trees like *Albizia amara, Anacardium occidentale, Borassus flabellfier, Diospyros chloroxylon, Ixora pavetta, Manilkara hexandra, Memecylon edule, Ficus religiosa, Maba buxifolia, Ochna squarrosa, Salvadora persica, Sapindus emarginata, Syzygium cumini, Strychnos nux-vomica, Phoenix sylvestris, Thespesia populnea, Ziziphus xylopyrus, Benkara malabarica* etc. Some of the shrubs include *Carissa spinarum, Cassia auriculata* etc.

II. Inland Forest Vegetation

The Inland Forests are again divided into 4 types. They are:

1. Southern Dry mixed Deciduous types
2. *Hardwickia binata* type
3. *Acacia nilotica* type
4. Dry scrub jungles

1. Dry Mixed Deciduous Type

These are found in fairly deep and fertile soils with mixed deciduous species forming two to three storeys. These are formed by a mixture of trees, all of them are deciduous during dry season, usually for several months.

The majority are not particularly gregarious so that more or less pure associations are a rare feature. The lower canopy is almost entirely deciduous. Bamboos are restricted to moist localities, but are not luxurient. Climbers are comparatively few but include large woody lianas. Epiphytes and ferns are quite inconspicuous. These forests occur in Venkatayapalem of Sattenapalle range, Bollapalle of Vinukonda range and Loyapalle of Macharla range.

The top storey consists of trees with more than 12 m height. They include: *Anogeissus latifolia, Cassia fistula, Chloroxylon swietenia, Hardwickia binata, Albizia lebbeck, Boswellia serrata, Soymida febrifuga,* etc. The middle storey attains a height of 9 metres and is composed to *Albizia amara, Acacia horrida, A. chundra, Dichostachys cinerea, Diospyros melanoxylon, Gardenia latifolia, G. resinifera* and *Ziziphus xylopyrus, Bauhinia racemosa, Gyrocarpus americanus*. The bottom storey is composed of shrubs like *Mundulea sericea, Dodonaea viscosa, Carissa spinarum, Erythroxylum monogynum, Helicteres isora* and *Tarenna asiatica*. The ground is almost bare of herbs and grasses on account of excessive grazing.

2. The Hardwickia binata Type

This type of forest occurs in Macherla, Vinukonda and Sattenapalle reserve Forests. It is confined to well-drained soils and overlying quartzite rocks on the lower and middle slopes of the low hill ranges. *Hardwickia binata* is nearly evergren and grows almost gregariously with *Chloroxylon swietenia, Albizia lebbeck* and *Anogeissus latifolia* as its chief associates which give it a character of its own.

This forest may be considered as a special type by itself on account of its gregarious habit and semi-evergreen nature. It is fairly prolific in reproducing itself by natural seedling regeneration and hence may be considered as local climax type. The species, *Hardwickia binata* is a boon to this district, where grass is rare and cattle are excessive, because it's leaf makes good fodder.

3. Acacia nilotica type

This is found in abandoned tank beds as in Bollapalle section. The localities are characterised by very fine black silt and are sumberged under water during rainy seasons. Very few associated species occur mainly, where the *Acacia* canopy is broken. There is very thin growth of grass and little undergrowth is noticed.

4. Dry scrub jungles

This type is most commonly met with on the local shallow, red, gravelly soils with little or no humus. It is a low forest of about 6 m to 10 m height. The thorny hard wooded species like *Acacia* are predominant. Two storeys are recognisable—the upper storey is composed of the truly xerophytic species like *Acacia chundra* and *A. horrida* in the main, and to a lesser extent *Albizia amara, Anogeissus latifolia, Grewia bracteata* and *Dichrostachys cinerea.* The lower storey is composed of thorny shrubs like *Acacia caesia, Ziziphus oenoplia, Maytenus emarginata, Toddalia asiatica, Carissa spinarum, Catunaregum spinosa, Mimosa intsia, Canthium parviflorum, Pleiospermum alatum, Helicteres isora* and *Cassia auriculata.* The floor bears fodder grasses such as *Sehima nervosum, Heteropogon contortus* and *Aristida mutica*. Climbers like *Cissus quadrangularis, Tinospora cordifolia, Cocculus hirsutus, Rhynchosia minima, Abrus precatorius, Pergularia daemia* and *Wattakaka volubilis* are common.

In addition to the above types, *Borassus flabellifer* springs up gregariously in localities which hold out little promise to other species. Such palmyrah occur both in coastal and inland areas. The occurrence of Bamboo, *Dendrocalamus stirictus* is mostly restricted to moist localities near Pulichintapayavagu in Vinukonda range. The Bamboo is mostly lean and lanky and does not serve the commercial needs.

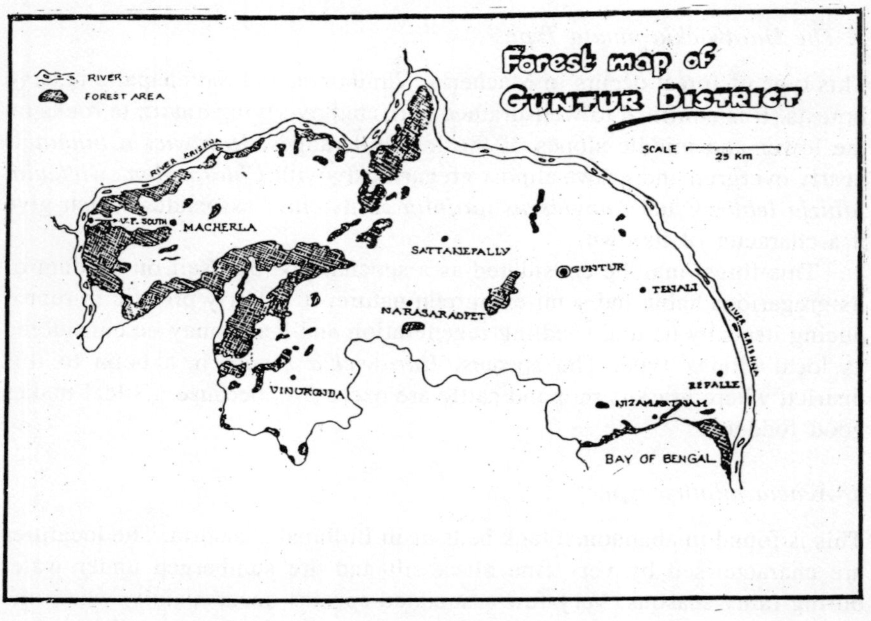

Fig. 3. Forest Map of Guntur District

Herbaceous Flora of the Forests

Herbs are commonly seen soon after the onset of monsoon and disappear by the beginning of January. The common herbs that are encountered in all the types of forests are:

Aerva lanata, Ageratum conyzoides, Indoneesiella echioides, Cassia tora, Chlorophytum tuberosum, Desmodium gangeticum, Enicostemma littorale, Goniogyna hirta, Glinus oppositifolius, Hemigraphis latebrosa, Indigofera linnaei, Polygala elongata and *Zornia gibbosa.*

The grasses that are encountered in the forest area are: *Aristida adscensionis, A. setacea, Chloris dolichostachya, Chrysopogon fulvus, Cymbopogon caesius, Dichanthium annulatum, Erogrostis coarctata, Heteropogon contortus* and *Oplismenus burmanii.*

The notable parasitic species found in the district are *Cassytha filiformis, Cuscuta reflexa, Dendrophthoe falcata, Orobanche cernua, Striga asiatca, S. gesnerioides* and *Viscum articulatum.*

Vegetation of Plains

The vegetation of plains is divided into two categories - Aquatic and Terrestrial.

Terrestrial Vegetation

Vegetation found along roadsides, waste places and weeds in cultivated fields comprise the terrestrial vegetation. This is again divided into two types 1. Coastal and 2. Inland.

1. Coastal vegetation

This is composed of the flora found on sand beaches of the coasts. This includes: *Ipomoea pes-caprae, Launaea sarmentosa, Portuluca oleracea, Sesuvium portulacastrum, Heliotropium curassavicum, Cressa cretica, Cyperus arenarius, Fimbristylis ferruginea, Kyllinga triceps, Aeluropus lagopoides, Spinifex littoreus* etc.

2. Inland Vegetation

The following trees are observed on roadsides and villages as avenue or shade giving trees. *Azadirachta indica, Delonix elata, Ficus benghalensis, F. religiosa, Mangifera indica, Melia azedarach, Samanea saman and Tamarindus indica.*

The common herbs observed in wastelands and roadsides include: *Acalypha indica, A. lanceolata, Acanthospermum hispidum, Achyranthes aspera, Amaranthus tricolor, A. spinosus, Boerhavia diffusa, Cassia auriculata, C. occidentalis, C. tora, Cleome viscosa, Corchorus aestuans, C. fascicularis, Croton bonplandianum, Echinops echinatus, Euphorbia hirta, E. indica, Crotalaria hebecarpa, Indigofera hirsuta, I. linnaei, Hedyotis puberula, Polygala erioptera, Solanum surattens, Sida acuta, S. cordata, S. cordifolia, Tephrosia pumila, T. purpurea, T. villosa, Tribulus terrestris, Tridax procumbens, Vernonia cinerea, Vicoa indica, Ziziphus mauritiana* etc.

Common weeds of cultivated dry and irrigated fields are: *Acalypha indica, Ageratum conyzoides, Alloteropsis cimicina, Alysicarpus rugosus, Amaranthus spinosus, A. viridis, Amischophacellus axillaris, Aristolochia bracteolata, Biophytum nervifolium, Cassia absus, Celosia argentea, Cleome gynandra, Crotalaria orixensis, C. retusa, Commelina benghalensis, Cynodon dactylon, Cyperus rotundus, Dactyloctenium aegyptium, Desmodium triflorum, Dicanthium annulatum, Digera muricata, Dinebra retroflexa, Eragrostis atrovirens, E. viscosa, Euphorbia heterophylla, E. hirta, Ischaemum pilosum, Lagascea mollis, Leucas aspera, Merremia tridentata, Panicum repens, Parthenium hysterophorus, Phyllanthus amarus, P. maderaspatana, Physalis minima, Portulaca oleracea, P. quadrifida* and *Zornia diphylla.*

In rice fields, the weeds commonly met with are: *Cyperus rotundus, C. compressus, Echinochloa colona, E. crusgalli, Fimbristylis bisumbellata, Leersia hexandra* and *Marsilia quadrifida.*

Weeds like *Coldenia procumbens, Glinus lotoides, Blumea eriantha* are found in harvested rice fields. Weeds like *Heliotropium strigosum, Glinus lotoides, G. oppositifolius* and *Sphaeranthus indicus* are found in dried up ponds and tanks.

Aquatic Vegetation

Guntur district is having many lakes, ponds, tanks, canals and streams. Rich and varied aquatic flora is met within these habitats. These aquatics are classified under the following heads depending on their relation with soil, water and air.

1. Floating Aquatics: Three types are included in this category.

a. *Free floating on the surface of water*: These are the plants which have no contact with soil and are drifted passively by water currents: These include: *Eichhornia crassipes, Nymphoides hydrophylla, Pistia stratiotes, Spirodela polyrhiza* etc.

b. *Rooted hydrophytes with floating shoots*: These are attached to the muddy floor by their roots but their shoots come afloat on the surface of water. These include: *Ipomoea aquatica*.

c. *Rooted hydrophytes with floating leaves*: These plants are attached to the muddy floor. Their stem remain under water while leaves float on water surface. These include *Ottelia alismoides, Nelumbo nucifera, Nymphaea nouchali, N. pubescens*.

2. Submerged Aquatics

These plants always remain under water surface and can be grouped into two categories

a. *Suspended submerged hydrophytes*: These plants remain in water but have no contact with soil. These include: *Ceratophyllum demersum, Najas graminea* etc.

b. *Attached submerged hydrophytes*: These plants remain in contact with soil and water. These include *Aponogeton echinatus, Hydrilla verticillata, Nechamandra alternifolia, Vallisneria natans* etc.

3. Emergent Aquatics

Plants which are attached to soil covered with water but most of their vegetative parts come out of water surface. They include: *Aeschynomene indica, Ammannia baccifera, Bacopa monnieri, Cyperus difformis, C. pangorei, Echinochloa colona, Eriocaulon quinquangulare, Hygrophila auriculata, Ischaemum rugosum, Limnophila indica, Polygonum glabrum, Phragmites karka* and *Typha angustata*.

4. Wetland Hydrophytes

These plants are rooted to the soils saturated with water, which may also survive in dried conditions in the later part of their life cycle: These include: *Alternanthera sessilis, Caesulia axillaris, Commelina ensifolia, Cynodon dactylon, Eclipta prostrata, Eleocharis atropurpurea, Eriochloa procera, Glinus lotoides, Hygrophila auriculata, Ipomoea carnea, Ludwigia perennis, Panicum trypheron, Phyla nodiflora, Saccharum spontaneum, Sphaeranthus indicus* etc.

Economic Botany

More than 80% of the geographical area of the district is under agriculture. The most important crop in the district is rice (*Oryza sativa*) cultivated

Economic Botany

Cultivated Plants

No.	English name	Telugu name	Botanical name
Cereals			
1.	Macaroni wheat	Semya godhuma	*Triticum durum* Desf.
2.	Bread wheat	Rotte godhuma	*Triticum aestivum* L.
3.	Emmer wheat	Upma godhuma	*Triticum dicoccum* (Schrank) Schubles
4.	Rice, Paddy	Vari, Biyyamu	*Oryza sativa* L.
Millets			
5.	Pearl millet	Sajja	*Pennisetum typhoides* Stapf & Hubb.
6.	Finger millet	Ragulu	*Eleusine coracana* Gaertn.
7.	Great millet	Jonnalu	*Sorghum bicolor* (L.) Moench
8.	Italian millet, Fox tail millet	Korralu	*Setaria italica* Beauv.
9.	Maize	Mokkajonna	*Zea mays* L.
Pulses			
10.	Black gram	Minumulu	*Vigna mungo* (L.) Hepper
11.	Green gram	Pesalu	*Vigna radiata* (L.) Wilczek
12.	Horse gram	Ulavalu	*Vigna unguiculata* (L.) *Walp.* ssp. *unguiculata* (= *Dolichos biflorus* L.)
13.	Cow pea	Alasandalu	*Vigna unguiculata* (L.) Walp. ssp. *cylindrica* L. Walp. (= *Vigna catjang* Wall.)
14.	Chick pea, Bengal gram	Senagalu	*Cicer arietinum* L.
15.	Pigeon pea, Red gram	Kandulu	*Cajanus cajan* Milsp.
Vegetables			
16.	White gourd	Budida gummadi	*Benincasa hispida* Cogn.
17.	Bitter gourd	Kakara	*Momordica charantia* L.
18.	Bottle gourd	Sorakai	*Lagenaria siceraria* (Molina) Standl. (= *L. vulgaris* Ser.)
19.	Little gourd	Dondakai	*Coccinia grandis* Voigt (= *Coccinia indica* Wt. & Arn.)
20.	Ridge gourd	Beerakai	*Luffa acutangula* (L.) Roxb.
21.	Smooth gourd	Nethibeerakai	*Luffa cylindrica* (L.) Roem.
22.	Pumpkin	Gummadi	*Cucurbita pepo* L.

No.	English name	Telugu name	Botanical name
23.	Snake gourd	Potlakai	*Trichosanthes anguina* L.
24.	Cucumber	Dosakai	*Cucumis sativus* L.
25.	Brinjal	Vankai	*Solanum melongena* L.
26.	Tomato	Tamota, Ramamulaga	*Lycopersicon esculentum* Mill.
27.	Cluster bean	Goru chikkudu	*Cyamopsis tetragonaloba* Taub.
28.	Sword bean	Thammakai	*Canavalia gladiata* DC. (= *C. ensiformis*)
29.	Drum stick	Munagakai	*Moringa oleifera* Lam.
30.	Lady's finger	Bendakai	*Abelmoschus esculentus* (L.) Moench
31.	Onion	Ullipayalu	*Allium cepa* L.
32.	Sweet potato	Chilagada dumpa	*Ipomoea batatas* L.
Oil seeds			
33.	Groundnut	Veru sanaga	*Arachis hypogaea* L.
34.	Coconut	Kobbari	*Cocos nucifera* L.
35.	Gingelly	Nuvvulu nune	*Sesamum orientale* L.
36.	Sunflower	Suryakanthi	*Helianthus annuus* L.
37.	Castor	Amudamu	*Ricinus communis* L.
38.	Mustard	Aavalu	*Brassica juncea* (L.) Czern & Coss.
Commercial crops			
38.	Tobacco	Pogaku	*Nicotiana tabaccum* L.
39.	Cotton	Pathi	*Gossypium herbaceum* L. *Gossypium arboreum* L. *Gossypium hirsutum* L. *Gossypium barbadense* L.
40.	Chillis	Mirchi	*Capsicum frutescens* L.
41.	Coriander	Dhaniyalu Kothimeera	*Coriandrum sativum* L.
Miscellaneous			
42.	Sunhemp	Janumu	*Crotalaria juncea* L.
43.	Betel	Thamalapaku	*Piper betel* L.

under irrigation from Nagarjunasagar Right canal and Krishna delta canal system. Black gram (*Vigna mungo*) is cultivated as Rabi crop after the harvest of rice during summer. The important pulses grown as rainfed crops include Red gram (*Cajanus cajan*), Groundnut (*Arachis hypogaea*), Bengal gram (*Cicer arietinum*), and Cow pea (*Vigna unguiculata*). The common cereal crops in drylands include *Sorghum vulgare* (Sorghum), *Pennisetum typhoideum* (Pearl millet) and *Zea mays* (Maize). Other important commecial crops in dry and irrigated lands include *Nicotiana tabaccum* (Tobacco), *Gossypium* sp (Cotton) and *Capsicum frutescens* (Chillies). The important fruit gardens in the district include: *Musa paradisiaca* (Banana), *Manilkara zapota* (Sapota), *Psidium guajava* (Guva), *Citrus limonia* (Lemon), *C.*

sinensis (Sweet lime), *C. aurantifolia* (orange) and *Cocos nucifera* (coconut). The common vegetable crops in the distrct under cultivation include *Abelmoschus esculantus* (Ladies finger), *Cyamopsis tetragonaloba* (Cluster bean), *Lagenaria siceraria* (Bottle gourd), *Luffa acutangula* (Ridge gourd), *Trichosanthes anguina* (Snake gourd), *Lycopersicon esculentum* (Tomato) and *Solanum melongena* (Brinjal).

Other crops in the district include: *Brassica juncea* (Mustard), *Coriandrum sativum* (Coriander), *Sesamum orientale* (Gingelli), *Piper betel* (Betal vine), *Manihot esculenta* (Tapioca) and *Allium cepa* (onions).

Muralidhara Rao and Pullaiah (1998) made an Ethnobotanical survey and reported about 50 medicinal plants used by the tribals, which are listed below. The main tribes of this district are Chenchus, Yerukalas, Yanadis and Sugalis. The tribal population is 1.82 lakhs which is 4.4% of the total population of the district.

Abrus precatorius, Aerva lanata, Aristolochia indica, Asparagus racemosus, Azadirachta indica, Butea monosperma, Centella asiatica, Chloroxylon swietenia, Datura metel, Eclipta alba, Gardenia latifolia, Gloriosa superba, Gmelina arborea, Helicteres isora, Hemidesmus indicus, Hygrophila auriculata, Momordica charantia, Ocimum tenuiflorum, Pongamia pinnata, Phyllanthus amarus, Pterocarpus marsupium, Smilax zeylanica, Solanum nigrum, S. surattense, Strychnos nux-vomica, Thevetia peruviana, Tylophora indica, Vitex negundo, Wattakaka volubilis, Wrightia tinctoria etc.

Chapter 4

PAST AND PRESENT WORK

a. Past Work

Guntur district, which is a part of Coastal Andhra Pradesh, has not been fully explored previously. Random collections were made by some eminent explorers from limited localities.

Gamble in 1887 made some collections from Vinukonda Forest and Guntur surroundings. The collections of Gamble were housed in Madras Herbarium, Coimbatore (MH), Central National Herbarium, Calcutta (CAL), Forest Research Institute Herbarium, Dehradun (DD) and Royal Botanic Gardens, Kew (K), England. Barber in 1902 and 1924 made extensive collections from Vinukonda, Bollapalle, Nakrekal, Kondavedu Fort and Madinepadu—all from Vinukonda Reserve Forest. The collections of Barber are housed in Madras Herbarium, Coimbatore (MH) and Central National Herbarium, Calcutta (CAL). Jacob in 1924 also made some collections from Kondavedu Fort, Madinepadu, Nakrekal, Bollapalle, Karempudi - all from Vinukonda range. His collections are housed in Madras Herbarium, Coimbatore (MH). Uhl, in 1929 made some collections from Guntur town and its surroundings. His collections are not available in any herbaria. He published a booklet entitled "Flora of Guntur" (Uhl, 1929). This is the first flora of a town in India.

Sebastine, in 1959, while making collections from Nalgonda district, collected extensively from Vijayapuri (Nagarjunasagar) which is partly included in Guntur district (V.P. South). The collections of Sebastine were housed in MH and CAL. Agarwal, Guha, Nanda, Patil and Madhuram, Scientists from ICAR, made some random collection of grasses from Macherla Reserve Forest in the year, 1955. Their collections are stored in Central National Herbarium, Cacutta (CAL) and Forest Research Institute Herbarium, Dehradun (DD).

Thothathri in 1961 made collections from Nagarjunakonda valley and the surrounding hills. A total of 251 species were collected from the valley and surrounding hills. The valley is now completely submerged under water of Nagarjunasagar Reservoir. Parts of Nagarjunakonda valley, now comes under Guntur jurisdiction. *Crotalaria paniculata* var. *nagarjunakondensis* is a new taxon collected from the valley. Thothathri's collections are deposited in CAL.

Wagh during 1958–60 has made some random collections from Mellavagu and Vinukonda for his doctoral work. His collections were housed in Blatter Herbarium, Bombay (BLAT). Sastry and Ananda Rao of Botanical Survey of India, Howrah, studied the flora and vegetation of coastal Andhra Pradesh in 1973, covering the entire coast of Andhra Pradesh including Guntur district. Lakshmana Reddy and Rao of Nagarjuna University, in 1985, made Eco-floristic studies of Nizampatnam mangroves. Their collections are being housed in Herbarium of Botany Department, Nagarjuna University, Guntur.

Scientists of Botanical Survey of India, Southern Circle, Coimbatore, while working on "Ethnobotany of Eastern Ghats" collected plants from Guntur district. These collections are incorporated in Madras Herbarium (MH).

A systematic enumeration of the weed flora of 75 Ha High Density Energy Plantations in Nagarjunanagar, was given by Ashakumari (1989; M.Phil. Thesis, NU).

Mohana Rao and Aasiya (1995) surveyed, Nagarjuna University campus and its surroundings and reported 388 species belonging to 80 families.

Recently Pullaiah (1997), Pullaiah and ali Moulali (1997), Pullaiah and Chennaiah (1997) and Pullaiah and Surya Prakash Babu (1998) brought out "Flora of Andhra Pradesh" in 4 volumes which included collections from Guntur district. The collections of the second author have been quoted in the text.

Pullaiah and Sandhya Rani (1999) documented the trees of Andhra Pradesh and including those from Guntur district.

b. Present Work

The present work comprised intensive field trips conducted in different parts of the district covering all the seasons of the year. Field and herbarium methods of Santhapau (1955) and Jain and Rao (1977) have been followed. The present enumeration includes all the specimens collected now and those collected by earlier workers. A total of 891 species have been enumerated.

Materials and Methods

Exploration trips were regularly conducted in every season and in all the representative localities of the district from floristic point of view, covering all topographical zones. Besides the plants in the forests, those of the waste lands, cultivated fields, roadsides, aquatic habitats etc., have also heen collected. At the outset, the site for collection was chosen after a thorough and detailed study of the density of vegetation. Each collection of individual species was labelled with field numbers in quadruplicates after making a critical observation on the habit, habitat, frequency of occurance, latex and other relevant ecological features and macroscopic characters which cannot be observed from the dried herbarium sheets. It case of plants in vegetative stages in a particular field trip, localities of these plants were carefully marked in the field note books, so that these could easily be spotted in the flowering and fruiting stages at later dates. In this way almost all the specimens were collected with flowers and fruits. Care was taken in collecting the bulbs, corms, rhizomes, tubers etc. if the plants possess them.

Immediately after cutting the twigs of about 30 cm in length, they were placed in a polythene bag to avoid further desiccation and to protect from withering. Collections were made in the forenoons and were restricted to 25 field numbers per day for the sake of convenience. The collected specimens were poisoned by dipping the whole twig in saturated mercuric chloride in ethyl alcohol solution. Immediately the poisoned specimens were placed between blotting papers with the help of foreceps. These were properly tied in iron pressors with the help of ropes. After 24 hours, the blotting papers were changed and the specimens were spread properly. Some leaves are placed facing upwards and some downwards to show the characters on both surfaces. These spread specimens were once again placed in other dried blotting sheets and tied again in iron pressors and this process is continued till the specimens were dried completely.

Flowers and fruits of some important and rare plants were fixed in formalin-acetic acid-alcohol (F.A.A) in specimen tubes for further study and reference. In case of succulents like Cacti and Euphorbias the plant parts are sliced longitudinally and transversely placed half upwards and the other downwards. Special care was taken for the aquatic and small herbaceous plants. The poisoned, pressed and dried specimens were pasted with glue on thick mount boards, measuring 42 × 28 cm, and stitched with thread. Labels with all relevant information such as name of the plant, family, locality, altitude, date of collection, salient notes and collector's name were affixed on the right hand bottom corner of the mounted sheets.

Provisional identifications were made with help of Gamble's 'Flora of Presidency of Madras' and Hooker's Flora of British India' (1872–1897). Later the identifications were confirmed with the help of latest monographs, floras and comparing with authentic specimens in the Madras Herbarium, Coimbatore (MH), Central National Herbarium, Calcutta (CAL) and Forest Research Institute Herbarium, Dehradun (DD). Flowering and fruiting seasons were accurately recorded for each species together with relative degree of abundance.

Dissections

Dissections of floral parts were made and studied under binocular stereoscope microscope for the minor details and the identifications were confirmed with reference to the floras.

After identification, the sheets were incorporated in the Herbarium of Botany Department of Sri Krishnadevaraya University. The herbarium specimens were arranged according to Bentham and Hooker's system of classification with certain exceptions to accommodate recent changes. The genera within the family and species within the genus are kept in alphabetical sequence. Nomenclature of each species has been checked according to International Code of Botanical Nomenclature (ed. 1994).

Plan of Enumeration

The families are broadly arranged according to Bentham and Hooker's (1862–1883) system with recent delimitation of certain families. The genera in a family and the species in a genus are arranged alphabetically. The treatment of each species includes the original citation of the valid names followed by relevant synonyms if any, thereby facilitating reference to 'Flora of British India' and 'Flora of Presidency of Madras' followed by short description, relative abundance and ecological features, distribution in the areas under study, flowering and fruiting periods, local names and uses wherever available and citation of the specimens examined. Nomenclatural notes have been given under the genera and the species wherever and necessary. New records and other observations of significance are also given. If the species is alien, its country of origin and probable time of introduction is given.

Abbreviations used

Common abbreviations used in the enumeration are as follows:

FBI : Hooker, J.D. *et al.*, *Flora of British India*, 7 volumes, London. Dates of Publications of the parts of the seven volumes are:

Vol. No.	Part No.	Pages	Date of publication
I	1	1–208	May, 1872
	2	209–464	January, 1874
	3	464–end	February, 1875
II	4	1–240	May, 1878
	5	241–496	July, 1878
	6	497–end	May, 1879

III	7	1–192	May, 1880
	8	193–448	March, 1881
	9	449–end	December, 1882
IV	10	1–256	June, 1883
	11	257–512	January, 1884
	12	513–end	August, 1885
V	13	1–240	August, 1886
	14	241–462	December, 1887
	15	463–686	December, 1888
	16	687–end	April, 1890
VI	17	1–224	December, 1890
	18	225–448	July, 1892
	19	449–672	September, 1893
	20	673–end	April, 1894
VII	21	1–end	November, 1897

Bor : Bor, N.L. 1960. *The Grasses of Burma, Ceylon, India and Pakisthan* (excluding *Bombusaeae*), London.
CAB : C.A. Barber
CAL : Central National Herbarium, Botanical Survey of India, Calcutta
DLR : D. Lakshmana Reddy
DMR : D. Muralidhara Rao
Eng. : English name
Fl. & Fr. : Flowering and Fruiting period
Gamble : Gamble, J.S. and Fischer, C.E.C., *Flora of the Presidency of Madras*; 3 volumes 1915–1935. Botanical Survey of India, Calcutta, reprinted 1957. Page number outside parenthesis is that of 1915–1935 edition while that in parenthesis is of 1957 edition.
GVN : G.V. Narayana
JSG : J.S. Gamble
KCJB: K.C. Jacob
KMS : K.M. Sebastine
KSM : K. Sri Rama Murthy
KTH : K. Thothathri
MH : Madras Herbarium, Botanical Survey of India, Coimbatore
NUH : Nagarjuna University Herbarium
N. Varam : Nagula Varam
PNR : P.N. Rao
RF : Reserve Forest
SSR : S. Sandhya Rani
TP : T. Pullaiah
Vern. : Vernacular name
V.P. South : Vijayapuri South

VRK : V. Rama Krishnaiah
p.p. : *pro-parte* or in part

Floristic analysis

Synopsis

Group	Families		Genera		Species	
	Total	Percentage	Total	Percentage	Total	Percentage
Dicotyledons						
Polypetalae	64	48.85	188	37.97	374	42.50
Gamopetalae	27	19.84	162	32.72	255	28.61
Monochlamydae	17	12.97	51	10.30	96	10.27
Monocotyledons	18	13.84	88	17.60	160	17.95
Ferns	6	4.58	6	1.41	6	0.67
Total	132	100.00	495	100.00	891	100.00

The ratio of Monocots to Dictos is; 1:4.53 of species: 1:4.56 of genera; 1:5.9 of families. The ratio of genera to species in Angiosperms in this study is 1:2.29. In Flora of British India, it is 1:7. This shows the small proportion of species to number of genera and families.

In order to get an insight into the floristic status of the Flora of Guntur district in contrast wih that of regional and country floras a comparative list of ten dominant families in Guntur district, with that of Madras Presidency (Gamble) and Flora of India (BSI) is given in order of their numerical dominance.

Rank	Guntur district (Present study)		Madras Presidency (Gamble & Fischer)		Flora of India (BSI)	
	Family	No. of species	Family	No. of species	Family	No. of species
1.	Leguminosae	139	Leguminosae	432	Poaceae	1225
2.	Poaceae	89	Poaceae	388	Orchidaceae	990
3.	Euphorbiaceae	45	Rubiaceae	226	Leguminosae	775
4.	Acanthaceae	36	Acanthaceae	202	Asteraceae	754
5.	Cyperaceae	29	Eurphorbiaceae	199	Rubiaceae	495
6.	Asteraceae	28	Orchidaceae	199	Cyperaceae	449
7.	Convolvulaceae	28	Asteraceae	189	Euphorbiaceae	419
8.	Rubiaceae	26	Cyperaceae	172	Lamiaceae	393
9.	Malvaceae	22	Lamiaceae	128	Acanthaceae	379
10.	Tiliaceae	19	Asclepiadeaceae	93	Scrophulariaceae	356

Note: The family Orchidaceae is represened by only 2 species in Guntur district.

Dominant Genera with 5 or more species

	Genera	No. of speices
1.	*Crotalaria*	17
2.	*Indigofera*	15
3.	*Ipomoea*	13
4.	*Cyperus*	12
5.	*Cassia*	11
6.	*Acacia*	11
7.	*Grewia*	10
8.	*Heliotropium*	9
9.	*Phyllanthus*	9
10.	*Commelina*	8
11.	*Fimbristylis*	8
12.	*Alysicarpus*	7
13.	*Capparis*	7
14.	*Barleria*	7
15.	*Sida*	7
16.	*Solanum*	7
17.	*Ficus*	7
18.	*Corchorus*	6
19.	*Eragrostis*	6
20.	*Rhynchosia*	6
21.	*Tephrosia*	6
22.	*Ziziphus*	6
23.	*Cleome*	5
24.	*Polygala*	5
25.	*Dalbergia*	5
26.	*Merremia*	5
27.	*Setaria*	5
28.	*Brachiaria*	5
29.	*Acalypha*	5

Plants of Botanical Interest

Little known taxa

1. *Sida spinosa*
2. *Tropaeolum majus*
3. *Pleiospermum alatum*
4. *Xylocarpus granatum*
5. *Abrus fruiticulosus*
6. *Crotalaria paniculata* var. *nagarjunakondensis*
7. *Indigofera astragalina*
8. *Ixora finlaysoniana*
9. *Sesamum alatum*
10. *Alpinia calcarata*
11. *Rhaphidophora pertusa*
12. *Fuirena ciliaris*
13. *Acrachne racemosa*
14. *Cleome angustifolia*
15. *Bergia ammannioides*

Endangered plants

1. *Entada pursaetha*
2. *Crotalaria paniculata* var. *nagarjunakondensis*
3. *Rostellularia vahlii* var. *rupicola*
4. *Habenaria roxburghii*
5. *Withania somnifera*
6. *Andrographis paniculata*
7. *Aristolochia indica*

All mangrove species are recommended for '*in situ*' conservation, where available, the 'vivipary' factor is to be suitably utilized for 'Coastal Farming' of these species.

1. *Avicennia officinalis*
2. *Avicennia marina*
3. *Exoecaria agallocha*
4. *Rhizophora mucronata*
5. *Rhizophora apiculata*
6. *Bruguiera cylindrica*
7. *Bruguiera gymnorhiza*
8. *Sonneratiq apetala*
9. *Aegiceras corniculata* and
10. *Lumnitzera racemosa*

Species recommended for *In situ* conservation other than mangroves to avoid illegal harvesting.

1. *Strychnos nux-vomica*
2. *Terminalia bellirica*
3. *Gyrocarpus americanus*
4. *Gloriosa superba*
5. *Pterocarpus marsupium*
6. *Soymida febrifuga*

Plants recommended for conservation by Micro propagation

1. *Withania somnifera*
2. *Andrographis paniculata*
3. *Hemidesmus indicus*
4. *Gymnema sylvestre*
5. *Aristolochia indica*
6. *Tylophora indica*
7. *Plumbago zeylanica*

which are till recently common but fast disappearing due to over exploitation and smuggling outside the country offering, typical examples of Botanical Piracy.

Local weeds as petrocorps in wastelands

Calvin (1979) suggested conversion of latex from wild plants to lower hydrocarbons which can substitute petroleum.

Various weeds are screened as potential sources of petrocrops. A large number of latex bearing plants of families Apocynaceae, Asclepiadaceae,

Convolvulaceae, Sapotaceae and Euphorbiaceae are being evaluated for production of biocrude. Active centres of work of this nature are Indian Institute of Petroleum (IIP) at Dehradun and NBRI, Lucknow (Bhatia, 1984). IIP conducted a workshop on petrocrops in 1986. The dwindling of existing petroleum resources and with the impending threat of a crisis all over the world paved way for serious consideration of options for producing liquid fuels from biomass. Plant biocrude is a complex mixture of liquids, triglycerides, waxes, terpenoids, phytosterols and other modified isoprenoid compounds that can be catalytically upgraded as liquid fuels. Bhatia (1984) listed 14 species as potential crops. Of these ten belong to Euphorbiaceae, three to Asclepiadaceae and one to Convolvulaceae. Raina (1984) mentioned that the biocrude from *Jatropha curcas*, a Euphorbiaceous member was tested for comparative performance of running a four stroke cycle diesel engine. The generation of pollutants like sulphur dioxide and carbon monoxide and smoke were found to be quite less in this experiment. The search for botanochemicals as substitutes or supplements to petrochemicals has been going on (Wang and Huffmann, 1982) and the literature was reviewed by Srivastava and Bhatia (1985). Inventories of petrocrops are being prepared for Rajasthan by Ashwini Kumar (1987) and for Andhra Pradesh by Seshagiri Rao and Prasad (1987). Petrocrops are being recommended for utilization of wastelands. Apart from being possible sources of hydrocarbons and other phytochemicals, the species involved are usually hardy, stress tolerant, disease free and can be put to multiple uses.

The following are the potential petrocrops of this area with some comments on their relative mertis and demerits.

Euphorbiaceae: Euphorbia hirta, E. indica, E. thymifolia, (3 species of the genus). These are all either prostrate or erect herbs of small size and their large scale exploitation as substitutes for petroleum fuels is doubtful.

Euphorbia tirucalli, E. nivulea, E. antiquorum, E. caducifolia, E. heterophylla, E. cyathophora. Of these six species of the genus; *E. cyathophora,* and *E. heterophylla* are herbaceous annuals and their cultivation is much simpler. The biocrude yield will be considerable. Their seed output and reproductive capacity are very high. The size of the natural populations of these weeds is very large. Grown in high densities, they improve the appearance of the landscape. The above two have crimson throated bracts and have ornamental value. They are not grazed. The latex of these species is considered as antiseptic. *E. tirucalli* is a spineless, leafless tree with enormous number of green, herabaceous, laticiferous tubular branches. This is one of the first among the plants to be identified as a potential source of hydrocarbons from the forests of South America by Calvin. It is a perennial, sizable crop of each tree can be periodically harvested. Other species of *Euphorbia* cited are bushy, dendroid, thorny xerophytic shrubs or trees that provide an impenetrable biofence. Their cultivation feasibility is expected to be tougher than the herbaceous large sized *Euphorbias.*

Local shrubby weeds *Jatropha curcas, J. gossypifolia* and *J. glandulifera* are useful for planned cultivation in wastelands.

Asclepiadaceae; Calotropis gigantea and *C. procera* are two self-spreading, sympatric species, already established in vast expanses of wasteland. If their potential as viable petrocrops could be established—they offer a good example of weeds turning worthies. These are generally established as poisonous plants and various parts including latex are put to diverse but regulated medicinal use.

There are others in this family already put to use in medicine. They are *Hemidesmus indicus, Gymnema sylvestre, Tylophora indica* and *Leptadenia reticulata.* Various other species of *Tylophora, Pergularia daemia* and *Cryptostegia grandiflora* are all climbers and hence their cultivation is to be planned on the lines of betel leaf crop supporting them on tall; stout poles or pillars. The biocrude output from these large climbers would repay the efforts of their planned cultivation in wastelands.

Apocynaceae: Catharanthus roseus, a wasteland weed of this family has already been inducted to extensive cultivation as a medicinal plant in Guntur district. Its roots are poisonous and contain alkaloids with purported similar properties as those of *Rauwolfia serpentina* and *R. tetraphylla;* sources of antihypertension drugs. It is also grown as an ornamental called the rose perwinkle. *Catharanthus pusillus* is a smaller annual herb and a common weed in moist places is easy to cultivate but its quantitative biocrude output is likely to be less economical. Both these valuable species deserve to be added to the petrocrop resources of Andhra Pradesh published by Seshagiri Rao and Prasad (1987) where they are suprisingly omitted. A total inventory of members of Apocynaceae may provide an impressive list of potential petrocrops but several of them are exploited more as confirmed medicinal species and ornamentals. For instance, the wood of *Wrightia tinctoria* is white, soft, easily manageable and extensively used in preparation of the famous Kondapalli toys. Species of *Nerium, Plumeria, Tabernaemontana, Vallaris, Thevetia, Allamanda* and others are cultivated extensively as ornamentals. Several of them are valuable medicinal plants. It is however, welcome to examine their petrocrop potential also since cultivation of the tree species here can easily be done by cuttings. The cultivation of the climbing weed *Ichnocarpus frutescens* on the lines of *Leptadenia* and *Pergularia* can be implemented.

Convolvulaceae: Several *Convolvulaceae* have already been identified and listed as weeds. Three members of the family were mentioned as potential petrocrops by Seshagiri Rao and Prasad (1987). In view of the restricted watery latex that oozes out from these species, they may not prove to be very useful material. *Ipomoea carnea* the erect, ascending, morning glory of the railway line bunds if proves biocrude worthy, will be rewarding. This species ca be multiplied by cuttings. It is in wide use as biofencing, ornamental, soil binder, preventive measure in soil erosion.

According to Wang and Huffmann's (1982) classification, potential liquid fuel source species are categorized into three types.

i. Seed oils,
ii. Essential oils, and
iii. Exudates and extracts.

The present examination brought forth the following categorization of the species.

Seed/Fruit oil-bearing species: *Jatropha curcas* (2500trees/could yield 2000 kg of crude oil/ha; Srivastava and Bhatia, 1985), *J. gossypifolia* and *J. glandulifera.* Benerji *et al.* (1983) estimated energy values of the seed oil of these species.

Apart from the above weeds, avenue trees very successfully established locally also may be examined for the fuel property of their oils. The main species are *Azadirachta indica, Pongamia pinnata* and *Calophyllum inophyllum.*

Essential oils: Members of Labiatae fall in this category. The local weeds here are *Ocimum americanum* L., *O. gratissimum* L., *Anisomeles malabarica* and *Hyptis suaveolens.* Due to the volatile nature of the essential oils, economic feasibility in fuel oil exploration is questionable. However, the weed *0. gratissimum* is recommended for cultivation in wastelands for its lemon—scented essential oil found to be a mosquito repellent.

Exudates and extracts: All the latex bearing species recorded above fall into this category.

Apart from cultivation of the potential local weed petrocorps, species recommended for wastelands at other arid and semi-arid zones are being successfully cultivated.

As the threat of depletion of fossil fuels always looms large, possession of a data base for alternative combustible resource material has futuristic value.

The people of the district are generally progressive, competitive and enterprising in their outlook. Former local rulers of Palnadu and Kondaveedu are known to be highly courageous and dashing. In fertile areas of the district, there is general inclination to modernise farming system. The quantum of use of fertilizers, herbicides, pesticides, rodenticides is generally higher. Agricultural practices embrace food, fodder, fuel, fiber and narcotics. The introduction of bole guard cotton by MNC and the suspicion of introduction of cotton seeds with 'terminator' gene hit headlines in the district. Recently, farmers suffered from overproduction of cotton. The district is a well known tobacco exporting centre.

The climate is overall of the hotter side and the area is referred to have 3 seasons in a year: hot, hotter and hottest.

Environmental Impact on Floristic Composition

During 1988–98, there are several changes in district for improvement of irrigation, laying of new Railway lines (the Nadikudi-Bibinagar), construction of additional bridges and initiation of 4-track highway through the district on NH5 and modification of arable crop land for building construction, enriched educational facilities and recreational spots (the Park Manasa Sarovar inauguration during 1999 near Takkellapadu village) laying of newer electrification activity into the interior villages and providing bus links to remote parts. These are some to have their impact on the floristic content, some might have been introduced from adjacent districts or states. Certain species might have disappeared especially due to increasing awareness and importance of effective herbal medicine. Plant introduction is always an activity influencing the floristic composition. The impact of harbour construction on Nizampatnam mangrove, has already been highlighted while enumerating mangroves.

The anthropogenic factors thus have an effective impact addressing to the need of frequent updation and revision, an aspect of which the authors are conscious about.

Chapter 5

KEY TO FAMILIES

1. Seeds present:
   2. Vascular bundles in a concentric ring, venation reticulate, flowers usually 4- or 5-merous, rarely 3-merous, cotyledons 2:
      3. Perianth 2-many seriate:
         4. Petals free, stamens often numerous:
            5. Segments of calyx distinct:
               6. Torus small or elongated, stamens hypogynous:
                  7. Plants aquatic, leaves peltate:
                     8. Leaves when mature raised above the water, carpels free, ovules 1 or 2 in each cell, stigma peltate .... *Nelumbonaceae*
                     8. Leaves floating, carpels connate, sunk in the receptacle, ovules numerous per cell, stigma radiating .... *Nymphaeaceae*
                  7. Plants terrestrial, leaves not peltate:
                     9. Carpels free when ripe:
                        10. Plants woody:
                           11. Stipules absent, connective of anthers produced into an oblong or truncate head .... *Annonaceae*
                           11. Stipules present, connectives not produced beyond the anthers .... *Magnoliaceae*
                        10. Plants herbaceous:
                           12. Leaves compound, terminal leaflet transformed into tendril, flowers bisexual, stamens many .... *Ranunculaceae*
                           12. Leaves simple, tendrils absent, flowers unisexual, stamens 4 to 6 .... *Menispermaceae*
                     9. Carpels not free when ripe:
                        13. Sepals 2 to 3:
                           14. Fruits and leaves prickly, latex yellow, ovary superior, petals 6 .... *Papaveraceae*
                           14. Fruits and leaves not prickly, latex colourless, ovary semi-inferior, petals 4 to 5 .... *Portulacaceae*
                        13. Sepals more than 3:
                           15. Placentation axile:
                              16. Leaves decussate, stipules persistent, sepals imbricate, petals persistent .... *Elatinaceae*
                              16. Leaves alternate, stipules deciduous, sepals valvate, petals deciduous:
                                 17. Filaments of anthers united:
                                    18. Stamens up to 15, anthers 2-celled .... *Sterculiaceae*
                                    18. Stamens more than 20, anthers 1-celled:
                                       19. Leaves simple or lobed, stamens monadelphous .... *Malvaceae*

19. Leaves palmately compound, stamens polydelphous .... *Bombacaceae*
17. Filaments of anthers free .... *Tiliaceae*
15. Placentation other than axile:
20. Leaves minute, scale like. Plants riparian .... *Tamaricaceae*
20. Leaves well developed:
21. Placentation free central .... *Caryophyllaceae*
21. Placentation parietal:
22. Flowers irregular, stamens connate or monadelphous:
23. Petals 3, stamens 8 .... *Polygalaceae*
23. Petals and stamens 5 each .... *Violaceae*
22. Flowers regular; stamens free:
24. Sepals persistent, plants thorny .... *Flacourtiaceae*
24. Sepals deciduous, plants not thorny:
25. Fruit an indehiscent berry .... *Capparaceae*
25. Fruit a dehiscent capsule:
26. Capsule 2-valved, flowers 0.5 to 6 cm across:
27. Small trees, sepals and petals 5 each; fruit echinate ... . *Bixaceae*
27. Herbs, sepals and petals 4 each; fruit glabrous or hairy, not echinate .... *Cleomaceae*
26. Capsule 3 to 5-valved, flowers ca. 15 cm across .... *Cochlospermaceae*
6. Torus expanded into a thickened or fleshy disc:
28. Plants herbs:
29. Leaves simple, flowers pale pink or pink .... *Balsaminaceae*
29. Leaves compound, flowers yellow or violet:
30. Leaves pinnate, flowers solitary, stamens 10 .... *Zygophyllaceae*
30. Leaves digitately 3- foliolate, flowers in umbels, stamens 5 .... *Oxalidaceae*
28. Plants trees, shrubs or twiners:
31. Leaves gland-dotted:
32. Fruits capsules .... *Flindersiaceae*
32. Fruits berries .... *Rutaceae*
31. Leaves not gland-dotted:
33. Leaves compound (rarely simple):
34. Placentation parietal, fruits elongated .... *Moringaceae*
34. Placentatin axile or axile-pendulous, fruits not elongated:
35. Leaves 2-foliolate, flowers in cymes .... *Balanitaceae*
35. Leaves 3-many foliolate, flowers in racemes or panicles:
36. Styles 5, fruit samarioid .... *Simaroubaceae*
36. Styles 1–2, fruit berries or drupes or capsules:
37. Stamens monadelphous, leaves opposite or subopposite, leaflets oblique; flowers bisexual .... *Meliaceae*
37. Stamens free, leaves alternate, leaflets usually not oblique; flower polygamous or polygamo dioecious or bisexual:
38. Flowers regular, sepals usually valvate, often persistent, fruit drupe with 2–5 pyrenes .... *Burseraceae*
38. Flowers regular or irregular, sepals imbricate, rarely valvate, not persistent, fruit loculicidal capsule .... *Sapindaceae*
33. Leaves simple:
39. Ovules 1 in each cell:
40. Stamens numerous, styles connate .... *Ochnaceae*
40. Stamens 5-10, styles not connate:
41. Leaves decussate, often with paired glands on petiole at leaf base .... *Malpighiaceae*

41. Leaves alternate, eglandular:
42. Stem with resin passages, flowers polygamodioecious or monoecious, in panicles or racemes; gynoecium of 1 carpel .... *Anacardiaceae*
42. Other than above:
43. Plants often armed, petals cucullate, not exceeding calyx, stamens enclosed by petals .... *Rhamnaceae*
43. Plants unarmed rarely armed, petals equal to or longer than sepals, stamens not enclosed by petals:
44. Ovary 1-celled .... *Opiliaceae*
44. Ovary 3-4-celled:
45. Drupe enclosed by calyx, stamens free or adnate to entire petals .... *Olacaceae*
45. Drupe not enclosed by calyx, stamens monadelphous, petals with an erect double ligule on the inner side .... *Erythroxylaceae*
39. Ovules 2 or more in each cell:
46. Stamens 10, sepals deciduous in fruit, styles 5, to 1 cm long .... *Linaceae*
46. Stamens 4 or 5, sepals persistent in fruit, style 1, short or 0:
47. Tendril bearing plants, inflorescence leaf-opposed, fruit berry .... *Vitaceae*
47. Plants not tendrillar, inflorescence axillary or terminal, fruit drupe or capsular .... *Celastraceae*
5. Segments of calyx united:
48. Leaf base pulvinous, fruits legumes:
49. Flowers actinomorphic .... *Mimosaceae*
49. Flowers zygomorphic:
50. Corolla papilionaceous, descendingly imbricate in bud, monadelphous or diadelphous stamens, all anthers fertile .... *Papilionaceae.*
50. Corolla not papilionaceous, ascendingly imbricate in bud, stamens free, often few sterile .... *Caesalpiniaceae*
48. Leaf base not pulvinous, fruits other than legumes:
51. Plants spiny:
52. Shrubs, stems succulents, flattened .... *Cactaceae*
52. Trees, stems not succulent, not flattened .... *Alangiaceae*
51. Plants not spiny:
53. Leaves compound or lobed, climbers .... *Passifloraceae*
53. Leaves simple, entire, non climbers:
54. Petals absent:
55. Plants mangroves .... *Sonneratiaceae*
55. Plants inland:
56. Ovules 1-3 per locul .... *Aizoaceae*
56. Ovules more than 3 per locule .... *Molluginaceae*
54. Petals present:
57. Leaves alternate, rarely clustered at the tip of branches:
58. Flowers in umbels; fruits of 2 indehiscent dorsally or laterally compressed mericaps .... *Apiaceae*
58. Flowers solitary, or in racemes, panicles, subcapitate spikes; fruit drupe or capsule:
59. Ovules on parietal placentae, styles 3, stigmas fringed .... *Turneraceae*
59. Ovules on other than parietal placentae, style single, stigmas simple or globose or lobed:

60. Flowers 6-12 cm across, the ligulate bundle of fertile, purple stamens arching over the pistil and the smaller bundle of yellowish stamens (mostly staminodes), fruit large, 15-20 cm dia .... ***Lecythidaceae***

60. Not as above characters:

61. Ovary 4-5-celled, ovules numerous per cell on axile placentae .... ***Onagraceae***

61. Ovary 1-2-celled, ovules 2–8 per cell on pendulous or horizontal placentae:

62. Foliar glands absent; inflorescence pendulous, petals scarlet, stamens numerous, pink, drupes not winged .... ***Barringtoniaceae***

62. Foliar glands sometimes present, inflorescence erect, petals cream or greenish yellow, stamens twice as many as calyx lobes, not pink, drupes winged, rarely entire .... ***Combretaceae***

57. Leaves opposite, decussate, rarely alternate:

63. Plant with vivipary, mangroves .... ***Rhizophoraceae***

63. Plant without vivipary, inland:

64. Bark thin, papery, brown, fruit usually indehiscent .... ***Myrtaceae***

64. Bark mostly absent, if present not papery, fruits dehiscent:

65. Erect herbs, viscid-pubescent ... ***Crassulaceae***

65. Erect herbs, trees or stragglers, not viscid-pubescent:

66. Calyx free from the ovary .... ***Lythraceae***

66. Calyx adnate to the ovary .... ***Melastomataceae***

4. Petals united, stamens often less than 12:

67. Ovary inferior:

68. Plants tendrillar climbers .... ***Cucurbitaceae***

68. Plants erect or ascending herbs:

69. Stipules absent, anthers often united around the style, flowers in heads .... ***Asteraceae***

69. Stipules present (intra or inter petiolar), anthers free, flowers not in heads .... ***Rubiaceae***

67. Ovary superior:

70. Plants aquatic:

71. Petal lobes contorted, stamens inserted in the tube of the petals .... ***Menyanthaceae***

71. Petal lobes imbricate, stamens inserted near the base of the petals .... ***Hydrophyllaceae***

70. Plants terrestrial:

72. Plants leafless, root parasites .... ***Orobanchaceae***

72. Plants leafy, autotrophs:

73. Leaves pinnate, seeds usually winged .... ***Bignoniaceae***

73. Leaves usually simple:

74. Flowers usually unisexual, stamens inserted on receptacle .... ***Ebenaceae***

74. Flowers usually bisexual, stamens inserted on corolla:

75. Ovary 1-celled:

76. Sepals, petals, stamens 4 each; ovules on basal placentation .... ***Salvadoraceae***

76. Sepals, petals, stamens 5 each; placentation other than above:

77. Ovules solitary on pendulous placentation .... ***Plumbaginaceae***

77. Ovules numerous on free central placentation .... ***Myrsinaceae***

75. Ovary 2-many-celled:

78. Branches strigose or scabrous .... *Boraginaceae*
78. Branches other than strigose or scabrous:
79. Inflorescence verticillaster or spicate, flowers zygomorphic, mature ovary separating into 1-seeded nutlets, style gynobasic .... *Lamiaceae*
79. Inflorescence other than verticillaster, mature ovary not separating into nutlets, style terminal:
80. Ovary 3- or more-carpelled, or if 2-carpelled, then stamens more than corolla lobes .... *Sapotaceae*
80. Ovary 2-carpelled, stamens as many as petals or fewer:
81. Corolla regular:
82. Leaves opposite, calyx deciduous in fruit (persistent in *Avicenniaceae*)
83. Stamens 2:
84. Fruits berries, or if capsules, then not compressed, woody .... *Oleaceae*
84. Fruit capsules, compressed, papery .... *Nyctanthaceae*
83. Stamens 4 or more:
85. Carpels and styles not always united, plants laticifereous:
86. Pollen aggregated into pollinia, gynostegium present .... *Asclepiadaceae*
86. Pollen not aggregated into pollinia, gynostegium absent, ovules on axile placentation .... *Apocynaceae*
85. Carpels and styles always united, plants not laticiferous:
87. Plants mangroves, roots with pneumatophores, ovary imperfectly 4-locular .... *Avicenniaceae*
87. Plants non mangroves, ovary 2-locular:
88. Plants trees or shrubs, ovules on axile placentation .... *Loganiaceae*
88. Plant herbs, ovules on peltate or parietal, intruded placentae .... *Gentianaceae*
82. Leaves alternate, calyx persistent in fruit:
89. Ovary cells 2-ovuled, rarely 4-ovuled, ovules erect or pendulous or on basal axile placentae:
90. Plants twiners or prostrate herbs, rarely erect shrubs, fruits capsules .... *Convolvulaceae*
90. Plants trees, fruits pulpy drupes .... *Cordiaceae*
89. Ovary cells many-ovuled on peltate axile placentae .... *Solanaceae*
81. Corolla irregular:
91. Carpels 2-many ovuled:
92. Fruits opening elastically from the apex of 2 loculicidal valves .... *Acanthaceae*
92. Fruits not opening elastically, rarely indehiscent:
93. Plants viscid-pubescent, seeds exalbuminous, ovules 2 or numerous on pendulous or axile placentae .... *Pedaliaceae*
93. Plants not viscid-pubescent, seeds albuminous, ovules numerous on swollen axile placentae .... *Scrophulariaceae*
91. Carpels 1-ovuled, stem 4-angled .... *Verbenaceae*
3. Perianth 1-seriate or absent:
94. Flowers bisexual:
95. Plants leafless parasitic:
96. Stems jointed .... *Viscaceae p.p.*
96. Stems not jointed .... *Lauraceae*
95. Plants leafy:
97. Ovary inferior:

98. Plants epiphytes, ovules solitary:
99. Perianth limb tubular .... *Loranthaceae*
99. Perianth limb campanulate .... *Viscaceae p.p.*
98. Plants not epiphytes:
100. Perianth irregular, ovules numerous on perietal connivent placentae, fruit a septicidal capsule .... *Aristolochiaceae*
100. Perianth regular, ovules 2-3, pendulous, fruit drupaceous .... *Santalaceae*
97. Ovary superior:
101. Leaves stipulate .... *Polygonceae*
101. Leaves estipulate:
102. Leaves succulent, glabrous, utricles enclosed in fleshy perianth .... *Basellaceae*
102. Leaves not succulent, pubescent, utricles not enclosed in fleshy .... perianth:
103. Perianth tubular, not scarious .... *Nyctaginaceae*
103. Perianth lobes not tubular:
104. Perianth scarious and dry, leaves membranous .... *Amaranthaceae*
104. Perianth membranous, leaves fleshy .... *Chenopodlaceae p.p.*
94. Flowers unisexual or polygamous:
105. Leaves reduced to scales .... *Casuarinaceae*
105. Leaves not as above:
106. Stipules absent:
107. Plants mangroves .... *Chenopodiaceae p.p.*
107. Plants other than above:
108. Plants terrestrial .... *Hernandiaceae*
108. Plants aquatics .... *Ceratophyllaceae*
106. Stipules present:
109. Ovary 3-celled, 3-locular .... *Euphorbiaceae*
109. Ovary1-celled, 1-locular:
110. Latex milky, fruits mostly figs .... *Moraceae*
110. Latex not milky, fruits other than figs:
111. Stamens inflexed in bud, fruit achenes .... *Urticaceae*
111. Stamens not inflexed in bud, fruit drupaceous or winged samara .... *Ulmaceae*

2. Vascular bundles scattered, venation parallel, flowers typically 3-merous, cotyledons 1:
112. Plants stout, arborescent:
113. Leaves pinnately or palmately divided, without spines stilt roots absent, perianth in two whorls .... *Arecaceae*
113. Leaves entire, margins spiny, stilt roots present, perianth absent .... *Pandanaceae*
112. Plants herbaceous:
114. Perianth 2-seriate:
115. Ovary 3-celled, axile placentation .... *Liliaceae*
115. Ovary 1- or 2-celled, ovules on other than axile placentation:
116. Seeds minute, exalbuminous:
117. Flowers unisexual, plant aquatics, outer perianth segments calycine .... *Hydrocharitaceae*
117. Flowers bisexual, plants epiphytes or terrestrial, outer perianth segments corolline .... *Orchidaceae*
116. Seeds conspicuous, albuminous:
118. Ovary inferior:
119 Flowers irregular, leaves usually distichous, stamen 1 .... *Zingiberaceae*

119. Flowers regular, leaves radical, stamens 6 .... *Amaryllidaceae*

118. Ovary superior:

120. Outer perianth calycine, inner perianth petaloid, ovules 1 or 2 in each cell, plants terrestrial .... *Commelinaceae*

120. Two whorls of perianth-petaloid, free floating or rooted aquatic herbs, ovules numerous per cell .... *Pontederiaceae*

114. Perianth 1-seriate or reduced to scales or absent:

121. Plants mostly terrestrial or marshy:

122. Inflorescence a solitary head .... *Eriocaulaceae*

122. Inflorescence a spike or panicle:

123. Leaves fleshy, spine tipped .... *Agavaceae*

123. Leaves not fleshy, not spine-tipped:

124. Stems solid, style 1, leaves 3-ranked .... *Cyperaceae*

124. Stems hollow except at nodes, styles 2, leaves 2- ranked ....*Poaceae*

121. Plants mostly aquatic:

125. Leaves in a rosette, roots tufted .... *Araceae*

125. Leaves not in rosette, roots not tufted:

126. Leaves 2-3, floating herbs .... *Lemnaceae*

126. Leaves more than 5, not floating:

127. Flowers axillary, solitary, or in clusters .... *Najadaceae*

127. Flowers in racemes or spikes:

128. Flowers unisexual, inflorescence drying ....*Typhaceae*

128. Flowers bisexual, inflorescence not drying .... *Aponogetonaceae*

1. Seeds absent:

129. Plants aquatic:

130. Plants rooted in the mud:

131. Leaflets four, sporangia in sporocarps at the base of stipes .... *Marsileaceae*

131. Leaflets many, pinnately arranged; sporangia on narrow leaflets with reflexed edges .... *Parkeriaceae*

130. Plants floating:

132. Leaves many, minute (below 1 × 0.75 mm), normal leaves covered by simple trichomes above .... *Azollaceae*

132. Leaves above 1.5 × 2 cm, normal leaves covered by branched hairs .... *Salviniaceae*

129. Plants terrestrial:

133. Fronds palmate, fan like .... *Polypodiaceae*

133. Fronds pinnatifid .... *Pteridaceae*

Chapter 6

SYSTEMATIC ENUMERATION

RANUNCULACEAE

Clematis Linnaeus

Clematis gouriana Roxb. ex DC. Syst. 1: 139. 1817; FBI 1: 4; Gamble 3 (2).

Woody climbers, up to 10 m long; leaves 2–3-pinnate, to 14 × 10 cm; leaflets oblong-lanceolate, 3–6 × 1–2.5 cm, 3–5-nerved, chartaceous, pubescent below, base obtuse or cordate, entire, acuminate; petiole twining; flowers cream, in dense panicles; achenes sessile, ovoid; style feathery; seeds linear.

Rare climber on hills on trees. Fl. & Fr.: Nov.–Feb. *Vern.:* Gowri-kuntala. Leaf paste is applied to wounds of cattle. Reported in Guntur Forest Division Work Plan.

MAGNOLIACEAE

Michelia Linnaeus

Michelia champaca L. Sp. Pl. 536. 1753; FBI 1: 42, Gamble 9(6).

Medium sized trees, up to 15 m tall. Leaves ovate-lanceolate, coriaceous, glabrescent, base subacute or cuneate, margin entire, apex long-acuminate; flowers solitary, axillary, pale yellow, very fragrant. Fruit an aggregate of follicles, fruitlets warty, seeds ca. 5, subglobose, red.

Cultivated in parks, gardens, NU campus, Mangalagiri. Flowers during Feb.–May. Fruits of two successive years occur together. *Vern.*: Chettu sampangi, Simhachalam sampengi.

ANNONACEAE

1. Fruit a ring of one-seeded berries. Inner petals prominent ...... ***Polyalthia***
1. Fruit of confluent carpels, fleshy. Inner petals scaly or 0 .... ***Annona***

Annona Linnaeus

1. Leaves obtuse, glaucous beneath. Fruit green, ovoid, with projecting ovoid areoles .... ***A. squamosa***
1. Leaves acuminate, green beneath. Fruit orange, subglobose with flat areoles .... ***A. reticulata***

Annona reticulata L. Sp. Pl. 537. 1753; FBI 1: 78; Gamble 20 (14).

Small deciduous trees, up to 6 m tall; leaves elliptic, oblong-lanceolate, 4.5–8 × 2.5–4 cm, thin coriaceous, glabrescent, base cuneate, apex acuminate; flowers in clusters or solitary, greenish-yellow; fruit cordate or ovoid, to 7 cm across; monocarps 1-seeded, brownish.

Cultivated for its fruits. Fl. & Fr.: Through the year. *Vern.*: Ramaphalam. Fruits are said to be used in curing biliousness and diseases of blood; ripe fruits are edible.

Annona squamosa L. Sp. Pl. 537. 1753; FBI 1: 78; Gamble 20 (14).

Small deciduous trees, up to 5 m tall; bark grey; leaves simple, elliptic-oblong, lanceolate, 5–8 × 7–3 cm, glabrous, acute to cuneate, entire; flowers greenish-yellow, solitary, or a few together, axillary or leaf-opposed fascicles; syncarps globose, green, grooved; pulp sweet; seeds many, angular, oblong, black, shining.

Occasional in forests, cultivated in gardens for its fruit. Introduced from tropical America by the Portuguese. Fl. & Fr.: Apr.–Nov.. *Vern.*: Seethaphalamu. Fruits edible. The powdered seed is believed to possess valuable insecticidal properties. N. Varam, *VRK* 3967.

Polyalthia Blume

Polyalthia cerasoides (Roxb.) Beddome Fl. Sylv. t. 1. 1869; FBI 1: 63; Gamble 17(12). *Uvaria cerasoides* Roxb. Pl. Cor. t. 33. 1795.

Semi evergreen trees, up to 10 m tall; bark light grey, rough, lenticellete; leaves simple, oblong, lanceolate, 3.5–15 × 2–5 cm, dark green, glabrous and glossy above, puberulous to pubescent beneath, apex acuminate, margin entire, base obtuse; flowers greenish, axillary or extra axillary; monocarps 35–40 in a cluster, reddish, ellipsoid to broadly ovoid, apiculate, seed one, brown, globose. .

Occasional in hill slopes, along ravines in dry deciduous forests. Fl. & Fr.: Apr.–Aug. *Vern*: Gutti. Davupalli, *SSR & KSM* 21652.

Artabotrys hexapetalus (L.f.) Bhandari is cultivated for its fragrant flowers. *Vern.*: Tiga sampangi. *Polyalthia longifolia* (Sonn). Thw. is planted as an avenue tree in the district in front of buildings and gardens. *P. longifolia* Hook. var. *pendula*, often treated as a different species, *P. pendula* cultivated as an avenue tree; monopodial with drooping branches. *Cananga odorata* (Lam.) Hook. f. & Thoms, tree, branches drooping, flowers greenish, scented. Occasionally cultivated. *Vern.*: Sakalaphala sampangi.

MENISPERMACEAE

1. Leaves peltate .... ***Cissampelos***
1. Leaves not peltate:
    2. Leaves villous, drupes compressed .... ***Cocculus***
    2. Leaves glabrous, drupes globose or ovoid:
        3. Leaves pinnately nerved, seeds horse-shoe shaped .... ***Tiliacora***
        3. Leaves palmately nerved, seeds globose or oblong .... ***Tinospora***

Cissampelos Linnaeus

Cissampelos pareira L. var. **hirsuta** (Buch.-Ham. ex DC.) Forman in Kew Bull.22: 356. 1968. *C. hirsuta* Buch.-Ham. ex DC. Syst. 1: 535. 1817. *C. pareira* L. Sp. Pl. 1031. 1753 *p.p.*; FBI 1: 103; Gamble 30 (21).

Climbing shrubs; leaves reniform or orbicular, 4.5 × 6.5 cm, peltate, chartaceous, 5–7-nerved, pubescent, retuse; base cordate; flowers unisexual, greenish, in corymbose cymes, in foliaceous bracts; drupes ovoid, 4 cm across.

Occasional in forests. Fl. & Fr.: Throughout the year. *Vern*: Adavi banka teega. Leaves and roots are used as an antidote for snake-bite and scorpion stings and for curing cough, dyspepsia, dropsy and diarrhoea. An indigenous substitute for curare, d-tubocurarine and other curariform drugs. Stem fibres are used for manufacturing ropes. Kondavedu Fort, *CAB* 4647 (MH).

Cocculus A.P. de Candolle

Cocculus hirsutus (L.) Diels in Engl. Pflanzenr. IV. 94. 276: 1910; Gamble 29 (21). *Menispermum hirsutum* L. Sp. Pl. 341. 1753. *Cocculus villosus* (Lam.) DC. Syst. 1: 525. 1817; FBI 1: 101. *Menisperum villosum* Lam. Encycl. 4: 97. 1797.

Straggling or climbing hirsute evergreen shrubs, leaves variable, ovate, ovate-oblong or sagittate, 2–6 × 1.5–5 cm, pubescent above, velvety-tomentose below, base cordate, obtuse or truncate, entire, apex obtuse or retuse mucronate; flowers greenish yellow, male in slender panicles, female in fascicles; drupes laterally compressed, endocarp transversely ridged, purplish-black.

Common climber on bushes and trees in forests and hedges. Fl. & Fr.: Nov.–Mar.. *Vern.:* Dusaraputiga. Koppanur. *VRK* 3245; Javuku, *VRK* 3297, Loyapalle, *VRK* 3824.

Tiliacora Colebrooke

Tiliacora acuminata (Lam.) Miers in Ann. Hist. Nat. Ser. 2. 7: 39. 1851; Gamble 28 (20); Forman in Kew Bull. 30: 90. 1975. *Menispermum acuminatum* Lam. Encycl. 4: 101. 1797. *Tiliacora racemosa* Colebr. in Trans. Linn. Soc. London 13: 67. 1822; FBI 1: 99.

Large woody, evergreen climbing shrub, stems from a woody base; leaves ovate to ovate-lancolate, cordate, 8–12 × 7–8 cm, dark green, glabrous, acute-acuminate, entire, base obtusely rounded or truncate to subcordate, prominently veined; flowers greenish-yellow, small, in axillary panicles; drupes subovoid, red.

Common along the streams in forests. Fl. & Fr.: Sept.–Jan. *Vern.*: Theendra Theega. Roots mixed with water given as a drink for cure in snake bite. Javuku, *VRK* 3861 & 3872; Kondavedu, *SSR & KSM* 21440.

Tinospora Miers

Tinospora cordifolia (Willd.) Miers ex Hook. f. & Thoms. Fl. Ind. 184. 1855; FBI 1: 97; Gamble 26(19). *Menispermum cordifolium* Willd. Sp. Pl. 4: 826. 1806.

Deciduous woody straggler up to 10 m high; leaves broadly ovate-cordate, 2–6 × 2.5–5 cm, glandular-papillose patches below, entire, apex acute, sinuate at base; flowers yellow, male in fascicles, female solitary; drupes globose.

Occasional in forests. Fl. & Fr.: Jan.–Aug. *Vern.*: Tippa teega. An ingredient in several Ayurvedic preparations; used for general debility, dyspepsia, fevers and urinary diseases and to heal fractures. Dried stem powder washed with water, mixed with honey is used to treat coughs. The juice of the plant is largely used for the treatment of stomach ailments. Javuku, *VRK* 3284; Kondavedu *SSR & KSM* 21481.

NELUMBONACEAE

Nelumbo Adanson

Nelumbo nucifera Gaertn. Fruct. 1: 73. 1788. *Nelumbium speciosum* Willd. Sp. Pl. 2: 1258. 1799; FBI 1: 116; Gamble 34 (24).

Large aquatic perennial herbs with stout creeping underground rhizomes; leaves rotund, 30–50 cm across, mature ones raised up to 70 cm above water surface, glaucous beneath, entire; flowers light rose; carpels numerous, free, loosely embedded on receptacle.

Abundant in tanks of Repalle and Tenali. Fl. & Fr.: Through the year. Sacred lotus of India. It is the national flower. *Vern.:* Thamara, Kamalam,

Padmam. Flowers for religious functions, tuberous rhizomes and ripe carpels are edible and are sold in market. The flowers are used in medicine and the leaves as plates. Nizampatnam, *VRK* 6820.

NYMPHAEACEAE

Nymphaea Linnaeus *nom. cons.*

1. Leaves glabrous on both surfaces .... *N. nouchali*
1. Leaves velvety pubescent below .... *N. pubescens*

Nymphaea nouchali Burm. f. Fl. Ind. 120. 1768. *N. stellata* Willd. Sp. Pl. 2: 1153. 1799; FBI 1: 114; Gamble 33 (24).

Aquatic rhizomatous herbs; leaves floating, elliptic-orbicular to sub rotund, glabrous, green above, pale purple below, repand to irregularly sinuate-dentate with broad obtuse teeth or sub-entire; flowers bluish-purple to pale pink; anthers brown, with creamy appendages; stigmas bright orange.

Occasional in ponds. Fl. & Fr.: Through the year. *Vern.*: Kaluva.The tuberous rhizomes and peduncles are eaten as vegetable, seeds on frying are eaten as puffed grain. Found in N.U. Campus (Mohan Rao & Aasiya, 1995).

Nymphaea pubescens Willd. Sp. Pl. 2: 1154. 1799; Gamble 34 (24). *N. lotus* auct. non L. 1753; FBI 1: 114. *N. lotus* L. var. *pubescens* (Willd.) Hook. f. & Thoms. Fl. Ind. 241. 1855; FBI 1: 114.

Aquatic fresh water rhizomatous herbs; leaves long petioled, peltate, orbicular, sagittate when young, 15–25 × 12–20 cm, pubescent below, sharply dentate, venation radiatingly reticulate; flowers white or pink, solitary, 4–15 cm across, slightly fragrant, arising from the rhizome; berries spongy; seeds minute.

Occasional in ponds. Fl. & Fr.: June–Nov. *Vern.:* Kaluva. Uses are similar to above. Repally RF, *VRK* 5777.

PAPAVERACEAE

Argemone Linnaeus

Argemone mexicana L. Sp. Pl. 508. 1753; FBI 1: 17; Gamble 35 (25).

Annual prickly herbs, up to 75 cm tall; divaricately branched; stems greyish-white with yellow sap; leaves sessile, oblong-obovate, 7–18 × 3–6 cm, glaucous, prickly, sinuate-pinnatifid, cordate or semi-amplexicaul at base; flowers yellow, terminal, solitary; capsules oblong, 4 × 2 cm, prickly; seeds blackish-brown.

Occasional in wastelands, roadsides and open forests. Native of Central America, introduced into India during 17th century. Fl. & Fr.: Throughout the year. *Vern.:* Yerrikusuma. The yellow juice of the plant is used as a medicine for dropsy, jaundice and cutaneous infections. It is also used as diuretic, to relieve blisters and ulcers. The seed is used as a medicine for ulcers and eruptions and also for burning in lamps. N. Varam, *VRK* 3593; Nallapadu tank, *VRK* 6768.

BRASSICACEAE

Brassica juncea (L.) Czern (Mustard, *Vern.:* Telugu : Avalu) is cultivated for its seeds. Varieties of *Brassica oleracea* L. viz., var. *botrytis* L. (*Vern.* Cauliflower), var. *capitata* L. (*Vern.* cabbage) and var. *gongylodes* (*Vern.* Knol-kohl) of Brassicaceae are cultivated for vegetables.

CAPPARACAE

1. Calyx tubular below .... *Maerua*
1. Calyx divided to the base into 4 sepals:
   2. Stamens 4 or 6, inserted at the middle of the gynophore .... *Cadaba*
   2. Stamens >20, inserted at the base of the gynophore:
      3. Leaves simple, sepals unequal .... *Capparis*
      3. Leaves digitately 3-foliolate, sepals equal .... *Crataeva*

Cadaba Forsskal

Cadaba fruticosa (L.) Druce Bot. Exch. Club. Soc. Brit. Isles 3: 415. 1914. *Cleome fruticosa* L. Sp. Pl. 671. 1753. *Cadaba indica* Lam. Encycl. 1: 544. 1785; FBI 1: 172; Gamble 43 (31).

Erect branched shrubs, up to 3 m tall; leaves oblong or oblong-lanceolate, 2.5–5.5 × 1–1.8 cm, green, glabrescent, glaucous, obtuse and mucronate, entire, base rounded; flowers greenish-yellow in terminal, corymbs; gynandrophore elongated; berries cylindrical, seeds embedded in orange yellow pulp.

Common in forests. Fl. & Fr.: Apr.–Aug. *Vern.*: Yenuka-parike, Adda morinke. Chenchu tribes use leaves for treating indigestion. Leaves are crushed and 3 spoonfuls of the filtrate taken daily for 3 days. Javuku, *VRK* 6758; Loyapalli *SSR & KSM* 21626; Nallapadu, *VRK* 6758.

Capparis Linnaeus

1. Mature branches leafless .... *C. decidua*
1. Mature branches leafy:
   2. Plants unarmed:
      3. Flowers in terminal corymbs .... *C. grandis*
      3. Flowers in axillary umbels .... *C. floribunda*
   2. Plants armed:
      4. Flowers solitary:
         5. Berry smooth .... *C. divaricata*

5. Berry corrugate .... *C. brevispina*
4. Flowers in racemes, umbels or clusters:
6. Flowers in umbels, berries 1 cm across. .... *C. sepiaria*
6. Flowers in axillary clusters, berries 4 cm across .... *C. zeylanica*

Note: Plants usually noctiflorus, with flowers opening at dusk and pollinated by butterflies, moths or other nocturnal insects. Seed dispersal is effected by mammals or birds. Sterility is prevalent in both sexes.

Capparis brevispina DC. Prodr. 1: 246. 1824; Gamble 45 (32). *C. zeylanica* auct. non L.; FBI 1: 194.

Shrubs to 4 m, branchlets stellate-pubescent; leaves elliptic or oblong, coriaceous, apex obtuse or rotund, mucronate, margin entire, base cuneate or subacute; thorns often straight; flowers cream, solitary, axillary; berry ellipsoid, beaked; seed 3 mm.

Occasional in Nagarjuna University campus (Mohan Rao and Aasiya, 1995). Fl. & Fr.: Feb.–Aug. The green fruits are sliced, dried, cooked and eaten in Sri Lanka (Trimen, Handb. Fl. Ceylon **1**: 61. 1893).

Capparis decidua (Forssk.) Edgew. in J. Linn. Soc. Bot. 6: 184. 1872. *Sodada decidua* Forssk. Fl. Aeg.-Arab. 81. 1775. *Capparis aphylla* Roth Nov. Pl. Sp. 238. 1821; FBI 1: 174; Gamble 45 (32).

Straggling shrubs; branches smooth, green; leaves in the young shoots only, small, less than 1 cm long, linear-oblong, acute, spinous-tipped; stipular thorns long, straight, sharp; flowers red in many-flowered corymbs from the old branches; fruit globular, 1.5 cm across, beaked, smooth, red when ripe.

Occasional in Nagarjuna University campus area (Mohan Rao and Aasiya, 1995). Fl. & Fr.: Feb.–Aug. Flower buds and fruits are used as vegetables and pickles. Young shoots, leaves, roots and fruits are powdered and used in treating boils, skin eruptions, rheumatism etc. Wood resistant to white ants and hence utilised for rafters, beams, ploughs, and country boats. Often planted as a sand binder to arrest spread of deserts.

Capparis divaricata Lam. Encycl. 1: 606. 1785; FBI 1: 174. *C. stylosa* DC. Prodr. 1: 246. 1824; Gamble 45 (32).

Small trees, up to 5 m tall; leaves variable, dimorphic, either linear to linear-oblong or elliptic-lanceolate or oblong, 4–6 × 1–2 cm, pubescent, apex acute-mucronate, entire, base cuneate or rounded, midrib somewhat raised; flowers yellow, axillary, solitary; gynophore 1.5–2 cm long, berry 5 × 2 cm, ribbed, apiculate, red when ripe.

Common in Loyapalle section. *Vern.:* Badreni. Fl. & Fr.: Mar.–July. Loyapalle, *VRK* 3997; Nagarjunakonda, *KTH* 9723 (CAL).

Capparis floribunda Wight Ill. 1: 33. t. 14. 1840; FBI 1: 177; Gamble 46 (33).

Straggling shrubs, unarmed; twigs straight, leaves simple, alternate, oblong or elliptic, ovate-lanceolate, 6–8 × 3–3.5 cm, base obtuse, entire, tip mucronate, glabrous; midrib slightly grooved above, flowers whitish-green in umbellate panicles; gynophore shorter than stamens; fruits fleshy, globose, green, smooth, many seeded.

Occasional along the streams. Fl. & Fr.: Jan.–Aug.. Loyapalle, *VRK* 3837; Davupalli, *SSR & KSM* 21420.

Capparis grandis L. f. Suppl. Pl. 263. 1781; FBI 1: 176; Gamble 46 (33).

Shrubs or trees up to 8 m tall; branchlets tomentose; leaves obovate, sub orbicular or elliptic-ovate, 3–5 × 2–3.5 cm, rusty to velutinous below, acute, entire, base cuneate; flowers creamish, fragrant, in terminal corymbs; berries globose, orange red to purple, seeds 2–6, embedded in pink pulp.

Occasional in forests. Fl. & Fr.: Mar.–July. *Vern.*: Reguti, Guli, Nallaupi. Kondavedu fort, *JSG* 17490 (MH).

Capparis sepiaria L. Syst. Nat. ed. 10. 2: 1071. 1759; FBI 1: 177; Gamble 46 (33).

Prickly shrubs, 2–3 m tall; leaves ovate-oblong, oblong-lanceolate, 1–2 × 0.6–1.2 cm, pubescent, emarginate or retuse, entire, base acute, rounded or rarely subcordate; flowers white in axillary corymbose umbels; berries globose, 1 cm across; seeds numerous.

Common in woodlands, hedges and open forests. Fl. & Fr.: Apr.–July. Vern.: Nalla uppi. It can be raised from cuttings, can be used as a hedge plant. Berries are edible. It is an alterative, tonic and febrifuge and for skin affections. Loyapalle, *VRK* 3976; Bollapalle, *VRK* 5746; Nagarjunakonda, *K. Thotharthri* 9727 (CAL).

Capparis zeylanica L. Sp. Pl. ed. 2. 720. 1762; Gamble 46 (33). *C. horrida* L. f. Suppl. Pl. 264. 1781; FBI 1: 178.

Large climbing armed shrubs, twigs densely covered with rufous, greyish stellate scurfy deciduous tomentum; leaves ovate or elliptic, 3–6 × 1.5–5 cm, glabrous or glabrescent, obtuse, entire, base subcordate or cuneate; stipules 2, modified into spines; flowers whitish-yellow, stamens many with violet filaments, berries subglobose, reddish brown; seeds many.

Common in open forests and hedges. Indicator of scrub jungle, deciduous low lying forests. Unripe fruits are pickled and mature ones are edible. Fl. & Fr.: Jan.–Apr. *Vern.:* Adavi donda. Sattenapally RF, *VRK* 6769; Vinukonda RF, *VRK* 6870. Loyapalli, *SSR & KSM* 21632.

Note: The nomenclatural problem in this species has been dealt with by Jacobs (in Blumea 12: 509. 1965). Unaware that Linneaus had already described the species based on Hermann's collection from Sri Lanka, the younger Linnaeus described *C. horrida* based on Koenig's collection, also from Sri Lanka. Wight & Arnott (Prodr. Fl. Pen. Ind. Or. 25. 1834, excl. synonym *pyrifolia* Lam.) had correctly interpreted both the species but failed to recognise that both are related to one and the same species.

The description of *C. zeylanica* in Hook. f. (Fl.Brit. India 1: 174. 1872) pertains to *C. brevispina* DC., an error repeated by many other Indian botanists, except by Dunn who could distinguish *C. zeylanica* from *C. brevispina*.

Crateva Linnaeus

Crateva magna (Lour.) DC. Prodr. 1: 243. 1824. *Capparis magna* Lour., Fl. Cochinch. 1: 331. 1790. *Crataeva nurvala* Buch.-Ham. in Trans. Linn Soc. London 15: 121. 1827. *C. religiosa* var. *nurvala* (Buch.-Ham.) Hook. f. & Thoms. in FBI 1: 172. 1372. *C. religiosa auct. non.* G. Forster; Gamble 47 (34).

Densely foliaceous, deciduous tree, up to 10 m tall, bark grey, fairly smooth with horizontal wrinkles; leaves 3-foliolate, leaflets chartaceous, ovate, lanceolate, glabrous, apex abruptly acuminate, margin entire, slightly undulate, base attenuate; flowers greenish-white in dense terminal corymbs; berry oblong-ellipsoid or oblong-ovate, woody, yellowish grey, seeds numerous, embedded in pulp, dark brown, tubercled.

Occasional, in deciduous forests, along streams, often planted in temples and gardens. Fl. & Fr.: Apr.–Aug.. *Vern*: Maga lingamu, utimidi, young berries are edible. Leaves are bitter and used in treating skin diseases. It is a valuable, species of sacred belief, hugging tree may create fertility in women as the belief goes. It is an alterative, astringent, antiperiodic, antipyretic, cholagogue, demulcent, diuretic, laxative, as antidote to snake bite and for stomach ache or liver complaints.

Maerua Forsskal

1. Stragglers, leaves simple, berries moniliform .... *M. oblongifolia*
1. Trees, leaves 3-foliolate, berries ovoid .... *M. apetala*

Maerua apetala (Roth) Jacobs in Blumea 12: 207. 1964. *Capparis apetala* Roth Nov. Pl. Sp. 238. 1821. *Niebuhria linearis* DC. Prodr. 1: 244. 1824; FBI 1: 171. *N. apetala* Dunn in Gamble Fl. Pres. Madras 41 (30), 1915.

Unarmed trees up to 10 m, leaves only on young shoots, 3-foliolate, leaflets linear-oblong, 4–7 × 1–1.5 cm, glabrous; flowers creamish in terminal corymbs; berries ovoid; seeds solitary.

Occasional in forests. Fl. & Fr.: July–Dec.. *Vern.*: Lukki chettu. A paste of the root bark is applied for leucoderma and the extract given orally for the same by Chenchus and Lambadis. Tender leaves ground with spices and the paste made into pills are given orally for nervous disorders and foot pains by Chenchus. N. Varam, *VRK* 3939.

Maerua oblongifolia (Forssk.) A. Rich. Tent. Fl. Abyss. 1: 32. t. 6. 1847. *Capparis oblongifolia* Forssk. Fl. Aegypt - Arab. 99. 1775. *Maerua* *arenaria* (DC.) Hook. f. & Thoms. in Hook. f. Fl. Brit. India 1: 771. 1872; Gamble 42 (30). *Niebuhria arenaria* DC. Prodr. 1: 244. 1824.

Bushy unarmed stragglers; leaves ovate, 2–5 × 1–2 cm, scabrous, mucronate, obtuse; flowers greenish in corymbs; berries moniliform, twisted; seeds globose, smooth.

Occasional in forests. Fl. & Fr.: Jan.–May. *Vern.:* Putta Tige. Leaves serve as food for camels and goats. Macherla RF, *VRK* 3837; Loyapalli *SSR* & *KSM* 21663; Guntur, *JSG* 18611 (DD).

Notes: An extremely variable species with reference to size, shape, texture, indumentum and venation of leaves.

CLEOMACEAE

Airy Shaw (Kew Bull. 18: 256. 1965) raised the section—Stanleyeae, of Pax (Pflanzenf. ed. 2. 17b. 146. 1936) to that of a family, stating that the family is intermediate between Capparaceae and Cruciferae differing from the former in the glandular covering, and in the fruit being a cruciferous siliqua with a replum; from the latter in the digitately 3–7-foliolate (rarely simple) leaves.

Cleome Linnaeus

1. Gynandrophore elongate, longer than petals .... *C. gynandra*
1. Gynandrophore absent:
    2. Plants viscous .... *C. viscosa*
    2. Plants not viscoous:
        3. Fruits compressed, flat, stamens more than 20 .... *C. felina*
        3. Fruits terete or subterete, stamens 6:
            4. Leaflets filiform, stems glabrous .... *C. angustifolia*
            4. Leaflets oblanceolate, stems and leaves asperous with minute scattered prickles .... *C. aspera*

Cleome angustifolia Forssk. Fl. Aeg. Arab. 120. 1775. *C. tenella* L. f. Suppl. Pl. 300. 1781; FBI 1: 169; Gamble 49 (29).

Herbs 10 cm, tall, leaves 3-foliolate, leaflets linear; flowers purplish in racemes; capsule sub terete, seeds with transverse ridges.

Rare in waste lands, rocky slopes in the under growth near Ananda Jyothi opposite to NU campus.

Cleome aspera Koen. ex DC. Prodr. 1: 241. 1824; FBI 1: 169; Gamble 41 (29).

Annual, much-branched ascending herb; leaves 3-foliolate, leaflets obovate or oblanceolate, 1–2 × 0.3–0.6 cm, obtuse, base cuneate; flowers yellow, axillary, solitary; capsules cylindrical; seeds 10.

Occasional in wastelands, sandy or gravelly soils and forests. Fl. & Fr.: July–Jan. Macherla RF, *VRK* 5780; N.U. Campus, *VRK* 6985; Nakkerikel, *KCJB,* 17467 (MH).

Note: Flowers are yellow but Gamble (l.c) mentions 'pink'.

Cleome felina L. f. Suppl. Pl. 300. 1781; FBI 1: 170; Gamble 41 (29).

Prostrate decumbent, hairy herbs, branches radiating from base, ascending, stems and leaves entirely covered with stiff appressed scale like hairs; leaves 3-foliolate, clothed with rigid scale-like hairs, leaflets obovate, 0.8–2 × 0.5–1 cm, glandular, obtuse, entire, ciliate, base cuneate; flowers pink, solitary, stamens usually 50 or more; capsules linear, flat, like legume; beaked at apex, glabrous, seeds reniform, yellowish brown.

Occasional in open forests and wastelands. Fl. & Fr.: July–Jan. N. Varam, *VRK* 3512; Mellavagu, *VRK* 5712; Bollapalle, *KCJB* 17437 (MH).

Cleome gynandra L. Sp. Pl. 671. 1753; Sant. & Wagh in BBSI 15(2): 106. 1933; Iltis in Brittonia 12(4): 284. 1960. *Gynandropsis pentaphylla* DC. Prodr. 1: 238. 1824; FBI 1: 171; Gamble 40 (29).

Erect branched herbs, up to 50 cm tall; leaves 5-foliolate, leaflets obovate, oblanceolate or rhombate, 0.5–2 × 0.2–3 cm, hispid, acute, entire, cuneate at base, flowers white or creamy yellow in corymbose racemes; capsules cylindrical, 4–6 × 0.2–0.3 cm, seeds numerous, black.

Rare weed of cultivated fields and roadsides. Fl. & Fr.: June–Dec.. *Vern.:* Atakura; Laja gandla, Totta vaminta, Vaminta. Plants and tender leaves used as herbage and as vegetables after boiling and salting. Leaves used in treating rheumatism and leaf juice for headache by applying to fore head and pouring into nostrils. Seed oil is efficaceous in killing lice. The medicinal properties of the herb is due to cleomine present in it. B.K. Palem, *VRK* 3252; Nagarjunakonda, *KTH* 9676 (CAL).

Note: Following Iltis (1960) C.P. De Wolf (Fl. Trop. East. Africa, 1964) treated *Gynandropsis* DC. under *Cleome* L. as *Cleome gynandra* L. But Hubbard & Milne—Red Head, the editors of Flora of Tropical East Africa (1964) kept the taxon under *Gynandropsis* only based on the presence of androphore while stating that Iltis admitted that this species occupies a very isolated position with in *Cleome* and no near relations are known. Krishna Mitra (BBSI 12(1–4): 7–31. 1975), Vijaya Kumar and Bir Bahadur (JIBS 57: 39–46. 1978) based on pollen and seed morphological studies respectively support the merging of *Gynandropsis* with *Cleome.*

The data on foliar epidermal characters presented by Balamani and Rolla S. Rao (Ind. Journ. Bot. 5: 1982) have clearly supported the position of *Cleome gynandra* positively in between *C. viscosa* and *C. monophylla* besides their interrelationship. Thus it is clearly evident that *C. gynandra* once considered under the genus *Gynandropsis* DC. is not only very close or rather identical to the genus *Cleome* L. but also takes the position in line with other species of *Cleome* and not at all an isolated position as it was considered earlier on the basis of few external morphological characters only.

Cleome viscosa L. Sp. Pl. 672. 1753; FBI 1: 170; Gamble 41 (29). *Polanisia icosandra* Wight & Arn. Prodr. Fl. Ind. Orient. 22. 1834. 2. 1838.

Erect herbs, up to 1 m tall; branchlets viscid to glandular; leaves 3–5-foliolate; leaflets obovate or elliptic, middle one 2.4–5 × 1–2 cm, laterals 1.5–3.5 × 0.8–1.5 cm, inequilateral, base cuneate, ciliate, acute; flowers yellow in terminal racemes; capsules terete, up to 7 cm, seeds ovoid-globose.

Key to the Varieties

1. Stamens 36–40; styles 5–8 mm long in flower, elongating to 20 mm long in fruits .... var. ***nagarjunakondensis***
1. Stamens 12–18; styles 1–3.5 mm long in flower, 4–5 mm long in fruits .... var. ***viscosa***

var. **nagarjunakondensis** Sund.—Ragh. in Bull. Bot. Surv. India 28: 187–191, t. 1–8. 1986 (1988); Flora of India 2: 318. 1993.

Herbs, up to 1–2 m tall, woody and sparsely branched from base. Leaves 3–5(–7) foliolate, 0.6–3 × 0.2–1.5 cm; flowers cream-coloured, 2–2.5 cm across, stamens 36–40, intermixed with staminodes; capsules linear to oblong, 6–8.5 cm long.

Rare, in deciduous forests of Guntur district (Flora of India). Fl. & Fr.: July–Sept. Endemic. Differs from var. *viscosa* in the conspicuous longer style, larger flowers and more number of stamens.

var ***viscosa***

Common weed in waste places and roadsides. Fl. & Fr.: July–Oct.. *Vern.:* Kukka vaminta. Three to four drops of leaf juice poured into ears relieves ear infections as used by the tribals. Nagarjuna Konda, *KTH* 9616 (CAL); V.P. South, *VRK* 6894; Madinepadu, *KCJB* 17458.

VIOLACEAE

Hybanthus N.J. Jacquin *nom. cons.*

Hybanthus enneaspermus (L.) F.V. Muell. Fragm. Phyt. Austr. 10: 81. 1876. *Viola enneasperma* L. Sp. Pl. 937: 1753. *Ionidium suffruticosum* (L.) Ging. in Ic. Prodr. 1: 311. 1824; FBI 1: 85; Gamble 49 (35). *Viola suffruticosa* L. Sp. Pl. 951. 1753.

Perennial branched erect woody herbs; leaves lanceolate, 2–3 × 0.3–0.5 cm, acute, serrate; flowers pink, axillary, solitary, capsule subglobose, yellowish; seeds many.

Common weed of wastelands. Fl. & Fr.: July–Jan. *Vern.*: Ratnapurusha. Root source of a aphrodisiac drug, named as Ratnapurusha Lehyam. Plant is considered to possess diuretic and demulcent properties. Decoction or powder of the whole plant taken to improve memory and vitality and as a remedy in asthma, fever, leprosy. Shampoo made from this plant removes dandruff. Bollapalle, *VRK* 5747; Vinukonda RF, *VRK* 6943; Kondavedu *SSR & KSM* 21655; Madinepadu *KCJB* 17460 (MH); Nagarjuna Konda, *KTH* 9837 (CAL).

Note: A polymorphic species, very variable in habit and leaf form. Dalzell & Gibson (Bombay Fl. 12. 1861) were the first to unite *Viola suffruticosa* L. (1753) (*Ionidium suffruticosum* (L.) Roem & Sch.) and *Viola enneasperma* L. (1953) and adopted the epithet '*enneaspermum*' for the combined taxon. Therefore the correct name for this species is *Hybanthus enneaspermus* (L.) F. Muell. according to Art. 57 of ICBN (1978).

BIXACEAE

Bixa Linnaeus

Bixa orellana L. Sp. Pl. 512. 1753; Bedd. Fl. Sylv. t. 79. 1871; FBI 1: 190; Gamble 51 (37).

Evergreen trees, up to 5 m tall; leaves alternate, cordate, glabrous, apex acuminate, margin entire, base cordate; flowers purplish, showy, fragrant, large in terminal panicles; capsule reddish brown, ovoid, 2-valved, soft bristly, seeds trigonous, covered with a red pulp.

Exotic American species cultivated in gardens for its dye. Fl. & Fr.: Aug.–Feb. *Vern.*: Jabaru kaya.

The red coating on the seeds yields the well known 'Annato' dye. Now, the seeds are much valued for extraction of bixin, which is great demand especially for colouring food stuffs, butter, cheese, chocolate, cosmetics, oils, varnishes, soap and in ancient times wool and silk.

Oncoba spinosa Forssk. cultivated as garden plant.

COCHLOSPERMACEAE

Cochlospermum Humboldt, Bonpland *et* Kunth *nom. cons.*

Cochlospermum religiosum (L.) Alston in Trimen Fl. Ceylon 6 (Suppl.): 14. 1931. *Bombax religiosum* L. Sp. Pl. 552. 1753. *Cochlospermum gossypium* DC. Prodr. 1: 527. 1824; FBI 1: 190; Gamble 50 (36).

Deciduous trees, up to 12 m tall; leaves orbicular in outline, 3–5-lobed, glabrous above, densely brownish-tomentose below, acute-acuminate at apex, lobes undulate or obscurely crenate along margins. Flowers yellow in terminal panicles; capsules obovoid, valves 5, striate-ribbed, seeds numerous, brown woolly.

Rare in forests. It grows mostly in dry gravelly soil. Fl. & Fr.: Feb.–Aug. Konda gogu. Bark yields an edible gum, called 'Kutara gum". Fibre covering seeds used for stuffing mattresses. It is sedative and stimulant and for cough and gonorrhoea. Reported in Guntur Forest Division Work Plan.

FLACOURTIACEAE

Flacourtia L'Herit.

Flacourtia indica (Burm. f.) Merr. Interpr. Rumph. Herb. Amb. 377. 1917.
Gmelina indica Burm. f. Fl. Ind. 132. t. 39. f. 5. 1768. *Flacourtia ramontchi* L'Herit. Strip. Nov. 3: 59. tt. 30 & 30 B. 1786; FBI 1: 194; Gamble 54 (39). *F. sepiaria* Roxb. Pl. Cor. t. 68. 1795.

Shrub or small tree, up to 5 m tall, leaves usually clustered at apex, basally sub-opposite or alternate, ovate or oblong, thin coriaceous, glabrous, usually cuneate and 3–5-pliveined at base, acute, obtuse or sometimes emarginate at apex, coarsely crenate to subentire along margins. Flowers yellowish green, solitary or few in axillary bracteate glabrous racemose clusters or terminating in short lateral 3–4 leaved thorny twigs; berries ellipsoid with 6–7 pyrenes, dark purple or red when ripe, seeds trigonous, pale yellow to brown.

In scrub jungles. Fl.: Dec.–Mar. Fr.: Ripens from May–Aug. *Vern.*: Puttelka, Chiki yelaka. Fruits are edible, also used medicinally for jaundice and enlarged spleen. Davupalli, *DMR* 18153.

POLYGALACEAE

Polygala Linnaeus

1. Flowers usually solitary .... *P. erioptera*
1. Flowers in racemes:
   2. Racemes more than 5 cm long .... *P. elongata*
   2. Racemes less than 5 cm long:
      3. Branching erect upwards; racemes and flowers erect; outer sepals obovate-oblong, mucronulate at apex .... *P. rosmarinifolia*
      3. Branching prostrate, decumbent or ascending, flowers drooping; outer sepals ovate falcate, acuminate at apex:
         4. Prostrate or decumbent weak herbs; leaves orbicular or oblong- obovate, up to 4 cm long, caruncle appendages subequal .... *P. arvensis*
         4. Erect stout herbs; stems woody, leaves linear-lanceolate or oblong- ovate to oblong lanceolate, 4–9 cm long; caruncle appendages equal .... *P. chinensis*

Polygala arvensis Willd. Sp. Pl. 3: 876. 1802; Chandrabose & Nair in Proc. Indian Acad Sci. B. 90: 123. 1981. *P. chinensis* auct. non L. FBI 1: 204; Gamble 58 (42).

Annual erect herbs, up to 30 cm tall; leaves linear, oblanceolate, 1.5–3.5 × 0.6–1 cm, chartaceous, glabrous to pubescent, obtuse-apiculate, base subacute; flowers yellow, in lateral racemes; capsules ovate-oblong, emarginate; seeds ellipsoid, villous.

Occasional in plains. Fl. & Fr.: Aug.–Jan. N.U. Campus, *VRK* 6984; Nagarjunakonda, *KTH* 9779 (CAL).

Polygala chinensis L. Sp. Pl. 704. 1753; FBI 1: 204; Gamble 58 (42).

Erect much branched herbs, up to 30 cm tall; leaves obovate, 1.5–3 × 0.5–1 cm, glabrescent, emarginate, entire; flowers yellow in lateral racemes; capsules orbicular or sub orbicular, flat; seeds 2, oblong.

Occasional in plains. Fl. & Fr.: June–Dec.. Eng: Indian milk wort. Macherla RF, *VRK* 6822.

Polygala elongata Klein ex Willd. Sp. Pl. 3. 379; FBI 1: 203; Gamble 58 (42); Chandrabose and Nair in Proc. Indian Acad. Sci. (Plant Sci.) 90: 125. 1981.

Herbs to 30 cm tall, erect, leaves linear, oblong or elliptic, 1.5–4 × 1.5–0.5 cm, chartaceous, base attenuate, margin entire, apex obtuse, apiculate; racemes lateral, leaf-opposed or terminal, lax, to 10 cm; flowers yellow; capsule rhomboid, flat, glabrous, seed solitary, dark, appressed downy, white pilose, caruncle pale brown, hood-shaped, bristly, 3-appendaged.

Less common, in Nagarjuna University Campus grass lands. Fl. & Fr.: Aug.–Jan.

Polygala erioptera DC. Prodr. 1: 326. 1824; FBI 1: 203; Gamble 59 (42).

Annual erect herbs, up to 25 cm tall; leaves lanceolate to linear, 1–2.5 × 0.1–0.2 cm, obtuse, revolute along margins, base narrow, glabrescent above, patently hairy near margins; flowers yellow, solitary or in axillary racemes; capsules elliptic-oblong, obliquely notched, pubescent; seeds 2, oblong, black.

Common weed along roadsides and wastelands. Fl. & Fr.: Aug.–Dec.. Macherla RF, *VRK* 5807; Sattenapally RF, *VRK* 6762.

Polygala rosmarinifolia Wight & Arn. Prodr. 27. 1834; FBI 1: 204; Gamble 58 (42); Chandrabose and Nair in Proc. Indian Acad. Sci. (Plant Sci.) 90: 122. 1981.

Herbs, erect, decumbent or prostrate up to 45 cm tall; branching upwards; branches densely appressed-pubescent; leaves linear, elliptic or very rarely oblong or oblanceolate, 10–40 × 2–4 mm, acute or acuminate and apiculate, attenuate base, thickened and recurved along margins, hairy on the upper surface, pubescent beneath, flowers yellow, rarely solitary or in lateral extra-axillary or axillary, erect up to 3 cm long, 3–8-flowered racemes; capsules rhombate or obliguely obcordate, narrowly winged and ciliate along margins; seeds truncate at apex, black, brownish, hirsute.

Open hilly grasslands, cleared areas and roadsides in Nagarjuna University campus area (Mohan Rao & Aasiya, 1995). Fl. & Fr.: Aug.–Jan.

Reproductive features and ecad differentiation in local populations of species of *Polygala* has been studied by Mary Roy (1999; doctoral work).

CARYOPHYLLACEAE

Polycarpaea Lamark *nom. cons.*

1. Leaves flat and with many leaves in the axils, stipules much fimbriate, stem tomentose or scattered hairy, sepals white .... *P. corymbosa*
1. Leaves with inrolled margins and with few or no leaf in the axils; stipules not much fimbriate, stem densely white tomentose, sepals with brownish tinge .... *P. aurea*

Polycarpaea aurea Wight & Arn. Ann. Nat. Hist. Ser. 1. 3291. 1834; Gamble 65 (49). *P. corymbosa* Lam. var. *aurea* (Wight & Arn.) Wight, Ill. 2: 44 t. 110. 1850; FBI 1: 345.

Much branched, prostrate herbs, up to 30 cm tall; stem densely white tomentose; leaves linear, 0.8–1 cm, margin inrolled, acute; flowers brownish-white in cymes, capsule oblong.

Rare in rocky places of dry deciduous forests (Saxena & Brahmam, 1994). Fls. & Frs.: Oct.–Nov.

Polycarpaea corymbosa (L.) Lam. Tabl. Encycl. 2: 129. 1797; FBI 1: 245; Gamble 64 (460. *Achyranthes corymbosa* L. Sp. Pl. 205. 1753.

Annual, erect, much branched herbs, up to 25 cm tall; leaves whorled, linear, 1–3 × 0.1–0.2 cm, minutely hairy, acute, stipules membranous, fimbriate; flowers reddish-white, in terminal and axillary cymes; capsules oblong; seeds compressed, brown.

Common weed in sandy places of wastelands, roadsides, cultivated fields and forests. Fl. & Fr.: Aug.–Dec.. *Vern.:* Rajuna. Vinukonda RF, *VRK* 5709; Macherla RF, *VRK* 5806; Vijayapuri, *KMS* 9831 (MH); Nagarjunakonda, *KTH* 9739 (CAL).

Ornamentals; *Dianthus*; the Carnations are under cultivation pots & as hedge plants in Guntur town.

PORTULACACEAE

Portulaca Linnaeus

1. Nodes without scales or hairs .... *P. oleracea*
1. Nodes with scales or hairs:

2. Leaves lanceolate to ovate; flowers solitary, half sunk in the enlarged end of the pedicel and surrounded by 4 involucral leaves .... ***P. quadrifida***
2. Leaves terete or linear, flowers clustered, abundant brown hairs round the flowers and usually at the nodes .... ***P. suffruticosa***

Portulaca oleracea L. Sp. Pl. 445. 1753; FBI 1: 246; Gamble 66 (47). var. ***olearacea.***

Prostrate succulent herbs, stems grey-reddish, glabrous; leaves fleshy, spirally arranged or sub-opposite, subsessile, oblong-obovate, 1.5–2.5 × 0.5–1 cm, flowers light yellow in terminal clusters; capsules oblong-ovoid; seeds numerous, black.

Common weed in moist localities. Fl. & Fr.: Through the year. *Vern.:* Pavalaku, Pappukura. Leaves used as vegetable. Koppanur, *VRK* 3273; Vinukonda RF, *VRK* 6927. World's worst weed No. 9. Population biology of Guntur area studied by K.S.R. Murthy (1981), M.Phil. Nagarjuna University.

Portulaca quadrifida L. Syst. Nat. ed. 12. 1766 & Mant. Pl. 1: 73. 1767; FBI 1: 247; Gamble 66 (47).

Herbs, branchlets creeping, fluffy at nodes, pinkish; leaves decussate, elliptic to oblong or ovate to lanceolate, succulent, 4–2 × 1–2.5 mm, glabrous, base cuneate, margin entire, apex acute. Flowers terminal, solitary, half sunk in the enlarged end of the pedicel; capsule ovoid, glabrous, seeds numerous, tubercled.

In moist grasslands of Nagarjuna University (Mohan Rao, 1995). Fl. & Fr.: Throughout the year. *Vern.*: Sanna payala.

Portulaca suffruticosa Wall. ex Wight & Arn. Prodr. 356. 1834; FBI 1: 247; Gamble 66 (47).

Suffruticose branched herbs with tuberous roots, up to 15 cm tall; stems with silky white hairs; leaves spiral, oblong, obtuse or rounded at apex, 1–2 cm long, subacute; flowers light pink in terminal clusters covered by hairs; capsules ovoid; seeds numerous, black, tuberculate.

Occasional in moist localities in open forests. Fl. & Fr.: June–Nov. *Vern.: Peddhapayilikura, Ganga-pavilikura.* Macherla RF, *VRK* 5784.

Portulaca grandiflora Hook. Sun plant cultivated as an ornamental for its attractive pink, yellow, red white flowers. *Portulaca pilosa* L. (Table rose) Introduced.

Talinum portulacifolium (Forssk.) Asch. ex Schweinf. *Vern.:* Seema bacchali, cultivated as vegetable and ornamental. Flowers rose or white. Running wild.

TAMARICACEAE

Tamarix Linnaeus

1. Trees, branchlets jointed, androecium haplostemenous, stamens 5 .... *T. aphylla*
1. Bushy shrubs, branchlets terete, androecium diplostemenous, stamens 10 .... *T. ericoides*

Tamarix aphylla (L.) Karsten, Deut. Fl. 641. 1882; Baum. Gen. Tamarix 81. 1978. *Thuya aphylla* L. Cent. Pl. 1: 32. 1755.

Trees, up to 5 m tall; branchlets jointed, enclosed by sheathing leaf bases; leaves triangular abruptly acute to acuminate at apex, flowers pinkish, fragrant, in terminal panicled spikes; capsules 3-gonous; seeds terete, pilose.

Introduced tree in plains, especially on Krishna river banks. Often cultivated as an ornamental, wind breakers and hedge. It is also recommended for planting in shifting sands. The timber is used in house building, for making agricultural implements and furniture etc. Baskets are made out of twigs and the wood is valued as cheap fuel. The bark and galls are used in tanning and as mordant in dyeing. Fl. & Fr.: Feb.–Apr. Krishna river, *VRK* 6733.

Tamarix ericoides Rottl. in Ges. Naturf. Freunde Berlin Neue Schriften 4: 214. 1803; FBI 1: 249; Gamble 69 (49).

Bushy shrubs, up to 2.5 m tall; bark dark brown or blackish, rough; leaves minute, ovate-lanceolate to triangular-ovate, scaly, sheathing at base, acuminate or acute at apex, upcurved, flowers pale-to-bright rosy, shortly pedicellate, in terminal branched spikes; capsules conical, glabrous; seeds oblong, pilose, hairs pale white.

Occasional on the Krishna river bed. Fl. & Fr.: Aug.–Mar.. The leaves are used in the treatment of cough in children and for treating enlarged spleen. The paste made from the tender shoots is applied on skin rashes by the Koyas. The galls are astringent. Krishna river at V.P. South, *KMS* 9801 (MH).

ELATINACEAE

Bergia Linnaeus

Bargia ammannioides Roxb. ex Roth, Nov. Pl. Sp. 219. 1821; Roxb. Fl. Ind. 2: 457. 1832, '*ammanoides*' ; FBI 1: 251. 1874; Gamble 69: 50; *Elatine ammannoides* (Roxb. ex Roth) Wight & Arn. Prodr. 41. 1834.

Annual, erect, much branched herb, 10–35 cm tall; stems terete, reddish-purple, glandular-pubescent or subglabrous, swollen at nodes; leaves oblanceolate or obovate-oblong, 15–25 × 3–5 mm, base attenuate, acute at apex, upper half sparsely serrate with gland-tipped teeth, lower half entire, hispid to glandular pubescent or glabrescent; flowers few to many in axillary fascicles; capsules subglobose, reddish, seeds numerous, minute, dark brown with shiny reticulations.

Grows in muddy areas, seasonally inundated fallow fields and some times along roadside ponds, river beds and in waste marshy lands. Fl. & Fr.: Aug.–Mar.. In Nagarjuna University Campus area (Mohan Rao & Aasiya, 1995).

Note: In general appearence, *Bergia ammannoides* resembles *Ammania baccifera*, hence the specific epithet '*ammannoides*'.

CLUSIACEAE

Calophyllum inophyllum L. known as Ponna chettu in Telugu is cultivated as garden plant for its foliage and showy flowers.

MALVACEAE

1. Fruits schizocarps:
   2. Epicalyx absent:
      3. Leaves cordate; carpels bursting irregularly, points not spreading ..... ***Abutilon***
      3. Leaves ovate, lanceloate or oblong; carpels dehiscing by a regular central line .... ***Sida***
   2. Epicalyx present:
      4. Shrubs, leaves entire, mericarps hooked .... ***Malvastrum***
      4. Herbs, leaves lobed; mericarps other than hooked:
         5. Mericarps 5; flowers solitary .... ***Pavonia***
         5. Mericarps 10; flowers in condensed racemes .... ***Malachra***
1. Fruits capsules:
   6. Style unbranched .... ***Thespesia***
   6. Style 3-branched:
      7. Calyx spathaceous, lobed or toothed at apex, splitting on one side during expansion of the corolla, adnate to the corolla and falling off with it .... ***Abelmoschus***
      7. Calyx not spathaceous, deeply 5-lobed or 5-parted, free from the corolla and persistent:
         8. Fruits winged .... ***Fioria***
         8. Fruits not winged .... ***Hibiscus***

Abelmoschus Medikus

Abelmoschus ficulneus (L.) Wight & Arn. Cat. No. 14. 1883; Borssum in Blumea 14: 106. 1966. *Hibiscus ficulneus* L. Sp. Pl. 695. 1753; FBI 1: 340; Gamble 97 (67).

Annual erect herbs or under shrubs; branchlets simple hairy; leaves 2–12 cm across, orbicular, palmately 3–5-lobed, base cordate, lobes 2–8 × 1–4 cm, obovate to spathulate, rounded to obtuse at apex, serrate, sparsely stiff simple hairy on both surfaces; also mixed with stellate hairs beneath;

flowers white become pink with deep purple centre, axillary, solitary or in terminal racemes, capsule ovoid, 5-angular, obtuse or rarely short acuminate, glabrescent; seeds ca. 3 mm across, blackish.

Rare in cultivated fields and waste lands. Fl. & Fr.: Sept.–Mar. *Vern.:* Nela benda, Parupu benda, seed yield an oil and are used for flavouring coffee. Stems yield an excellent fibre. Davupalli, *SSR* & KSM 21684.

Note: This species can be easily recognised by its flask-shaped calyx in buds and blunt, hairy capsules. The leaves resemble those of *Ficus cairica* L. hence the specific epithet.

Abutilon P. Miller

1. Mericarps rounded at apex .... *A. pannosum*
1. Mericarps acute at apex:
    2. Corolla with a purplish centre, mericarps 20–25, leaves woolly below ..... *A. hirtum*
    2. Corolla without a purplish centre, mericarps up to 18, leaves glaucous below .... *A. indicum*

Abutilon hirtum (Lam.) Sweet Hort. Brit. ed. 1. 53. 1826; Gamble 91 (65); Borssum in Blumea 14: 168. 1966. *Sida hirta* Lam. Encycl. 1: 7. 1783. *S. graveolens* Roxb. ex Hornem. Hort. Bot. Hafn. (Suppl.) 77. 1819. *Abutilon graveolens* (Roxb. ex Hornem.) Wight & Arn. var. *hirtum* (Lam.) Mast. in Hook. f. Fl. Brit. India 1: 327. 1874.

Erect, hirsute, odorous undershrubs, up to 2.5 m tall; leaves cordate, 6–8 × 5–6.5 cm, woolly below, acute, serrate; flowers yellowish-cream, axillary, solitary; schizocarp globose, mericarps awnless; seeds black, minute, white hairy.

Occasional in Sattenapalle range. Fl. & Fr.: July–Jan. *Vern.:* Pala Benda, Nela Benda. Sattenapally RF, *VRK* 5841.

Abutilon indicum (L.) Sweet, Hort. Brit. ed. 1. 54. 1826; FBI 1: 326; Gamble 91 (65). *Sida indica* L. in Torner, Cent. Pl. 2: 26. 1756.

Key to the subspecies

1. Calyx shorter than shizocarp, lobes spreading at maturity; mericarps shortly acuminate at apex .... sub sp. ***indicum***
1. Calyx about as long as schizocarp, lobes appressed; mericarps obtuse or rounded or acute at apex, with an erect acumen:
    2. Apex of the leaves obtuse to acute or shortly acuminate, coarsely stellate hairy; mericarps long acute at apex .... sub sp. ***guineensis***
    2. Apex of the leaves long acuminate, velutinous with short stellate hairs, mericarps usually rounded or obtuse, rarely long acute at apex .... sub. sp. ***albescens***

Abutilon indicum (L.) Sweet ssp. **albescens** var. **australiense** Hochr. in Ann. Cons. Jard. Bot. Geneva 6: 20. 1902; Borssum in Bumea 14: 175.

t. 19 (a.b.). 1966; Chandrabose in Bull. Bot. Surv. India 12: 276. 1970. *Abutilon indicum* sensu Span. in Linnaea 15: 171. 1841.

Under shrubs, stem terete, branched, stellate hairy, leaves orbicular, cordate at base, apex obtusely acuminate, minutely crenate to irregularly dentate, velutinous on both surfaces; schizocarp depressed globose, indented at apex, mericarps reniform; seeds 2 or more in each mericarp, reniform, covered with stellate hairs.

Occasional. Fl. & Fr. Nov.–Jan. Nagarjuna Sagar, Krishna river bank, *KMS* 9807 (MH).

Abutilon indicum (L.) Sweet ssp. **indicum**—Borssum in Blumea 14: 171. 1966.

Densely tomentose undershrubs, up to 1 m tall; leaves cordate, 2–6 × 1.5–4 cm, chartaceous, glaucous below, acute, acuminate, crenate-dentate; flowers orange-yellow, axillary, solitary; schizocarps globose, mericarps acute; seeds black, tubercled.

Occasional weed of the open forests. Fl. & Fr.: July–Jan. *Vern.:* Thuthura Benda. Leaves, bark, roots and seeds are medicinal. The medicinal importance of this plant is known among the local people; since long it is referred in Ayurveda and Unani systems of medicine. N. Varam. *VRK* 3590; Kondavedu Fort. *CAB* 4674 (MH).

Abutilon indicum (L.) Sweet, Hort. Brit. ed. 1. 54. 1826; FBI 1: 326; Gamble 91 (65); ssp. ***guineensis*** (Schumach.). Borssum in Blumea 14: 175. 1966. *Sida guineensis* Schumach. in Kongel. Danske vidensk. Selek. Ska. 4: 81. 1829. *Abutilon asiaticum* (L.) Sweet, Hort. BRIT. ed. 1: 53. 1826; FBI 1: 326; Gamble 91 (66). *Sida asiatica* L. Cent. Pl. 2: 26. 1756

ssp. *guineense* is allied to ssp. *indicum* but differs in having stem with dense woolly hairs, leaves more or less scabrid on the upper surface, large corolla, mericarps long acute with woolly stellate hairs on the dorsal surface.

Occasional in dry areas. Fl. & Fr.: Oct.–Nov. Anupu, *VRK* 3590.

Abutilon pannosum (Forst. f.) Schlect. in Bot. Zeitung (Berlin) 9: 828. 1851. *Sida pannosa* Forst. f. Comm. Phys. Soc. Phys. Soc. Reg. Goett. 1787: 62. 1789. *Abutilon glaucum* (Cav.) Sweet. Hort. Brit. ed. 1. 54. 1826; Gamble 91 (66). *Sida glauca* Cav. Icon. 1: 8. t. 11. 1791. *Abutilon muticum* (Del. ex DC.) Sweet Hort. Brit. ed. 2. 65. 1830; FBI 1: 327. *Sida mutica* Del. ex DC. Prodr. 1: 470. 1824.

Undershrubs or shrubs, up to 2 m tall; stems slender, tomentose by stellate and some simple hairs; leaves orbicular, rounded-cordate, apex acute, irregularly dentate, 5–9-nerved at base, densely stellate-hairy on both surfaces; flowers axillary, yellow with deep brown centre, solitary, sometimes in terminal racemes by reduction of the upper leaves; shizocarps subglobose, depressed at the top, mericarps 20–25, reniform with awn compressed on sides, hairy on the dorsal margin, 2–3 seeded, seed brownish.

Occasional in waste places of Nagarjuna University campus and Nallapadu. Fl. & Fr.: Oct.–Mar. Stem yields fibre of poor quality. Mucilaginous leaves used as pectoral.

Fioria Mattei

Fioria vitifolia (L.) Mattei Bot. R. Orte. Bot. Palermo 2: 71. 1916. *Hibiscus vitifolius* L. Sp. Pl. 696. 1753; FBI 1: 338; Gamble 98 (70).

Annual, rarely perennial, erect, hispid shrubs, up to 1 m tall; leaves broadly ovate-orbicular or slightly 3–5-lobed, 2.5–12 × 2–10 cm, tomentose, acute, serrate, base subcordate to rounded; flowers yellow with pink streaks and purplish inside base, axillary, solitary; epicalyx lobes 7, linear; capsules shorter than calyx, shortly beaked, globose, 5-winged, hairy; seeds 2–4 in each locule, reniform, tubercled, glabrous, brownish-black.

Occasional shrub in the forest. Fl. & Fr.: May–Jan. *Vern.:* Karu-patti. Koppanur, *VRK* 3899; Davupalli, *SSR & KSM* 21433. Kondavedu Fort, *CAB* 4672 (MH).

Note: The species was treated previously by Masters (1874) and Borssum Walkes (1966) under *Hibiscus*. Mattei (l.c) transferred it to *Fioria* based mainly on fruit with conspicuous, scarious, strongly veined wings—a unique character absent in the genus *Hibiscus.*

Hibiscus Linnaeus

1. Leaves entire, not lobed .... *H. ovalifolius*
1. Leaves lobed:
    2. Epicalyx absent .... *H. lobatus*
    2. Epicalyx present:
        3. Epicalyx segments basally connate .... *H. panduriformis*
        3. Epicalyx segments free .... *H. sabdarifa*

Hibiscus lobatus (Murr.) Kuntze Rev. Gen. Pl. 3: 19. 1898; Borssum in Blumea 14: 77. 1966. *Solandra lobata* Murr. in Commentat. Soc. Regiae Sci. Gott. 6: 20. t. 1. 1785. *Hibiscus solandra* L' Herit. Stirp. Nov. 1: 103. t. 49. 1789; FBI 1: 326; Gamble 98 (70).

Annual erect pubescent herbs, up to 1 m tall; lower leaves 3-lobed, lobes linear, lanceolate, ovate, upper ones lanceolate cordate, rounded or truncate and 3–5-nerved at base, 3–5 × 0.6–1.5 cm, hairy above, woolly below, obtuse, acute to acuminate at apex, crenate or coarsely serrate, flowers whitish-yellow, axillary, solitary, rarely in terminal racemes by decrescence of upper leaves; capsules ovoid, seeds numerous, black, tubercled.

Occasional in forests and common in plains. Fl. & Fr.: July–Nov., V.P. South, *VRK* 6710; Ethipothala, *VRK* 5797; Loyapalli, *SSR & KSM* 21665.

Hibiscus ovalifolius (Forssk.) Vahl Symb. Bot. 1: 50. 1790. *Urena ovalifolia* Forssk. Fl. Aeg.—Arab. 124. 1775. *Hibiscus micranthus* L. f. Suppl. Pl. 308. 1781; FBI 1: 335; Gamble 97 (70).

Erect woody shrubs, up to 2 m tall, branches slender, terete, scabrid or with scattered stellate hairs; leaves ovate, 0.5–2 × 0.3–1 cm, stellate-tomentose, acute, serrate, base obtuse, stipules linear, narrow; flowers white, axillary, solitary; epicalyx lobes more than 3, capsules globose, dehisces into 5 valves, smooth outside, glossy inside; seeds reniform, black, hirsute with long white silky hairs.

Abundant weed in rocky areas in the forests. Fl. & Fr.: Jan.–Dec.. *Vern.:* Nityamalli. Javuku, *VRK* 3285; Nagarjuna University Campus, *SSR & KSM* 21659.

Hibiscus panduriformis Burm. f. Fl. Ind. 151. L. 47. f. 2. 1768; FBI 1: 338 "*panduriformis*," Gamble 98 (70).

Subshrubs, up to 1 m tall; branchlets densely stellate tomentose generally with large, silky shiny prickly stellate hairs; lower leaves cordate-ovate, lobed or angular, 2.5–5 × 2–4 cm, 7-nerved at base, thin coriaceous, stellate-velutinous above, cinereously stellate-velutinous tomentose beneath, crenate-serrate, subentire, acute; flowers axillary solitary, golden yellow; capsules ovoid to globose, acute to acuminate, smooth and shiny inside, seeds reniform, angular, brown.

Occasional in foot hills, especially near streams and ponds. Fl. & Fr.: Oct.–Jan. Guntur *ARS* 565 (MH).

Hibiscus sabdariffa L. Sp. Pl. 695. 1753; FBI 1: 340; Gamble 99 (71).

Glabrous shrubs, up to 1 m tall; leaves palmately lobed; lobes oblong-lanceolate, 3–4 × 1–2 cm, base cuneate, apex acute, serrate, mid rib reddish-purple or green with a nectary beneath, petiole green or reddish purple, flowers pink or yellow with—pink shades, fleshy; capsules ovoid, beaked, densely strigose, seeds reniform, furfuraceous.

Occasional, in forests. Fl. & Fr.: Dec.–Apr.. *Vern.:* Pundicura. The stem yields the Rozella hemp of commerce which is strong and silky. The fibre is used as substitute to jute. The fleshy calyces are edible as vegetable and also used for the preparation of jelly and soft drinks. Guntur (*Uhl* 1929).

Malachra Linnaeus

Malachra capitata (L.) L. Syst. Nat. ed. 12. 2. 458. 1767; FBI 1: 329; Gamble 102 (73). *Sida capitata* L. Sp. Pl. 685. 1753.

Herbs or undershrubs, up to 1.8 m tall, leaves orbicular or slightly ovate, palmately 3- or 5-lobed or angled, 7-nerved at base, enitre-slightly lobed, apex obtuse, base cordate, sparsely hairy; flowers yellow and white, axillary or terminal, 2–7 in a cluster or in dense capitate head; capsule subglobose, wrinkled, mericarps indehiscent, 3-gonous, prominently nerved, glabrous; seeds reniform, brown, smooth, glabrous.

Rare, in Bapatla road and Power house.

Malavastrum A. Gray *nom. cons.*

Malvastrum coromandelianum (L.) Garcke in Bonplanida 5; 295. 1857; Gamble 88 (64). *Malva coromandeliana* L. Sp. 687. 1753. *Malvastrum tricuspidatum* (R. Br.) A. Gray Pl. Wright 1: 16. 1852; FBI 1: 321. *Malva tricuspidata* R. Br. in Ait. Hort. Kew. ed. 2.4: 210. 1812.

Erect, hairy herbs, up to 50 cm tall; leaves ovate to oblong, sometimes lanceolate, 3–4.5 × 1–2 cm, scattered hairy, obtuse-acute, crenate-serrate, base truncate, 5-nerved; flowers yellow, axillary or terminal, solitary or 2–4 in clusters; mericarps 10–14, hairy, indehiscent, 1-seeded; seeds solitary, glabrous, brownish-black.

Common weed in wastelands, roadsides, fallow fields and open forests. Native of South America. Fl. & Fr.: Through the year, V.P. South *VRK* 6723; Davupalli, *SSR & KSM* 21670.

Note: This species is often confused with the *Sida* species especially *S. acuta* but can be easily distinguished from the latter by its epicalyx and strigose indumentum on stems, petioles and pedicels.

Pavonia Cavanilles *nom. cons.*

1. Upper leaves lobed, mericarps winged at angles .... *P. zeylanica*
1. Upper leaves not lobed, mericarps wingless .... *P. odorata*

Pavonia odorata Willd. Sp. Pl. 3: 837. 1800; FBI 1: 331; Gamble 93 (67).

Erect, perennial, odorous, hairy herbs, up to 1 m tall; leaves, orbicular-ovate, occasionally upper ones lanceolate, 1.5–3 × 1–2 cm, stellately hairy, acute, cuneate-serrate, base cordate, or more or less truncate, obscurely 3–5-lobed, mericarps 5, more or less reniform, unarmed, glabrous, flowers creamish or pink, solitary; seeds brown, minutely papillose.

Common weed in wastelands and occasional in forests. Fl. & Fr.: Throughout the year. Chenchu Colony, *SSR & KSM* 21402. Nakkarekal, *KCJB* 17469 (MH).

Pavonia zeylanica (L.) Cav. Diss. 3: 34. 1787; FBI 1: 730; Gamble 93 (67). *Hibiscus zeylanicus* L. Sp. P. 697. 1753.

Erect, perennial, branched, viscid herbs, up to 1.2 m tall; stems stout and woody at base, profusely branching towards tip, leaves 3–5-lobed, lobes oblong, 1–2.5 × 1–2 cm, sparsely hirsute, acute or rounded, serrate, base cordate; flowers pinkish-white, axillary, solitary or in terminal clusters; fruits globose, pubescent; mercarps 5, 4 × 3 mm, oblong, obtuse, slightly winged, seeds ascending, reniform, minutely pubescent.

Occasional weed in wastelands, cultivated fields and rare in dry deciduous forests. Fl. & Fr.: May–Jan. *Vern.*: Karu benda. N.U. Campus, *SSR & KSM* 21496. Madinapadu, *KCJB* 17462 (MH); Nagarjuna Sagar, *KMS* 9770 (MH), NU Campus, *PNR* 2298 (NUH).

Sida Linnaeus

1. Plants velvety, sepals shorter than awns of carpels .... *S. cordifolia*
1. Plants not velvety, sepals longer than awns of carpels:
   2. Mericarps 5:
      3. Mericarps with prominent reticulation, awned; seeds dispersed by apical dehiscense; leaves penninerved .... *S. spinosa*
      3. Mericarps smooth, mucronate, seeds dispersed by withering of wall; leaves palminerved:
         4. Erect herbs or under shrubs; flowers in axillary or terminal racemes or panicles .... *S. glutinosa*
         4. Prostrate or diffuse herbs; flowers usually solitary or in few-flowered racemes .... *S. cordata*
   2. Mericarps 6–12:
      5. Stipules of each pair dissimilar, one linear to lanceolate and other near to filiform; leaves sparsely hairy, soon glabrescent .... *S. acuta*
      5. Stipules of each pair similar; leaves densely pubescent:
         6. Mericarps dehiscent, glabrous .... *S. rhombifolia*
         6. Mericarps indehiscent, pubescent .... *S. ovata*

Sida acuta Burm. f. Fl. Ind. 147. 1768, emend. K. Schum. in C. Martius, Fl. Bras. 12. 3: 326. 1891; Gamble 90(64); Borssum in Blumea 14: 186. 1966. *S. carpinifolia* non L. f. 1781. FBI 1: 323.

Erect branched herbs up to 75 cm tall; leaves lanceolate to linear, elliptic-lanceolate or ovate-oblong, sparsely stellate and simple hairy on both surfaces, soon glabrescent, acute, serrate, base rounded, 3-nerved; flowers light yellow, axillary, solitary or 2–8 in clusters, schizocarps subglobose, mericarps 6–10, tetrahedral, reticulately striate, 1-seeded, 3-angular, brown, awned.

Common weed of open forest and roadsides. Fl. & Fr.: Aug.–Jan. *Vern.:* Vishaboddi. The twigs are used for brooms. N. Varam, *VRK* 3226; Mellavagu, *VRK* 5734; N.U. campus, *PNR,* 2392 (NUH).

Sida cordata (Burm. f.) Borssum in Blumea 14: 182. 1966. *Melochia cordata* Burm. f. Fl. Ind. 143. 1768. *Sida veronicifolia* Lam. Encycl. 1: 5. 1783; Gamble 89 (64). *S. humilis* Cav. Diss. 5: 277. 1788; FBI 1: 322.

Herbs, prostrate or ascending, up to 1 m tall; branchlets pubescent with scattered, long, patent, simple and minute hairs, leaves cordate, 4–7 × 2–5 cm, hairy, acute-acuminate, crenate-serrate, 5–9-nerved at base; flowers yellow, axillary, solitary, ultimately in few-flowered racemes; mericarps 5, slightly longitudinally keeled on the back, awnless, seeds trigonous, brownish black.

Occasional in open forests. Fl. & Fr.: Sept.–Jan. Stems, leaves and root bark are medicinal. Stem yields good fibre. *Vern.:* Gayapaku, Tirunala benda. Javuku, *VRK* 3876. N.U. campus, *PNR*, 2366, 2385 (NUH).

Note: The species shows variation in indumentum and inflorescence. It has the general appearence of *S. mysorensis* Wight & Arn. but differs from the later in having semi prostrate habit and absence of glandular hairs.

Sida cordifolia L. Sp. Pl. 684. 1753; FBI 1: 324; Gamble 89(64); Borssum in Blumea 14: 199. 1966.

Much-branched undershrubs, up to 1 m tall with an unpleasant smell, leaves ovate to oblong or orbicular, 2–5 × 2–4 cm, tomentose, obtuse or acute, serrate-crenate, shallowly cordate at base; flowers pale yellow, axillary, solitary or 2–5 in clusters; schizocarps exceeding calyx; mericarps 8–10, awned; seeds trigonous, dark brown or black.

Occasional weed of the open forests. Fl. & Fr.: July–Jan. Vern.: Chiru benda, Suvarnamu, Tella Antisa. The seeds are used for improving appetite. A decoction of the root with ginger is helpful in the treatment of certain types of fever. The root bark is ground into a powder and administered in leucorrhoea and nervous disorders. The root juice helps in healing. Nizampatnam, *VRK* 6817; V.P. South, *VRK* 3528.

Sida mysorensis Wight & Arn. Prodr. 59. 1834; FBI 1: 322; Borsum in Blumea 14: 180. 1966. *Sida glutinosa* Roxb. Fl. Ind. 3: 1. 72. 1832. non Cav. 1785. Gamble 89 (64); Borssum in Bumea 14: 190. 1966.

Erect undershurbs, up to 1 m tall, much branched, odorous, branchlets stellate-tomentose; lower leaves ovate, upper ones lanceolate, 3–10 × 2–8 cm, stellate-tomentose, acute-acuminate, crenate, base sub-cordate or obtuse, 5–11-nerved; flowers yellow, solitary or in racemes; schizocarps smooth, not exceeding calyx, mericarps tetrahedral with rounded angles; seeds 3-gonus, glabrous, brown-black, dispersed by withering of wall.

Occasional weed of the forest. Fl. & Fr.: Sept.–Apr. Loyapalle, VRK 3839.

Note: Most of the botanists like Cooke (1901), Gamble (1915), Haines (1921) etc. have treated *Sida mysorensis* Wight & Arn. as synonym of *Sida glutinosa* Comm. ex Cav, but it can be easily distinguished from the latter by its awnless mericarps. Borssum-Waalkes (1966) placed *S. mysorensis* Wight & Arn. and *S. glutinosa* Cav. in two different sections like *Nela vaga* and *Sida* respectively.

Sida ovata Forssk. Fl. Aeg.-Arab. 124. 1775. *S. grewioides* Guill. & Perr. Fl. Seneg. Tent. 1: 71. 1831; FBI 1: 323.

Erect stellately hairy undershrubs; leaves ovate-oblong, 1.5–4 × 1.2–2.7 cm, stellately grey tomentose on both surfaces, rounded at base, apex obtuse, crenate; flowers pale yellow or white, in axillary, soilitary or clustered at the end of branches; shizocarps subglobose, indehiscent, stellately hairy; seeds rounded-reniform, glabrous, uniformly smooth or minutely pitted, blackish.

Very rare in the district. A new record for South India. Fl. & Fr.: July–Dec.. Nagarjuna Konda, KTH 9724 (CAL).

Sida rhombifolia L. Sp. Pl. 684. 1753; emend. Mast. in FBI 1: 323; Gamble 90 (65). *S. rhomboidea* Roxb. ex Fleming in Asiat. Res. 6: 178. 1810; Gamble 90 (65).

Annual, much branched, undershrubs, up to 60 cm tall; branchlets tomentose; leaves ovate-elliptic or rhomboid; 1.5–3 × 0.5–3 cm, stellate-hairy above, woolly below, obtuse, entire; flowers yellow, axillary, solitary or in clusters; schizocarps rugulose, mericarps 7–10; seeds solitary, ovoid.

Occasional in forests. Fl. & Fr.: July–Dec.. Kondavedu Fort, *CAB* 4634 (MH), Nagarjuna University, *PNR s.n.*

Sida spinosa L. Sp. Pl. 683. 1753; FBI 1: 323; Gamble 89 (64); Borssum in Blumea 14: 191. 1966.

Erect hispid shrubs, up to 1 m tall; leaves elliptic-ovate-oblong, 1.5–4 × 1–2.5 cm, tomentose, obtuse to acute, serrate, 3–5 nerved at base, flowers yellow, axillary, solitary; schizocarps wrinkled, exceeding calyx, mericarps 5; seeds ovoid, slightly 3-angled.

Common weed along the roadsides. Fl. & Fr.: Sept.–Mar.. *Vern.:* Nagabala, Mayilu-Manikyam. Nizampatnam, *VRK* 6791; Guntur, *CAB* 4771 (MH).

Thespesia Solander ex Correa *nom. cons.*

Thespesia populnea (L.) Soland. ex Correa in Ann. Mus. Nat. Hist. Par. 9: 290. 1807; FBI 1: 345; Gamble 101 (72). *Hibiscus populnea* L. Sp. Pl. 694. 1753.

Small trees of 10 m tall; younger parts clothed with peltate scales; leaves entire, cordate, 8–10 × 10–15 cm, acuminate, base deeply cordate, 7-nerved, flowers axillary, solitary, yellow with purple base inside, older ones turn brownish-black; fruits fleshy, 2–3.5 cm across, hardly dehiscent; seeds obovoid, angular, covered with yellowish-brown, long simple hairs, veined.

Occasional roadside trees, often cultivated for shade. Fl. & Fr.: Aug.–Dec.. *Vern.:* Gangaravi. Nizampatnam, *VRK* 6818.

Urena Linnaeus

Urena lobata L. Sp. Pl. 692. 1753; FBI 1: 327; Gamble 72(66). ssp. *sinuata* (L.) Borssum in Blumea 14: 142. 1966. *Urena sinuata* L. Sp. Pl. 692. 1753; Gamble 92 (66).

Erect woody shrubs; leaves entire-3 lobed, upper ones ovate, 5–8 x 3–7 cm, tomentose above, rusty woolly below, apex obtuse-acuminate, margin, crenate, dentate, base obtuse-acute; flower pale pink, solitary or 2–3 in axillary clusters; schizocarp globose, mericarp 5, trigonous, glochidiate; seeds reniform, angular, glabrescent.

Occasional in waste places and forests. Fl. & Fr.: July–Dec. *Vern.:* Nalla Benda, Pedda Benda, N.U. Campus, *SSR & KSM* 21498.

Note: Santapau following Hochreutiner and Merril has united *U. sinuata* with *U. lobata* for the reason that the characters of the leaf namely "deeply lobed" and "shallowly lobed" cannot be considered as distinctive, as both types with intermediate forms are found on the same plant.

Abelmoschus esculentus Moench (Ladies Fingers), *Hibiscus cannabinus* L. (famous in the Andhra Pradesh State) are cultivated as vegetables. *Gossypium herbaceum* L., *G. arboreum* L. and *G. barbadense* L. are extensively cultivated for cotton in the district. The district is well known as cotton grower and exporter. Species of *Gossypium* occur as escapes at several places, grown as ornamentals in residences. *Hibiscus rosa-sinensis* L. is cultivated as an ornamental.

BOMBACACEAE

Ceiba R. Miller

Ceiba pentandra (L.) Gaertn. Fruct. 2, 244. t. 133. 1791. *Bombax pentandrum* L. Sp. Pl. 511. 1753. *Eriodendrum anfractuosum* DC. var. *indicum* DC. Prodr. 1: 479. 1824; FBI 1: 350. *E. pentandrum* (L.) Kurz in Journ. Asiat. Soc. Bengal 43: 113. 1874; Gamble 100 (72).

Large handsome deciduous trees forming a spreading crown of branches; buttressed at base, bark grey, smooth, or covered with hard sharp conical prickles when young; branches horizontal in whorls of 5–7 in all directions at intervals of 60–150 cm; leaves large, with 5–9 leaflets, entire, lanceolate, 10–15 cm, glabrous; petioles 10–15 cm long; flowers large, white, usually in clusters of 3–10, axillary or grouped, near the ends of the branches, appearing before the new leaves; capsules woody, 10–15 cm, ellipsoid-fusiform; seeds numerous, globose, enveloped in copious floss.

Occasional in open country, often cultivated. Fl. & Fr. February–May. Eng.: White silk cotton tree. *Vern.:* Buruga. Cotton used for stuffing pillows. Gum exuded from the trunk is an astringent used as tonic in bowel complaints. Roots are reported to be used in treating diarrohea, as an antidote for scorpion sting and as a cure for diabetes. V.P. South, *VRK* 3937.

STERCULIACEAE

1. Trees or shrubs:
   2. Petals absent, flowers unisexual or polygamous .... ***Sterculia***
   2. Petals 5, flowers bisexual:
      3. Staminal column adnate to gynophore forming stalk, dialated above into a cup, fertile anthers alternating with staminodes towards margin. Follicles spirally twisted, shrubs .... ***Helicteres***
      3. Staminal column free from carpels; anthers alternating with staminodes along the margin of the cupular column at its apex. Follicles not spirally twisted, trees .... ***Guazuma***
1. Herbs or undershrubs:
   4. Flowers pink in terminal cymes .... ***Melochia***
   4. Flowers yellow in axillary clusters:
      5. Staminodes absent; ovary 1-celled, bracteoles deciduous .... ***Waltheria***
      5. Staminodes present, ovary 5-celled, bracteoles persistent .... ***Melhania***

Guazuma P. Miller

Guazuma ulmifolia Lam. Encycl. 3: 52. 1789; Robyns in Ann. Missouri Bot. Gard 51: 102. 1964. *Theobroma guazuma* L. Sp. Pl. 782. 1753. *Guazuma tomentosa* Kunth Nov. Gen. Sp. 5: 320. 1823; FBI 1: 376; Gamble 111 (79).

Trees, up to 10 m tall; branchlets stellate tomentose; leaves ellipticlanceolate, 10–16 × 3–6 cm, glabrous above, tomentose below, acute to acuminate, serrulate, base obliquely cordate; flowers yellow, numerous, small in terminal cymes; capsules subglobose, woody, indehiscent, tubercled; seeds many, small, globose.

Introduced and naturalised; rare. Fl. & Fr.: May–Sept.. Bhadraksha, Rudraksha. Seeds mixed with Rudraksha in rozaries. Guntur (*Uhl.* 1929).

Helicteres Linnaeus

Helicteres isora L. Sp. Pl. 963. 1753; FBI 1: 365; Gamble 107 (77).

Stellately tomentose, arborescent shrubs, up to 5 m tall; leaves broadly elliptic, elliptic-abovate, 5–12 × 3–6 cm, pubescent beneath, sub-acute, crenate-serrate, base slightly obliquely cordate; flowers red and yellow in axillary cymes; follicles 5, spirally twisted with an apical beak, seeds many, tubercled, tomentose.

Common in Javuku. Fl. & Fr.: July–Dec.. *Vern.*: Guba tada, source of fibre. Macherla RF, *VRK* 3567; Vinukonda RF, *VRK* 6970; Bollapalle, *KCJB* 17461 (MH).

Melhania Forsskal

1. Leaves elliptic-lanceolate; involucral bracts linear-oblong, slightly shorter than sepals .... *M. incana*
1. Leaves ovate-cordate; involucral bracts ovate-lanceolate, longer than the sepals .... *M. hamiltoniana*

Melhania hamiltoniana Wall. Pl. As. Rar. 1: 69. t. 77. 1830; FBI 1: 372; Gamble 113 (81).

Evergreen shrubs with spreading branches, tomentose; leaves ovate to ovate-lanceolate, acute-acuminate, serrate, base subcordate, 12–10 x 6–10 cm; flowers orange red on axillary peduncles, bracteoles cordate, ovate, longer than the calyx, capsules globose, woolly, loculicidally dehiscent; seeds 4 or more in each locule, 4-sided, tubercled.

Occasional in V.P. South Section. Fl. & Fr.: Aug.–Sept.. N. Varam, *VRK* 3511; Kondavedu, *SSR & KSM* 21446; Nagarjunakonda, *KTH* 9642 (CAL).

Melhania incana Heyne ex Wight & Arn. Prodr. Fl. Ind. Orient. 68. 1834; FBI 1: 372; Gamble 113 (91).

Erect, woody herbs, up to 30 cm tall; branchlets densely tomentose; leaves elliptic-lanceolate, 2–5 × 0.6–1.5 cm, glabrescent above, grey pilose below, obtuse, crenate-serrate; base subcordate; flowers yellow in axillary peduncled cymes; capsules globose-ovoid, hairy; seeds 2–4 in each locule, angled, tubercled.

Common along edges of tanks and streams. Fl. & Fr.: Aug.–Jan. N. Varam, *VRK* 3961; Papaiah palem, *VRK* 5855; Nakkarikal, *KCJB* 17468 (MH); Nagarjunakonda, *KTH* 9696 (CAL).

Melochia Linnaeus

Melochia corchorifolia L. Sp. Pl. 675. 1753; FBI 1: 374; Gamble 110 (79).

Robust, much branched, stout herbs, up to 50 cm tall; leaves variable, ovate, ovate-lanceolate, 1.5–5 × 0.5–2 cm, glabrous above, thinly stellate-hairy below, acute, serrate, base rounded or cordate; flowers pink or white with yellow near the base in terminal cymes; capsules subglobose, hispid, loculicidally 5-valved, seeds angular, brown, one in each locule.

Occasional weed in forests. Fl. & Fr.: Aug.–Jan. *Vern.:* Sithakura. Javuku, *VRK* 3294, 3574; N.U. Campus, *SSR & KSM* 21488.

Sterculia Linnaeus

Sterculia urens Roxb. Pl. Cor. t. 24. 1795; FBI 1: 355; Gamble 106 (75).

Large deciduous trees, up to 10 m tall; bark smooth, grey, outer papery, exfoliating in large thin irregular papery flakes; leaves crowded at the ends of branches, tomentose beneath, glabrous above, lobes 5, caudate-acuminate, entire, base cordate; flowers small, yellowish green, in much-branched, glandular pubescent, crowded panicles, densely pubescent with stinging hairs; seeds black, oblong.

Common in deciduous forests. Fl. & Fr.: Dec.–Apr. *Vern.*: Tapsy. Kovila chettu. It yields a gum used in medicines and textiles; seeds roasted and eaten by natives. Jayanthipuram, *VKR* 6877.

Waltheria Linnaeus

Waltheria indica L. Sp. Pl. 673. 1753; FBI 1: 374; Gamble 111 (79).

Erect branched pubescent herbs, up to 1.5 m tall; stems terete, softly pubescent; leaves ovate, elliptic, cordate-ovate or oblong, 1.5–2 × 1–1.5 cm, velvety tomentose, obtuse, crenate-serrate or dentate, base rounded to cordate or truncate; flowers yellowish, sessile in axillary clusters; capsules membranous; seeds solitary, kidney-shaped, black.

Common weed of open forests, waste places, cultivated fields and amidst boulders on rocky slopes. Fl. & Fr.: July–Jan. *Vern.*: Nalla benda. N. Varam, *VRK* 3228; Javuku, *VRK* 3290, 3598; Mellavagu, *VRK* 5723; Nagarjunakonda, *KTH* 9743 (CAL).

Remarks: Hutchinson (Fam. Fl. Plants, 1926) disagreeing with Engler and Prantl in assigning the genus *Waltheria* Linn. to the Bombacaceae has retained it in the Sterculiaceae. Venkata Rao (Journ. Indian Bot. Soc. 29: 163–176. 1950), working on the embryology of *Waltheria* has concluded that the genus *Waltheria* is appropriately placed in the Sterculiaceae. Interestingly Hutchinson (1969) is rather doubtful about the distinction between Sterculiaceae and Tiliaceae. He states (Genera Pl. P. 496. 1967), '........ the somewhat slender differences used to separate them have largely broken down. ... a future monographer may combine them into one and probably recast the rather unsatisfactory tribes'.

Note: The choise of this name in preference to *Waltheria americana* L. is given in the Intern. Code of Bot. Nomencl. ed. 1956, as an example of the applicaiton of Art. 57 which says "the author who first unites taxa bearing names or epithets of the same date has the right to choose one of them and his choice must be followed.' In this case Robert Brown seems to have been the first one to unite these two in 1818 and he choose *Waltheria indica* L.

Sterculia foetida L. is cultivated in NU campus near Engineer's office.. Planted in 1975. *Vern.:* Gurrapu badam.

Pentapetes phoenicia L. is cultivated in gardens, flowers red.

TILIACEAE

1. Trees or shrubs, fruit a drupe .... *Grewia*

1. Herbs or undershrubs, fruit a capsule:
   2. Capsules globular, echinate .... *Triumfetta*
   2. Capsules linear, not echinate .... *Corchorus*

Corchorus Linnaeus

1. Capsules angled:
   2. Capsule 6-angled, 3-angles prominently winged. Capsules ending in 3 radiating points .... *C. aestuans*
   2. Capsule 3-angled, capsules ending in 1 point:
      3. Capsule neither falcate nor strigose, seeds not wrinkled .... *C. trilocularis*
      3. Capsule falcate, strigose; seeds wrinkled .... *C. urticifolius*
1. Capsules terete:
   4. Capsules ending in a single beak:
      5. Capsule 1.3 cm long, cylindric, pubescent, 3-valved .... *C. fascicularis*
      5. Capsule 2.5 to 8 cm long, 10-ribbed, glabrous, 5-valved .... *C. olitorius*
   4. Capsules ending in a 3-radiatng points .... *C. tridens*

Corchorus aestuans L. Syst. Nat. ed. 10. 1079. 1759. *C. acutangulus* Lam. Encycl. 2: 104. 1786; FBI 1: 398; Gamble 121 (86).

Annual branched, erect hairy herbs, up to 50 cm tall; leaves sub-orbicular or ovate-oblong, 3–6 × 2–3 cm, sparsely hairy, acute, serrate, base rounded; flowers yellow, axillary, solitary or in pairs; capsules 6-angled, 3 of the angles winged, septate; seeds numerous, dark brown, truncate at both ends.

Common weed of roadsides and forest undergrowth. Fl. & Fr.: May–Dec.. *Vern.:* Kalasa. N. Varam, *VRK* 3229; Koppanur, *VRK* 3271; Nagarjunasagar, *KMS* 9728 (MH).

Corchorus fascicularis Lam. Encycl. 2: 104. 1786; FBI 1: 398; Gamble 121 (87).

Annual glabrous herbs to undershrubs, up to 60 cm tall; stems woolly with scaly bark; leaves oblong-lanceolate, 2–4 × 0.7–2 cm, glabrous, obtuse, serrate, 3-nerved; flowers greenish-yellow, 2–5 (8-flowered), in leaf-opposed, pedunculate cymes; capsules triangular, densely strigose, shortly beaked; seeds numerous, wedge shaped, truncate at one end and obliquely produced at the other, black.

Occasional weed in wastelands. Fl. & Fr.: Sept.–Feb. Nagarjunakonda, *VRK* 6706; Krishna river, *KMS* 9822 (MH); Nagarjunakonda, *KTH* 9849 (CAL).

Corchorus olitorius L. Sp. Pl. 529. 1753; FBI 1: 397; Gamble 122 (87).

Sub shrubs to 80 cm long, stem glabrous; leaves elliptic-lanceolate, 6–11 × 2.2–5 cm, glabrous, acute to acuminate, serrate, the lower serratures on each side prolonged into a filiform appendage, base obtuse or rounded, 3–5-nerved; flowers pale yellow, solitary or 2–3 fascicled, axillary or extra axilllary; capsules linear, cylindric, 10-ribbed, beaked, 5-valved; seeds trigonous, black.

Occasional in moist places and hedges of the fields. Fl. & Fr.: Throughout the year. Eng.: Red jute or Desi jute. *Vern.:* Parintakura. The well known fibre jute is obtained from this plant. Sirigiripadu, *SSR & KSM* 21463.

Corohorus tridens L. Mant. Pl. 566. 1771; FBI 1: 198; Gamble 121 (87).

Erect subshrubs, up to 70 cm, branchlets glabrous; leaves oblong-lanceolate to linear-lanceolate, or elliptic, 3–9 × 0.8–2 cm, glabrescent, serrate, acute, base cuneate; flowers yellow in leaf-opposed cymes; capsules terete, beak 3-fid, spreading, glabrous, seeds obliquely truncate at both ends, black.

Occasional at foot hills and weed in cultivated fields. Fl. & Fr.: Through the year. Nagarjunakonda, *KTH* 9738 (CAL).

Corchorus trilocularis L. Mant. Pl. 77. 1767; FBI 1: 397; Gamble 122 (87).

Annual, erect, much branched herbs up to 80 cm tall, pubescent; leaves oblong-lanceolate, 3–7 × 1.5–3 cm, bluntly serrate, lower serratures modified into filiform appendages, sparsely hairy on both surfaces, stipules lanceolate; flowers yellow in axillary, short peduncled cymes; capsules 3-angled, 3-valved, muricate, beaked, hairy when young, glabrous at maturity; seeds trigonous, blackish brown.

Common along the edges of streams and tanks and as a weed in cultivated fields. Fl. & Fr.: Aug.–Jan. *Vern.:* Bankitutturu. V.P. South, *VRK* 3527; *VRK* 5833; Kondaveedu, *SSR & KSM* 21454; Duggirala, *JSG* 12662 (MH); Nagarjunakonda, *KTH* 6788 (CAL).

Corchorus urticifolius Wight & Arn. Prodr. 73. 1834; '*urticaefolius*', FBI 1: 397; Gamble 122 (87).

Herb to subshrubs to 70 cm, branchlets pubescent; leaves ovate-lanceolate, 4.5–5.5 × 3–3.3 cm, sparsely appressed strigose, base obtuse, margin serrate, apex acute; flowers yellow, in leaf-opposed, 2–4-flowered cymes; capsule falcate, 3-angled, 2.3–3 cm long, beak short, strigose, seeds numerous, 3-gonous.

Occasional in the forests. Fl. & Fr.: Throughout the year. Davupalli, *SSR & KSM* 21421.

Grewia Linnaeus

1. Stipules auriculate, medium-sized tree .... *G. tiliifolia*
1. Stipules not auriculate, shrubs or small trees:
    2. Leaves 3-ribbed:
        3. Flowers polygamous .... *G. hirsuta*
        3. Flowers bisexual:
            4. Peduncle shorter than pedicels .... *G. orientalis*
            4. Peduncle equal to or longer than pedicels:
                5. Cymes axillary:
                    6. Fruits unlobed, leaves conspicuously serrate, peduncles slender, more than 1.5 cm long .... *G. rothii*

6. Fruits 2-parted, leaves finely serrate, peduncles stout, up to 1.5 cm long .... *G. damine*
5. Cymes leaf-opposed or terminal:
7. Androgynophore absent .... *G. rhamnifolia*
7. Androgynophore prominent:
8. Leaves glabrous; peduncles not capillary, 3-many-flowered .... *G. bracteata*
8. Leaves woolly or stellate-pubescent; peduncles capillary, 1–2-flowered .... *G. tenax*
2. Leaves 5-ribbed:
9. Leaf serratures with woolly tufts; peduncles much shorter than or as long as petioles .... *G. villosa*
9. Leaf serratures without woolly tufts; peduncles longer than petioles ..... *G. orbiculata*

Grewia bracteata Heyne ex Roth Nov. Pl. Sp. 243. 1821; Wight & Arn. Prodr. Fl. Ind. Orient. 76. 1834. in part; FBI 1: 389. *G. obtusa* Wall. ex Gamble Fl. Madras 1: 117 (83). 1915.

Small trees, up to 6 m, branchlets sparsely stellate-tomentose; leaves broadly-ovate, 5.5–9 × 3.5–5.5 cm, thin coriaceous, 3-nerved, glabrous, base rotund to obtuse, margin irregularly crenate-serrate, apex acuminate or acute; flowers light yellow, leaf-opposed; drupes broadly ovoid or depressed globose, 4-lobed.

Common in the deciduous forests. Fl. & Fr.: June–Oct.. Loyapalle, *VRK* 3984; Jayantipuram, *VRK* 6959. A new record for Circars.

Grewia damine Gaertn. Fruct. 2. 113. 1791; Gamble 118 (84). *G. salvifolia* Heyne ex Roth Nov. Pl. Sp. 240. 1821; FBI 1: 386, *p.p.*

Woody shrubs or small trees up to 3.5 m tall; branchlets pubescent; leaves lanceolate or ovate-oblong, glabrous above, densely-tomentose below, base rounded or oblique, apex obtuse or subacute, finely serrate; flowers yellow in axillary umbels; drupes globose, distinctly 2-lobed, sparsely pubescent, seeds stony.

Frequent in scrub jungles. Fl. & Fr.: Apr.–Dec.. *Vern.:* Adavi Jana, Chitti jana. Loyapalle, *VRK* 3983; J. Puram, *VRK* 6961; Karempudi, *KSJB* 17454 (MH).

Grewia hirsuta Vahl Symb. Bot. 1: 34. 1790; FBI 1: 391; Gamble 119 (85).

Villous shrubs, up to 2 m tall; leaves ovate-lanceolate, or ovate-elliptic, 2.5–7 × 1–2 cm, pubescent above, woolly below, acuminate, serrate, base oblong; flowers white, in axillary cymes; polygamous, male and bisexual; drupes subglobose, obscurely 4-lobed, fleshy, wrinkled, hirsute, nuts 4.

Common in the forests. Fl. & Fr.: May–Oct.. *Vern.:* Jibilika, Tella jana. Fruits are edible. Loyapalle, *VRK* 3988; Mellavagu, *VRK* 5731.

Grewia orbiculata Rottl. in Ges. Naturf. Freunde Berlin Neue Schriften 4: 205. 1803; FBI 1: 386. *G. rotundifolia* Juss. in Ann. Mus. Paris 4: 50. 92. 1804; Gamble 118 (84).

Unarmed shrubs or small trees, up to 4 m tall; young branches softly pubescent; leaves broadly elliptic or obovate, ovate-oblong, suborbicular or obovate, 3–8.5 × 2–5 cm, 5-nerved, serrate, velvety above, base rounded; stipules lanceolate, caducous; flowers yellow in umbellate cymes; drupes subglobose, obscurely 2-lobed, pilose.

Occasional in scrub jungles. Fl. & Fr.; Mar.–July. *Vern.:* Jana. Jayanthipuram, *VRK* 6892; Nagarjunakonda *KTH* 9626 (CAL).

Grewia orientalis L. Sp. Pl. 964. 1753; Gamble 117 (84). *G. columnaris* J.E. Smith in Rees, Cyclop. 17: 5. 1811; FBI 1: 383.

Shrubs or small trees, up to 2.5 m tall; leaves elliptic-ovate, 4–10 x 2–4.5 cm, pubescent, acute or acuminate, crenulate, base obtuse or obliquely cordate; flowers white in axillary cymes; drupes 2–4-lobed, yellow, hairy.

Common in deciduous forests. Fl. & Fr.: July–Dec.. *Vern.:* Tegali. Anupu, *VRK* 3501; Loyapalle, *VRK* 3998; Papaiah palem, *VRK* 5850.

Grewia rhamnifolia Heyne ex Roth Nov. Pl. Sp. 244. 1821; Gamble 117 (83). *G. orientalis* auct. non L. Roxb. Fl. Ind. 2: 586. 1832; FBI 1: 383. *p.p.*

Shrubs, up to 3 m tall; branchlets densely tomentose; leaves ovate, rhomboid-ovate, or lanceolate, 4.5–6.5 × 3–4 cm, thin coriaceous, glabrous, base subcordate, obtuse, acuminate or acute; flowers in axillary or terminal, umbellate cymes; drupes globose, 4-lobed.

Occasional, in forests. Fl. & Fr.: June–Oct.. Nakkerikal, *KCJB* 17482 (MH); Bollapalle, *KCJB* 17498 (MH); Vinukonda, *KCJB* 17494 (MH).

Grewia rothii DC. Prodr. 1: 509. 1857; Gamble 117 (84). *G. excelsa* auct. non Vahl 1790; FBI 1: 385. *p.p.*

Small trees, up to 3 m tall; young branchlets tomentose, leaves ovate-oblong, 2–4 × 0.8–1.5 cm, densely tomentose below, serrulate up to the acuminate tip; flowers yellow in axillary clusters; drupes entire, villous hairy.

Occasional in forests. Fl. & Fr.: February–Aug.. *Vern.:* Peddacipuru, Muvvanchi. Fruits edible. Anupu, *VRK* 3957.

Grewia tenax (Forssk.) Fiori in Agric. Colon. 5. Suppl. 23. 1912. *Chadara tenax* Forssk. Fl. Aegypt-Arab, 105–1775. *Grewia populifolia* Vahl Symb. Bot. 1: 33. 1799; FBI 1: 385. *G. betulifolia* Juss. in Ann. Mus. Natl. Hist. Nat. 4: 92. t. 50. f. 1. 1804; Gamble 117 (83).

Much branched shrubs, up to 3 m tall; apical branchlets stellate-tomentose; leaves irregular in shape, broadly ovate, rotund or elliptic, scabrid above, glabrous or sparsely hairy beneath, obtuse, 2-dentate; flowers white in axillary cymes; drupes 4-lobed, glabrescent, yellow when ripe.

Common in forests. Fl. & Fr.: July–Dec.. *Vern.:* Kadara. N. varam, *VRK* 3958 & 3503; Koppanur, *VRK* 3998; Anupu, *VRK* 3507; Bollapalle, *KCJB* 17429; Kondaveedu, *CAB* 4622 (CAL).

Grewia tiliifolia Vahl, Symb. Bot. 1790; FBI 1: 386; "*tiliaefolia*". Gamble 118 (84). *G. arborea* Roth, Nov. Pl. Sp. 24. 1821.

Moderate sized tree; bark peeling off; leaves ovate-elliptic, 7–15 × 4–9.5 cm, glabrescent, acuminate or rounded, margin serrate to crenate-serrate, base obliquely cordate, 3–5-nerved; flowers yellow in axillary or extra-axillary cymes; drupe globose, didymous, black, entire to 2–4-lobed.

Occasional in dry deciduous forests. Fl. & Fr.: Apr.–Sept. *Vern.:* Tada chettu, Pedda jana, Nulli jana. Stem bark is acrid with a sweetish taste and employed for digestion; bark yield a fibre used for cordage. Fruit edible and has an agreeable acid flavour. The wood has emetic properties, used as an antidote to opium poisoning in a powder form; it is strong and flexible used in agricultural implements, bent parts of carts, carriages, spokes etc. Davupalli, *SSR & KSM* 21458.

Grewia villosa Willd. Ges. Naturf. Freunde Berlin Neue Schriften 4: 205. 1803; FBI 1: 388; Gamble 119 (85).

Shrubs or small trees, 3 to 5 m tall; branchlets rusty-velvety; leaves ovate, orbicular or cordate, 5–9 × 4–7 cm, scabrid above, villous below, rounded to abruptly acuminate at apex, serrate, base cordate; flowers yellow, in axillary or leaf-opposed cymes; drupes entire or obscurely 4-lobed.

Occasional in forests. Fl. & Fr.: Sept.–Dec.. *Vern.:* Banta. Ethipothala, *VRK* 3795; Karempudi, *KCJB* 17486 (MH).

Triumfetta Linnaeus

1. Leaves entire: fruit spines pubescent, stamens 10–25 .... ***T. rotundifolia***
1. Leaves lobed:
    2. Leaves tomentose, capsules globose, spines glabrous .... ***T. rhomboidea***
    2. Leaves not tomentose, capsules ovoid, spines with a line of hairs .... ***T. pentandra***

Triumfetta pentandra A. Rich. in Guill. & Perr. Fl. Seneg. Tent. 93. t. 19. 1931; Gamble 120 (86). *T. neglecta* Wight & Arn. Prodr. Fl. Ind. Orient. 75. 1834; FBI 1: 396.

Herbs to undershrubs; branchlets stellately tomentose; basal leaves rhomboid-ovate, entire or palmately, 3–5-lobed, 2–7 × 1–7 cm, strigose, acuminate; flowers yellow in terminal racemes; capsules ovoid, hispid.

Occasional, Fl. & Fr.: July–Nov. *Vern.:* Chirusitrika. Loyapalle, *VRK* 3979.

Triumfetta rhomboidea Jacq. Enum. Carib. 22. 1760; FBI 1: 395; Gamble 120 (86). *Bartramia indica* L. Sp. Pl. 389. 1753. *nom. illeg. Triumfetta bartramia* L. Syst. Nat. (ed. 10) 1044. 1759. *nom. illeg.*

Erect, much-branched, pubescent undershrubs, up to 1 m tall; lower leaves broadly orbicular, upper leaves ovate, 1–2.5 × 0.5–2 cm, pubescent, 2-serrate, base rounded to cuneate, apex acute to acuminate; flowers yellow in terminal cymes; capsules globose, greenish-brown, clothed with smooth, hooked spines; seeds 1–2 in each locule.

Common in undergrowth of forest. Fl. & Fr.: Aug.–Jan. Eng.: Paroquet burg. *Vern.:* Sirusitrika. It is an astringent and demulcent and for diarrhoea, dysentery, gonorrhoea, leprosy, intestinal ulcers; also reported to facilitate child birth and hasten inception or delayed parturition. Loyapalle, *VRK* 3854.

Note: Triumfetta and *Bartramia* were dealt separately by Linnaeus in his Species plantarum (1753) and combined the two genera under *Triumfetta* in Systema Naturae (1759). *Bartramia indica* L. (1753) although the oldest name for this taxon cannot be used as a base for the plant in the genus *Triumfetta* as the combination *Triumfetta indica* is already used by Lamarck in 1791 for another plant. The combination *Triumfetta bartramia* is a *nom. illeg.* since Linnaeus (1759) gave a reference to Rumphius. Merrill disclosed that the Rumphius illustrations are drawn from different elements (1917). Therefore any name based on these combinations becomes *nom. ambigum.* So the earliest validly published name by Jacquin 1760 is used here.

Triumfetta rotundifolia Lam. Encycl. 3: 421. 1792; FBI 1: 395; Gamble 120 (86).

Suffruticose herbs, up to 1 m tall; branchlets stellately densely hairy, roughish; leaves orbicular-rhomboid, 1.5–2.5 × 1–2 cm, scabrid above, woolly below, base obtuse, 3–5-nerved, irregularly serrate; flowers yellow in leaf opposed cymes arranged in interrupted racemes; capsules ovoid, prickly; seeds beaked, black.

Common weed in open forests. Fl. & Fr.: Sept.–February. *Vern.:* Bankituthuru. V.P. South, *VRK* 3532; Kondavedu Fort. *CAB* 4635 (MH).

Muntingea calabura L. Elaeocarpaceae, the evergreen foilage, dwarf habit and umbrella like contour make the plant an excellent tree for lawn. The red cherry like fruits contain a sweet pulp and numerous poppy like seeds, relished by children. It is commonly planted in gardens for ornament. Native of central and tropical South America.

LINACEAE

Hugonia Linnaeus

Hugonia mystax L. Sp. Pl. 675. 1753; FBI 1: 413; Gamble 126 (90).

Stragglers, up to 6 m; branchlets yellow tomentose, leaves elliptic-obovate, or obovate-oblong, 1–2.5 × 0.5–1.5 cm, coriaceous, glabrous, nerves pubescent below, base tapering, entire, obtuse or subacute, lateral nerves prominent on both surfaces; flowers golden yellow in terminal or axillary cymes, lower peduncles spirally hooked; drupes globose; seeds compressed, pendulous.

Occasional straggler in scrub jungles. The yellow root bark is aromatic and is employed as an antidote in poisons. The bruised root is applied to inflammatory swellings. The root powder is administered as an anthelmintic and febrifuge. Fl. & Fr.: Through the year. *Vern.:* Kakibera. Nakkarekal, *KCJB* 17472 (MH).

ERYTHROXYLACEAE

Erythroxylum P. Browne

Erythroxylum monogynum Roxb. Cor. t. 88. 1798; FBI 1: 415; Gamble 127 (91).

Dense evergreen shrub or small trees, up to 3 m tall, bark brown, very rough, lenticellate, wood scented; leaves elliptic-ovate, 2-5 × 1–2 cm, glabrous, obtuse, entire, base cuneate; flowers pale green with white disc, scented, dimorphous, axillary, solitary or in clusters; drupes oblong-ellipsoid, triangular, 3-loculed, green turning to blood red when ripe, 1-seeded.

Common in forests. Grows gregariously even under extreme drought conditions. Fl. & Fr.: June–Nov. *Vern.:* Deva daru. The dried leaves are cooked and eaten during famine. Sometimes the leaves are also used as fodder for cattle. An infusion of the wood and bark is used in treating stomachic and as diaphoretic and diuretic. Loyapalle, *VRK* 3985; Mellavagu, *VRK* 5702; Kondavedu, *SSR & KSM* 21659; Bollapalle, *KCJB* 17430 (MH).

MALPIGHIACEAE

Hiptage J. Gaertner

Hiptage benghalensis (L.) Kurz in J. Asiat. Soc. Bengal. Pt. 2, Nat. Hist. 43: 136. 1874. *Banisteria benghalensis* L. Sp. Pl. 427. 1753. *Hiptage madablota* Gaertn. Fruct. Sem. Pl. 2: 69. t. 116. t. 4. 1790; FBI 1: 418; Gamble 128 (91).

A large straggling climbing shrubs; wood rough, reddish; branchlets densely woolly; leaves opposite, ovate to elliptic-oblong, 7-9 × 4-6 cm, coriaceous, glaucous below, glabrous above, glands 2 at leaf-base, base obtuse, margin entire, apex acute to acuminate; flowers with 4 white and one coloured petal, fragrant, in terminal and axillary racemes; samaras with 3 unequal wings, median larger than laterals; seed solitary, globose.

Occasional along the streams in Macherla RF. Fl. & Fr.: Dec.–Mar. *Vern:* Madhavilatha, Koppanur, *VRK* 3990.

Galphimia glauca Cav. cultivated in gardens for its showy yellow flowers.

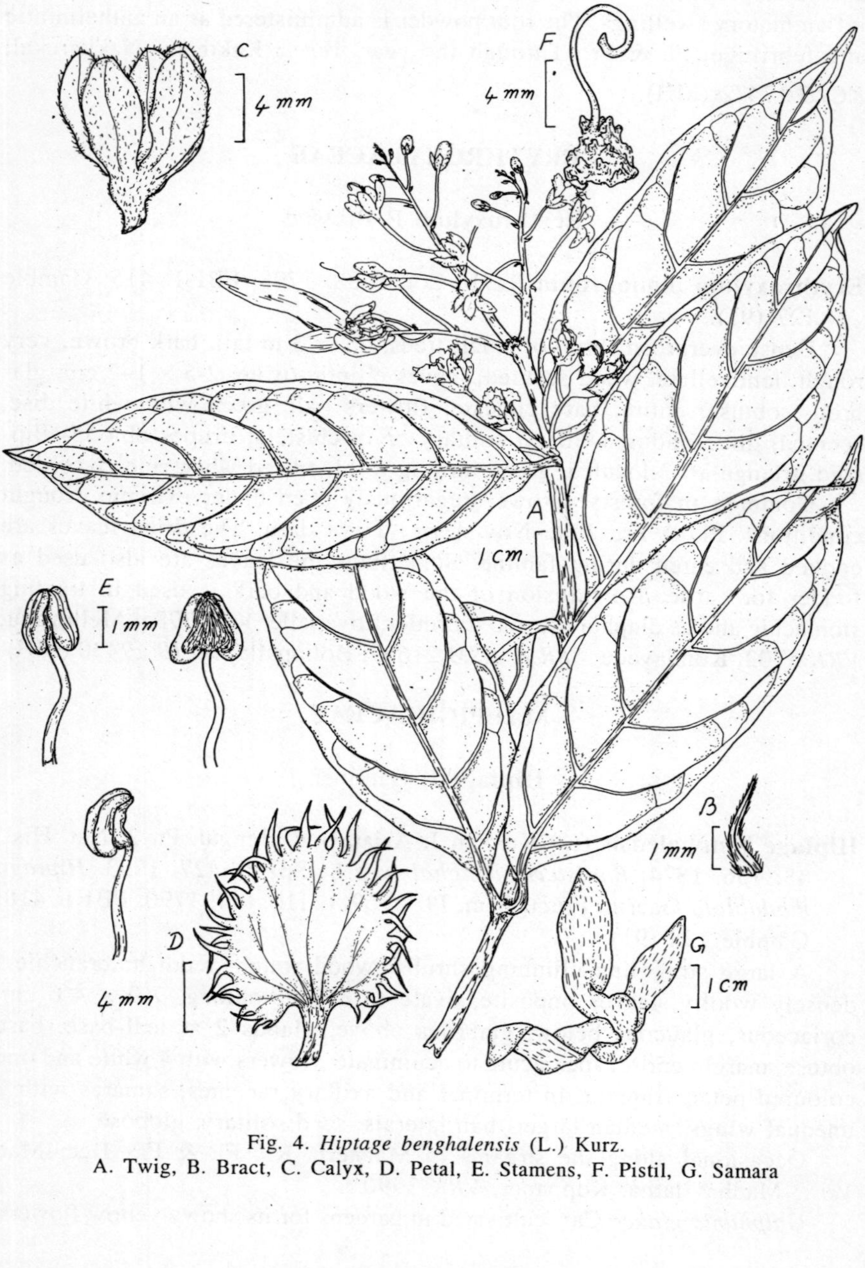

Fig. 4. *Hiptage benghalensis* (L.) Kurz.
A. Twig, B. Bract, C. Calyx, D. Petal, E. Stamens, F. Pistil, G. Samara

ZYGOPHYLLACEAE

Tribulus Linnaeus

Tribulus terrestris L. Sp. Pl. 387. 1753; FBI 1: 423; Gamble 130 (92).

Prostrate, branched, pubescent herbs; leaves paripinnate, 2-5 cm, leaflets 4 or 5 pairs, usually pairs unequal, base obtuse-cuneate, entire, apex acute; flowers yellow, axillary, solitary; capsule hairy, breaking into 3-5 cocci, woody, each with two sharp spines, seed solitary per coccus.

Common weed in open forests, roadsides and rocky places. Those in shady and moist places are robust and others in dry regions have stunted growth and are densely hairy. Fl. & Fr.: Throughout the year. Eng.: Puncture vine. *Vern.:* Palleru. Fruits are helpful in urinary and kidney complaints. It promotes urination. It is a stimulant and styptic and for chest complaints, convulsion, cramps, tumor, stomachache, diabetes, asthma and phythisis. N. Varam. *VRK* 3239 & 3520.

OXALIDACEAE

1. Leaves digilately 3-(4)-folioate, valves of capsules adnate to central column .... *Oxalis*
1. Leavés even-pinnate, valves of capsules spreading .... *Biophytum*

Biophytum A.P. de Candolle

Biophytum nervifolium Thw. Enum. Pl. Zeyl. 64. 1858. *B. sensitivum* DC. var. *nervifolium* (Thw.) Edgew. & Hook. f. in Fl. Brit. India 1: 43. 1874.

Unbranched, annual herbs, 15-20 cm long; leaves radical, pari pinnate, leaflets 8–10 pairs, oblong, 0.8–1.5 × 0.4–0.7 cm, glabrous, base obtuse, apex shortly apiculate; flowers yellowish cream in terminal umbels; capsules oblong-ovoid; seeds 2–4 per cell.

Common weed in black cotton soils, rare in forest. Fr. & Fr.: Sept. – Dec. *Vern.:* Talapushpa, Flowers red, orange, yellow; heterostylous; a good example for Diallelic system of incompaibility, Sattenapally, *VRK* 5863; Dachepalli, *KCJB* 17465 (MH); Nagarjunakonda, *KTH* 9849 (CAL).

Oxalis Linnaeus

Oxalis corniculata L. Sp. Pl. 435. 1753; FBI 1: 436; Gamble 172 (94).

Diffuse herbs; branchlets creeping, rooting at nodes; leaves digiltately 3-foliolate, leaflets obcordate, apex emarginate; flowers in pseudo-axillary umbels; capsule oblong, seeds numerous.

In shady, moist localities, in NU campus. Fl. & Fr.: Throughout the year. *Vern.:* Puli chinta. Leaves are said to be used in curing dysentery, improving appetite and digestion of dyspetic patients.

Oxalis latifolia Kunth, flowers rose, cultivated in gardens. Native of central and tropical America.

BALSAMINACEAE

Impatiens Linnaeus

Impatiens balsamina L. Sp. Pl. 939. 1753; var. ***balsamina***; FBI 1: 453; Gamble 142 (101).

Erect, glabrescent herbs, up to 1 m tall; leaves ovate-lanceolate, 2.5–5 × 1.5–3 cm, sparsely pubescent, acute, serrate; flowers pale pink, solitary or in clusters; capsules oblong, tomentose; seeds ovoid, smooth.

Rare in forests, cultivated as an ornamental. Fl. & Fr.: Sept.–Mar. *Vern.*: Kasi Tummi, Mudda Gorinta. Guntur (*Uhl*, 1929).

RUTACEAE

1. Plants unarmed .... ***Murraya***
1. Plants armed:
   2. Leaves 1-foliate .... ***Atalantia***
   2. Leaves 3- or more-foliolate:
      3. Berries smooth, fleshy .... ***Toddalia***
      3. Berries with woody rind:
         4. Petioles winged .... ***Naringi***
         4. Petioles not winged:
            5. Flowers polygamous, stamens 10–12, seeds exalbuminous .... ***Limonia***
            5. Flowers bisexual, stamens 20–50, seeds albuminous .... ***Aegle***

Aegle Correa

Aegle marmelos (L.) Corr. in Trans. Linn. Soc. London 5: 223. 1800; FBI 1: 516; Gamble 161 (115). *Crateva marmelos* L. Sp. Pl. 444. 1753.

Armed deciduous trees, up to 10 m tall; branchlets pubescent; spines axillary, straight, single or paired; leaves 3-foliolate; leaflets elliptic, lanceolate or oblong-obovate, 2.5–6 × 1–3.2 cm, chartaceous, glabrous, subcrenulate, obtuse to acute, base cuneate or rounded; flowers greenish-white, in axillary panicles; berries ovoid, rind woody; seeds numerous, oblong, embedded in aromatic pulp.

Occasional in deciduous forests, also cultivated. Fl.: Mar.–Apr. Fr.: Ripening after one year. Trees planted in compounds of Siva temples. *Vern.:* Gudu Bilvamu, Sriphalamu. Eng.: Basi. Fruits of economic value or snuff boxes. Fresh, ripe or partially ripe fruits of this tree yield the drug called Bel. Bel drug is used for treating chronic diarrhoea and dysentry particularly in cases where spells of constipation occur between prolong periods of diarrhoea and dysentry. Drinks prepared from the pulp of fruits soothe the intestine of patients recovering from bacillary dysentry. Fruits improve appetite and digestion. Pulp of fruit used locally as crude gum. Davupalli, *SSR & DMR* 21400; Kondavedu, *SSR & KSM* 21480.

Atalantia Correa

Atalantia monophylla (L.) Corn. in Ann. Mus. Natl. Hist. Nat. 6: 383. 1805; FBI 1: 511; Gamble 159 (113). *Limonia monophylla* L. Mant. Pl. 237. 1771.

Densely foliaceous armed trees, up to 8 m tall; branchlets velutinous; base obtuse, leaves simple, lanceolate or elliptic, 3–7 × 1–2.5 cm, glabrous, emarginate, base obtuse, entire or crenulate; flowers creamish in axillary racemes; berries globose; seeds ovoid.

Occasional in Macherla RF. Fl. & Fr.: May–Oct. Eng.: Wild lime. *Vern.:* Adavi nimma, Karu nimma. The seeds are said to yield a sweet smelling oil, used in rheumatism and paralysis. Javuku, *VRK* 3868.

Limonia Linnaeus

Note: Swingle (in Journ. Wash. Acad. Sci. 4: 328. 1914) is of the view that the generic name *Limonia* L. (1764) is illegitimate being a mere variant of *Limonium* Miller (1754) of Plumbaginaceae. He therefore took the next available name *Feronia* Corr. for the genus which has been subsequently followed by many. Airy Shaw (in Kew Bull. 1939: 293. 1939) states that *Limonia* is not an orthographic variant of *Limonium* of Plumbaginaceae and the legitimate name *Limonia* L. must be retained.

Limonia acidissima L. Sp. Pl. ed. 2: 554. 1762; Panigr. in Taxon 26: 576. 1977. *Feronia elephantum* Corr. in Trans. Linn. Soc. London 5: 224. 1800; FBI 1: 516; Gamble 160 (114).

Armed deciduous trees, up to 10 m tall; bark deeply fissured, black; leaves 1–3 in a cluster, imparipinnate, 3–7-foliolate, rachis narrowly winged, leaflets obovate, 2–4.5 × 1–2 cm, glabrous, scented, obtuse, entire, base narrow-acute; flowers creamish in axillary and terminal racemes, anthers red; berries woody, globose, seeds numerous.

The Wood Apple common in the forests. Fl. & Fr.: Mar.–Dec. *Vern.:* Velaga. Fruits edible. Fruit and bark used for digestive troubles. Loyapalle, *VRK* 3981 & 3847.

Note: Panigrahi (loc. cit.) has argued for the rejection of the name *Limonia acidissima* L. as *nomen ambiguum*. However Stone and Nicolson (in Taxon 27:551–552. 1978) have argued for retaining the original name. The concept of Stone and Nicolson is followed here.

Murraya J.S. Koenig ex Linnaeus *nom. cons.*

1. Leaflets with aromatic oil glands, leaflets more than 8 pairs, flowers in long corymbose panicles, fruit purplish black when ripe .... *M. koenigii*
1. Leaflets without aromatic glands, leaflets less than 8 pairs, flowers in short corymbs, fruit red when ripe .... *M. paniculata*

Murraya koenigii (L.) Spreng. Syst. Veg. 2: 315. 1826; FBI 1: 503; Gamble 156 (111). *Bergera koenigii* L. Mant. Pl. 565, 1771.

Small aromatic deciduous trees; up to 5 m tall; leaves odd-pinnate, leaflets 15–25, oblong-lanceolate, 2–4 × 1–2 cm, glabrous, acute, entire and serrate, base obtuse, asymmetical, flowers white, fragrant in terminal corymbose panicles; berries sub-globose, purplish-black when ripe apiculate, rough with glands, seeds 1–2, globose.

The curry leaf tree cultivated for its aromatic leaves, occasional in forests. Fl. & Fr.: Mar.–Aug. *Vern.:* Karivepa, Karepaku. Leaves used in culinary preparations, source of 'C' vitamin. V.P. South, *VRK* 3931.

Murraya paniculata (L.) Jack, Malayan Misc. 1 (No.5); 31. 1820. *Chalcas paniculata* L. Mant. Pl. 68. 1767. *Murraya exotica* auct. non L. 1771, FBI 1: 502; Gamble 155 (110).

Small evergreen trees, up to 5 m tall, trunk short; bark greyish-white, fragrant, leaves with 3–5 leaflets, leaflets ovate or ovate-elliptic, entire or crenate, glabrous, shining, apex obtusely apiculate, base oblique to acute, flowers white in terminal panicles; fruits ovoid, 1–2-seeded.

Commonly cultivated and rarely wild. Fl. & Fr.: Mar. –Sept. Eng.: Orange Jasmine. *Vern.:* Golimi, Naga Golugu, Golung, N.U. campus, *VRK* 6991.

Naringi Adanson

Naringi alata (Wall. ex Wight & Arn.) Ellis in Bull. Bot. Surv. Ind. 22: 193. 1980. *Limonia alata* Wall. ex Wight & Arn. Prodr. Fl. Ind. Orient. 92. 1834; FBI 1: 508; Gamble 157 (112). *Pleiospermum alatum* (Wall. ex Wight & Arn.) Swingle in Journ. Wash. Acad. Sci. 6: 427. 1916.

Armed shrubs or trees, up to 8 m tall; branchlets tomentose; spines solitary, axillary, to 3 cm, straight; leaves 3-foliolate, 3.5–7 cm long; leaflets obovate, ovate or elliptic, 2.5–6.5 × 2–4 cm, thin, coriaceous, glabrous, subentire, obtuse, base cuneate; petiole winged; flowers green in terminal or axillary panicles; berries globose, up to 2.5 cm across; seeds up to 4.

Common in scrub forests. Fl. & Fr. Jan.–Aug. *Vern.:* Munkudu, Nalla Munkudu. Loyapalle, *VRK* 3847, 3981.

Toddalia A.L. Jussieu *nom. cons.*

Toddalia asiatica (L.) Lam. Tabl. Encycl. 2: 116. 1797; var. *gracilis* Gamble Fl. Madras 1: 152 (107). 1915. *Paullinia asiatica* L. Sp. Pl. 365. 1753. *Toddalia aculeata* Pers. Syn. 1: 348. 1805; FBI 1: 497.

Slender, very prickly climbing shrub; branchlets glandular-tomentose, prickles recurved; leaves 3-foliolate; leaflets-sessile, elliptic, ovate-oblong or lanceolate; 3–4.3 × 1.5–2.2 cm, entire to crenulate, obtusely acute, glabrous, midnerve spiny below; flowers creamish in axillary or terminal panicles, berries 1-seeded, globose-much lobed.

Occasional in scrub forests. Fl. & Fr.: Nov.–Jan. *Vern.:* Konda Kashinda, Veru patta. Root paste is used as antidote to snake bite and scorpion sting. Vulusalapenta, *DMR* 18190.

Citrus maxima, Citrus medica L. (Lungamu), *Citus limonia* Osbeck (Nimma), *Citrus sinensis* Osbeck. (Narangi), *Citrus aurantifolia* Swingle (Bathayi) and *Citrus grandis* Osbeck. (*Vern.:* Pampala masam) are cultivated for their fruits.

SIMAROUBACEAE

Ailanthus Desfontaines *nom. cons.*

Ailanthus excelsa Roxb. Pl. Cor. t. 23. 1795 & Fl. Ind. 2: 450. 1832; FBI 1: 518; Gamble 163 (116).

Medium sized deciduous trees, up to 15 m tall; branchlets with persistent leaf-scars, densely fluffy; leaves odd-pinnate, 15–40 cm; leaflets 8–14 pairs, foetid, oblong-lanceolate, 4–8 × 2–4 cm, glabrous, sublobate, acuminate, base inequilateral; flowers yellowish, polygamous, in axillary or terminal panicles; samaras oblong, flat, membranous, narrowed or pointed at both ends, twisted at the base, seed one in the middle of the samara.

Occasional semievergreen foliaceous trees in forests. Fl. & Fr.: Aug.–Sept. Eng.: Tree of heaven. *Vern.:* Pedda Manu, Pedda Vepa. The bark is bitter, astringent, febrifuge and anthelmintic. Sirigiripadu, *SSR & KSM* 21402.

BALANITACEAE

Balanites Delile *nom. cons*

Balanites aegyptiaca (L.) Delile Fl. Egypte 77 & 221. t. 2B. f. 1. 1813; Basak in Fasc. Fl. Ind. 4: 20. 1986. *Ximenia aegptiaca* L. Sp. Pl. 1194. 1753. *B. roxburghii* Planch. Ann. Sci. Nat. Bot (ser. 4). 2: 258. 1854; FBI 1: 522; Gamble 164 (117).

Armed trees, up to 6 m tall; branchlets tomentose, ending in sharp, straight thorns; leaves 2-foliolate, leaflets oblanceolate-elliptic to obovate, 1.5–3 × 0.5–1.5 cm, glabrescent, obtuse, apiculate, entire, base acute, flowers white or greenish white in axillary pseudo umbels, drupes oblong-ovoid, woody; seed solitary, embedded in pulp.

Occasional in forests and plains. Fl. & Fr.: Mar.–Jan. *Vern.:* Gara chettu. Wood is used for walking sticks, seeds and leaves are used in native medicine. The fruit contains a soapy substance and is used by local people in some areas for washing their clothes. Macherla RF, *VRK* 5890, Bellamkonda forest, *DMR* 18120.

Note: Sands, Kew. Bull. (1989), has shown that the Indian plant is distinct from the African *B. aegyptica* (L.) Del. to which Roxb. Fl. 2: 253. 1832, had first referred the Indian plant.

The plant is included in family Simaroubaceae by most of the authors; Chavan and Oza (1966) in their "Flora of Pavagadh" have followed Blatter (in JBNHS 34: 895) in placing it in family Zygophyllaceae; while the genus has been raised to the rank of family (Balanitaceae Endlicher, Enchiridion 547, 1841) also.

OCHNACEAE

Ochna Linnaeus

1. Leaves elliptic or obovate, flowers in terminal and axillary racemes .... ***O. obtusata***
1. Leaves oblong-lanceolate, flowers in axillary cymes .... ***O. lanceolata***

Ochna lanceolata Spreng. in L. Syst. Veg. (ed. 16) 2: 597. 1825. *O. wightiana* Wall. ex Wight & Arn. Prodr. Fl. Ind. Orient. 152. 1834; FBI 1: 524; Gamble 166 (119).

Trees up to 8 m tall; leaves lanceolate, 4–5 x 1–2 cm, thin-coriaceous, glabrous, serrate, acute, apiculate, base obtuse; flowers yellow, in axillary cymes; drupes 1–3-lobed; seeds up to 5 mm across.

Occasional in forests. Fl. & Fr.: Feb.–July. Nagarjunakonda. *KTH* 9731 (CAL).

Ochna obtusata DC. var. ***gamblei*** (King ex Brandis) Kanis in Blumea 16: 34. 1968. *O. gamblei* King ex Brandis Indian trees 128. 1906; Gamble 165 (118). *O. beddomei* Gamble in Kew Bull. 1916. 34. 1916 & Fl. Pres. Madras 165 (118). 1915.

Deciduous trees, up to 8 m tall; leaves elliptic or obovate, 5–13 × 2–5 cm, glabrescent, obtuse, serrulate; flowers yellow, fragrant, in terminal and axillary racemes; drupes lobed, shining black.

Occasional in forests. Fl. & Fr.: Feb.–Aug. *Vern.:* Sunari, Tammi. Guntur (*Uhl*, 1929).

Ochna squarrosa L. cultivated in NU Botanical garden, small tree, flowers red.

BURSERACEAE

1. Drupes trigonous, dehiscent, seeds winged .... ***Boswellia***
1. Drupes globose, indehiscent, seeds not winged:
   2. Leaflets 3–7, disc free from calyx; ovary 2-celled .... ***Commiphora***
   2. Leaflets 6–15 pairs, disc adnate to calyx, ovary 5-celled .... ***Garuga***

Boswellia Roxburgh ex Colebrooke

Boswellia serrata Roxb. ex Colebr. in Asiat. Res. 9: 379. t. 5. 1807; FBI 1: 528; Gamble 168 (120). *B. glabra* Roxb. Pl. Cor. t. 207. 1811; Gamble 168 (120). *B. serrata* Roxb. var. *glabra* (Roxb.) Bennett in Fl. Brit. India 1: 528. 1815.

Deciduous trees, up to 10 m tall; branchlets pubescent; bark greenish, ashy–grey exfoliating in thin flakes; leaves odd-pinnate, apically clustered, leaflets oblong-lanceolate, 2–6 × 0.6–2.5 cm, tomentose, acute, slightly crenate, base obliquely acute, flowers white in axillary panicles; drupes trigonous, brown, pyrenes 3, each 1-seeded, enclosed in heart shaped pyrenes attached to the inner angles.

Occasional in Macherla RF. Fl. & Fr.: Mar.–Aug. *Vern.:* Guggilam, Anduga. This tree yields a golden yellow coloured gum resin, which is used in treating skin diseases and rheumatism. The fresh stem bark is chewed in toothache and to remove pus formation in the mouth. The wood is soft and of reddish tinge, and is used in making News-print paper. Bellam Konda forest, *DMR* & *SSR* 18112.

Commiphora N.J. Jacquin *nom. cons.*

1. Armed arboreous shrubs, flowers in fascicles, drupes apiculate .... ***C. berryi***
1. Unarmed deciduous trees, flowers in lax, dichotomous panicles; drupes entire .... ***C. caudata***

Commiphora berryi (Arn.) Engl. in A. & C. DC. Monogr. Phan. 4: 17. 1883; Gamble 170 (122). *Balsamodendron berryi* Arn. Ann. Nat. Hist. 3: 86. 1839; FBI 1: 529.

Armed, arboreous shrubs; branchlets spine-tipped; leaves in alternate clusters, 3-foliolate; leaflets ovate-suborbicular, 1.5–2 × 1–2.5 cm, glabrous, crenate, obtuse, base cuneate; flowers in axillary clusters; drupes ovoid-subglobose; seeds 2.

Occasional in scrub jungles. Planted as hedge near Kopparavuru, Nambur, Kanteru. Fl. & Fr.: Mar.–Aug. Karempudi, *KCJB* 8806; *CAB* 17443 (MH).

P 967, Curr. Sci. 73, 1997 gives notes on its economic and therapeutic values.

Commiphora caudata (Wight & Arn.) Engl. in DC. Monogr. Phan. 4: 17. 1883; Gamble 171 (122). *Protium caudatum* Wight & Arn. Prodr. 176. 1834; FBI 1: 530.

Deciduous, unarmed trees, up to 18 m tall; leaves 3–7-foliolate, leaflets elliptic-ovate, orbicular to abruptly long-acuminate, 3–10 × 1.5–6 cm, glabrescent, base unequal; flowers red, scented in axillary cymes; drupes globose, pyrenes 4, bony, seeds solitary, membranous.

Occasional in forests. Fl. & Fr.: Mar.–Oct. *Vern.:* Konda regu, Konda mamidi, metta mamidi. Fruits are pickled. The bark and leaves have the odour of mangoes and the oleoresin obtained from the tree is sometimes used as incense. Kondavedu Fort, *KCJB*, 17498 (MH).

Garuga Roxburgh

Garuga pinnata Roxb. Pl. Cor. t. 208. 1811 & Fl. Ind. 2: 400. 1832; FBI 1: 528; Gamble 169 (121).

Trees, up to 20 m tall; branchlets pubescent; leaves even-pinnate; leaflets 6–8 pairs, opposite, ovate-obovate or oblong, 3–6 × 1–2 cm, puberulous, subentire, abruptly caudate-acuminate, base unequal, flowers cream-yellow in axillary panicles; drupes oblong; pyrenes 2 or 3, seed 1 per pyrene.

Occasional, deciduous trees in forests. Fl. & Fr.: Feb.–Oct. *Vern.:* Garuga. The fruit is edible. It is a digestive and cooling agent and for asthma, conjuctivitis and stomachache. Reported in Forest Division Work Plan.

FLINDERSIACEAE

Chloroxylon A.P. de Candolle

Chloroxylon swietenia DC. Prodr. 1: 625. 1824; FBI 1: 569; Gamble 152 (109).

Trees, up to 8 m tall; bark thick, yellowish, corky, spongy, deeply and characteristically fissured; wood yellowish; leaves pinnate, leaflets oblong-lanceolate, 2–3.5 × 0.5–1.5 cm, gland dotted, obtuse, entire, base oblique, flowers white in terminal panicles; capsules ovoid-oblong, 3-loculicidal; seeds compressed, apically broad winged.

Common in forests. Fl. & Fr.: Mar.–Aug. *Vern.* Billudu, Billu. Wood hard, valuable and has great demand for agricultural implements, furniture, walking sticks etc. The stem bark pounded with *Azadirachta indica*, *Pongamia pinnata* bark administered orally after delivery as an antiseptic and healthy as general tonic. Loyapalle, *VRK* 3829.

MELIACEAE

1. Leaves 2 or 3-pinnate .... *Melia*
1. Leaves 1-pinnate:
   2. Seeds winged:
      3. Seeds winged at both ends, leaflets with obtuse-retuse apex .... *Soymida*
      3. Seeds winged at the base, leaflets with acuminate apex .... *Swietenia*
   2. Seeds not winged:
      4. Fruits berries, leaflets odd-pinnate, non-mangroves, flowers 5-merous .... *Azadirachata*
      4. Fruits capsules, leaflets even-pinnate, mangroves, flowers 4-merous .... *Xylocarpus*

Azadirachta A.H.L. Jussieu

Azadirachta indica A. Juss. Mem. Hist. Nat. 19: 231. t.2. f. 5. 1832; Gamble 177 (127). *Melia azadirachta* L. Sp. Pl. 385. 1753; FBI 1: 544.

Moderate sized trees, up to 10 m tall; leaves odd- and even pinnate, leaflets obliquely lanceolate, falcate, 3–5.5 × 1–2 cm, glabrous, acute-acuminate, serrate; flowers white in axillary panicles; drupes oblong, yellow when ripe; seed solitary, ellipsoid.

The Neem tree. Common in plains, roadsides and rare in forests. Fl. & Fr.: Mar.–June. *Vern.:* Vepa. Every part of the tree useful. It is one of the Kalpa vrikshas. Wood in agricultural implements, and minor timber; leaves pesticidal, insecticidal, antibacterial, antifungal, antidiabetic (young leaves) antimalarial, dermatitis. *Azadirachtin* is the active principle, microbicide and insecticide. Internaitonal controversy arose when US obtained patent for this Indian native and obtained a legal stay on patent. The ripe fruits are largely collected for the oil, which is used in parasitic skin diseases and also applied for hair for destroying lice. The twigs are used as tooth brushes. Occasionally a sap exudes continuously from the tree and it is also medicinally important. The insecticidal Azadirachtin is sold as Econeem in Guntur district. V.P. South, *VRK* 3952.

An outstanding example of revival, confirmation and implementation of old health tradition; attempts of micropropagation of plus neem trees (Elite trees with maximum Azadirachtin content) is carried out (Kavi Kishore, 1998).

Melia Linnaeus

Melia azedarach L. Sp. Pl. 384. 1753; FBI: 1: 544; Gamble 177 (126).

Medium sized trees, up to 5 m tall; leaves 2–or 3 pinnate, 30 cm long, leaflets elliptic-lanceolate, 3–5.5 × 1–2 cm, glabrous, acuminate, serrate, base oblique; flowers bluish-purple, fragrant in long panicles; drupes ellipsoid, fleshy; seed solitary, elliptic.

Commonly cultivated in backyards and fields. Rarely as an escape. Fl. & Fr.: Apr.–Dec. *Vern.:* Thuraka vepa. The fruits are poisonous and used in leprosy. V.P. South, *VRK* 3930.

Soymida A.H.L. Jussieu

Soymida febrifuga (Roxb.) Adr. Juss. Mem. Mus. Hist. Nat. 19. 251, t. 11. f. 26. 1830; FBI 1: 567; Gamble 185 (133). *Swietenia febrifuga* Roxb. Pl. Cor. t. 17. 1795 & Fl. Ind. 2: 398. 1832.

Unarmed deciduous trees, up to 10 m tall; branchlets with persistent leafscars; leaves even-pinnate; leaflets 6 pairs, opposite, ovate-oblong, 7–9 × 3–5 cm, thick-coriaceous, glabrous, entire, obtuse, base rounded-obtuse; flowers greenish-white in axillary or terminal panicles; capsules obovoid, woody; seeds numerous, oblong, compressed, winged at both ends.

Rare in forests. Fl. & Fr.: Feb.–July. *Vern.:* Somi, Cheva manu. Bark and other parts as febrifuge. The wood is extremely hard, very durable and is used in furniture making. Davupalli, *DMR & SSR* 18160; Davupalli, *SSR & KSM* 21403.

Swietenia N.J. Jacquin

Swietenia mahagoni (L.) Jacq. Enum. Pl. Carib. 20. 1760; Gamble 187 (134). *Cedrus mahagoni* L. Syst. Nat. (ed. 10) 2: 940. 1759.

Large trees, up to 20 m tall; leaves even-pinnate, to 15 cm; leaflets 4 or 5 pairs, oblong-lanceolate, slightly falcate, 4.5–7 × 1–2 cm, coriaceous, glabrous above, puberulous below, base oblique, entire, acuminate; flowers cream in panicles; capsules oblong; seeds winged, numerous.

The Mahogony tree. Introduced and naturalised as avenue trees. Fl. & Fr.: Apr.–July. Timber very durable, suitable for radio, television cabinets and other high grade work. Guntur (*Uhl* 1929).

Xylocarpus Koenig

Xylocarpus granatum J. Koenig Natur forschen. 20: 2. 1784; Noamexi Rev. Xylocarp. 107. 1964. *X. obovatus* (Blume) Adr. Juss. Mem. Mus. Nat. 19: 244. t. 9. f. 22. 1830; Gamble 185 (132). *Carapa obovata* Blume Bijdr. 179. 1825. *C. moluccensis* auct. non Lam.: FBI 1: 567.

Medium trees, up to 8 m tall; leaves even pinnate; leaflets 2–3 pairs, opposite, obovate, 4–6 × 2.5–3.5 cm, thick coriaceous, puberulous, base cuneate, entire, obtuse or round; flowers white in axillary panicles; capsules subglobose, 4–valved.

Common in Nizampatnam mangroves associated with *Avicennia* populations. Fl. & Fr.: Sept.–Dec. Wood hard and valuable. Nizampatnam, *VRK* 6913.

OLACACEAE

1. Armed small trees, stamens 8–10, calyx not enclosing drupes .... *Ximenia*
1. Unarmed shrubby climbers, stamens 3, calyx enclosing drupes to half the length .... *Olax*

Olax Linnaeus

Olax scandens Roxb. Pl. Cor. t. 102. 1799 & Fl. Ind. 1: 163. 1820; FBI 1: 575; Gamble 190 (136).

Small unarmed trees, up to 5 m tall; branchlets sparsely pubescent; leaves oblong or ovate-lanceolate; 3.5–8 × 2.5–3 cm, thin coriaceous, base cuneate, margin entire, ciliate, obtuse-acute; flowers creamish-yellow in axillary racemes; drupes globose-oblong, 1-seeded.

Occasional in scrub forests. Fl. & Fr.: Mar.–Oct. *Vern.:* Murki matte. Jayanthipurm, *VRK* 6894; 6958.

Ximenia Linnaeus

Ximenia americana L. Sp. Pl. 1198. 1753; FBI 1: 574; Gamble 189 (135).

Armed straggler, to 3 m tall; leaves elliptic-ovate, 3–5.2 × 2–3.3 cm, thin coriaceous, base rounded to acute, entire, apex retuse; flowers pale green in terminal or axillary subumbellate racemes; drupe globose, orange when ripe, pulpy, seed solitary.

Occasional in dry deciduous forests on stony ground. Fl. & Fr.: Mar.–July. *Vern.:* Natara. Fl. & Fr.: Round the year. Ripe fruits are eaten raw. The leaves are made into a paste and applied on externally to treat scabies. Davupalli, *SSR & KSM* 21412.

OPILIACEAE

1. Inflorescence axillary spikes, stigma 4-lobed .... *Cansjera*
1. Inflorescence axillary racemes, stigma simple .... *Opilia*

Cansjera A.L. Jussieu

Cansjera rheedii J. Gmelin in L. Syst. Nat. (ed. 13) 2(1): 280. 1791; FBI 1: 582; Gamble 193 (139). *C. scandens* Roxb. Pl. Cor. t. 103. 1799.

Unarmed woody lianas; branchlets tomentose; leaves ovate or lanceolate, 4–9 × 2–3.5 cm, thin-coriaceous, glabrous, base obtuse, entire, acuminate; flowers yellowish in axillary spikes; drupes scarlet, ovoid, endocarp thin, hard, seeds solitary.

Common in scrub jungles and hilly areas. Fl & Fr.: July–Dec. Jayanthipuram, *VRK* 6895.

Opilia Roxburgh

Opilia amentacea Roxb. Pl. Cor. t. 158. 1802 & Fl. Ind. 2: 87. 1832; FBI 1: 583; Gamble 192 (138).

Unarmed stragglers; branchlets pubescent; leaves elliptic or oblong, lanceolate, 2–9 × 1–3 cm, thin coriaceous, glabrous, enitre, acuminate, base obtuse; flowers greenish-yellow, in axillary racemes; drupes orange-yellow, obovoid-globose, pedicelled; seeds ovoid.

Occasional straggler in the forests. Fl. & Fr.: Apr.–Sept. *Vern.:* Adavi bandaldhi. Bollapalle, *KCJB* 17435 (MH).

Remarks: C.A. Barber (in Memoirs of the Department of Agriculture in India, Bot. Ser. Vol. 2, Part 5) has indicated that this is a root parasite.

CELASTRACEAE

1. Leaves decussate, trees .... ***Cassine***
1. Leaves alternate, shrubs or stragglers:
    2. Plants unarmed, flowers polygamous .... ***Celastrus***
    2. Plants armed, flowers bisexual .... ***Maytenus***

Cassine Linnaeus

Cassine glauca (Rottb.) Kuntze, Rev. Gen. Pl. 1: 114. 1891; var. **glauca** Ding Hou in Steenis Fl. Males; Ser. 1. 6: 287. 1962. *Mangifera glauca* Rottb. Nye Saml. Kongel. Danske Vidensk. Selsk. Skr. 2: 534 t. 4. f. 1. 1783. *Elaeodendron glaucum* (Rottb.) Pers. Syn. B1. 1. 341. 1805; FBI 1: 623; Gamble 211 (152).

Unarmed trees, up to 20 m tall; branchlets glabrous; leaves decussate, ovate-oblong, orbicular, 8–11 × 1–6 cm, thin coriaceous, glabrous, serrate-crenate, acute-acuminate, base obtuse; flowers green in divaricately branched, corymbose cymes; drupes oblong, 1-seeded.

Common avenue tree. Fl. & Fr.: Mar.–Sept. *Vern.:* Neridi. The fresh stem bark made into a fine paste, is applied on fore head for quick relief from head ache. Gum is roasted and eaten. Bollapalle, *KCJB* 17434 (MH); Davu palle, *SSR & KSM* 21419.

This taxon resembles Euphorbiaceae members in fruiting stage due to its 3-celled capsule.

Note: Ding Hou (op. cit.) has traced the history of this genus since the time of Linnaeus. The main difference between the two genera viz. *Cassine* and *Elaeodendron* apparently is that in *Cassine* the drupe is supposedly juicy with a thin crust like putamen, whereas in *Elaeodendron,* it is rather dry with a hard ligneous putamen. He states (op. cit. p. 285) Bentham & Hooker (Gen. Pl. 1, 1802, 363, 367) kept *Cassine* and *Elaeodendron* separate, but

Baillon (Hist. Pl. 1, 1877, 33) recognised only the latter. In 1892, Loesener reduced *Elaeodendron* to *Cassine* and distinguished them as two different sections of *Cassine* subg. *Elaeodendron*, adding that sect. *Elaeodendron* has vessels with scalariform, rarely also simple perforations, and *Cassine* simple round or elliptic perforations.. Later, however, he reinstated *Cassine*, *Elaeodendron* and *Mystroxylon* as distinct genera....'. Taking into cognisance the additional anatomical evidence of Metcalf & Chalk (1950), he concludes (p. 285): ' ...The recognition of only one genus, *Cassine*, seems therefore to be final'.

Celastrus Linnaeus

Celastrus paniculatus Willd. Sp. Pl. 1: 1125. 1798; FBI 1: 617; Gamble 208 (150).

Unarmed stragglers, up to 8 m; branchlets puberulous; leaves alternate, ovate, obovate or orbicular, 4–7 × 4–6 cms, thin-coriaceous, base obtuse to cuneate, crenate, serrulate, acuminate to caudate; flowers greenish-white, in terminal panicles; capsules globose, orange when ripe, 0.5–1 cm; seeds 2, elliposoid to ovoid, enclosed in arils.

Common stragglers on large trees in the forests. Fl. & Fr.: Mar.–Oct. *Vern.:* Kasara, Kora surrgudee, Eriki. Seeds yield an oil used in medicine as aphrodisiac and in rheumatism. Jayanthipuram, *VRK* 6899.

Maytenus Molina

Maytenus emarginata (Willd.) Ding Hou in Steenis, Fl. Males. Ser. 1. 6: 241. 1962. *Celastrus emarginatus* Willd. Sp. Pl. 1: 2: 1798. *Gymnosporia montana* (Roth) Benth. Fl. Austral. 1: 400. 1863; FBI 1: 621; Gamble 209 (150). *Celastrus montanus* Roth in Roem. & Schult. Syst. Veg. 5: 427. 1819. *Gymnosporia emarginata* (Wlld.) Thw. Enum. Zeyl. 409. 1864; FBI 1: 621; Gamble 219 (151).

Armed erect woody shrubs, up to 4 m tall; stem short, crooked, spines on the stem bearing leaves and flowers, leaves obovate, rarely broadly elliptic, coriaceous, 3.5–8 × 1.5–4 cm, emarginate, base cuneate, minutely serrate; flowers greenish-white in axillary panicles; capsules 2-valved, pyriform or subglobose, concave at tip; purplish when ripe; seeds rugose, orange brown or reddish.

Common in scrub forests, often in wastelands. Fl. & Fr.: Sept.–Jan. *Vern.:* Danti. Unripe fruits are cooked as vegetable. Mellavagu, *VRK* 5703; Sattenapally RF, *VRK* 5851.

RHAMNACEAE

1. Erect shrubs or trees .... *Ziziphus*
1. Stragglers or lianas:
   2. Fruits winged, unarmed lianas .... *Ventilago*
   2. Fruits not winged, armed stragglers .... *Scutia*

Scutia (Commerson ex A.P. de Candolle) A.T. Brongniart

Scutia myrtina (Burm. f.) Kurz in J. Asiat. Soc. Bengal. Pt. 2. Nat. Hist. 44. 168, 1875; Gamble 233 (160). *Rhamnus myrtinus* Burm. f. Fl. Ind. 60. 1768. *R. circumscissus* L. f. Suppl. Pl. 152. 1781. *Scutia indica* Brongn. Ann. Sci. Nat. (Paris) 10: 363, 1827; FBI 1: 640.

Armed stragglers, up to 5 m tall; branchlets glaucous; spines recurved; leaves obovate-orbicular, 1.5–3 × 1.5–2 cm, coriaceous, glossy above, entire, obtuse, base cuneate; flowers green in axillary umbellate clusters; drupes subglobose-obovoid; green, ultimately turning dark blue, seeds subglobose; compressed.

Common in scrub jungles. Fl. & Fr.: June–Oct. *Vern.:* Pariki, Nalla Korinta. Reported in Guntur Forest Division Work Plan.

Ventilago Gaertner

Ventilago maderaspatana Gaertn. Fruct. Sem. Pl. 1: 223. t. 49. f. 2. 1788; FBI 1: 631; Gamble 218 (156).

Large much-branched woody climbers; young branchlets pubescent; leaves ovate or ovate-lanceolate, elliptic, oblong, 4–6.5 × 1.5–3 cm, glabrous, usually acuminate, crenate, base rounded; flowers greenish in large drooping, axillary and terminal panicles; samaras subglobose; seeds solitary.

Occasional in the forest. Fl. & Fr.: Jan.–Aug. *Vern.*: Surala tiga, Surudugudu, Surali chekka. Root bark yields an important dye and forms important minor forest produce. It also contain the drug 'emodin' a stomach tonic and stimulent. Koppanur, *VRK* 3869.

Ziziphus P. Miller

Note: The spelling of the generic name of the taxon has raised controversy in the past (Santapau in JBNHS 51: 501, 1953). The present spelling, however, has been adopted from Dandy (1967) in *Regnum vegetabile*, Vol. 5).

1. Plants unarmed .... *Z. glabrata*
1. Plants armed:
   2. Inflorescence up to 30 cm long, petals absent .... *Z. rugosa*
   2. Inflorescence up to 3 cm long, petals present:

3. Trees, prickles paired, recurved .... ***Z. maurtitiana***
3. Shrubs, prickles paired, one straight, one recurved:
   4. Leaves acute or acuminate, drupes 2 mm across .... ***Z. oenoplia***
   4. Leaves obtuse or rounded, drupes 2 cm across:
      5. Styles 3, free, peduncles of cymes mostly longer than the pedicels, drupes green, leaves pubescent beneath .... ***Z. xylopyrus***
      5. Styles 2, connate, peduncles of cymes shorter than the pedicels, drupes red, leaves grey-pubescent on both sides .... ***Z. nummularia***

Ziziphus glabrata Heyne ex Roth Nov. Pl. Sp. 159. 1821; FBI 1: 633. *Z. trinervia* Roxb. Fl. Ind. 1: 606. 1832; Gamble 220 (158).

Unarmed trees, up to 8 m tall; leaves ovate to oblong-elliptic, 4–7 × 2.5–5 cm, glabrous or pubescent on the nerves beneath, obtuse or acute, crenate, base obtuse; flowers white in axillary cymes; drupes globose-obovid, 8 mm across, rugose; seed solitary.

Occasional in Macherla RF. Fl. & Fr.: Apr.–June. *Vern.:* Korupala. Loyapalle, *VRK* 3833; Jayanthipuram, *VRK* 6942.

Ziziphus mauritiana Lam. Encycl. 3: 319. 1788. var. **mauritiana**. *Z. jujuba* (L.) Gaertn. Fruct. 1: 203. 1788. non Mill. 1768; FBI 1: 632; Gamble 219 (157). *Rhamnus jujuba* L. Sp. Pl. 194. 1753.

Prickly trees with spreading crowns, up to 10 m tall; bark rough, grey, spines paired, recurved; leaves ovate-elliptic, 2–3.5 × 1–2 cm, rounded at both ends, slightly in-equilateral at the base, crenate; flowers reddish-green or yellowish green, in axillary cymes; drupes globose or ovoid, red or brownish pink, edible; seeds solitary, pulp jelly like.

The jujube tree. Common in wastelands, roadsides and scrub jungles. Fl. & Fr.: Oct.–Jan. *Vern.:* Regu or Reni. Ripe fruits are edible. Javuku, *VRK* 3919.

Various cultivars are released with larger fruits round or oval and marketed in season.

Note: The correct name of this taxon has been discussed by Santapau in J. Bombay Nat. Hist. Soc. 51. 807. 1959. *Z. jujuba* Gaertn. is a later homonym of *Z. jujuba* Mill. and hence invalid. *Z. jujuba* was first employed by Miller for a different tree.

Ziziphus nummularia (Burm. f.) Wight & Arn. in Wight. Cat. 31. 1873 & Prodr. 163. 1834; FBI 1: 158; Gamble 220 (158). *Rhamnus nummularia* Burm. f. Fl. Ind. 61. 1768.

Small armed shrubs; branchlets shortly zig-zag; tomentose; leaves alternate, 3–5-ribbed, orbicular, elliptic-oblong, 1.5–2.5 × 0.5–2 cm, serrulate, grey pubescent on both sides, thorns straight and recurved, slender; flowers creamish in axillary cymes; drupes globose, stones bony.

Occasional in scrub forests. Fl. & Fr.: Dec.–Feb. *Vern.:* Nela regu. The leaves are collected, dried and stored and are a good source of fodder for goats during winter season. The fruits are eaten and when dried it is used medicinally in Ayurveda. Vinukonda, *CAB* 4762 (MH).

Ziziphus oenoplia (L.) Mill. Gard. Dict. ed. 8. n. 3. 1768; FBI 1: 634; Gamble 220 (158). *Rhamnus oenoplia* L. Sp. Pl. 194. 1753.

Straggling, prickly, tomentose shrubs, up to 3 m tall; prickles paired, one straight and other recurved; leaves ovate, ovate-lanceolate 2–4 × 1–2.5 cm, glabrous or pubescent above, densely silky with appressed rufous hairs below, acute, denticulate; flowers yellowish green in axillary fascicles; drupes minutely globose, dark-pink; seed solitary, woody.

Common in scrub forests. Fl. & Fl.: Oct.–Jan. *Vern.:* Banka, Pariki. Ripe fruits are eaten raw. Loyapalle, *VRK* 3975; Loyapalle, *SSR & KSM* 21649. Nagarjunakonda, *KTH* 9697 (CAL).

Ziziphus rugosa Lam. Encycl. 3: 318. 1789; FBI 1: 636; Gamble 220 (158).

Woody, armed stragglers; thorns stout; branchlets rusty-tomentose; leaves orbicular, broadly elliptic, 8–15 × 5–10 cm, glabrous above, woolly below, obtuse, serrulate, base subcordate; flowers greenish in long peduncled tomentose cymes arranged along usually leaflets, forming panicles; drupes globose; seeds solitary, stony.

Occasional in scrub forests. Fl. & Fr.: July–Dec. *Vern.:* Goti tige, Pindu parigamu. Reported in Guntur Forest Division Work Plan.

Ziziphus xylopyrus (Retz.) Willd. Sp. Pl. 1: 1104. 1798; FBI 1: 634; Gamble 220 (158). *Rhamnus xylopyrus* Retz. Obs. Bot. 2: 11. 1781.

Straggling, prickly tree-like shrubs, up to 5 m tall; prickles paired, one straight, one recurved; leaves elliptic-oblong or sub orbicular, 2–8 × 1.5–3 cm, glabrous above, pubescent beneath, obtuse, very oblique and sometimes cordate at the base, crenate-serrate; flowers greenish-yellow, in axillary panicles; drupes globose, hard, woody, tomentose; seeds oblong.

Common along the hill slopes and forests. Fl. & Fr.: May–Sept. *Vern.:* Gotti, Gotika. Bark is used for tanning. Javuku, *VRK* 3913; Nagarjunakonda, *VRK* 6718; Karempudi, *CAB* 4754 (MH); Nagarjunakonda, *KTH* 9796 (CAL).

VITACEAE

Note: Planchon in DC. Monogr. Phan. divided *Vitis* into number of genera, they were noticed to belong to the following. 1. *Cissus* petals 4, style long. 2. *Cyphostemma*, branchlets bristly setose, 3. *Cayratia,* cymes axillary.

1. Leaves simple .... ***Cissus***
1. Leaves pedately or digitately 3–7-foliolate:
    2. Berres 1-seeded, cymes leaf-opposed .... ***Cyphostemma***
    2. Berres 2–4 seeded, cymes axillary .... ***Cayratia***

Cayraita A.L. Jussieu *nom. cons.*

1. Leaves 3-foliolate, seeds pyriform .... *C. trifolia*
1. Leaves 5–7-foliolate, seeds hemispheric .... *C. pedata*

Cayratia pedata (Lam.) Juss. ex Gagnep. in Lecomte, Notul. Syst. (Paris) 1: 346. 1911; Gamble 236 (169). *Cissius pedata* Lam. Encycl. 1: 31. 1783. *Vitis pedata* (Lam.) Wall. ex Wight & Arn. Prodr. 128. 1834; FBI 1: 661.

Climbing shrubs, branches striate, puberulous, leaves pedately, 5–7-foliolate, leaflets elliptic, oblong or ovate-lanceolate, 3–12 × 1.6–5 cm softly pubescent, acuminate, base inequilateral, subentire to serrate, tendrils leaf opposed; flower pale green in axillary branched cymes; fruit subglobose, depressed, 2–4 seeded; seeds convex and concave, the convex side closed by a membrane.

Occaisonal in forests. Fl. & Fr.: Jan.–Nov. *Vern.*: Edakul mandula-marri, Gummadi tiga. Kondavedu, *SSR & KSM* 21478.

Note: Nomenclaturally this taxon is interesting, for, as pointed out by Furtado (in Taxon 13; 240. 1964.) the specific epithet, *pedata* originated in 1783 as *Cissus pedata* Lamk. and not in 1791 by Loureiro, as indicated in the 1961 Botanical code. Not withstanding, the actual type of the genus *Cayratia* Juss. is based on Loureiro's specimen.

Cayratia trifolia (L.) Domin. Biblioth. Bot. 89: 371. 1971. *Vitis trifolia* L. Sp. Pl. 203. 1753. *Cayratia carnosa* (Wight & Arn.) Gagnep. in Lecomte, Notul. Syst. (Paris) 1: 347. 1911; Gamble 237 (169). *Vitis carnosa* Wight & Arn. Prodr. Fl. Ind. Orient. 127. 1834; FBI 1: 654.

Climbing shrubs, rather succulent; leaves 3-foliolate, leaflets ovate-orbicular, 2–5 × 1–3 cm, pubescent, acute, dentate, base obtuse, flowers greenish-white, in corymbose cymes; berries obovoid; seeds 2.

Occasional in forests, roadsides and plains. Fl. & Fr.: June–Nov. *Vern.:* Kanupu tige, Pulla mada. Roots of the plant are said to be used in curing tumors, pains and spleen complaints. Kondavedu, *SSR & KSM* 21454.

Note: Though some earlier authors treated *Cissus carnosa* Lamk. as distinct taxon from Linnaeus *Vitis trifolia*, it is now evident from the critical studies by Merril (inter-Herb.Amb.) and subsequently by Backer & Bakhuizen f. (Fl. Jav. 1965) that both are conspecific and hence its current name under the genus *Cayratia* should be *C. trifolia* (L.) Domin.

Cissus Linnaeus

1. Stem 4-angular, 4-winged .... *C. quadrangularis*
1. Stem terete, not winged .... *C. vitiginea*

Cissus quadrangularis L. Mant. Pl. 39. 1767; Gamble 234 (167). *Vitis quadrangularis* (L.) Wall. ex Wight Cat. 36. 1833; FBI 1: 645.

Climbing herbs with stout root stock; stems fleshy, quadrangular, angles slightly winged, odorous with much constricted nodes; leaves simple, ovate or reniform, 2–5 × 1.5–4 cm, crenate; tendrils leaf-opposed, highly coiled; flowers greenish-red in leaf-opposed umbels; berries ovoid, reddish-brown; seed solitary, obovate, smooth.

Common in scrub jungles and hedges. Fl. & Fr.: May–Dec. *Vern.*: Nalleru. Tender shoots and leaves are cooked as vegetable. Plant is used in medicine for fractured bone setting after crushing and adding with coconut oil. N. Varam, *VRK* 3971; Nagarjunakonda, *KTH* 9634 (CAL).

Cissus vitiginea L. Sp. Pl. 117. 1753; Gamble 334 (167). *Vitis linnaei* Wall. ex Wight & Arn. Prodr. Fl. Ind. Orient. 26. 1834; FBI 1: 649.

Vines; branchlets densely pubescent; leaves simple, broadly cordate, 5–8 × 4.5–8.5 cm, 5-angular or deeply lobed, thick coriaceous, pubescent, dentate, acuminate, base cordate; flowers pale yellow, in umbellate cymes; berries pyriform, rugose, apiculate, purplish blue when ripe, seeds tessellate.

Occasional in plains. Fl. & Fr.: June–Dec. *Vern.:* Adavi draksha. Nagarjunakonda, *KTH* 9665 (CAL).

Cyphostemma A.H.J. Alston

Cyphostemma setosum (Roxb.) Alston in Trimen, Handb. Fl. Ceylon 6 (Suppl.): 53. 1931. *Cissus setosa* Roxb. Fl. Ind. 1: 410. 1820; Gamble 235 (168). *Vitis setosa* (Roxb.) Wall. ex Wight & Arn. Prodr. 127. 1834; FBI 1: 804.

Fleshy herbaceous vine to 5 m, branchlets bristly setose, leaves 3-foliolate, leaflets oblong-obovate, 3.5–7.5 × 3–5 cm, subsucculent, glandular puberulous, apex and base rounded, margin sharply dentate-serrate; petiole 0; tendrils forked; flowers green in dichasial, divaricate, glandular pubescent cymes; berries bright red when ripe, seeds deeply pitted and crenate at margins.

Plains to the foothills. Fl. & Fr.: Nov.–Jan. *Vern.:* Pullabachali. Kondavedu, *SSR* & KSM 21654.

SAPINDACEAE

1. Tendril climbers .... *Cardiospermum*
1. Shrubs or trees:
    2. Flowers irregular, disc unilateral:
        3. Leaves 3-foilolate .... *Allophylus*
        3. Leaves pinnate .... *Lepisanthes*

2. Flowers regular, disc annular:
   4. Drupes smooth .... *Sapindus*
   4. Drupes tubercled or muricate .... *Dodonaea*

Allophylus Linnaeus

Allophylus serratus (Roxb.) Kurz. in Asiat. Soc. Bengal. 44: 185. 1876; Gamble 246 (175). *Ornitrophe serrata* Roxb. Pl. Cor. t. 61. 1796. *Allophylus cobbe* forma *serratus* (Roxb.) Hiern in Hook. f. Fl. Brit. India 1: 673. 1875.

Small trees, up to 4 m tall; leaves 3-foliolate, leaflets elliptic, ovate-lanceolate, 3–9 × 2–4 cm, woolly when young, glabrescent, acute, crenate-serrate, cuneate at base, flowers pale green or white in pubescent, axillary racemes; schizocarps 2-lobed, lobes globose, orange-red, seeds solitary.

Occasional in forests. Fl. & Fr.: June–Dec. *Vern.:* Jutti tiga, Erravalu. Guntur (*Uhl* 1919).

Cardiospermum Linnaeus

1. Capsules winged at the angles, seeds with large hilum, leaflets glabrous, acuminate apex .... *C. halicacabum*
1. Capsules not winged at the angles, seeds with small hilum, leaflets usually obtuse, the end one mucronate .... *C. canescens*

Cardiospermum canescens Wall. Pl. As. Rar. 1: 14. t. 14. 1839; FBI 1: 680; Gamble 244 (175).

Climbing herbs; stems deeply furrowed; leaves biternate; leaflets ovate, 1–5.6 × 1–2.5 cm, pubescent, dentate, serrate, acute or cuneate, flowers white in long-peduncled umbellate cymes on tendrils; capsules inflated, subglobose, triangled; seeds globose with orbicular hilum.

Common along the streams in open scrubs. Fl. & Fr.: July–Dec. *Vern.:* Pedda budda busara. Loyapalle, *VRK*, 3845; Davupalli, *SSR & KSM* 21668.

Cardiospermum halicacabum L. Sp. Pl. 366. 1753; FBI 1: 670; Gamble 244 (175).

Pubescent climbing herbs; leaves bipinnate, leaflets ovate-lanceolate, 1–3 × 0.5–2 cm, dark green; deeply incised, coarsely dentate, acuminate, cuneate; flowers white in axillary umbels;capsules pyriform, winged at angles, pubescent, loculicidally dehisced, seeds black, globose, with a heart-shaped black aril.

Common in open places, often climbing over hedges, thickets. Fl. & Fr.: July–Nov. Eng: Baloon vine. *Vern.;* Buddakakara. Koppanur, *VRK* 3254; Papaiah palem, *VRK* 5838; Kondavedu Fort, *C.A. Barber* 4630 (MH).

Dodonaea Miller

Dodoneae viscosa (L.) Jacq. Enum. Pl. Carib. 19. 1760; FBI 1: 697; Gamble 253 (181). *Ptelea viscosa* L. Sp. Pl. 118. 1753.

Evergreen shrubs, up to 2 m tall; leaves oblong or oblanceolate, 5–8 × 2–3 cm, shining, viscid, obtuse with short apicule, entire, base tapering into a short petiole; flowers greenish yellow, in axillary cymes; capsules 2–3-winged; yellowish green or yellowish brown, wings prominent, membranous; seeds subglobose, black.

Common in open scrubs, wastelands and hill slopes. Fl. & Fr.: May–Jan. Eng.: Jamaica switch sorrel. *Vern.:* Bandheru. Wood used for walking sticks. The leaves are made into paste and applied in the form of plaster in injuries for healing wounds. V.P. South, *VRK* 3933, Mellavagu, *VRK* 5708; Davupalli, *DMR* 13124.

Lepisanthes Blume

Lepisanthes tetraphylla (Vahl) Radlk. in Sitzumgsber. Math. Phys. Cl. Koenigl. Bayer. Akad. Wiss. Muenchen 8: 276. 1878; Gamble 247 (176). *Sapindus tetraphylla* Vahl, Symb. Bot. 3. 54. 1794. *Hemigyrosa canescens* (Roxb.) Bl. in Rumphia 3: 166. 1849; FBI 1: 671. *Molinaea canescens* Roxb. Pl. Cor. t. 60. 1796. *Lepisanthes deficiens* (Wight & Arn.) Radlk. in Sitzungsber. Math. Phys. Cl. Koenig. Bayer Akad. Wiss. Mueneben 8: 276. 1878, Gamble 247 (176). *Hemigyrosa deficiens* (Wight. & Arn.) Bedd. Fl. Sylv. t. 231. 1822; FBI 1: 671. *Sapindus deficiens* Wight & Arn. Prodr. 111. 1834.

Evergreen tree, up to 10 m tall; bark grey, fairly rough; branchlets conspicuously striate when young; leaves paripinnate, leaflets 2-pairs, thick coriaceous, variable, narrowly elliptic-oblong to broadly elliptic oblong, 3.2–20 × 1.1–9 cm, glabrous, base cuneate, entire, apex obtuse; flowers white, in dense flowered racemiform panicles; drupe woody, yellow when ripe, seed 1, oblong.

Rare in dry forests. Fl. & Fr.: Apr.–May. *Vern.;* Salikunkudu, Korivi. Wood hard and used as good fuel. Occasionally used by the natives for building purpose. Davupalli, *SSR & KSM* 21424.

Sapindus Linnaeus

Sapindus emarginatus Vahl Symb. Bot. 3: 54. 1794; Gamble 250 (178). *S. trifoliatus* non L. 1753; FBI 1: 682. *p.p.*

Trees, up to 8 m tall; leaves paripinnate; leaflets 2–3 pairs, elliptic-oblong, 5–10 × 2–5 cm, glabrous above, pubescent below, emarginate, entire, base cuneate-obtuse; flowers greenish-yellow in terminal racemes; drupes deep brown, ovoid-tomentose, wrinkled when ripe; seeds black, smooth, one in each locule.

The soapnut tree; occasionally cultivated, rare in forests. Fl. & Fr.: Oct.–Feb. Eng.: Soap berry. *Vern.:* Kunkudu. Dried fruit powder, used as a detergent for hair wash and washing of delicate and valuable silks. Fruit possess emitic, tonic, astringent, anthelminitic properties and are used in treatment of asthma, colic due to indigestion, diarrhorea etc. Fruit as well as root bark are used as fish poision. Javuku, *VRK* 3921; Kondavedu Fort, *CAB*, 4631 (MH).

ANACARDIACEAE

1. Leaves compound .... ***Lannea***
1. Leaves simple:
   2. Drupes with a basal hypocarp .... ***Anacardium***
   2. Drupes without a basal hypocarp:
      3. Drupes oblong or globose, with fleshy pericarp, flowers polygamous .... ***Mangifera***
      3. Drupes lenticular, not fleshy, flowers bisexual .... ***Buchanania***

Anacardium Linnaeus

Anacardium occidentale L. Sp. Pl. 383. 1753; FBI 2: 20; Gamble 260 (185).

Small evergreen trees, up to 8 m tall; branchlets glabrous; leaves simple, obovate, 8–15 × 6–9 cm, thick-coriaceous, glabrous, entire, obtuse, base cuneate-round; flowers unisexual, greenish with red stripes, in terminal or axillary panicles; nuts reniform, hypocarp pyriform; seeds reniform, ascending, 1–5.2 × 1 cm.

Introduced and planted along the coastline. Fl. & Fr.: Jan.–Aug. *Vern.:* Jidi mamidi, Munta mamidi. Eng.: Cashewnut. The seeds are roasted or eaten raw. The pericarp of the fruit yields on roasting a black acrid oil, hypocarp is fleshy and edible. Reported in Guntur Forest Disivion Work Plan.

Buchanania Sprengel

Buchanania lanzan Spreng. in J. Bot. (Schrader) 2: 234. 1800; Gamble 258 (184). *B. latifolia* Roxb. Fl. Ind. 2: 385. 1832; FBI 2: 23.

Small trees, 6–8 m tall; branchlets with prominent leaf-scars, rusty tomentose when young; leaves elliptic obovate, broadly oblong, 15–25 × 6–10 cm, glabrescent above, slightly villous beneath, obtuse-retuse, entire, base acute-round; flowers creamish in axillary and terminal panicles; drupes globose, obliquely lentiform, black, stony; seeds 1 or 2, biconvex.

Occasional in forests. Fl. & Fr.: Nov.–May. *Vern.:* Pedda sarapappu. The fruit is largely eaten and also used in the preparation of sweets. Reported in Guntur Forest Division Work Plan.

Lannea Richard *nom. cons.*

Lannea coromandalica (Houtt.) Merr. in J. Arnold Arbor. 19: 353. 1938.
Dialium coromandalicum Houtt. Nat. Hist. ser. 2. 2: 39. t. 5. f. 2. 1774.
Odina wodier Roxb. Fl. Ind. 2: 293. 1832; FBI 2: 29; Gamble 263 (187).

Deciduous trees, up to 8 m tall; branchlets pubescent; leaves odd-pinnate, leaflets ovate, 3–8 × 2–4 cm, chartaceous, acuminate-acute, unequally rounded at base, entire; flowers greenish-yellow in axillary panicles; drupes fleshy, 1–2 × 0.5–6 cm, compressed, seed solitary.

Occasional in forests. Fl. & Fr.: Mar.–May. *Vern.:* Gumpena, Oddi. Wood fairly good but not much used. It yields in considerable quantity. Paste of the stem bark together with egg paste is applied in bone fracture and tied with the help of bamboo sticks for one week. Vinukonda RF, *VRK* 6876; Javuku, *VRK* 3884.

Mangifera Linnaeus

Mangifera indica L. Sp. Pl. 200. 1753; FBI 2: 13; Gamble 259 (185).

Large evergreen trees, with a dense crown, up to 15 m tall; leaves crowded at the tip of branches, lanceolate, 10–20 × 3–5 cm, glabrous, apex acute to shortly or obtusely acuminate, entire, base cuneate; flowers yellowish-green in terminal panicles; drupes fleshy, orange yellow when ripe; seed solitary, oblong, compressed.

The Mango tree; cutivated as commercial fruit crop, often wild. Fl. & Fr.: Dec.–Aug. *Vern.:* Mamidi. Fruits edible; used to prepare pickles, dishes, juice as drinks and Jams; wood is used for building constructions. V.P. South, *VRK* 3932.

MORINGACEAE

Moringa Adanson

1. Leaves 3-pinnate .... ***M. oleifera***
1. Leaves 2-pinnate .... ***M. concanensis***

Moringa concanensis Nimmo ex Dalz. & Gibson Bombay Fl. 311. 1861; FBI 2: 45; Gamble 270 (192).

Deciduous trees, up to 10 m tall; branchlets tomentose; leaves 2-pinnate, secondary rachis articulated, and with a gland at the articulations, leaflets broad, ovate-elliptic, 1–3 × 1–2 cm, glabrous, pale beneath, obtuse, entire, base rotund; flowers pale yellow in terminal panicles; capsules beaked; seeds numerous, winged, white or pale yellow.

Rare in Vinukonda RF. Fl. & Fr.: Feb.–Aug. *Vern.:* Karu munaga, Konda munaga. Fruits are not edible. Stem brittle, wood soft. Jayanthipuram, *VRK* 6966.

Moringa oleifera Lam. Encycl. 1: 398. 1785; Gamble 270 (192). *M. pterygosperma* Gaertn. Fruct. 2; 314. 1791; FBI 2: 45.

Medium sized trees, up to 8 m tall; trunk grey-white with longitudinal wrinkles; leaves 3-pinnate, up to 40 cm long, terminal odd leaflet obovate or elliptic, 1.5–2.5 × 1–1.5 cm, glabrous, obtuse; flowers white in terminal panicles, sweet scented; capsules long-linear, loculicidally dehiscent, woody; seeds winged.

The Drumstick tree. Cultivated in gardens and house compounds. Fl. & Fr.: Feb.–Sept. *Vern.*: Munaga, Mulaga. Fruits and leaves eaten as vegetables. The roots have the flavour of horse radish and the seeds are eaten in curries and yields a valuable oil; wood is used as floater. Javuku, *VRK* 3901.

Note: Lamarck (1785) cited *Balanus myrepsica* Garacult (1764) under *Moringa oleifera*. But *B. myrepsica* being a distinct species, was later reduced to *Moringa myrepsica* (Gara.) Thell. Some authors doubting the circumscription of *M. oleifera* as given by Lamarck include two species *M. oleifera* and *M. pterygosperma* Gaertn.

FABACEAE (*nom. alter.* PAPILIONACEAE)

1. Trees, shrubs and erect or prostrate herbs:
  2. Trees:
    3. Leaflets alternate:
      4. Corolla golden yellow, pods broad, winged, rounded .... ***Pterocarpus***
      4. Corolla other than golden yellow, pods narrow, winged, or without wings .... ***Dalbergia***
    3. Leaflets opposite:
      5. Leaves 3-foliolate:
        6. Branchlets prickled .... ***Erythrina***
        6. Branchlets tomentose .... ***Butea***
      5. Leaves more than 3-foliolate:
        7. Branchlets glandular pubescent, pods up to 18 cm long, seeds more than 5 .... ***Gliricidia***
        7. Branchlets glabrescent, pods up to 5 cm long, seeds 1–2 .... ***Pongamia***
  2. Shrubs, herbs, erect or prostrate:
    8. Leaves simple:
      9. Leaves ovate-cordate .... ***Indigofera*** (*p.p*)
      9. Leaves obovate, oblong, linear-lanceolate:
        10. Pods globose, seeds solitary .... ***Indigofera*** (*p.p.*)
        10. Pods linear or oblong, seeds more than 2:
          11. Pods not constricted, dehiscent:

12. Pods flattened, stamens diadelphous .... *Tephrosia* (*p.p*)
12. Pods turgid or inflated, stamens monadelphous .... *Crotalaria* (*p.p*)
11. Pods constricted, indehiscent:
13. Joints of pods turgid .... *Alysicarpus*
13. Joints of pods flattened .... *Desmodium* (*p.p*)
8. Leaves compound:
14. Leaves even-pinnate:
15. Pods enclosed by calyx-tube .... *Smithia*
15. Pods exceeding the calyx-tube:
16. Pods prickly .... *Zornia*
16. Pods glabrous .... *Sesbania*
14. Leaves odd-pinnate:
17. Leaves 3-foliolate:
18. Procumbent herbs or undershrubs:
19. Bracts dimorphic .... *Stylosanthes*
19. Bracts unimorphic:
20. Pods 2-seeded:
21. Flowers yellow, pod turgid, straight .... *Crotalaria* (*p.p*)
21. Flowers red, pod torulose, deflexed .... *Indigofera* (*p.p*)
20. Pods more than 2-seeded:
22. Flowers yellow, solitary or in 3-flowered fascicles .... *Rothia*
22. Flowers pink or red, many in racemes:
23. Pod flat, compressed, hooked pubescent, non-septate ... . *Pseudarthria*
23. Pod torulose, not compressed, glabrous or slightly hairy, septate .... *Indigofera* (*p.p*)
18. Erect herbs or under shrubs:
24. Flowers yellow .... *Cajanus (cajan)*
24. Flowers pink, purplish or blue .... *Desmodium* (*p.p*)
17. Leaves more than 3-foliolate:
25. Flowers in terminal or leaf-opposed corymbose racemes:
26. Stamens monadelphous .... *Mundulea*
26. Stamens diadelphous .... *Tephrosia* (*p.p*)
25. Flowers in axillary or extra axillary racemes:
27. Pods compressed .... *Aeschynomene*
27. Pods not compressed .... *Indigofera* (*p.p*)
1. Climbers, stragglers or vines:
28. Leaves even-pinnate.. *Abrus*
28. Leaves odd-pinnate:
29. Leaflets 5-15:
30. Pods winged, stamens monadelphous .... *Derris*
30. Pods not winged, stamens diadelphous.. *Clitoria*
29. Leaflets 3:
31. Pods 1–2-seeded (rarely 3) .... *Rhynchosia*
31. Pods more than 3-seeded:
32. Flowers yellow:
33. Leaves gland dotted beneath .... *Cajanus (scarabaeoides)*
33. Leaves not gland dotted beneath .... *Vigna*
32. Flowers other than yellow:
34. Stamens monadelphous:
35. Twining herbs, standard petal simple .... *Teramnus*
35. Twining woody shrubs, standard petal appendaged .... *Canavalia*

34. Stanmens diadelphous:
    36. Leaflets elliptic-lanceolate or linear- oblong:
        37. Prostrate or twining herbs, branchlets glabrescent, leaves villous beneath .... *Galactia*
        37. Climbers, branchlets and leaves densely villous .... *Macrotyloma*
    36. Leaflets cordate-ovate .... *Lablab*

Abrus Adanson

1. Seeds subglobose, blood red, with a black blotch at the hilum, leaflets glabrous .... *A. precatorius*
1. Seeds flat, black, leaflets silky pubescent .... *A. fruticulosus*

Abrus fruticulosus Wall. ex Wight & Arn. Prodr. 236. 1834; FBI 2: 176. 1876; Gamble 350 (247).

Twining shrubs, leaves even pinnate; up to 6 cm long; leaflets opposite, up to 12 pairs, 1 × 0.3 cm, silky pubescent, entire, acute, oblong-linear; flowers pink in axillary racemes; pods flat, linear-oblong, 3 × 1 cm; seeds 3–5, flattened, black.

Common twiner in hedges of open forests and roadsides, Fl. & Fr.: Jan.–Oct. Nagarjunakonda, *VRK* 6708; V.P. South, *VRK* 6751; Krishna river, *KMS* 9853 (MH).

Note: Raghava Swamy and Rao (1987) reported this species from Nalgonda for the first time from Andhra Pradesh (J. Econ. Tax. Bot. 9: 489–492).

Abrus precatorius L. Syst.Nat. ed. 12. 2: 473. 1767; FBI 2: 175; Gamble 349 (247); Verdc. in Kew Bull. 24: 240. 1970. *Glycine abrus* L. Sp. Pl. 753. 1753.

Woody twiners, leaves even pinnate, 3–8 cm long, leaflets 7–12 pairs, oblong, 1–2.5 × 0.4–0.8 cm, glabrous, obtuse or mucronate, base rounded; flowers rose to pink in axillary racemes; pods oblong, 4–5 × 0.8–1.2 cm, pilose, wrinkled; seeds blood red, with a black lateral blotch around the hilum.

Occasional in forests. Fl. & Fr.: July–Dec. Eng. Indian liqueries. *Vern.:* Guruginja. Seeds are poisonous. Its seeds have been in use since ancient times for weighing commodities in very minor quantities viz. gold and medicines. Fresh leaves when chewed with sugar relieve hoarseness of voice strengthens gums; paste of leaves is applied in skin diseases (leucoderma). Loyapalle, *VRK* 3974; Mellavagu, *VRK* 5704; Nagarjunakonda, *KTH* 9654 (CAL).

Aeschynomene Linnaeus

Aeschynomene indica L. Sp. Pl. 713. 1753; FBI 2: 151; Gamble 331 (234).

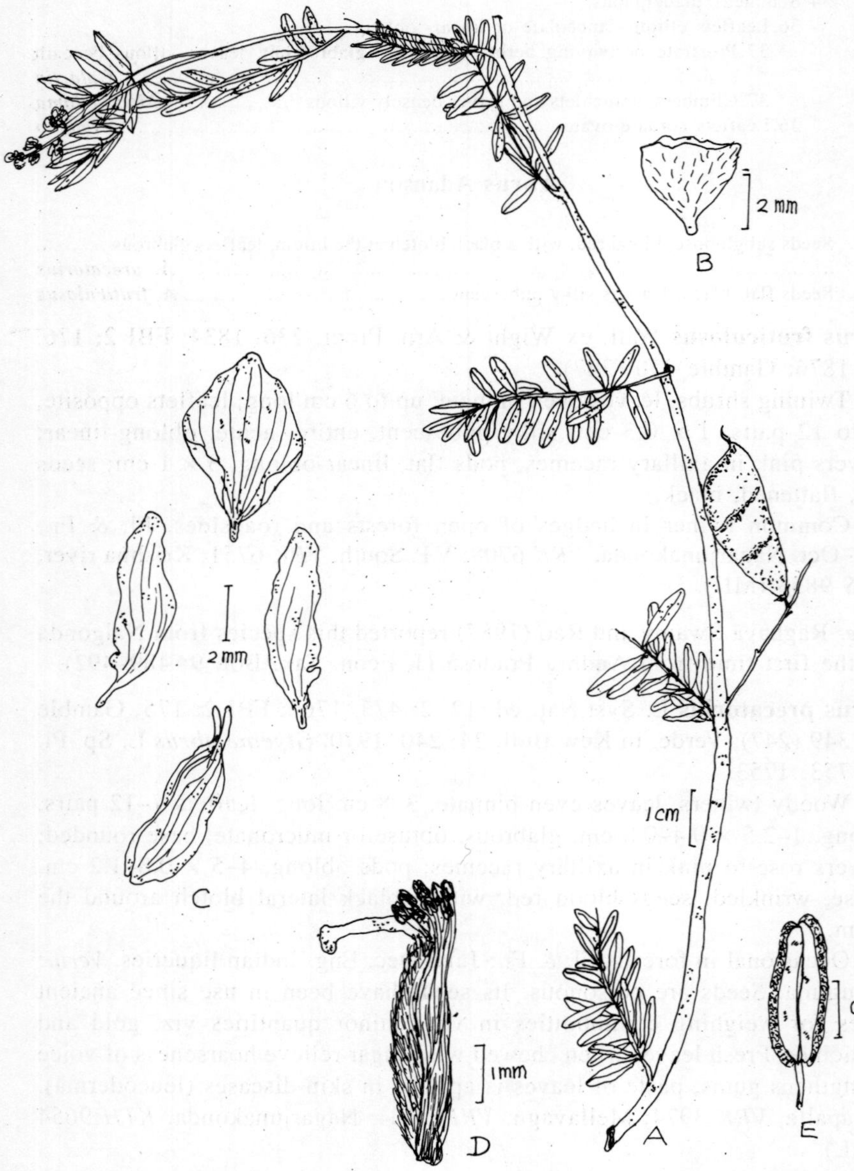

Fig. 5. *Abrus fruticulosus* Wall.
A. Twig, B. Calyx, C. Corolla (spread out), D. Pistil with stamens, E. Anther.

Annual erect, branched herbs, up to 1 m tall; leaflets oblong, 2.5–5 × 1–1.5 cm, glabrous, base and apex obtuse, entire; flowers yellowish cream in axillary, solitary or in 2–4-flowered racemes; pods jointed, 3–4 × 0.3–0.5 cm, lower suture indented; seeds brown, smooth.

Common marshy herbs along the edges of streams and ponds. Fl. & Fr.: July–Nov. *Vern.*: Teega Jeeluga. Vinukonda RF, *VRK* 6938; NU Campus *SSR & KSM* 21678; Kondavedu Fort, *CAB*, 4648 (MH).

Note: Rudd (1959) says "TYPE LOCALITY" "INDIA" Collector not known. It seems that *A. indica* is native of America and was introduced into the old world.

Alysicarpus Desvaux *nom. cons.*

1. Erect, stout, woody herbs .... ***A. longifolius***
1. Prostrate or ascending herbs:
   2. Upper leaves 3-foliolate:
      3. Pods puberulous; 4–8-jointed .... ***A. heterophyllus***
      3. Pods glabrous, 2–5-jointed .... ***A. scariosus***
   2. All leaves 1-foliate:
      4. Pods not moniliform:
         5. Calyx shorter than the first joint of the pod, pod compressed, shallowly indented on margins .... ***A. hamosus***
         5. Calyx equalling or slightly longer than the first joint of the pod, pod turgid, not indented on margins .... ***A. vaginalis***
      4. Pod moniliform:
         6. Calyx shorter or very slightly longer than the first joint of pod .... ***A. monilifer***
         6. Calyx much longer than the first joint of pod .... ***A. bupleurifolius***

Alysicarpus bupleurifolius (L.) DC. Prodr. 2: 352. 1825; FBI 2: 158; Gamble 338 (239). *Hedysarum bupleurifolium* L. Sp. Pl. 745. 1753.

Prostrate or ascending herbs, up to 50 cm tall; leaves linear-lanceolate, 2–7 × 0.2–0.5 cm, glabrous, acute, enitre; flowers yellow, with pink tinge in racemes; pods jointed, smooth; seeds 3–4, ovoid, smooth shiny, light-brown.

Occasional weed in wastelands and cultivated fields, rare in forests. Fl. & Fr.: Aug.–Oct. Kondavedu Fort, *CAB* 4692 (MH).

Alysicarpus hamosus Edgew. in Journ. Asiat. Soc. Beng. 21: 171. 1853; FBI 2: 157; Gamble 338 (238).

Slender diffuse herbs, clothed with spreading hairs; leaves orbicular, mucronate, hairy on both surfaces, stipules sheath like, base slightly cordate; flowers purple-violet in pairs along the rachis in axillary racemes; pods compressed, clothed with hairs, 2–8-jointed.

Common weed of cultivated fields and wastelands. Fl. & Fr.: July–Dec. Nagarjunakonda, *KTH* 9825 (CAL).

Note: It is an interesting species in the genus and it resembles *Desmoidum* in the presence of compressed pod and on the other hand it is similar to

Alysicarpus is having 1-foliate leaves and ebracteolate flowers. This combination of characters required experimental studies. It is an important problem in the Bio-systematic studies.

Alysicarpus heterophyllus (Benth. ex Baker) Jafri & Ali in Biologia (Lahore) 12: 13. 1966. *A. vaginalis* (L.) DC. var. *heterophyllus* Benth. ex Baker in Hook. f. Fl. Brit. India 2: 158. 1876.

Glabrous prostrate herbs; leaves vary in shape, lower simple, ovate-oblong, upper 1–3-foliolate, oblong-lanceolate to linear-lanceolate, 1–2.5 × 0.5–1 cm, glabrescent; base and apex subacute; flowers deep blue with red tinge in racemes, pods moniliform, longitudinally rugose, seeds orbicular.

Occasional in forests. Fl. & Fr.: July–Dec. Kondavedu Fort, *CAB* 4690 (MH).

Alysicarpus longifolius.Wight & Arn. Prodr. 233. 1834; FBI 2: 159; Gamble 338 (239).

Stout woody glabrescent herbs to undershrubs, up to 90 cm tall; leaves oblong-lanceolate, 2–11 × 1–2.5 cm, obtuse, base rounded; pods 3–6-jointed, compressed; seeds brown.

Common weed in fields and rare in forests. Fl. & Fr.: Sept.–Dec. *Vern.:* Peddakandikaraku. Vinukonda, *CAB* 4768 (MH).

Alysicarpus monilifer (L.) DC. Prodr. 2: 353; FBI 2: 157; Gamble 338 (238). *Hedysarum moniliferum* L. Mant. 1: 102. 1767.

Prostrate herb, up to 50 cm long, branchlets with deciduous pilose hairs; leaves ovate or elliptic-oblong, 0.7–3.5 × 0.5–2 cm, sparsely pubescent, obtuse, entire, cordate; flowers violet in 4–10-flowered axillary racemes; pod moniliform, nerves obscure, finely downy with minutely hooked hairs, articles 5–7, turgid.

Common in deciduous and open forests. Fl. & Fr.: Throughout the year. *Vern.*: Amera. NU Campus, *SSR & KSM* 21500, 21680.

Alysicarpus scariosus (Rottl. ex Spreng) Garh. ex Thw. Enum. Pl. Zey. 88. 1858. *A. rugosus* DC. var. *styracifolius* Baker in Hook. f. Fl. Brit. India 2: 159. 1876; Gamble 338 (239).

Annual erect herbs, up to 40 cm tall, stipules papery, yellowish, hairy; leaves simple, lower small, elliptic-oblong, 1–1.8 × 0.5–0.8 cm, obtuse, entire; flowers pink in dense racemes; pods enclosed in calyx lobes, 3–5-jointed; seeds sub-orbicular.

Occasional in open forests. Fl. & Fr.: Aug.–Jan. *Vern.:* Chinna Kandikaraku. Koppanur,*VRK* 3261.

Alysicarpus vaginalis (L.) DC. Prodr. 2: 353. 1825. var. ***vaginalis***; FBI 2: 158; Gamble 338 (239). *Hedysarum vaginalis* L. Sp. Pl. 746. 1753. *Alysicarpus vaginalis* (L.) DC. var. *nummularifolius* Miq. Fl. Ind. Bot. 1: 232. 1855; FBI 2: 158; Gamble 338 (238).

Procumbent herbs, leaves obovate-oblong, 0.6–2 × 0.4–1.2 cm, glabrous, pale beneath, obtuse, entire, base obtuse-cordate; flowers purplish in terminal racemes; pods terete, 1–8 × 0.2 cm, jointed; seeds 5.7 × 0.4 mm, smooth.

Occasional along the streams in forests. Fl. & Fr.: Nov.–Mar. It is used for cough. Javuku, *VRK* 3908; Achampet, *VRK* 5861; Madinepadu, *KCJB* 17469 (MH).

Butea Roxburgh ex Willdenow *nom. cons.*

Butea monosperma (Lam.) Taub. in Engler & Prantl Pfl. Fam. 3 (3): 366. 1894. *Erythrina monosperma* Lam. Encycl. 1: 391. 1785. *Butea frondosa* Koen. ex Roxb. in Asiat. Res. 3: 369. 1792; FBI 2: 194; Gamble 357 (252). *Plaso monosperma* (Lam.) O. Kuntze, Rev. Gen. Pl. 2: 202. 1891, incl. var. *rubra* & *flava*.

Trees, up to 8 m tall; branchlets densely tomentose; leaves 3-foliolate, leaflets rhomboid-ovate, 8–12 × 6–10 cm, glabrous above, grey tomentose or pubescent beneath, obtuse to retuse, entire, base cuneate; flowers yellowish-red in racemes on leafless branches; pods oblong, 15 × 4 cm, seeds obovate.

Occasional in forests. Fl. & Fr.: Feb.–Sept. *Vern.:* Moduga. It is commonly called as flame of the forest due to its flower in absence of leaves. Flowers yield yellow dye. A red juice exudes from natural cracks and artificial incisions in the bark and it hardens into a vitrious, ruby red gum, known as 'Butea gum' or 'Bengal kino'. It is an astringent, anthelmintic, depurative, aphrodisiac, rubifacient, antidote to snake bite and for diarrhoea, dysentery, menstrual disorder, piles, tumor and herpes. Seeds had 18% yellow taste-less oil and three alkaloids toxic to earthworms, besides proteolitic and lipolytic enzymes.

Note: The taxon is usually described under the combination *Butea frondosa* Koenig ex Roxb. (1792) but its specific epithet is to be accepted from Lamarck's (1783) combination (*Erythrina monosperma*). Some of the modern authors attribute this combination to O. Kuntze (1891 l.c) who is reported not to have made it (Sanjappa Fl. Saur. 1962 in Bull. Bot Surv. India 3: 12, 1961). The present combination thus should be attributed to Taubert (1894 l.c) who made it and published for the first time.

Cajanus A.P. de Candolle *nom. cons.*

1. Flowers in axillary clusters, slender twining shrub .... *C. scarabaeoides*
1. Flowers in terminal panicles, erect much branched bushy shrubs .... *C. cajan*

Cajanus cajan (L.) Millsp. in Publ. Field Columbian Mus. Bot. Ser. 2: 53. 1900. *Cytisus cajan* L. Sp. Pl. 739. 1753. *Cajanus indicus* Spreng. Syst. Veg. 3: 248. 1826; FBI 2: 217; Gamble 369 (261).

Erect much-branched bushy shrubs, up to 2 m tall; leaves 3-foliolate, leaflets elliptic-oblong, 3–7.5 × 1.5–3 cm, soft pubescent, entire, base obtuse; flowers yellow in terminal panicles; pods straight, compressed between seeds; seeds 3–5, discoid, smooth.

Important cultivated pulse crop, rarely as an escape. Fl. & Fr.: Oct.–Mar. Eng.: Pigeon pea. *Vern.:* Kandi. Seeds as pulses. Ethipothala, *VRK* 3918.

Cajanus scarabaeoides (L.) du Petit-Thou., Dict. Sci. Nat. 6: 617. 1817. *Dolichos scarbaeoides* L. Sp. Pl. 726. 1753. *Atylosia scarabaeoides* (L.) Benth. in Miq. Pl. Jungh. 242. 1852; FBI 2; 215; Gamble 369 (261).

Densely pubescent, spreading woody herbs; leaves 2.5 cm long, leaflets obovate-oblong, 1–2.5 × 1–1.5 cm, velvetty-pubescent, obtuse, entire, base cuneate; flowers yellow in axillary clusters; pods densely pubescent; seeds 4–6, oblong.

Occasional in Vinukonda RF. Fl. & Fr.: Dec.–Oct. *Vern.:* Adavi pesara. Seeds are cooked and eaten as pulse. Mangalagiri, *SSR* & *KSM* 21610.

Canavalia A.P. de Candolle *nom. cons.*

1. Leaves oblong, pods glabrescent .... *C. gladiata*
1. Leaves ovate-oblong; pods pubescent .... ***C. africana***

Canavalia africana Dunn in Pipers & Dunn in Bull. Misc. Inf. Kew 1922; 154. 1922; Verdc. in Kew bull. 2: 653, 1987. *C. virosa* (Roxb.) Wight & Arn. Prodr. 253. 1834; Gamble 359 (253). *C. ensiformis* DC. var. *virosa* (Roxb.) Baker in FBI 2: 176. 1879.

A long climbing perennial, up to 8 m long, branchlets glabrous, reddish; leaves 3-foliolate, leaflets ovate-oblong, chartaceous, acute-acuminate, entire, base subacute, glabrous above, sparsely puberulous below; flowers purple in long racemes; pod pubescent, apically obtuse; seeds ca. 10, ellipsoid or ovoid.

Rare, in the scrub forests. (Mohan Rao and Aasiya, 1995). Fl. & Fr.: Aug.–Mar. *Vern.:* Adavi Thamba

Canavalia gladiata (Jacq.) DC. Prodr. 2: 404. 1825. *Dolichos gladiatus* Jacq. Coll. Bot. 2: 276. 1788. *Canvalia ensiformis* sensu Baker in Hook. f. Fl. Brit. India 2: 195. 1876, non DC. 1825; Gamble 359 (253).

Stout, woody, perennial twining shrubs; leaves 3-foliolate, leaflets oblong, 5–12 × 3.5–7 cm, glabrous, acute, entire, base rounded; flowers lilac in axillary racemes; pods linear-oblong, 10 × 4 cm, glabrescent, flattened; seeds 6–12, ovate, reddish-brown or white.

Common twiner on hedges. This is a wild variety. Fl. & Fr.: Sept.–Mar. *Vern.:* Thammakaya. Eng.: Horse bean, Jack bean, Sword bean. The seeds and sliced pods are cooked and eaten. Nagarjunakonda, *VRK* 6701; Kondavedu Fort, *CAB* 4663 (MH); Nagarjunakonda, *KTH* 9748 (CAL).

Note: Chatterjee (1948 p. 86) has traced the history of this species and is of the opinion that the specific epithet, *ensiformis* which rightly belongs to an American plant has wrongly been used for the Indian plant. The old world plant should be known as *Canavalia gladiata* (Jacq.) DC., true *ensiformis* has white seeds, whereas *C. gladiata* has red seeds.

Clitoria Linnaeus

Clitoria ternatea L. Sp. Pl. 1753; FBI 2: 208; Gamble 365 (258).

Twining herbs; leaves odd-pinnate, leaflets 5–7, broadly ovate or elliptic, 2–4 × 1.5–3 cm, glabrous or with a few short appressed hairs, obtuse; flowers bluish violet with yellow streaks, axillary, solitary; pods flattened or sub-cylindrical, beaked, hairy; seeds 6–10, reniform, smooth, yellowish-brown.

Commonly grown in the gardens and often found as an escape in hedges and thickets. Fl. & Fr.: Sept.–Feb. Eng.: Conch flower creeper. *Vern.:* Sankupoolu. It is used as a cathartic, diuretic and antidote against snake-bite. Mallavagu, *VRK* 5726; Kondavedu, *CAB* 4715 (MH).

Crotalaria Linnaeus

1. Leaves 3-foliolate:
    2. Pods oblong, seeds 7–20:
        3. Herbs, standard petal < 6 mm across, corolla hardly exceeding calyx, bracts persistent .... ***C. orixensis***
        3. Shrubs, standard petal >2 cm across, corolla twice as long as calyx, bracts deciduous .... ***C. laburnifolia***
    2. Pods globose, seeds 1–4:
        4. Pods globose, tomentose .... ***C. trifoliastrum***
        4. Pods sub-globose, pubescent .... ***C. medicaginea***
1. Leaves simple:
    5. Pods flattened, branches dichotomous .... ***C. hebecarpa***
    5. Pods turgid, branches not dichotomous:
        6. Pods glabrous:
            7. Seeds up to 10 .... ***C. albida***
            7. Seeds more than 10:
                8. Branchlets pubescent, undershrubs .... ***C. retusa***
                8. Branchlets grey tomentose, erect herbs .... ***C. mysorensis***
        6. Pods hairy or pubescent:
            9. Flowers in panicles:

10. Seeds 10–12 .... *C. pulchra*
10. Seeds 1–2:
11. Erect woody herbs .... *C. ramosissima*
11. Erect shrubs .... *C. paniculata*
9. Flowers in racemes:
12. Racemes 2–3-flowered:
13. Standard petal blue within .... *C. bifaria*
13. Standard petal yellow:
14. Prostrate herbs, leaves hispid pubescent on both sides, seed 15–20 .... *C. angulata*
14. Erect herbs, leaves sparsely hairy on both surfaces, seeds 6–10 .... *C. hirsuta*
12. Racemes 3-many-flowered:
15. Flowers blue, racemes leaf-opposed .... *C. verrucosa*
15. Flowers yellow, racemes terminal or lateral:
16. Erect shrub, leaves 3–6 × 1–2 cm, apex acute, apiculate, flowers 2 cm across .... *C. juncea*
16. Prostrate herb, leaves 1.5–2.5 × 0.5–1 cm, apex obtuse, flowers 1 cm across .... *C. evolvuloides*

Crotalaria albida Heyne ex Roth Nov. Pl. Sp. 333. 1821; FBI 2: 71; Gamble 295 (208).

Erect annual herbs, up to 30 cm tall; branchlets pubescent; leaves simple, alternate, oblanceolate or oblong-elliptic, 0.6–1.5 × 0.2–0.4 cm, chartaceous, glabrous above, pubescent below, base cuneate, entire, obtuse; flowers yellow in terminal racemes; pods oblong, 1–2 cm, glabrous; seeds up to 10.

Occasional herb in dry deciduous forests. Fl. & Fr.: Oct.–Mar. Vinukonda RF, *VRK* 6939.

Crotalaria angulata Mill. Gard. Dict. ed. 8. n 9. 1768. *C. biflora* (L.) L. Mant. Pl. 570. 1771; FBI 2: 66; Gamble 292 (266).

Prostrate herbs, up to 50 cm long, branchlets hispid, leaves ovate, 1–1.6 × 0.5–1.2 cm, chartaceous, base rounded-obtuse, apex obtuse, margin entire, appressed hispid on both surfaces; racemes 1–2-flowered; corolla yellow; pod globose, turgid with golden hairs, seeds 15–20, grey.

Occasional, in forests. Fl. & Fr. Nov.–Feb.

Crotalaria bifaria L.f. Suppl. Pl. 322. 1781; FBI 2: 69; Gamble 293 (207).

Spreading herbs, up to 60 cm tall; branches 4-angled when young; leaves simple, lower ones orbicular, upper ones ovate-lanceolate, 2.5–5 × 1.5–3.5 cm, glabrescent above, puberulous below, subacute-obtuse; margin entire, base cuneate-obtuse; flowers exerted in 2-flowered racemes, standard petal yellow without, blue within; pods oblong, terete, turgid, puberulous, seeds ca. 10.

Rare, in N.U. Campus (Mohan Rao and Aasiya, 1995). Fl. & Fr.: Aug.–Jan.

Crotalaria evolvuloides Wight ex Wight & Arn. Prod. 138. 1834. excl. syn. var. *evolvuloides*; FBI 2: 68; Gamble 1. 292 (206).

Prostrate herbs, up to 60 cm tall, branchlets shortly hispid; leaves very variable, ovate, oblong-lanceolate, base and apex obtuse, entire, tomentose beneath; flowers yellow in 3–8-flowered racemes; pods oblong, pubescent; seeds 8–10.

Rare, in wastelands (Mohan Rao & Aasiya, 1995). Fl. & Fr.: Sept.–Apr.

Crotalaria hebecarpa (DC.) Rudd. in Phytologia 54: 28. 1983. *Goniogyna hirta* (Willd.) Ali in Taxon 16: 463. 1967. *Hallia hirta* Willd. Sp. Pl. 3(2): 1169. 1803. *Heylandia latebrosa* sensu Baker in FBI 2: 65. 1876; non (L.) DC. 1825; Gamble 280 (190).

Prostrate annual herbs, leaves simple, sessile, cordate-ovate, 1–1.5 × 0.5–0.8 cm, pubescent, acute, base cordate; flowers yellow, axillary, solitary; pods flat, oblong, compressed; seed solitary, reniform, brownish.

Occasional in open forests and weed in cultivated fields. Fl. & Fr.: Sept.–Mar. Mellavagu, *VRK* 5711; Sattenapally, *VRK* 5854; Vinukonda, *VRK* 6972; Kondavedu, *CAB* 4627.

Crotalaria hirsuta Willd. Sp. Pl. 3: 978. 1802; FBI 2: 68; Gamble 293 (207).

Hairy shrubs, up to 1 m tall, branchlets shortly hispid. Leaves ovate-elliptic, 3–6 × 2–4 cm, tomentose beneath, acute, entire, base acute or obtuse; flowers yellow in slender racemes; pods obovoid, hirsute.

Rare in the forest. Fl. & Fr.: Sept.–Apr. Mangalagiri, *SSR & KSM* 17632. Kondavedu Fort, *CAB* 4687 (MH).

Crotalaria juncea L. Sp. Pl. 714. 1753; FBI 2: 79; Gamble 297 (210).

Erect annual herbs; branchlets silky pubescent; leaves simple, narrowly oblong-elliptic, 3–6 × 0.8–1.5 cm, chartaceous, base and apex acute, margin entire, ciliate; flowers yellow in terminal racemes; pods sessile, oblong, pubescent, 3 × 1 cm; seeds 3 to 5.

Cultivated as green manure crop, often as an escape, which yields important fibre. Fl. & Fr.: Sept.–Dec. V.P. South, *VRK* 3946.

Crotalaria laburnifolia L. Sp. Pl. 715. 1753; FBI 2: 84; Gamble 301 (213).

Erect shrubs, up to 2.5 m tall; branchlets terete, glabrous; leaves 3-foliolate, leaflets ovate, 2–5 × 1.5–3 cm, glabrous, base and apex obtuse-subacute, margin entire, flowers yellow in 25 cm long terminal and leaf-opposed racemes; pod cylindric, with a stipe of 1–2 cm long, oblong, glabrous; seeds 15–20.

Rare, in wastelands and open forests (Mohan Rao & Aasiya, 1995). Fl. & Fr.: Sept.–Jan. *Vern.:* Pedda-Giligicha.

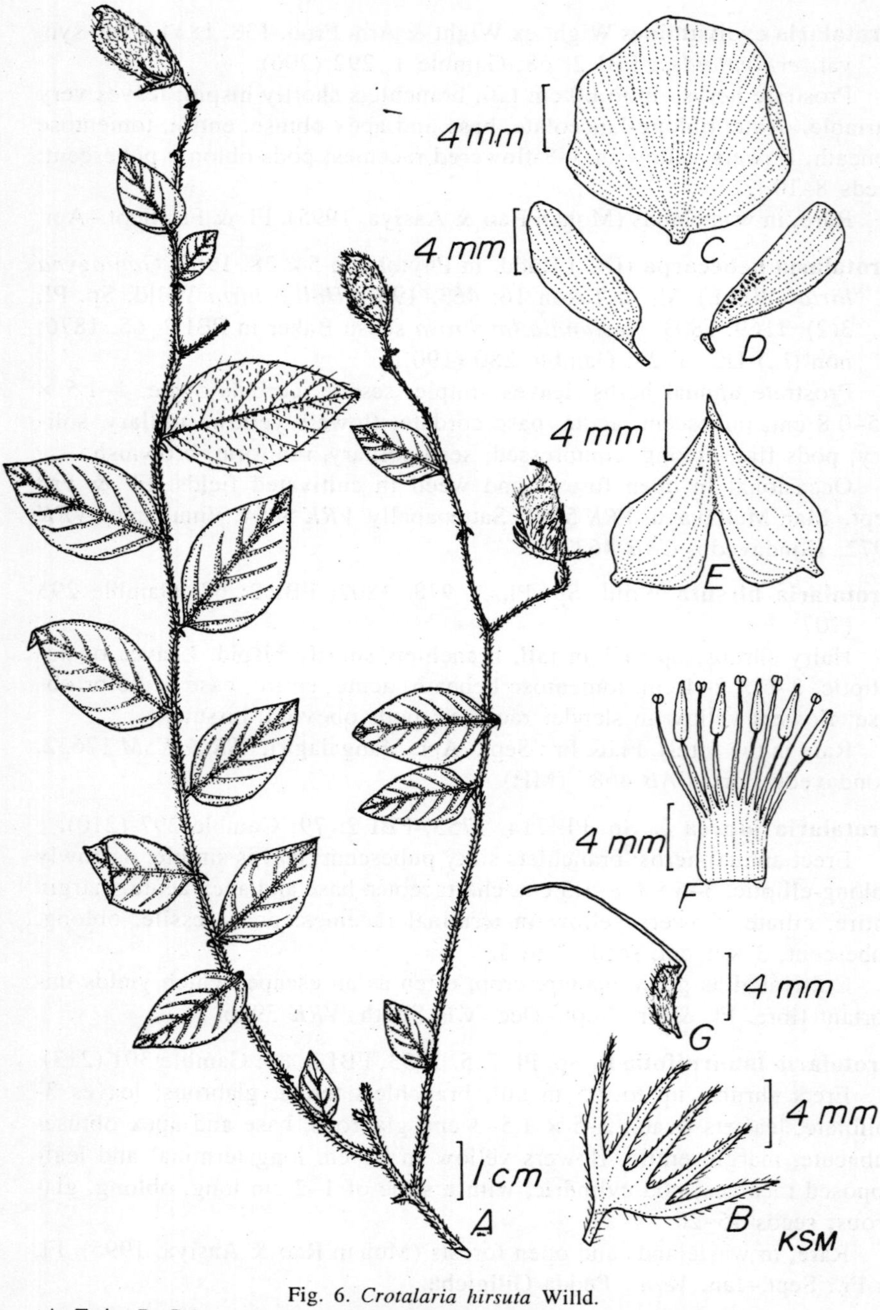

Fig. 6. *Crotalaria hirsuta* Willd.
A. Twig, B. Calyx split open, C. Standard petal, D. Wing petals, E. Keel petals, F. Androecium (split open), G. Pistil

Crotalaria medicaginea Lam.

1. Leaves pubescent below, seeds up to 2 .... *C. medicaginea* var. *medicaginea*
1. Leaves glabrous, seeds up to 4 .... *C. medicaginea* var. *neglecta*

Crotalaria medicaginea Lam. Encycl. 2: 201. 1790. FBI 2: 81; Gamble 298 (211); var. **medicaginea**; Munk in Reinwardtia 6: 208. 1962.

Diffuse, sparsely-pubescent divaricate herbs; leaves 3-foliolate, leaflets obovate, 0.5–1 × 0.2–0.5 cm, glabrous above, pubescent below, emarginate, entire; flowers yellow with maroon streaks in terminal racemes; pods subglobose, 4 mm long, slightly exceeding calyx; seeds 2.

Common in open forests. Fl. & Fr.: July–Dec. Mellavagu, *VRK* 5715; Nizampatnam, *VRK* 6917.

Crotalaria medicaginea Lam. Encycl. 2: 201. 1790; var. **neglecta** (Wight & Arn.) Baker in Hook. f. Fl. Brit. India 2: 81. 1876; Gamble 299 (212). *C. neglecta* Wight & Arn. Prodr. 192. 1834.

Low diffuse perennial herbs; branchlets ascending, sparsely appressed pubescent beneath, leaves 3-foliolate, leaflets 0.5–1 × 0.3–1 cm, obovate, obtuse, glabrous, entire; flowers yellow in short leaf-opposed racemes; pods sub-globose, glabrous; seeds 2–4, black, triangular.

Common weed in cultivated fields and wastelands. Fl. & Fr.: July–Dec. Bollapalle, *VRK* 5763; Sattenapalle, *VRK* 5843; Nallapadu, *VRK* 6772; Kondavedu Fort, *CAB* 4696 (MH); Nagarjunakonda, *KTH* 9845 (CAL).

Crotalaria mysorensis Roth Nov. Pl. 338. 1821; FBI 2: 70; Gamble 295 (208). Munk in Reinwardtia 6: 210. 1962.

Annual grey-tomentose herbs, or undershrubs, up to 1.5 m tall; branchlets densely hairy; leaves simple, oblong-elliptic, 1.5–6.5 × 1–3 cm, sericeous above, dense below, base and apex obtuse, entire; flowers yellow with maroon streaks in terminal elongated racemes; pods oblong; seeds 20–30, globose.

Rare in Vinukonda RF, near copper mines. Fl. & Fr.: Aug.–Dec. Bollapalle, *VRK* 5751; Nagarjunakonda, *KTH* 9761 (CAL).

Crotalaria orixensis Rottl. ex Willd. in Ges. Naturf. Freunde Berlin Neue Schriften 4: 217. 1803; FBI 2: 83; Gamble 300 (212).

Procumbent, pubescent herbs, branchlets pubescent, leaves 3-foliolate, leaflets obovate, 1–3 × 0.5–1.5 cm, pubescent below, obtuse, entire; flowers yellow in lateral lax racemes, pedicels filiform; pods oblong, bulged, turgid, up to 1.4 cm long, light brown; seeds 7–10, reniform, flattened, glabrous.

Occasional weed in fields. Fl. & Fr.: Oct.–Mar. Guntur, *G.V. Narayana* 16800 (MH).

Crotalaria paniculata Willd. Sp. Pl. 3: 982. 1802; FBI 2: 81; Gamble 299 (211); Munk in Reinwardtia 6: 211. 1962.

1. Bracts lanceolate, stipules linear, 1 cm .... *var. paniculata*
1. Bracts cordate, stipules 0 .... *var. nagarjunakondensis*

Crotalaria paniculata *var.* **paniculata**

Viscid, pubescent undershrubs, up to 1.5 m tall; leaves oblong to oblanceolate, 2–5 × 0.5–1.5 cm, velvety, mucronate, entire, base narrowed; flowers reddish brown in terminal racemes; pods ellipsoid, pubescent, horned, seeds 1–2, black.

Rare in the forests and occasional on river banks. Fl. & Fr.: Through the year. Satrasala, *VRK* 6781; Davupalli, *SSR & KSM* 21692.

Crotalaria paniculata Willd. Sp. Pl. 3: 982. 1802; var. ***nagarjunakondensis*** Thoth. in Bull. Bot. Surv. India 6: 67–68, 1964.

Undershrubs, up to 1 m tall, fulvous villous; leaves simple, alternate, estipulate, ovate or elliptic to obovate, 3.5–6.5 x 1.5–2.5 cm, villous, acute, entire, base cuneate; flowers yellow in 13–20 cm long panicles; pod ovoid, 10 × 6 mm, inflated with a recurved persistent style, pilose, 1–2-seeded; seeds black.

Occasional in forests. Fl. & Fr.: July–Dec. Nagarjunakonda valley, *KTH* 9764 (CAL).

Note: C. paniculata Willd. var *nagarjunakondensis* Thoth. differs from *C. paniculata* Willd. var. *paniculata* in having distinctly cordate bracts, shortly petiolate leaves and in the absence of stipules.

K. Thothathri has erected a new variety of *C. paniculata* in Bull. Bot. Surv. India 6: 67–68. 1964. The plant could not be collected now as the area is submerged under Nagarjunasagar reservoir.

Crotalaria pulchra Andr. Bot. Repos. 10. t. 601. 1810. *C. pulcherrima* Roxb. Fl. Ind. 2: 267. 1832; FBI 2: 180; Gamble 1: 295 (211).

Erect handsome shrubs; branchlets densely sericeous; leaves simple, obovate-oblanceolate, 4–8 × 2.5 cm, sericeous on both sides, entire, round-obtuse, apiculate, cuneate; sessile; flowers yellow in terminal or leaf opposed panicles, calyx sericeous, corolla scarcely exerted, standard golden silky outside; pod oblong, pubescent, slightly exerted over calyx, 7–9 seeded.

Occasional, in deciduous forests. Fl. & Fr.: Sept.–Mar. *Vern.:* Nela Janumu. Tender leaves are cooked as vegetable. Bollaka bavi, *DMR* 21101.

Crotalaria ramosissima Roxb. Fl. Ind. 3: 268. 1832; FBI 2: 80; Gamble 299 (211).

Much branched woody herbs; up to 3.5 cm tall; branchlets clothed with silky hairs; leaves linear or oblanceolate, 1–2.8 × 0.1–0.5 cm, silky hairs on both sides, entire, obtuse, base cuneate; flowers yellow tinged with purple in panicled racemes, pods sessile, oblong, as long as calyx, silky; seeds solitary, brownish-black.

Common in open and dry deciduous forests and NU Campus. Fl. & Fr.: Sept.–Feb. N. Varam, *VRK* 3596; Bellamkonda, *JSG* 18616 (MH); Nagarjunakonda, *KTH* 9741 (CAL).

Crotalaria retusa L. Sp. Pl. 715. 1753; FBI 2: 75; Gamble 293 (207).

Erect pubescent herbs, up to 1 m tall; branchlets terete; leaves oblanceolate, 2.5–5 × 1–2 cm, glabrous above, silky pubescent below, obtuse, retuse, entire, base cuneate; flowers yellow in terminal racemes; pods oblong, stalked; seeds 15–20.

Occasional weed in cultivated fields and wastelands. Fl. & Fr.: Nov.–Apr. *Vern.:* Potti-Gilligicha. Nizampatnam Port, *VRK* 6803; Guntur, *JSG* 18602 (MH).

Crotalaria trifoliastrum Willd. Sp. Pl. 3: 983. 1802; FBI 2: 82. Gamble 300 (212).

Erect herbs, up to 1.5 m tall; leaves 3-foliolate, leaflets 1.5–2 x 0.5–1 cm, obovate to obcordate, entire, mucronate, glabrous above, glabrescent below; flowers yellow in axillary, solitary and terminal racemes; pods globose, tomentose; seeds 1–2.

Occasional weed in cultivated fields. Fl. & Fr.: Aug.–Jan. Achempet, *VRK* 5858; Narsaraopet, *CAB* 4716 (MH).

Crotalaria verrucosa L. Sp. Pl. 715. 1753; FBI 2: 77; Gamble 297 (210); Munk in Reinwardtia 6: 217. 1962.

Erect subshrubs, up to 1 m tall; branchlets 4-angled, appressed pubescent; stipules semi-lunate, leaves ovate or broadly rhomboid, 4–9 × 3–4.5 cm, apex obtuse-acute, margin entire, base alternate; flowers bluish-white in leaf-opposed racemes; pod oblong, cylindric, sparsely pubescent; seed ca. 15.

Rare, in forests. Fl. & Fr.: July–Feb. *Vern.:* Gilligicha, Gheli-Gherinta, Davupalli, *SSR & KSM* 21414.

Dalbergia Linnaeus *nom. cons.*

1. Branchlets spine tipped, shrubs or small trees .... ***D. spinosa***
1. Branchlets without spines, medium-sized to large trees:
    2. Stamens isodiadelphous (5 + 5):
        3. Leaflets 9–13, oblong-obovate, standard longer than broad; pod short stipitate .... ***D. paniculata***
        3. Leaflets 11–15, ovate-oblong or orbicular, standard broader than long; pod long stipitate .... ***D. lanceolaria***
    2. Stamens monadelphous:
        4. Leaflets 2 pairs, apex abruptly acuminate, leaf rachis zig zag .... ***D. sissoo***
        4. Leaflets 3 or 4 pairs, apex rounded or emarginate, leaf rachis straight .... ***D. latifolia***

Dalbergia lanceolaria L.f. Suppl. Pl. 316. 1: 181; FBI 2: 235; Gamble 383 (270).

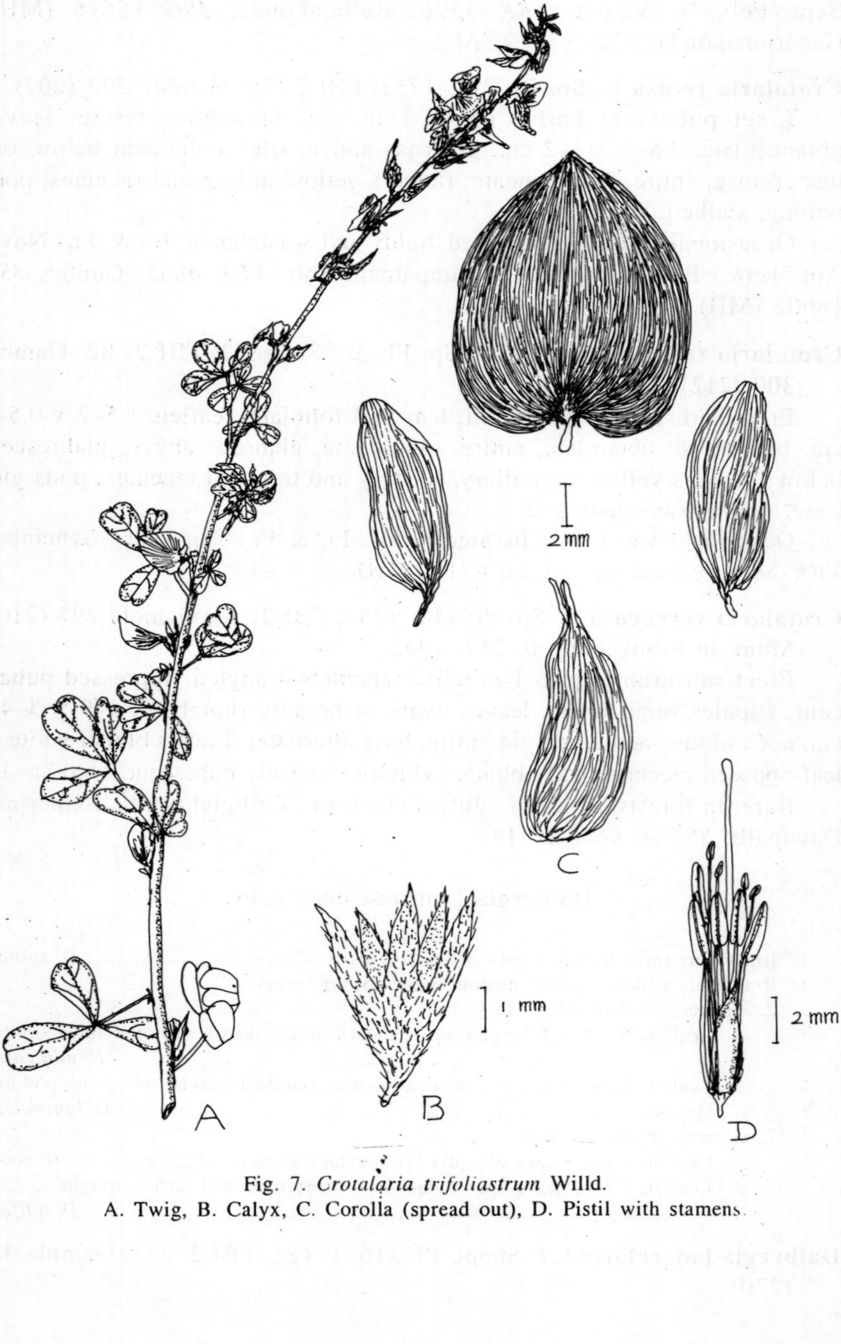

Fig. 7. *Crotalaria trifoliastrum* Willd.
A. Twig, B. Calyx, C. Corolla (spread out), D. Pistil with stamens.

Large, decidous, densely foliaceous trees, up to 1.5 m tall; bark smooth, grey, leaves imparipinnate, leaflets 11–17, chartaceous, elliptic-oblong, 1.8–4.3 × 1–2.8 cm, base cuneate-obtuse, entire, apex obtuse; flowers pale pink in terminal and axillary in short unilateral racemes to large panicles; pod purplish-brown, lanceolate, 9 × 2 cm, seeds 1–3.

Occasional in deciduous slopes. Fl. & Fr.: July–Jan. *Vern.*: Patsari, Yerra patsari, Nagulla patchari. The decoction of the bark is used in dyspepsia. The seed oil is applied in rheumatism. Wood moderately hard and of no great value, used for tool handles and agricultural implements. Javuku *VRK* 3881; Davupalli *SSR & KSM* 21409.

Dalbergia latifolia Roxb. Pl. Cor. t. 113. 1799; FBI 2: 231; Gamble 383 (270).

Deciduous trees, up to 10 m tall; leaves odd pinnate, leaflets 3–4 pairs, alternate, obovate-orbicular, 3–5 × 2.5–4 cm, glabrous above, puberulous below, base subacute-obtuse, apex rounded or emarginate, entire; flowers white with pale yellow tinge, scented, in axillary panicles; pods oblong-lanceolate, abruptly narrowed to a stipe, flat, glabrous, seeds 1–3.

Occasional in forests. Fl. & Fr.: July–Feb. It is drought resistent, good coppicer and produces root suckers. Eng.: Black wood, Rose wood. *Vern.:* Jittegi, Virugudcheva Iridi. Bark contains tannins. Oil from the seed is used as antidote for poison. Wood extremely hard and valuable. Reported in Forest Division Work Plan.

Dalbergia paniculata Roxb. Pl. Cor. t. 114. 1799; FBI 2: 236; Gamble 383 (270).

Deciduous trees, up to 10 m tall; branchlets tomentose; bark smooth grey, leaves odd-pinnate, leaflets 9–13, alternate, oblong-obovate, 2–3.5 × 1–2 cm, pubescent, base cuneate-obtuse, apex obtuse, entire; flowers white in terminal and axillary panicles; pods lanceolate, apex acute, faintly nerved, seeds 1–2, reniform, compressed.

Occasional in deciduous forests. Fl. & Fr.: May–Dec. *Vern.:* Porla-patchari. The wood is not durable. It is not fit for fuel even. Javuku, *VRK* 3881.

Dalbergia sissoo Roxb. Fl. Ind. 3: 223. 1832; FBI 2: 237; Gamble 384 (270).

Deciduous trees, up to 10 m tall; leaves odd-pinnate, leaflets 2 pairs, alternate, ovate or broadly elliptic, 4–6 × 3–5 cm, chartaceous, abruptly acuminate, entire, base rounded, flowers creamish in terminal panicles; pods yellowish-green, oblong, up to 7 × 1 cm; seeds 1–4, kidney-shaped, thin, flat.

Cultivated in forests and as avenue trees. Strong light demander. Fl. & Fr.: Jan.–Sept. Eng.: Sissoo tree. *Vern.*: Sissoo, Wood is valuable and used for furniture and ordinary purpose. The leaves are bitter and stimulent; decoction of the leaves is said to be useful in gonorrhoea. Nallapadu, *VRK* 6760.

Dalbergia spinosa Roxb. Fl. Ind. 3: 233. 1832; FBI 2: 238. Gamble 381 (269).

Shrub or small trees, 3–6 m tall, branches stiff and horizontal, ending in sharp spines; leaves crowded at the nodes of the spinous branchlets, leaflets alternate, 1 × 0.5 cm, elliptic ovate, base cuneate, margin entire, apex obtuse or emarginate; flowers purplish-white, small in short dense corymbose clusters in the leaf-axils; pod flat, kidney shaped, brownish, 2-seeded.

Core mangrove. *Vern.*: Chillingi. Nizampatnam estuary, *DLR* 138 (NUH).

Derris Loureiro *nom. cons.*

1. Pods up to 1 cm broad, acute at both ends, seeds 1–5 .... ***D. scandens***
1. Pods more than 2 cm broad, obtuse at both ends, seeds 1–2 .... ***D. trifoliata***

Derris scandens (Roxb.) Benth. in J. Linn. Soc. Bot. 4 (Suppl.): 103. 1860; FBI 2: 240; Gamble 387 (273). var. **scandens;** Thoth. in Fasc. Fl. Ind. 8: 1982. *Dalbergia scandens* Roxb. Pl. Cor. t. 192. 1805.

Large woody twiners; branchlets closely covered with reddish coloured, round lenticels; leaves odd-pinnate, leaflets 4–6 pairs, sub-opposite, oblong-obovate, 1–4.5 × 0.5–2 cm, glabrous above, puberulous below, obtuse, entire, base cuneate-acute; flowers pale pink, in terminal or axiallary panicles; pods oblong, winged, tapering at both ends; seeds 1–5, compressed.

Occasional in forests. Fl. & Fr.: July–Jan. Loyapalle, *VRK* 3835; *Vern.*: Nalla tiga, Mainapu tiga. Vinukonda, *KCJB* 17398 (MH).

Derris trifoliata Lour. Fl. Cochinch. 2: 433. 1790: Thoth. in Bull. Bot. Surv. India 3: 181. 1961. *D. uliginosa* (Roxb.) Benth. in Miq. Pl. Jungh. 1: 252. 1852; Baker in Hook. f. F. Brit. India 2: 241. 1878; Gamble 387 (273).

Climbing shrubs along the coastal mangroves; leaves up to 15 cm, leaflets 2 pairs, opposite, oblong-lanceolate, 4–9 × 2–4 cm, glabrous, base obtuse, entire, acuminate; flowers rose in terminal racemes; pods ovoid, winged; seeds 1–2.

Occasional among Kothapalem mangroves among *Rhizophora* populations. Fl. & Fr.: Aug.–Oct. *Vern.*: Tige Kanuga, Nacha tige. Kothapalem, *VRK* 5770. Nizampatnam estuary, *DLR* 129.

Desmodium Desvaux *nom. cons.*

1. Leaves simple .... ***D. gangeticum***
1. Leaves trifoliolate:

2. Prostrate or trailing herb; leaflets 1–2 × 0.8–1.3 cm; pod stalked, straight on the dorsal, indented on the ventral suture .... ***D. triflorum***
2. Diffuse much branched herb, leaflets 3–4 × 1.6–2.5 cm; pod sessile indented on both margins .... ***D. dichotomum***

Desmodium dichotomum (Willd.) DC. Prodr. 2: 236. 1825. *Hedysarum dichotomum* Willd. Sp. Pl. 3: 1180. 1802. *Desmodium diffusum* (Willd.) DC. Prodr. 2: 336. 1825 *nom. illeg* non. DC. 1825 quest *D. laxiflorum* FBI 2: 169. 1876; Gamble 346 (244).

Stout much branched herbs, up to 1 m tall, branchlets angular, grooved, clothed with long spreading hairs; stipules amplexicaul, foliaceous; leaves 3-foliolate, leaflets ovate-oblong, elliptic-oblong or obovate, 3–4 × 1.6–2.5 cm, obtuse, apiculate, ciliate with long white hairs on both surfaces, entire, stipels prominent; flowers bluish in 20 cm long racemes; pod sessile, 5–6-jointed, indented on both margins, clothed with hooked hairs.

Rare in dry deciduous forests. Fl. & Fr.: Dec.–Feb. Kothajupudi, *P.C. Nanda* 171 (DD).

Desmodium gangeticum (L.) DC. Prodr. 2: 327. 1925: FBI 2: 168; Gamble 345 (244). *Hedyasarum gangeticum* L. Sp. Pl. 746. 1743.

Erect herbs, up to 60 cm tall; branchlets sericeous; leaves 1-foliate, oblong-obovate or lanceolate, 4–10 × 2–5 cm, chartaceous, puberulous above, appressed tomentose below, base obtuse-subacute, entire, acute; flowers lilac in terminal racemes; pods moniliform, margin deeply undulate; seeds 6–8, 2 mm.

Common weed in open forest. Fl. & Fr.: Nov.–Mar. *Vern.:* Nalla-Nelapanki, Geetanaramu. Jayanthipuram, *VRK* 6900.

Desmodium triflorum (L.) DC. Prodr. 2: 234. 1825; FBI 2: 173; Gamble 347 (245); Van Meeuwen in Reinwardtia 6: 261. 1962. *Hedysarum triflorum* L. Sp. Pl. 749. 1753.

Much-branched, prostrate hairy herbs, rooting at nodes; leaves 3-foliolate, leaflets obcordate or obovate, 1–2 × 0.8–1.3 cm, glabrous above, pubescent below, emarginate, base cuneate; flowers purplish pink or pinkish blue in axillary clusters; pods linear, 1.5 × 0.3 cm, curved, flat, 3–6-jointed, straight on the dorsal, indented on the ventral suture, pubescent; seeds smooth, comprssed, brown.

Common in hedges and often in the forest along the streams. *Vern.:* Muntamandu. Fl. & Fr.: Aug.–Jan. Javuku, *VRK* 3907; Vinukonda RF, *VRK* 69534.

Erythrina Linnaeus

Erythrina stricta Roxb. Fl. Ind. 3: 251. 1832; FBI 2: 189; Gamble 353 (250).

Yellowish prickled deciduous trees, up to 12 m tall; leaves 3-foliolate, leaflets rhomboid-ovate, 7–10 × 8–12 cm, glabrescent, base deltoid or truncate, apex subacute, margin entire, flowers red in terminal racemes; pods sub-turgid, slightly curved, stipe slender; seeds 3–4, oblong.

Occasional in forests and common on canal banks. Fl. & Fr.: Feb.–Aug. *Vern.*: Mulla Moduga. It is used as fence plant. The wood is soft and light, but tough, fairly durable. The bark in powdered form is used in biliousness, fever, rheumatism, itch, asthma, leprosy, epilepsy. The flowers are said to be given as an antidote to poison. V.P. South, *VRK* 3936.

Galactia P. Brown

Galactia tenuifolra (Klein ex Willd.) Wight & Arn. Prodr. 206. 1834; FBI 2: 192; Gamble 356 (251). *Glycine tenuiflora* Klein ex Willd. Sp. Pl. 3: 1059. 1802.

1. Leaflets 2.5–3 cm long, lateral nerves 6 pairs, base subcordate, inflorescence to 1.5 cm, pod 6-seeded .... ***var. minor***
1. Leaflets 4–9 cm long, lateral nerves 8 pairs, base obtuse, inflorescence to 8 cm, pod 7–8-seeded .... ***var. tenuiflora***

Galactia tenuiflora (Klein ex Willd.) Wight & Arn. var. **minor** Baker in Hook. f. Fl. Brit. India 2: 192. 1876. *Leucodictyon malvensis* Dalz. & Gibs. Bombay Fl. 73. 1865.

Annual prostrate herbs, branchlets slender, appressed pubescent; leaves 3-foliolate, leaflets oblong, 2.5–3 × 1–1.7 cm, thin coriaceous, subcordate, apex obtuse-apiculate, margin entire, puberulous above, sericeous pubescent beneath; flowers purple, twin on slender rachis; pod straight, faintly puberulous, obliquely constricted between the seeds, margin thick; seeds 6.

Occasional, in Mangalagiri associated with grasses. Fl. & Fr.: Oct.–Dec. Mangalagiri RF, *KSM* 17630.

Galactia tenuiflora (Klein ex Willd.) Wight & Arn.var. **tenuiflora**

Twiners; leaves 3-foliolate, leaflets elliptic, 4–9 × 2.5–4.5 cm, thin-coriaceous, glabrous above, silky below; flowers pale pink, leaf-opposed, solitary or paired; pods slightly falcate, thinly grey silky.

Occasional twiners in Vinukonda RF. Fl. & Fr.: July–Dec. Jayanthipuram, *VRK* 6934; Kondavedu Fort, *CAB* 4697 (MH).

Gliricidia Humboldt, Bonpland et Kunth

Gliricidia sepium (Jacq.) Kunth ex Walp. Rep. 1: 679. 1843. *Robinia sepium* Jacq. Enum. Pl. Carib. 28. 1760.

Small deciduous trees, up to 4 m tall, branchlets glandular-pubescent; leaves odd-pinnate, grey-pubescent when young; leaflets 6–7 pairs, elliptic-

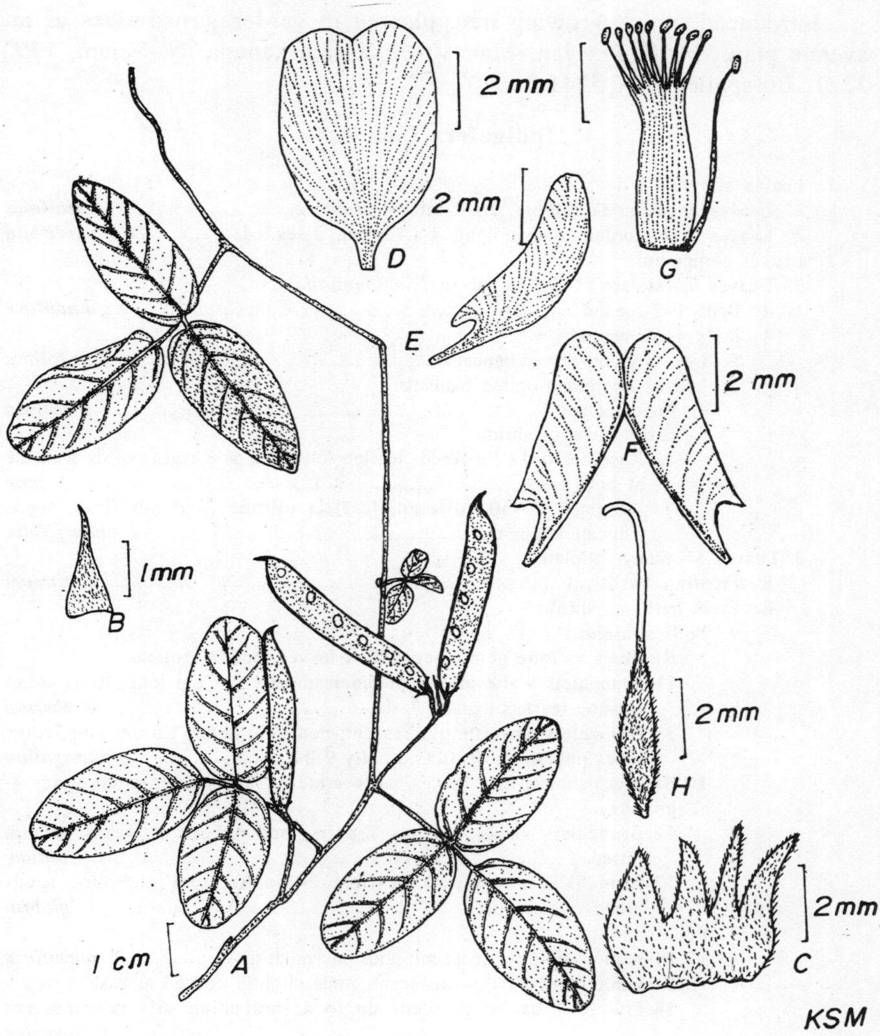

Fig. 8. *Galactia tenuflora* (Klein ex Willd.) Wt. & Arn. var. *minor* Baker.
A. Twig, B. Bract, C. Calyx split open, D. Standard petal, E. Wing petal, F. Keel petals, G. Androecium (split open), H. Pistil.

ovate, 3–6 × 1–2.5 cm, glabrous, green above, greyish beneath, acuminate, base obtuse; flowers rosy-white or yellowish-white in axillary racemes; pods thick, flat, linear, narrow at both ends, 18 cm long, glabrous, seeds 8–12, compressed.

Introduced, quick growing tree, planted in gardens, road sides as an avenue plant. Fl. & Fr.: Jan.–Mar. *Vern.*: Sasma kanuga. N. Varam, *VRK* 3221; Bolapalle, *VRK* 5741.

Indigofera Linnaeus

1. Leaves simple:
    2. Leaves linear, pods globose, 2 x 2 mm, 1-seeded .... ***I. linifolia***
    2. Leaves ovate-cordate, pods oblong, 4 x 1.5 mm, 2-seeded. .... ***I. cordifolia***
1. Leaves compound:
    3. Leaves 3-foliolate (1–5 foliolate in *I. oblongifolia*):
        4. Pods 1–2-seeded .... ***I. glandulosa***
        4. Pods 4–10 seeded:
            5. Leaves gland-dotted beneath .... ***I. trifoliata***
            5. Leaves not gland-dotted beneath:
                6. Prostrate herbs .... ***I. prostrata***
                6. Erect herbs or shrubs:
                    7. Racemes 6–12-flowered, leaflets obovate, pod rigid, seeds truncate at both ends .... ***I. trita***
                    7. Racemes 20–50-flowered, leaflets elliptic; pod not rigid, seeds truncate at one end .... ***I. oblongifolia***
    3. Leaves 5—many-foliolate:
        8. Prostrate herbs .... ***I. linnaei***
        8. Erect herbs or shrubs:
            9. Pods straight:
                10. Racemes as long as or longer than leaves, pods 4-gonous:
                    11. Branchlets white pilose, peduncle mostly 2.5 cm long, calyx lobes hirsute; leaflets usually 7–9 .... ***I. hirsuta***
                    11. Branchlets purplish, glabrescent, peduncle under 2.5 cm long, calyx lobes pubescent; leaflets usually 9–11 .... ***I. astragalina***
                10. Racemes shorter than leaves, pods terete or subclyindric to obscurely 4-gonous:
                    12. Branchlets viscid, leaflets 7 pairs, pod pubescent, seeds cuboid, pitted .... ***I. colutea***
                    12. Branchlets not viscid, leaflets 2–3-pairs, pod glabrous, seeds globose, not pitted .... ***I. glabra***
            9. Pods curved:
                13. Branchlets silvery-canescent, pods curved at tip .... ***I. parviflora***
                13. Branchlets not silvery-canescent, pods slightly curved at base:
                    14. Pod to 1 cm long, seeds up to 4; branchlets silvery canescent .... ***I. caerulea***
                    14 Pod to 2.5 cm long, seeds up to 12, branchlets glabrous .... ***I. tinctoria***

Indigofera astragalina DC. Prodr. 2: 228. 1825. *I. hirsuta* sensu Baker in FBI 2: 98. 1876. *p.p,* non L. 1753; Gamble 312 (221).

Erect herbs or small shrubs, up to 1 m tall; branchlets purplish, 4-angular, sparingly pubescent; leaves up to 13 cm long, imparipinnate; leaflets 7–11, opposite, elliptic-oblong, to 3.5 × 2 cm; pilose on both surfaces, apiculate, base acute; flowers pink, in axillary dense racemes; pods tetragonous, 1.5–2 cm long, deflexed, shortly pointed at apex, white, long hairy, seeds 6, blackish-brown, glabrous.

Occasional in forests. Fl. & Fr.: Aug.–Dec. Kondavedu, *SSR & KSM* 21444, Nagarjunakonda, *KTH* 5776 (CAL).

Indigofera caerulea Roxb. Fl. Ind. 3: 377. 1832; Ali in Bot. Not. 111: 564. 1958. *I. argentea* L. var. *caerulea* (Roxb.) Baker in Hook. f. Fl. Brit. India 2: 99. 1876. *I. articulata* auct. non Gonan; Gamble 311 (221).

Much-branched woody herbs to under shrubs; branchlets silvery canescent; leaves odd-pinnate, leaflets obovate, 1–3 × 1–2 cm, canescent, obtuse, cuneate; flowers pink in axillary racemes; pods turgid, deflexed, 1–1.5 cm long, silvery pubescent, shortly mucronate, seeds 3–4, striate.

Common in forests. Fl. & Fr.: Sept.–Mar. *Vern.:* Karu Neelia, Tender leaves are consumed as vegetable. Javuku, *VRK* 3582; N.U. campus, *SSR & KSM* 21497; Karempudi, *CAB* 4760 (CAL).

Indigofera colutea (Burm. f.) Merr. in Philipp. J Sci. 19: 355. 1921. *Galega colutea* Burm. f. Fl. Ind. 172. 1768. *Indigofera viscosa* Lam. Encycl. 3: 247. 1789; FBI 2: 95; Gamble 311 (220).

Subshrubs, up to 70 cm tall; branchlets viscid; leaves odd-pinnate; leaflets 7 pairs, opposite, obovate-elliptic, 1–1.5 × 0.4–0.6 cm, sparsely clothed with white appressed hairs, more or less beneath, base and apex obtuse, apiculate, entire; flowers orange-red in glandular 6–12-flowered racemes nearly as long as the leaves; pods straight, to 2.5 cm long, subcylindric, faintly torulose, with white appressed hairs; seeds ca. 15, pitted.

Common weed in wastelands. Fl. & Fr.: Sept.–Jan. *Vern.:* Nakkenara. Nagarjunakonda, *VRK* 6704, Phirangipuram, *CAB* 4603 (MH); Karempudi, *CAB* 4727 (MH).

Indigofera cordifolia Heyne ex Roth Nov. Pl. Sp. 357. 1821; FBI 2: 93; Gamble 309 (218).

Prostrate herbs; branchlets densely white-tomentose; leaves ovate-cordate, 0.5–1.2 × 0.5–0.8 cm, white hairy, obtuse, mucronate, base cordate, entire, sessile, flowers red in small condensed racemes, pods oblong, straight, 4x1.5 mm, pubescent, beaked, seeds globose.

Occasional in forests and wastelands. Fl. & Fr.: June–Oct. Kondavedu, *SSR & KSM* 21475, Bollapalle, *KCJB* 17411 (MH).

Indigofera glabra L. Sp. Pl. 751. 1753; Gamble 309 (219). *I. pentaphylla* Murr. Syst. Veg. ed. 13. 564, 1774; FBI 2: 95.

Annual erect herb up to 75 cm tall; branches numerous, ascending, glabrous or with a few spreading deciduous eglandular hairs, leaflets 5–7, opposite, obovate-elliptic, 1–2 × 8–1.2 cm, hairy, obtuse, apiculate, base cuneate-acute; flowers red, small, in 2–4-flowered short axillary racemes; pod straight, sub-cylindric, or obscurely 4-gonous, glabrous, not torulose; seeds 10–12, globose.

Occasional, in wastelands. (Mohan Rao & Aasiya, 1995). Fl. & Fr.: Sept.–Apr. *Vern.:* Pulichintaku.

Note: The specific epithet, *glabra* of Linnaeus has precedence over the specific epithet, *pentaphylla* of Murray. Further it is interesting to note that by error the accredition of the specific epithet, *pentaphylla* has been given to Linnaeus by most of the workers including De Candolle, Baker and Cooke (Fl. Pres. Bombay, 1902). It is, in fact, Murray who, in the 13th edition of Linnaeus *Systama Naturae* which renamed as *Systema Vegetabilium*, described the plant as a distinct species.

Indigofera glandulosa Roxb. ex Wild. Sp. Pl. 3: 1227. 1803; FBI 2: 94; Gamble 309 (218).

Pubescent undershrubs, up to 75 cm tall; leaves 3-foliolate, leaflets oblanceolate, 1.5–3 × 0.5–1 cm, obtuse, entire, base cuneate; flowers red in axillary capitate racemes, standard petal oblong-obovate, glandular-hirsute; pods oblong, 5 × 2 mm, slightly winged, deflexed, pubescent, turgid, seeds 2, rarely 1, ovoid.

Rare along the streams in forests. Fl. & Fr.: Oct.–Jan. Jayanthipuram, *VRK* 6979; Karempudi, *CAB* 4728 (MH); Guntur, *GVN* 16804 (MH).

Indigofera hirsuta L. Sp. Pl. 751. 1753; FBI 2: 96; Gamble 312 (221).

Densely pubescent undershrubs, up to 1 m tall; branches covered with brown hairs; leaves odd-pinnate, up to 15 cm long; leaflets subopposite, elliptic-oblanceolate, obtuse, entire, base subacute; flowers pink to brick red in axillary racemes; pods up to 2 cm long, deflexed, tomentose; seeds ca. 5, cuboid, angular.

Common in open forests, roadsides and wastelands. Fl. & Fr.: Sept.–Dec. *Vern.:* Kolapattitulu. Nagarjuna konda, *VRK* 5819 & 6705.

Indigofera linifolia (L.f.) Retz. Obs. Bot. 4: 29. 1786; Gamble 309 (218); Ali in Bot. Not. 111: 546. 1958. *Hedysarum linifolium* L.f. Suppl. Pl. 331. 1781.

1. Leaves linear, internodes long, racemes 6–8-flowered .... **var. *linifolia***
1. Leaves obovate,internodes short, racemes 1–4-flowered .... **var. *campbelli***

var. **campbelli** Wight ex Baker in Hook. f. Fl. Brit. India 2: 93. 1876; Gamble 209 (215).

Diffuse herbs, silvery canescent, stem prominently 2-edged with short internodes; leaves simple, ovate-elliptic or obovate, obtuse, subacute; flowers bright red in racemes, pod as in var. *linifolia.*

Occasional in open forests, grows on rocky area. Fl. & Fr.: July–Feb. Loyapalli, *SSR & KSM* 21438. (Population biology studies by B.V.N. Reddy 1978, M.Phil. Thesis, Nagarjuna Univ.).

var. **linifolia**

Diffuse annual herbs; branchlets silvery; leaves simple, linear, 1.5–3 × 0.1–0.2 cm, pubescent, acute; flowers red in axillary racemes; pods globose, 2x2 mm, silvery pubescent, seed solitary, globose.

Abundant weed in cultivated fields, roadsides, wastelands and occasional in forests. Fl. & Fr.: Sept.–Dec. *Vern.:* Gudlaku. Roots are used as good soil fertilizers. V.P. South, *VRK* 5829; Sattenapally *VRK* 5757; Karempudi, *CAB* 4757 (MH); Nagarjunakonda, *KTH* 9768 (MH).

Indigofera linnaei Ali in Bot. Not. 111: 549. 1958. *Hedysarum prostratum* L. Mant. Pl. 102. 1767. *Indigofera enneaphylla* L. Mant. Alt. 2: 271. 1771. *nom. superfl.;* FBI 2: 94; Gamble 309 (218).

Profusely branched ascending herbs; leaves odd-pinnate up to 3 cm long, leaflets 5-pairs, alternate, oblanceolate, 0.8–1 × 0.3–0.5 cm, densely hirsute below, mucronate, base cuneate-narrowed, apex obtuse; flowers pink in axillary racemes; pods ellipsoidal, 5 × 2 mm, white hirsute; seeds 2, ovate, truncate.

Abundant weed in plains, roadsides and wastelands. Fl. & Fr.: May–Dec. *Vern.:* Chalapachi. Nagarjuna konda, *VRK* 6712; Vinukonda RF, *VRK* 6921; Nagarjuna konda, *KTH* 9605 (CAL).

Note: The earliest valid name for this taxon is *Hedysarum prostratum* L. According to article 55 of the ICBN ed. 1978 when transferred to *Indigofera* the correct name should be *I. prostrata.* As Linnaeus did not make this combination his new name *I. enneaphylla* is illegitimate. Nor can the earliest epithet be reinstated now, as it would be a later homonym of Willd. 1803. Hence Ali has chosen the new name *Indigofera linnaei.*

Indigofera oblongifolia Forssk. Fl. Aeg. Arab. 137. 1775; Gamble 310 (219), *I. paucifolia* Del. Descr. Egypt. Hist. Nat. 251. 1312; FBI 2: 97.

Erect woody under shrubs, up to 1.5 m tall; branches numerous, stout, woody, silvery grey pubescent; leaves 1–5-foliolate, oblanceolate or elliptic-oblong, more or less hairy above, hoary with dense, fine white hairs beneath, base acute; flowers small, 20–50 in spicate racemes, rather close; pods numerous, along the whole length of the rachis, slightly curved, 1.3–2 cm long, pubescent; seeds 6–8, obscurely 4-gonous.

Rare, in wastelands (Mohan Rao & Aasiya 1995) . Fl. & Fr.: Sept.–Jan.

Indigofera parviflora Heyne ex Wight & Arn. Prodr. Fl. Ind. Orient. 201. 1834; FBI 2: 97; Gamble 311 (220).

Subshrubs, up to 60 cm tall; branchlets silvery-canescent; leaves odd-pinnate, 3–4.5 cm long, leaflets 7–9, opposite, oblong-oblanceolate, 1.2–5 × 0.2–0.4 cm, chartaceous, thinly hairy, entire, obtuse-mucronate, base cuneate; flowers rose, sessile in short close, 6–12-flowered, axillary racemes; pods terete, linear, pubescent, recurved at the tip, deflexed; seeds 15–20, scabroid, obscurely striate.

Occasional weed in wastelands. Fl. & Fr.: July–Dec. *Vern.*: Bapangu. Nagarjunakonda, *KTH* 9846 (CAL).

Indigofera prostrata Willd. Sp. Pl. 3: 1226. 1802; Gamble 310 (219); Santapau in Rec. Bot. Surv. India 16. 5. 1967. *I. trifoliata* auct. non L. FBI 2: 96. 1876.

Prostrate herbs, up to 60 cm long; branchlets appressed pubescent; stipules subulate; leaves 3-foliolate,leaflets obovate or oblanceolate, 1.5–2.3 × 0.3–0.5 cm, glabrous above, appressed pubesent beneath, apex obtuse with apicule, entire, base cuneate; flowers purple in short, 3–5-flowered racemes; pod filiform, to 1.7 cm long, sparsely hairy, deflexed, slightly winged at the sides of sutures; seed ca. 8.

Rare, in wasteland and dry deciduous forests. Fl. & Fr.: Nov.–Feb. Davupalli, *SSR & KSM* 21672; Mangalagiri, *KSM* 17831.

Indigofera tinctoria L. Sp. Pl. 751. 1753; FBI 2: 99; Gamble 311 (220). *I. sumatrana* Gaertn. Fruct. 2: 317. t. 148. f. 4. 1791; Gamble 311 (220).

Woody undershrubs, up to 1 m tall; branchlets terete, appressed pubescent, leaflets bluish-green, 7–11, oblong, or oblanceolate, 1–3 × 0.5–1.5 cm, glabrescent, apiculate, entire, base narrowed; flowers pale pink in spike like racemes; pods paired, cylindric, thick, turgid, recurved, shortly mucronate, brown; seeds 10–12, globose.

Rare in forests. *Vern.:* Nili, Kondanili. N.U. Campus, *SSR & KSM* 21490. Jayanthipuram, *C.A Barber* 4731 (MH); 4748 (MH); Vinukonda, *CAB* 4761 (MH), NU Hostel, *PNR* 2380 (NUH).

Indigofera trifoliata L. Cent. Pl. 2: 29. 1756; FBI 2: 98, *p.p.*; Gamble 310 (279); Ali in Bot. Not. 111. 552. 1958.

Erect undershrubs, up to 60 cm tall; branchlets brownish pubescent; leaves 3-foliolate, to 1.5 cm, leaflets oblanceolate, 1–1.7 × 0.2–0.4 cm, appressed pubescent, gland dotted, obtuse, cuneate, entire; flowers flame coloured in 6–10-flowered axillary, dense racemes; pod linear, deflexed, narrowly winged along margins; seeds 4, cuboid angular.

Occasional, in open and dry deciduous forests (Mohan Rao & Aasiya. 1995). Fl. & Fr.: Dec.–Feb. *Vern.:* Baragudamu, Muripindi.

Note: Gamble treats *I. trifoliata* and *I. prostrata* Willd. as distinct species. Ali after checking the specimens of the latter identified by Gamble and kept in Kew Herbarium concluded that these are the same as *I. trifoliata* L. But here Gamble's view is being followed and treated them as separate species. *I. prostrata* is distinct from *I. trifoliata* L. as pods are filiform, narrowly winged, more than 5-seeded.

Indigofera trita L.f. Suppl. Pl. 1781. ssp. ***trita***. FBI 2: 96; Gamble 310 (219).

Pubescent undershrubs, up to 1 m tall, branches hoary with fine pubescent hairs; leaves 3-foliolate, 4 cm long, leaflets obovate-oblong, 1.5–3 × 0.8–1.5 cm, silvery pubescent, emarginate, acute; flowers pink in axillary spicate racemes; pods oblong, 3 cm long, deflexed, rigid, 4-gonous, spine-tipped, hairy, seeds 6–10, oblong.

Occasional weed of the forests. Fl. & Fr.: Sept.–Jan. *Vern.:* Jidi vempali, Nakkanam, Loyapalle, *VRK* 3846; V.P. South, *VRK* 6722; Mangalagiri, *CAB* 4601 (MH); Nagarjunakonda, *KTH* 9671 (CAL); NU Hostel, *PNR* 2309, Kondavedu, *SSR & KSM* 21445.

Note: The pointed 4-angled divaricate pods render identification easy.

Lablab Adanson

Lablab purpureus (L.) Sweet. Hort. Brit. ed. 1.481. 1827 var. ***purpureus***; Verdc. in Kew Bull. 24: 410. 1970. *Dolichos lablab* L. Sp. Pl. 725. 1753; FBI 2: 209. *p.p.;* Gamble 367 (259).

Twining shrubs; leaves trifoliolate, leaflets cordate-ovate, 2.5–3 × 2–3 cm; entire, acute to acuminate, glabrous; flowers white in terminal racemes; pods flattened; seeds 2–3.

Commonly cultivated and also run willd. Fl. & Fr.: Dec.–Feb. Macherla RF, *VRK* 3856.

Macrotyloma (R. Wight et Arnott) Verdcourt

Macrotyloma unifolrum (Lam.) Verdc. in Kew Bull. 24: 322 & 401. 1970. *Dolichos uniflorus* Lam. Encycl. 2: 299. 1786. *D. biflorus* auct. non L. 89. 2: 210; Gamble 367 (259).

Densely villous climbers; leaves 3-foliolate, leaflets elliptic-lanceolate, 1.5–5 × 1–1.5 cm, villous, acute, entire, base obtuse; flowers yellow with purplish black blotch in the centre, axillary, solitary; pods oblong pubescent; seeds 4–6, oblong.

Commonly cultivated herb, occasionally wild. Fl. & Fr.: Aug.–Mar. *Vern.:* Wulawalu. V.P. South, *VRK* 5826.

Mundulea (A.P. de Candolle) Bentham

Mundulea sericea (Willd.) A. Cheval in Compt. Rend. 180. 1521. 1925. *Cytisus sericeus* Willd. Sp. Pl. 3: 121. 1802. *Mundulea suberosa* (DC.) Benth. in Miq. Pl. Jungh. 248. 1852; FBI 2: 110; Gamble 314 (222).

Silky tomentose shrubs, 2–4 m tall; bark thick, pale yellow, spongy, deeply fissured, leaves odd-pinnate, leaflets elliptic-oblong or lanceolate, 1.5–6 × 0.5–1 cm, glabrous above, silky below, acuminate, apiculate, ciliate, base obtuse; flowers pinkish in terminal corymbose racemes; pods linear, compressed, yellowish, thickened at the sutures; seeds 4–8.

Common, clumps on fully sun-exposed slopes in forests. Fl. & Fr.: June–Dec. *Vern.*: Verri Billu, Neela Murrai, Tella Billu. Seeds are used for poisioning fish. Plant possesses inseciticidal and piscicidal properties. All parts of the plant, particularly the bark and seeds are toxic and used as dust. It is toxic to caterpillers, cockroaches, pulse beetles and house flies. Davupalli, *SSR & KSM* 21409, Kakirala, *DMR* 21126.

Pongamia Ventenat *nom. cons.*

Pongamia pinnata (L.) Pierre Fl. Forest. Cochinch. Sub. t. 385. 1899; Thoth. in Bull. Bot. Surv. India 3: 418. 1962. *Cytisus pinnatus* L. Sp. Pl. 741. 1758. *Pongamia glabra* Vent. Jard. Mahm. 28. t. 28. 1803; FBI 2: 240; Gamble 385 (272). *Derris indica* (Lam.) Bennet in J. Bombay Nat. Hist. Soc. 68: 603. 1971. *Galedupa indica* Lam. Encyl. 2. 594. 1788.

Trees, up to 10 m tall; leaves odd-pinnate, 10–18 cm; leaflets 5–7, opposite, ovate-oblong, 6–12 × 4–8 cm, glabrous, acuminate, entire, base truncate; flowers pinkish white in axillary racemes; pods obliquely oblong or elliptic, up to 4 × 2 cm, compressed, indehiscent, seeds 1–2, pale red.

Common along the streams, forests and cultivated as avenue tree. Fl. & Fr.: Feb.–Sept. Eng.: Pongam oil tree. *Vern.:* Kanuga. Seeds for oil, leaves as green manure. The dried flowers are used in dyeing. Seeds are used as fish poison and give an oil used for burning and skin diseases. V.P. South, *VRK* 3940.

Pseudarthria Wight & Arnott

Pseudarthria viscida (L.) Wight & Arn. Prodr. 209. 1834; FBI 2: 154; Gamble 334 (236). *Hedysarum viscidum* L. Sp. Pl. 747. 1753.

Herb to under shrub, up to 1 m tall; branchlets viscid-pubescent; leaves 3-foliolate, terminal leaflets rhomboid, 5–7 × 4–9 cm, laterals obliquely ovate, 3–6 x 2–4 cm, base cuneate-obtuse, acute mucronate at apex, margin ciliate; flowers rose in slender axillary racemes; pod linear, 1.5 × 0.5 cm, hooked, pubescent, flattened, apex apiculate; seeds ca. 4, compressed, reniform.

Occasional in dry deciduous forests. Fl. & Fr.: Aug.–Dec. *Vern.*: Nayaku ponna. Fruits are eaten raw. Kondavedu, *SSR & KSM* 21482.

Pterocarpus Jacquin *nom. cons.*

Pterocarpus marsupium Roxb. Pl. Cor. t. 116. 1799; FBI 2: 239; Gamble 385 (271).

Large deciduous trees, up to 12 m tall; leaves odd-pinnate, leaflets alternate, elliptic-oblong or ovate, 6–10 × 2–3.5 cm, glabrescent, shining, emarginate, entire, base rounded to truncate; flowers golden yellow, scented in axillary panicles; pods winged, wings broad.

Occasional in forests. Fl. & Fr.: June–Dec. *Vern.:* Yerra Yegisa, Yegisa, Pedda yegi. The tree yields a 'kino' gum. The bark is used as an astringent and in toothache. The bruised leaves used in external application for boils, sores and skin diseases. The leaves are used as fodder. Davupalli, *SSR & KSM* 21696, Bollapalli RF, Vinukonda, *N.S. Veerachandra Rao* 65771 (MH).

Rhynchosia Loureiro *nom. cons.*

1. Pods usually 1-seeded .... ***R. rufescens***
1. Pods usually 2-seeded:
   2. Flowers solitrary or in pairs; branchlets viscid pubescent .... ***R. suaveolens***
   2. Flowers more than 5 in racemes; branchlets not viscid:
      3. Racemes capitate .... ***R. capitata***
      3. Racemes slender, lax:
         4. Under shrubs; leaves 4–7 × 2–4 cm, velvetty beneath, calyx accrescent; seeds strophiolate .... ***R. heynei***
         4. Trailing herbs; leaves 1–2.5 × 0.8–1.8 cm, glabrous or pubescent beneath; calyx deciduous; seeds estrophiolate .... ***R. minima***

Rhynchosia capitata (Roth) DC. Prodr. 2: 386. 1825; Gamble 373 (264). *Glycine capitata* Heyne ex Roth Nov. Pl. Spec. 346. 1821. *Rhynchosia aurea* sensu Baker in Fl. Brit. India 2: 221. 1876. *p.p non DC. 1825.*

Slender climbers; leaves 3-foliolate; leaflets obovate-suborbicular, 2–5 × 2.5–5 cm, scattered hairy, base subacute, apex rounded, flowers yellowish, seeds 2, oblong, brownish.

Occasional weed in wastelands, hedges and rare in forests. Fl. & Fr.: Sept.–Feb. Mellavagu, *VRK* 5725.

Rhychosia heynei Wight & Arn. Prodr. 240. 1834; FBI 2: 220; Gamble 374 (263). *R. codoorensis* Bedd. Ic. Pl. Ind. Or. 1: 70 t. 27. 1868–74.

Under shrubs up to 2 m tall; branchlets slender, velvety, leaves 3-foliolate, leaflets elliptic to oblong-ovate, 4–7 × 2–4 cm; pubescent above, velvety below, acute, entire, cuneate; flowers yellow in lax racemes; pod oblong, obliquely septate, 2-seeded, strophiolate.

Rare, in deciduous forests. Found in Nagarjuna University campus. Fl. & Fr.: Nov.–Apr. *Vern.:* Adavi kandi. Fruits are consumed as vegetable.

Rhynchosia minima (L.) DC. Prodr. 2: 1825; 2: 223; Gamble 375 (264); Van Meeuwen in Reinwardtia 5: 439. 1961. *Dolichos minimus* L Sp. Pl. 726. 1753.

Twining herbs; leaves 3-foliolate, leaflets obovate-rhombid, 1–2.5 × 0.8–1.8 cm, hairy on nerves below, gland dotted, obtuse-rounded; base cuneate, flowers yellow, in axillary racemes; pods falcate, 1.8 × 0.5 cm, glabrescent, horned; seeds 2 (rarely 1), subreniform.

Occasional twiner in forests. Fl. & Fr.: Sept.–Mar. *Vern.:* Nela Alumu, Gaddi chikkudu. Loyapalle, *VRK* 3860; V.P. South, *VRK* 5814, 6842; Jayanthipuram, *VRK* 6854; Krishna river, *VRK* 6753; Kondavedu, *SSR & KSM* 21443.

Rhynchosia rufescens (Willd.) DC. Prodr. 2: 387 1825; FBI 2: 200; Gamble 373 (263). *Glycine rufescens* Willd. in Ges. Naturf. Freunde Berlin Neue Schriften 4: 222. 1803.

Slender, pubescent twining shrubs; leaves 3-foliolate, leaflets elliptic-ovate. 2–5 × 0.5–3.5 cm, pubescent, acute, base rounded, margin entire; flowers yellow, with maroon streaks at base, in long racemes; pods ovoid, pointed at apex; seeds solitary.

Occasional in forests. Fl. & Fr.: July–Jan. Karempudi, *CAB* 4750 (MH).

Rhynchosia suaveolens (L.) DC. Prodr. 2: 387. 1825; FBI 2: 221, Gamble 374 (264). *Glycine suaveolens* L.f. Suppl. Pl. 326. 1781.

Subshrubs; branchlets viscid, glandular; leaves up to 30 cm; terminal leaflets ovate-oblong, lateral narrow-ovate, 3.5–5 × 1.5–2.5 cm, thin coriaceous, entire, acuminate, base obtuse; flowers yellow, in pairs or solitary; pods oblong, pubescent, septate between seeds; seeds 2, strophiolate.

Occasional in forests. Fl. & Fr.: Sept.–Feb. *Vern.:* Adavi kandi, Karra vulavalu. N.U. Campus, *SSR* & KSM 21493. Nagarjunakonda, *KTH* 9760 (CAL).

Rothia Persoon *nom. cons.*

Rothia indica (L.) Druce in Bot. Exch. Club Soc. Brit. Isles 3: 423. 1914. *Trigonella indica* L. Sp. Pl. 778. 1753. *Rothia trifoliata* Roth, Parr. syn 2: 302. 1807; FBI 2: 63; Gamble 279 (199). *Dillwynia trifoliata* Roth, Cat. Bot. 3: 71. 1806.

Annual prostrate, villous herbs, up to 60 cm tall; leaves 3-foliolate, obovate-oblong to oblanceolate, 1–2 × 0.4–0.6 cm, hirsute below, apex obtuse, margin ciliate, base acute to narrowed, sessile; flowers yellow, small, solitary or in short axillary racemes; pod linear, slender, compressed, appressed villous, apex acute; seeds to 20.

Occasional weed in cultivated fields and waste places. Fl. & Fr.: Oct.–Feb. *Vern.:* Vuchakura. N.U. Campus, *SSR & KSM* 21679.

Sesbania Scopoli *nom. cons.*

1. Erect annual prickly herbs .... ***S. procumbens***
1. Prickly or glabrous shrubs:
   2. Brchchlets with short prickles .... ***S. bispinosa***
   2 Branchlets glabrous
      3. Flowers 1.5–3.5 cm across, buds straight .... ***S. sesban***
      3. Flowers 10 cm long; buds falcate .... ***S. grandiflora***

Sesbania bispinosa (Jacq.) Wight in USDA Bur. Pl. Industr. Bull. no. 137. 15. 1909. *Aeschynomene bispinosa* Jacq. Icon. Pl. var. 3: 13. t. 564. 1792. *Sesbania aculeata* (Willd.) Poir. in Lam. Encycl. 7: 128. 1806. *nom. illegit.* FBI 2: 114; Gamble 323 (228).

Shrubs, up to 2 m tall; branchlets with short prickles; latex watery; leaves even-pinnate, up to 25 cm; leaflets 45–50 pairs, oblong, 1–2 × 0.4 cm, chartaceous, glabrous, enitre, obtuse-apiculate, base rounded; flowers yellow-brown in erect racemes; pods subterete, up to 30 cm long, beaked, torulose, seeds up to 40.

Occasional near cultivated fields. Fl. & Fr.: Nov.–Mar. *Vern.:* Ettejanga. Nagarjunakonda, *KTH* 9798 (CAL); V.P. South *VRK* 5825; Loyapalli, *SSR & KSM* 21631.

Sesbania grandiflora (L.) Poir. in Lam. Encycl. 7: 127. 1806; FBI 2: 115; Gamble 323 (228). *Robinia grandiflora* L. Sp. Pl. 722. 1753.

Trees, up to 12 m tall, branchlets pubescent, wood soft; leaves pinnate, leaflets to 20 pairs, oblong, 1–4 × 0.5–1 cm, emarginate, entire, obtuse; flowers white, 5 cm across in racemes; pod linear, 50 x 0.5 cm; seeds ca. 25.

Cultivated. (Mohan Rao & Aasiya, 1995). Fl. & Fr.: Dec.–Mar. *Vern.:* Avise.

Sesbania procumbens (Roxb.) Wight & Arn. Prodr. 215. 1834; FBI 2: 115; Gamble 123 (228). *Aeschynomene procumbens* Roxb. Fl. Ind. 3: 337. 1832.

Erect prickly annual herbs, 1–1.5 m tall; branchlets glabrous, striate; leaves even-pinnate, up to 15 cm long; leaflets linear, 1–1.5 × 0.2–0.5 cm, obtuse, base rounded; flowers yellow, in axillary pairs; pods straight, erect, much torulose, up to 25 cm long, beaked; seeds small, many.

Common weed of cultivated fields and wastelands. Fl. & Fr.: Sept.–Apr. Ethipothala, *VRK* 5800.

Sesbania sesban (L.) Merr. in Philipp. J. Sci. 7: 235. 1912. *Aeschynomene sesban* L. Sp. Pl. 714. 1753. *Sesbania aegyptiaca* Poir. in Lam. Encycl. 7: 128. 1806; FBI 2: 114; Gamble 323 (228).

Shrubs, up to 5 m tall; branchlets glabrous; leaves bipinnate; leaflets 15–20 pairs, opposite, oblong, 0.9–2 × 0.2–0.4 cm, glabrous, base and apex obtuse, entire; flowers purple to golden-yellow in pendent lax racemes; pods sub-terete; seeds oblong.

Occasional in plains, planted. Fl. & Fr.: Nov.–Mar. Guntur (*Uhl*, 1929).

Smithia W. Aiton

Smithia conferta Sm. in Rees. Cyclop. 33: n–2. 1819; Gamble 329 (232). *S. geminiflora* Roth Nov. Pl. Sp. 352. 1821; FBI 2; 149. 1876 incl. var. *conferta* (Sm.) Baker; Gamble 329 (232).

Annual diffuse herbs; leaves pinnate; leaflets 4–6 pairs, to 0.8 x 0.2 cm, linear oblong, apex rounded or acute, entire, midrib ending in setae; flowers yellow, in condensed head like spikes; calyx rigid with numerous parallel veins, bristles along the mid rib and top; fruit lomentum with turgid joints; seeds many, reniform.

Occassional on the banks of rice fields and tanks. Fl. & Fr.: July–Feb., Vinukonda RF, *VRK* 6874.

Stylosanthes Swartz

Stylosanthes fruticosa (Retz.) Alston in Trimen Handb. Fl. Ceylon 6 (Suppl.): 77. 1931. *Arachis fruiticosa* Retz. Obs. Bot. 5: 26. 1789. *Stylosanthes mucronata* Willd. Sp. Pl. 3: 1166. 1802. *nom. illeg.;* FBI 2: 148; Gamble 326 (230).

Rigid, diffuse, tomentose undershrubs, up to 60cm tall; leaflets, oblong-elliptic, 0.5–1.5 × 0.2–0.3 cm, coriaceous, sub-acute, obtuse, entire, flowers yellow, with maroon streaks in terminal heads; pods oblong, 4 mm beaked, compressed, hooked at the apex, seeds 2, compressed.

Occasional in wastelands and rare in forests. Fl. & Fr.: Sept.–Apr. *Vern.*: Salle kampa. Vinukonda RF, *VRK* 6955; Kotappakonda *KCJB* 392 (MH); Plots raised for source of seed by Forest Department during 1985–90 in NU B gardens—now found as an escape in NU Botanical gardens.

Tephrosia Persoon *nom. cons.*

1. Leaves simple .... *T. strigosa*
1. Leaves compound:
    2. Low shrubs, stipules spinescent .... *T. spinosa*
    2. Erect or prostrate herbs, stipules not spinescent:
        3. Prostrate herbs:
            4. Pod margin hispid, seeds 10, subovate .... *T. pumila*
            4. Pod margin glabrous, seeds 15, orbicular .... *T. maxima*
        3. Erect herbs:
            5. Inflorescence extra-axillary, pods glabrescent .... *T. purpurea*
            5. Inflorescence axillary, pods villous .... *T. villosa*

Tephrosia maxima Pers. Syn. 2: 329. 1807; Gamble 329 (225). *T. purpurea* (L.) Pers. var. *maxima* (Pers.) Baker in Hook. f. FBI 2: 113. 1876.

Prostrate herb, 70 cm long, branchlets appressed pubescent to sparsely pilose or glabrescent; leaves imparipinnate, to 9 cm, leaflets 6–9 pairs, 1–1.5 × 0.3–0.6 cm, puberulous, truncate-obtuse, retuse, mucronate, cuneate, entire; flowers bright purple in leaf-opposed axillary racemes; pods straight; sparsely pubescent, 5 × 0.5 cm; seeds ca. 15, orbicular, strophiolate.

Occasional in dry deciduous and open forests. Fl. & Fr.: Nov.–Jan. Davupalli, *SSR & KSM* 21669; Kondavedu fort, *CAB* 4682 (MH).

Tephrosia pumilla (Lam.) Pers. Syn. 2: Pl. 330. 1807. *Galega pumilla* Lam. Encycl. 2: 599. 1788. *Tephrosia purpurea* (L.) Pers. var. *pumila* (Lam.) Baker in Hook. f. Fl. Brit. India 2: 113. 1876. *T. procumbens* (Buch.-Ham.) Benth. Gen. Index to Trans. of Linn. Soc. 101. 1856; Gamble 320 (226)..

Perennial prostrate herbs; leaves odd-pinnate, up to 4 cm long; leaflets obovate-oblanceolate, 1–2 × 0.2–0.6 cm, tomentose, truncate or emarginate-mucronate, entire, base cuneate; flowers white or purplish-violet in leaf-opposed racemes; pods 4 × 0.4 cm, pubescent, seeds ca. 10.

Occasional weed of wastelands, cultivated fields and forests. Fl. & Fr.: Sept.–Mar. *Vern.:* Vempali. Jayanthipuram, *VRK* 6978; Kondavedu Fort, *CAB* 4681 (MH).

Tephrosia purpurea (L.) Pers. Syn. 2: 329. 1807; FBI 1: 112. *p.p.* Gamble 320 (226). *Cracca purpurea* L. Sp. Pl. 752. 1753. *Tephrosia hamiltonii* Drumm. in Gamble Fl. Pres. Madras 320. 1918, 226. 1957 (repr. ed.).

Erect undershrubs, up to 1 m tall; leaves odd-pinnate, up to 8 cm long, leaflets obovate to oblanceolate, 1–2.5 × 0.5–1 cm, pubescent below, obtuse, mucronate, base cuneate; flowers bluish-pink to purple in axillary (leaf opposed) racemes; pods up to 4 × 0.3 mm, glabrescent; seeds 5–8, ovoid.

Common weed of open forests. Fl. & Fr.: Throughout the year. *Vern.:* Yempali. Plants as green manure. Root juice mixed with hot water is given orally for the treatment of stomach ache. N. Varam, *VRK* 3599, Loyapalle, *VRK* 3991; Nagarjuna Sagar, *KMS* 9771 (MH).

Tephrosia spinosa (L.f.) Pers. Syn. Pl. 2: 230. 1807; FBI 2: 112; Gamble 320 (226). *Galega spinosa* L. f. Suppl. Pl. 335. 1782.

Low shrubs, up to 50 cm tall; branchlets grey; leaves to 3 cm; leaflets 4–5 pairs, obovate, 0.7–1.5 × 0.3–0.5 cm, chartaceous, pubescent, base cuneate, entire, truncate-emarginate, mucronate, flowers red, axillary, solitary; pods pubescent; seeds ellipsoid.

Rare in forests. Fl. & Fr.: July–Dec. Davupalli, *SSR & KSM* 21431.

Tephrosia strigosa (Dalz.) Sant. & Mahesh. in J. Bombay Nat. Hist. Soc. 54: 805. 1957. *Macronyx strigosus* Dalz. in Hooker's J. Bot. Kew Gard.

Misc. 2: 35. 1850. *Tephrosia tenuis* Wall. ex Dalz. & Gibs. Bombay Fl. 61. 1861; FBI 2: 111; Gamble 318 (224).

Annual spreading herbs; branchlets appressed pubescent; leaves simple, linearly oblong, 4–6 × 0.2–1 cm, glabrous above, pubescent below, obtuse; apiculate; flowers pale violet, axillary, solitary or paired; pods up to 2 × 0.3 cm, compressed, seeds 5–10.

A grass like herb, occasional weed of wastelands and cultivated fields. Vinukonda RF, *VRK* 6963; Mangalagiri *SSR & KSM* 21619; Narasaraopet, *KCJB,* 14103 (MH).

Note: This plant has been referred to, in Indian Floras under the name *Tephrosia tenuis* Wall. but like other names published in Wallich's catalogue, this too is a *nomen nudum* (*vide* J.K. Maheswari in JBNHS 54: 804, 1957). The oldest, validly published name is *Macronyx strigosus* of Dalz. (l.c.) and based on this the combination was proposed by Santapau and Maheshwari (l.c.)

Tephrosia villosa (L.) Pers. Syn. 2: 329. 1807; FBI 2: 113; Bosman & Hags. in Blumea 28: 476. 1983. *Galega villosa* L. Syst. Nat. ed. 10. 2: 172. 1753. *Tephrosia hirta* (Buch.-Ham.) Benth. Gen. Index to Trans. Linn. Soc. Lond. 101. 1866; Gamble 319 (225). *Galega hirta* Buch.-Ham. in Trans. Linn. Soc. Lond 13: 546. 1822.

Erect villose herbs, up to 1 m tall; leaves odd-pinnate, up to 6 cm long, leaflets obovate-oblanceolate, 2–2.5 × 0.5–0.8 cm, glabrous above, villous below, obtuse, entire, base cuneate; flowers violet or pink in terminal or extra-axillary racemes; pods 3 x 0.5 cm, sericeous, falcately curved upwards, densely woolly, seeds 4–8, subquadrate.

Common roadside weed, occasional in forest. Fl. & Fr.: July–Jan. *Vern.*: Nuguvempalli, Javuku, *VRK* 3525; Bollapalle, *VRK* 5754; Karempudi, *KCJB*, 17445 (MH); Nagarjunakonda, *KTH* 9774 (CAL).

Teramnus P. Browne

Teramnus labialis (L.f.) Spreng. Syst. Veg. 3: 235. 1826; FBI 2: 184; Gamble 352 (249). *Glycine labialis* L. f. Suppl. Pl. 335. 1781.

Twining herbs; stems hairy; leaves 3-foliolate, up to 6 cm long, leaflets obovate-lanceolate, 3–5 × 1–3 cm, glabrous above, pubescent below, obtuse, entire, base narrowed, flowers pink or purple in axillary racemes; pods linear, up to 3 × 0.2 cm, compressed, hairy; seeds 8–12, oblong.

Occasional, it grows at all elevations on hedges and bushes. Fl. & Fr.: May–Nov. *Vern.*: Masha parini. Pods are used as an astringent and in nervous disorders. Javuku, *VRK* 3556; Loyapalle, *VRK* 3825; Koppanur, *VRK* 3883; Vinukonda, *VRK* 6962, NU Campus, *PNR* 419 (NUH).

Vigna Savi

1\. Leaflets more than 3-lobed; stipules narrow-lanceolate, bracts 4 mm long, poduncle to 10 cm .... *V. aconitifolia*
1\. Leaflets 3-lobed, stipules broad ovate-oblong, bracts 7 mm long, peduncle to 30 cm .... *V. trilobata*

Vigna aconitifolia (Jacq.) Marech. in Bull. Jard. Bot. Nat. Belge. 39: 160. 1969; Verdc. in Kew Bull. 23: 464. 1969. *Phaseolus aconitifolius* Jacq. Obs. 3: 2. t. 52. 1768; FBI 2: 202; Gamble 363 (256).

Diffuse hairy twining herbs; leaves 3-foliolate, leaflets palmately lobed, lobes lanceolate-elliptic, up to 4 × 0.4 cm, glabrescent, acute, base obtuse; flowers yellow in axillary racemes; pods 2–4 × 0.2–0.6 cm, turgid, glabrescent; seeds 7–10, oblong.

Common weed in rice fields and occasional in moist waste lands. Fl. & Fr.: Sept.–Jan. Macherla, *VRK* 5895; Kondavedu. *CAB* 4695 (MH).

Vigna trilobata (L.) Verdc. in Taxon 17: 172. 1968 & in Kew Bull. 24: 560. 1970. *Dolichos trilobatus* L. Mant. Pl. 1: 101. 1767; FBI 2: 262. *Phaseolus trilobatus* (L.) Schrad. in Nov. Actorum Acad. Caes. Leop. Carol. Nat. Cur. 4: 132. 1770; Gamble 363 (256).

Diffuse prostrate or twining herbs; leaves 3-foliolate, leaflets entire or palmately 3-lobed, obovate, spathulate, 2–5 × 3–5 cm, membranous, base and apex obtuse, margin ciliate, flowers yellow in axillary capitate racemes; pods up to 5 × 0.2 cm, sparsely pubescent; seeds 7–12, oblong, smooth.

Occasional in cultivated fields and wastelands. Fl. & Fr.: Dec.–May. *Vern.*: Pillipesara. Plants as green manure and fodder. Sattenapalle, *VRK* 5870; Macherla, *VRK* 5892.

Zornia J.F. Gmelin

1\. Lower and upper leaflets lanceolate-ovate, stipules and bracts epunctate to sparsely punctate; branchlets pubescent .... *Z. diphylla*
1\. Lower leaflets obliquely oblong-ovate, upper lanceolate to narrowly lanceolate, stipules and bracts profusely punctate, branchlets glabrescent .... *Z. gibbosa*

Zornia diphylla (L.) Pers. Syn. Pl. 2: 318. 1807. *Hedysarum diphyllum* L. Sp. Pl. 247. 1753. *Zornia zeylonensis* Pers. l.c.; Gamble 325 (229). *Z. diphylla* (L.) Pers. var. *zeylonensis* (Pers.) Baker in Hook. f. Fl. Brit. India 2: 148. 1876.

Ascending or prostrate herbs; branchlets glabrescent; leaves broadly ovate, 1–1.5 × 0.5–1 cm, glabrous, base and apex obtuse, entire, apiculate; flowers yellow in axillary racemes; pods 2–5 cm, apically with glochidiate prickles.

Occasional in plains and rare in forests. Fl. & Fr.: Oct.–Mar. Nagarjunakonda, *KTH* 9793 (CAL).

Zornia gibbosa Span. in Linnaea 15: 192. 1841. *Z. diphylla* auct. non. (L.) Pers. 1807; FBI 2: 147. 1876. *p.p.;* Gamble 325 (229).

Annual pubescent prostrate herbs; leaves 2-foliolate, lanceolate, 1–2.5 × 0.2–0.5 cm, pubescent, obtuse, apex acute-apiculate, entire, gland dotted beneath, flowers yellow in long racemes; pods with scabrid prickels.

Occasional in forests. Fl. & Fl.: Sept.–Jan. N. Varam, *VRK* 3968; V.P. South, *VRK* 5779, 6836; Kondavedu, *CAB* 4694 (MH).

Note: The species *Zornia gibbosa* Span. due to its extensive pantropic distribution in Africa, India and East Asia, growing in a variety of habitat and climate, presents extremely variable forms and has confused several earlier workers. Noteboom (1961) accepted a composite species of the genus *Zornia; Z. diphylla* (L.) Pers. which includes *Z. gibbosa* Span. also. But Mohlenbrock (1961) in his monograph on the genus *Zornia* considered *Z. diphylla* (L.) Pers. as a perennial and separate form *Z. gibbosa* Span. which is an annual form, former growing in Ceylon and the latter in India, Burma, Malaysia, China, Australia and New guinea. Subsequent to Mohlenbrock's contention, what occurs in India was *Z. gibbosa*, S.K. Wagh (1964) has definitely pointed out that in India both *Z. diphylla* and *Z. gibbosa* occur. Recently, Chakravarthi and Jain have summarised stating that 4 species viz., *Z. walkeri, Z. diphylla, Z. gibbosa* and *Z. quilonensis* occur in India. Ravi (1979) stated that there are three species of *Zornia* in South India viz., *Z. diphylla, Z. gibbosa* and *Z. quilonensis* endemic to Kerala only. So in Andhra Pradesh two species viz. *Z. diphylla* and *Z. gibbosa* occur.

Arachis hypogaea L. commendable work has been done on Groundnut pathology in NU, Department of Botany in collaboration with ICRISAT. The Black gram and green gram in Guntur district especially in Kaza and Nambur are severely effected by the holoparasite, Goldenvine-China dodder, *Cuscuta chinensis* Lamk. Pusa Naubahar and Sada bahar C.V.s are recommended for inhibition of china dodder by Rao & Reddy (1987), Rao and Neelai (1997) and Neelai (1998). Soyabean and other cultivars of Black and Green grams are intensively studied in Agriculture Farm at Lamfarm, Guntur and several new varieties locally designed are released.

Seed stocks of all the pulses cultivars are with Lam Agri. Station of AP Agri. Univ.

CAESALPINIACEAE

1. Leaves simple .... ***Bauhinia***
1. Leaves pinnately compound:
    2. Leaves 1-pinnate:
        3. Leaflets adnate at their length up to middle and above the middle, palmately neved .... ***Hardwickia***
        3. Leaflets 4-numerous, not adnate, reticulately nerved:
            4. Petals 3, pods with pulpy mesocarp .... ***Tamarindus***

4. Petals 5, pods without pulpy mesocarp .... *Cassia*
2. Leaves 2-pinnate:
5. Plants armed:
6. Stragglers or lianas, pods winged .... *Pterolobium*
6. Shrubs or small trees, pods not winged:
7. Rachis flattened with very small leaflets; pod moniliform ...... *Parkinsonia*
7. Rachis rounded with large leaflets; pod not moniliform .... *Caesalpinia (p.p)*
5. Plants unarmed:
8. Sepals valvate, pods woody, long .... *Delonix*
8. Sepals imbricate, pods not woody, short:
9. Inflorescence rusty tomentose, stigma peltate .... *Peltophorum*
9. Inflorescence glabrous, stigma not peltate .... *Caesalpinia (p.p.)*

Bauhinia Linnaeus

1. Large climbers with circinate tendrils .... *B. vahlii*
1. Shrubs or trees, tendrils 0:
2. Fertile stamens 10:
3. Flowers yellow, more than 2.5 cm long; pod not turgid, flat ...... *B. tomentosa*
3. Flowers white, up to 1.5 cm long; pod turgid, falcate .... *B. racemosa*
2. Fertile stamens 4 .... *B. purpurea*

Bauhinia purpurea L. Sp. Pl. 375. 1753; FBI 2: 284; Gamble 407 (288).

Medium sized trees, up to 10 m tall; branchlets warty, glabrous, leaves oblong, 5–10 ´ 6–10 cm; leaflets connate about half-way up, base subcordate, entire, obtuse-subacute; flowers purple, showy in terminal or axillary racemes; pods linear-oblong, compressed, apiculate, seeds ovoid 12–15, flat, dark-brown, smooth.

Often cultivated in gardens and roadsides. Fl. & Fr.: Throughout the year. Eng.: Camel's foot tree. *Vern.:* Kanchanam, Devakasia. Wood hard and durable and used for agricultural implements. The bark gives fibre. The leaves are eaten as vegetables and also used as fodder. The tree yields gum, barks, tannins and seeds—non drying oil. Quercetin, Isoquercitrin, astragalin besides pelargodinidin, 3-glucoside and pelargonidin-3-triglucoside have been isolated from flowers. N.U. campus, *VRK* 6994; Vulasalapenta, *DMR* 13197.

Bauhinia racemosa Lam. Encycl. 1: 390. 1785; FBI 2: 276; Gamble 407 (288). *Piliostigma racemosa* (Lam.) Benth. in Miq. Pl. Jungh. 262. 1852.

Small deciduous bush like trees, up to 5 m tall, crown compact, leaves 2-lobed, fused to two thirds of their length, 2–4 x 3–5 cm, glabrous above, slightly pubescent and pale glaucous below, obtuse, entire, base cordate; flowers white or yellowish white in leaf-opposed, lax racemes; pods oblong, up to 16 × 2.5 cm, brown, woody, turgid, indehiscent, seeds 10, ovoid, compressed.

Common in forests. Fl. & Fr.: Apr.–Feb. *Vern.:* Are, Pacchare. Bark gives fibre. Wood is used as timber. The fruit is used in cough, asthma and diarrhoea and the bark is astringent in dysentry. Javuku, *VRK* 3577; Jayanthipuram, *VRK* 6993; Vinukonda RF, *KCJB* 17402 (MH), Nagarjunakonda, *KTH* 9735 (CAL).

Bauhinia tomentosa L. Sp. Pl. 375. 1753; FBI 2: 275; Gamble 406 (288).

Arboreous shrubs, up to 5 m tall; branchlets densely pubescent; leaves orbicular, 3.5–5.5 × 3.5–6.5 cm, leaflets connate above half their length, chartaceous, 7-nerved, glabrous above, densely pubescent below, base truncate, entire, obtuse or subacute; flowers yellow, large in terminal racemes; pods oblong, compressed; apex horned, base subacute-cuneate, dehiscent, seeds ovoid, 8–12, horizontal.

Occasional along river banks. Fl. & Fr.: Throughout the year. *Vern.:* Kanchini, Guntur (*Uhl,* 1929).

Bauhinia vahlii Wight & Arn. Prodr. 297. 1834; FBI: 2: 279; Gamble 406 (288).

Massive large leaved stragglers; leaves auriculate, 10–25 × 10–30 cm, grey or rusty-pubescent, bilobed, lobes connate one third of the way, obtuse; flowers yellowish-white in corymbose racemes; pods 15–30 × 2–4 cm, flat, rusty-tomentose, woody, indehiscent; seeds 7–9, flat, ovate, brightish colour.

Occasional gaint climber in forests. Fl. & Fr.: Apr.–Dec. *Vern.*: Addaku. Seed pulp is eaten raw. Stem bark fibre is used for making ropes. Reported in Guntur Forest Division work plan. Leaves source of minor forest produce. NWFP—leaves as plates.

Caesalpinia Linnaeus

1. Unarmed shrubs or trees:
    2. Shrubs, flowers yellow; pods flat .... *C. pulcherrima*
    2. Trees, flowers cream; pods spirally twisted .... *C. coriaria*
1. Armed stragglers or lianas .... *C. decapetala*

Caesalpinia coriaria Willd. Sp. Pl. 2: 532. 1799; Gamble 394 (279).

Unarmed trees, up to 10 m tall; branchlets warty; leaves even-pinnate; leaflets 20 pairs, oblong, 8 × 2 mm, punctate, glabrous, base oblique, cordate, entire, emarginate; flowers cream, in axillary and terminal panicles; pods twisted; seeds flattened, oblong.

Introduced, planted by Forest Department. Fl. & Fr.: Aug.–Nov. *Vern.:* Divi Divi tree. Fruit, bark in curing of leather. Large areas cultivated for source of tannins from fruits. Kondavedu *SSR & KSM* 21477.

Caesalpinia decapetala (Roth) Alston in Trimen, Fl. Ceylon 6 (Suppl.); 89. 1931. *Reichardia decapetala* Roth Nov. Pl. Spec. 212. 1821. *Caesalpina sepiaria* Roxb. Fl. Ind. 2: 360. 1832; FBI 2: 256; Gamble 394 (279).

Armed stragglers, up to 5 m tall; branchlets tomentose; prickles recurved; leaves bipinnate; pinnae 10–12 pairs; leaflets 10 pairs, elliptic or oblong-obovate, 1–1.5 × 0.3–0.5 cm, puberulous above, pubescent below, base cuneate, entire, obtuse-rotund; flowers golden yellow in axillary or terminal racemes; pods flat, turgid, tomentose; seeds ca. 8, oblong-globose.

Occasional in forests. Naturalized. Fl. & Fr.: Oct.–Feb. Reported in Guntur Forest Division Work Plan.

Caesalpinia pulcherrima (L.) Sw. Obs. Bot. 166. 1791; FBI 2: 255; Gamble 394 (279). *Poinciana pulcherrima* L. Sp. Pl. 380. 1753.

Glabrous shrubs, up to 3 m tall; leaves pinnate, pinnae 4–12 pairs, leaflets obliquely oblong, 0.5–2 × 0.2–0.5 cm, glabrous; flowers orange in terminal racemes; pods broadly linear, glabrous.

Native of tropical America. Cultivated in gardens, roadsides and rare in forests. Fl. & Fr.: Apr.–Dec. *Vern.:* Ratna gandhi. It is an abortifacient, febrifuge, purgative, tonic, emmenagogue, stimulant and for asthma, bronchitis and malarial fevers. Javuku, *VRK* 3924.

Cassia Linnaeus

1. Trees:
    2. Inflorescence erect. Pods long, flat, strap-shaped, up to 25 cm long .... *C. siamea*
    2. Inflorescence pendent. Pods long, round, cylindric up to 50 cm long ...... *C. fistula*
1. Shrubs, undershrubs or herbs:
    3. Prostrate herbs:
        4. Branchlets villous tomentose, leaflets up to 18 pairs; flowers small, on very short pedicels .... *C. pumila*
        4. Branchlets pubescent, leaflets up to 10 pairs, flowers large on long pedicels .... *C. kleinii*
    3. Erect undershrubs or shrubs:
        5. Leaflets 2–3 pairs:
            6. Leaflets 2 pairs, plants viscid pubescent, pods up to 4.5 cm long .... *C. absus*
            6. Leaflets 3 pairs, plants glabrous, pods up to 10 cm long .... *C. tora*
        5. Leaflets more than 3 pairs:
            7. Foliar glands present:
                8. Stipules foliaceous, leaflets pubescent; racemes to 10 cm, pod turgid, seeds ca. 6–10 .... *C. auriculata*
                8. Stipules not foliaceous, leaflets glabrous, racemes to 3 cm, pod compressed, seeds ca. 20–25 .... *C. occidentalis*
            7. Foliar glands absent:
                9. Pods curved, compressed; not porous:
                    10. Leaflets obovate-oblong, Pods crested .... *C. italica*
                    10. Leaflets linear-lanceolate, pods not crested .... *C. senna*
                9. Pods straight or slightly curved, turgid, porous .... *C. sophera*

Cassia absus L. Sp. Pl. 376. 1753; FBI 2: 265; Gamble 403 (285).

Viscid pubescent, bushy shrubs, up to 25 cm tall; leaves 1-pinnate, 5 cm long; leaflets 2 pairs, broadly ovate, obovate, inequilateral, 2–3.5 × 1.5–2.5 cm, pubescent, obtuse-subacute, entire, cuneate, rachis with subsessile glands

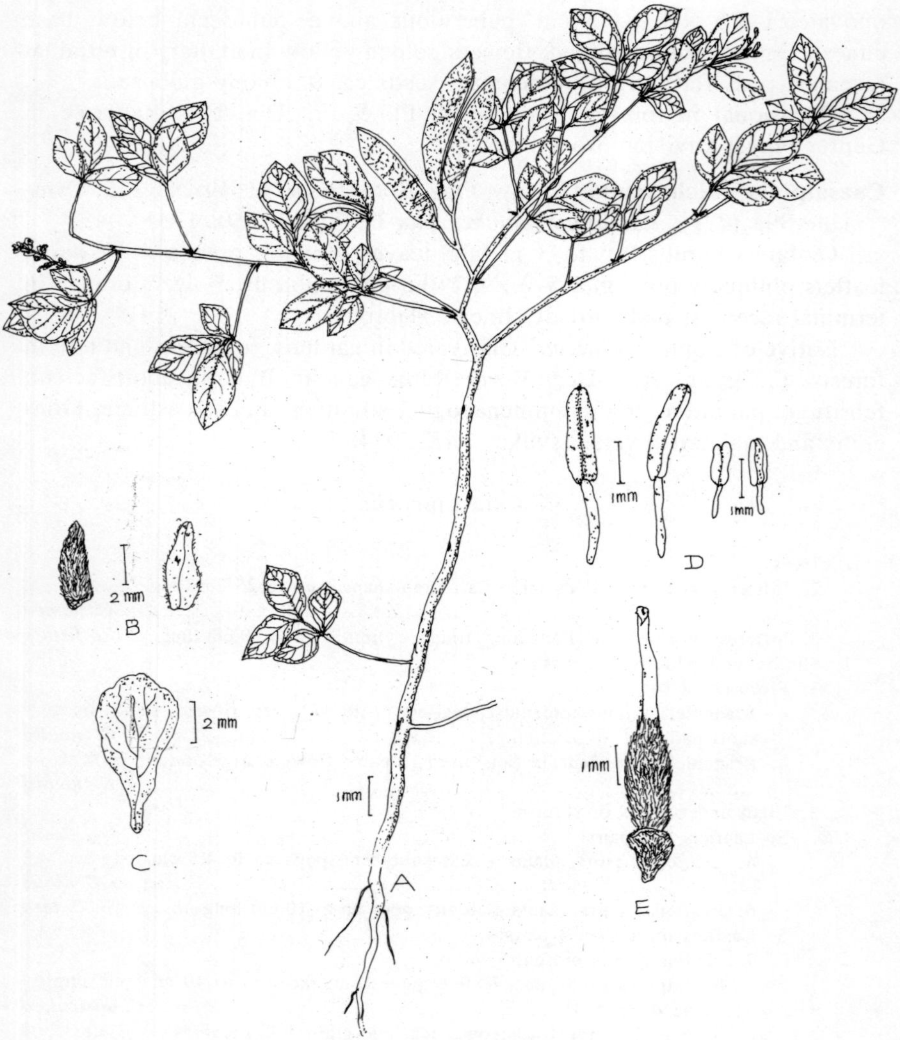

Fig. 9. *Cassia absus* L.
A. Twig, B. Sepals, C. Petal, D. Stamens, E. Pistil.

between leaflets, flowers red or pinkish-red, in terminal or leaf-opposed racemes; pods short-stipitate, falcate, 4.5 × 0.5 cm, flat, strigose; seeds 5–7, ovoid.

Occasional in forests and common as a weed in cultivated fields. Fl. & Fr.: Sept.–Feb. *Vern.*: Chanupala vittulu. Powdered seeds used in sore eyes. V.P. South, *VRK* 5815; N.U. campus, *VRK* 6992.

Cassia auriculata L. Sp. Pl. 379. 1753; FBI 2: 263; Gamble 402 (284).

Much-branched shrubs, up to 3 m tall; leaves pinnate, 8 cm long; glabrous above, puberulous below, obtuse-retuse, mucronate, entire, cuneate, or rounded, glands present, stipules foliaceous and reniform; flowers yellow in terminal and axillary corymbose racemes; pods long, stipitate, flat, thin, 10–13 × 1–1.5 cm, apex obtuse, mucronate, seeds 6–10, ovoid.

Abundant in open forests. Fl. & Fr.: Throughout the year. Eng.: Tannery cassia. *Vern.:* Thangedu. Bark is a source of tannin. All parts used in medicine and as green manure. It is an anthelmintic and astringent and used for conjuctivitis, diabetes, ophthalmia, skin affections and urinary complaints. The flower extract is used in the treatment of asthma. N. Varam, *VRK* 3237; Nagarjunasagar, *K.M. Sebastine* 9754 (MH); Nagarjunakonda, *KTH* 9623 (CAL).

Cassia fistula L. Sp. Pl. 377. 1753; FBI 2: 261; Gamble 400 (283).

Medium sized deciduous trees, up to 6 m tall; leaves pinnate, up to 3.5 cm long, leaflets 4–6 pairs, sub-opposite, ovate or ovate-oblong, 5–12 × 3–5 cm, glabrous above, pubescent below, acute-emarginate, entire, subacute-cuneate; flowers yellow in terminal, drooping, racemes; pods oblong, indehiscent, pendulous, cylindrical, 35–50 cm long, dark brownish-black when ripe; seeds numerous, immersed in a dark coloured sweetish pulp, and completely separated by transverse dissepiment.

Common deciduous tree in forests and planted along roadsides. Fl. & Fr.: Apr.–Sept. Eng.: Indian laburnum. *Vern.:* Rela. Bark is a source of tannin, wood for agricultural implements. The pulp of the fruit furnishes an excellent and safe purgative. Loyapalle, *VRK* 4000.

Cassia italica (Mill.) Lam. ex Andr. Fl. Pl. Anglo. Egypt. Sudan 2: 117. 1952. *Senna italica* Mill. Gard. Dict. ed. 8.772. 1768. *Cassia obtusa* Roxb. Fl. Ind. 2: 344. 1824 *nom. nud.* Gamble 402 (285). *C. obovata* Collad. Hist. Cass. 92. t. 15 A. 1816; FBI 2: 264.

Diffuse perennial undershrubs, up to 75 cm tall; stems striate; leaves pinnate, leaflets 4–8 pairs, pubescent, elliptic-oblong, 1.5–2.5 × 0.8–1.5 cm, glabrous, obtuse, mucronate, base cuneate; glands absent, flowers yellow in axillary racemes; pods flat, thin, much curved, 2–3 × 1–1.5 cm, flexible, glabrous, transversely veined and with a line of prominent longitudinal crests down the middle of the pod over the seeds; seeds 5–8.

Common in wastelands, roadsides and rare in forests. Fl. & Fr.: Oct.–May. V.P. South, *VRK* 6828; Bollapalle, *KCJB* 17424 (MH).

Note: Roxburgh's (1814) name *Cassia obtusa* is a *nomen nudum*. Two years later Colladon (1816) independently validated it by naming it *C. obovata* which has been followed by Bentham, and later by Baker in Flora of British India. However, *Cassia italica* (Mill.) Lam. ex Steud. Nom. ed. 1, a transfer based on Miller's *Senna italica* is the correct name to be used, as has been done by Andrews.

Cassia kleinii Wight & Arn. Prodr. Fl. Ind. Orient. 293. 1834; FBI 2: 266; Gamble 403 (285).

Prostrate herbs, up to 60 cm long; branchlets pubescent; leaves even pinnate; leaflets 8–10 pairs, oblong-elliptic, 5–9 × 3 mm, chartaceous, base and apex obtuse, entire; flowers yellow, in axillary racemes; pods flat, compressed, seeds flat, ovoid.

Occasional in forests. Fl. & Fr.: Throughout the year. *Vern.*: Nela tangedu. Kondavedu Fort, *CAB* 4695 (MH).

Cassia occidentalis L. Sp. Pl. 377. 1753; FBI 2: 262; Gamble 401 (284).

Annual herbs to undershrubs, up to 2 m tall; stipule prominent, lanceolate; leaves pinnate, up to 15 cm long, leaflets 3–6 pairs, ovate-lanceolate, glabrous, very acute, entire, base rounded, rachis with a single gland near its base; flowers yellow, in axillary and terminal racemes; pods flat, compressed, margins thickened, glabrous, 8–12 × 0.4–0.8 cm, seeds 20–25, ovoid.

Common weed of wastelands, occassional in cultivated fields and forests. Fl. & Fr.: July–Dec. The bark, leaves and seeds used medicinally. *Vern.*: Kasivinda. Eng.: Negro coffee. Bollapalle, *VRK* 5755.

Cassia pumila Lam. Encycl. 1: 651. 1785; FBI 2: 266; Gamble 403 (285).

Prostrate herbs, branchlets villous-tomentose, leaves pinnate, up to 4 cm long, leaflets 10–18 pairs, linear-oblong, 1–3 × 0.5–1 mm, glabrous above, villous below; rounded and apiculate at apex, flowers yellow, axillary, solitary; pod flat, 2–3.5 × 0.2–0.4 cm, compressed, linear, glabrous or with few scattered hairs, horned, seeds 5–7, rounded, brown.

Occasional weed in cultivated fields. Fl. & Fr.: Throughout the year. *Vern.:* Nalla jeeluga. V.P. South, *VRK* 5785.

Cassia senna L. Sp. Pl. 377. 1753. Brenan in Kew Bull. 13: 243. 1957. *C. angustifolia* Vahl Symb. Bot. 1: 29. 1790; FBI 2: 264; Gamble 404 (286).

Undershrubs, up to 1.5 m tall, leaves pinnate, leaflets 4–8 pairs, ovate-lanceolate, 1.1–3.6 × 0.5–0.6 cm, glabrous above, pubescent below, apiculate, entire, cuneate; flowers yellow in axillary racemes, pod short stipitate, slightly curved outside, compressed, flattened, strongly nerved, 3–5 × 1.5–2 cm, glabrous, greenish black, seeds 5–8, ovoid, membranous.

Rare in open forests. Fl. & Fr.: Oct.–May. Javuku, *VRK* 3522; V.P. South, *VRK* 6837.

Cassia siamea Lam. Encycl. 1: 648. 1785; FBI 2: 264; Gamble 402 (285).

Medium sized trees up to 10 m tall; branchlets glabrescent; leaves even-pinnate; leaflets 7–10 pairs, ovate-oblong, 2.5–6 × 1.5–2 cm, coriaceous, glabrous above, pilose below, base and apex obtuse, tipped with small sharp mucro, entire; flowers yellow, in terminal corymbose panicles; pods strap-shaped, flat, thickened at sutures; seeds 20–30, longitudinal.

Avenue trees, densely foliaceous, cultivated in gardens. Recommended for covering the denuded areas, hill slopes, flat terrain provided the drainage is good. Fl. & Fr.: Throughout the year. *Vern.*: Seema Tangedu. N.U. Campus, *VRK* 6986.

Cassia sophera L. Sp. Pl. 379. 1753; FBI 2: 262; Gamble 401 (284).

Erect shrubs, up to 2 m tall; leaves even pinnate, leaflets 5–8 pairs, 2–3.5 × 1–1.5 cm, elliptic-oblong, acute, entire, cuneate-obtuse, greenish; flowers yellow in terminal or axillary corymbose racemes; pods slightly curved, oblong, transversely septate, apically horned, 6–10-seeded.

Occasional in wastelands and forest out skirts. Fl. & Fr.: Nov.–Apr. *Vern.*: Paiditangedu. It is an antidote to snake-bite, diuretic and purgative and used for bronchitis and ringworm. Koppanur, *VRK* 1293.

Cassia tora L. Sp. Pl. 376. 1753; FBI 2: 263; *p.p.*; Gamble 401 (284).

Annual herbs or undershurbs, up to 1 m tall; leaves pinnate, leaflets obovate, 2–4.5 × 1–2.5 cm, gland present between lowest pair, obtuse-acute, entire, cuneate; flowers golden yellow, solitary or in pairs; pods linear-acicular, flat, up to 10 cm long, glabrous, obliquely septate; seeds 20–25, oblong.

Common, along roadsides and in wastelands. Fl. & Fr.: June–Nov. *Vern.*: Tantipu. Blue dye is prepared from seeds. It is a purgative, laxative, and antidote to snake-bite and for ring worms, itches and other skin affections. Javuku, *VRK* 3530.

Delonix Rafinesque

Delonix elata (L.) Gamble Fl. Pres. Madras 396. 1919 & 280. 1957 (repr. ed.). *Poinciana elata* L. Cent. Pl. 2: 16. 1756; FBI 2: 260.

Trees, up to 12 m tall; leaves 2-pinnate, up to 16 cm long, pinnae 7–9 pairs, leaflets 10–25 pairs, oblong-elliptic, 0.6–1 × 0.2–0.4 cm, puberulous, obtuse, shortly apiculate, base cuneate; flowers yellowish-cream in terminal corymbose racemes; pods linear or oblanceolate, 10–20 × 1.5–2 cm, horned, sutures grooved; seeds more than 10.

Cultivated as avenue tree and occasionally found as an escape. Fl. & Fr.: Sept.–May. Eng.: White Gold Mohur. *Vern.*: Chitti kesara, Sunkesala. The branches and leaves are used as leaf manure. Wood is light and used in preparing combs and matches etc. The leaves are said to be used in rheumatism and flatulense. Javuku, *VRK* 3583.

Hardwickia Roxburgh

Hardwickia binata Roxb. Pl. Cor. t. 209. 1811; FBI 2: 270; Gamble 412 (292).

Deciduous trees, up to 10 m tall; trunk straight, cylindrical, branchlets glaucous, slender, drooping; leaves 2-foliolate, leaflets obliquely ovate, 2–4 × 1–2 cm, glabrous, obtuse, entire, base truncate; flowers light yellowish-green in terminal or axillary panicled racemes; pods 5–8 × 1–1.5 cm, flat, compressed, oblong with parallel longitudinal veins; seeds solitary, near the top, to 2 × 0.8 cm.

Common in dry deciduous forests. It is drought resistant and reproduces vegetatively by root suckers. Fl. & Fr.: Aug.–Nov. *Vern.:* Yepi. Wood extremely hard, very heavy and durable. Bark yields a red brown strong fibre extensively used for well ropes and other agricultural purposes. Gum exudate from the bark used in varnishes. Javuku, *VRK* 3541.

Parkinsonia Linnaeus

Parkinsonia aculeata L. Sp. Pl. 375. 1753; FBI 2: 260; Gamble 397 (281).

Small armed trees, up to 4.5 m tall; spines woody, erect, leaves 2-pinnate, 25–30 cm long, secondary rachis flattened; leaflets minute. 30–35 pairs, lanceolate, 2–6 × 1–1.15 mm, reduced to scales; flowers yellow in axillary long, lax racemes; pods 12 × 0.6 cm, linear, moniliform, beaked, dehiscent, glabrous; seeds more than 6, oblong, brown.

Occasional in wastelands and in hedges. Long row of plants along NH5 near Chinakakani diversion Road. Native of tropical America. It is recommended for plantations in black cotton soil and sandy waste lands. Fl. & Fr.: Throughout the year. Eng.: Jew's thorn. *Vern.:* Seemathumma. N. Varam, *VRK* 39456; Loyapalle, *VRK* 3814.

Peltophorum (Vogel) Bentham *nom. cons.*

Peltophorum pterocarpum (DC.) Baker ex Heyne Nutt. Pl. Ned. Ind. 2: 255. 1927. *Inga pterocarpum* DC. Prodr. 2: 441. 1825.

Densely foliaceous trees, up to 10 m tall; crown round, spreading; bark brown, leaves 2-pinnate, pinnae 8–10 pairs, 15 cm long, leaflets 15–20 pairs, elliptic-oblong, 1–2 × 0.4–0.8 cm, glabrescent, retuse, entire, base oblique, sub-cordate; flowers golden yellow in terminal rusty tomentose panicles; pods flat, compressed, dark brown, closely longitudinally veined, indehiscent, seeds 3–4.

The copper pod; The rusty shield bearer. Cultivated as an avenue and ornamental tree. Branches very brittle and easily broken or damaged by monsoon winds. *Vern.*: Konda chinta, Pacha Sunkesula. Fl. & Fr.: Feb.–Aug. Native of Sri Lanka, Andamans, Malay peninsula, Archipelago and North Australia. V.P. South, *VRK* 6832.

Pterolobium R. Brown ex Wight et Arnott *nom. cons.*

Pterolobium hexapetalum (Roth) Sant. & Wagh in Bull. Bot. Surv. India 5: 108. 1963. *Reichardia hexapetala* Roth, Nov. Pl. Sp. 210. 1820. *Pterolobium indicum* A. Rich. Tent. Fl. Abyss. 1: 247. 1847; FBI 2: 259; Gamble 395 (280).

Armed stragglers, up to 8 m tall; thorns strong, woody, recurved; leaves 2-pinnate, 15 cm long; pinnae 5–6 pairs, leaflets 6–9 pairs, oblong-lanceolate, 0.8–1 × 0.3–0.5 cm, obtuse, entire, base truncate; flowers white in axillary racemes; pods samara like, winged at apex, pinkish purple, indehiscent, 3 × 0.3 cm, seeds solitary, obovoid.

Common in forests. Fl. & Fr.: Mar.–Aug. *Vern.*: Korinta, Erra siki, Loyapalli, *VRK* 3990; Nagarjunakonda, *KTH* 9667 (CAL).

Tamarindus Linnaeus

Tamarindus indica L. Sp. Pl. 34. 1753; FBI 2: 273; Gamble 409 (290).

Large evergreen trees, up to 15 m tall; crown umbrella shaped, leaves pinnate, leaflets 10–16 pairs, oblong, 1.5–2 × 0.2–0.6 cm, glabrous, obtuse, entire; flowers yelowish orange in terminal racemes; pods oblong, up to 10 × 1 cm, compressed, indehiscent with crustaceous brittle epicarp, pulpy mesocarp; seeds 2–8, compressed, shiny, black.

The tamarind tree, common on roadsides, also in forests. Fl. & Fr.: May–Nov. *Vern.:* Chinta. Tamarind is the fruit pulp. Fruit pulp is made into preserves and also exported as a drug. Ripe fruit regarded as refrigerent, digestive, carminative and laxative and useful in intoxication caused by alcohol or any poisonous substances. Young leaves are eaten. Wood is used for many agricultural and household purposes. Extract of bark is used as tonic to restore sensation after a paralytic stroke; leaf juice is used as an antiseptic to wash wounds. Loyapalle, *VRK* 3832.

Cassia javanica L. with pink flowers and filaments swollen at the middle, *Cassia spectabilis* DC. with acute-acuminate leaflets and yellow flowers, *C. alata* L. with showy oblong-elliptic to obovate-elliptic leaves and large bright yellow flowered racemes are common ormantal trees.

Saraca asoka (Roxb.) de Wilde, *Bauhinia variegata* L., *Delonix regia* Hook. f. in several floral colours are cultivated as ornamentals.

MIMOSACEAE

1. Plants armed:
   2. Pinnae 1 or 2 pairs:
      3. Stamens indefinite, connate, pods spirally coiled without pulp, seed arillate .... ***Pithecellobium***
      3. Stamens 10, free, pods straight or curved, mesocarp pulpy, seeds exarillate .... ***Prosopis***
   2. Pinnae 3 or more pairs:
      4. Flowers 4-merous, stamens 4–8, pods splitting transversely into 1- seeded segments .... ***Mimosa***
      4. Flowers 5-merous, stamens 10 or more, pods not splitting into 1-seeded segments:
         5. Flowers in bicoloured spikes, pods clustered, coiled, indehiscent, branchlets spine-tipped .... ***Dichrostachys***
         5. Flowers in heads or single coloured spikes, pods not clustered, straight or curved, dehiscent, branchlets not spine-tipped .... ***Acacia (p.p)***
1. Plants unarmed:
   6. Stamens indefinite:
      7. Stamens free .... ***Acacia (p.p)***
      7. Stamens connate below or monadelphous:
         8. Pod straight, compressed, thin, mesocarp scanty or 0 .... ***Albizia***
         8. Pod curved, turgid, woody, mesocarp copious .... ***Samanea***
   6. Stamens usually 10:
      9. Large climbers, flowers in spikes, fruits large up to 1 m long .... ***Entada***
      9. Non climbers, flowers in globose heads, fruits small to medium sized, 2–15 cm long:
         10. Prostrate diffuse herbs, pinnae 1–3 pairs .... ***Neptunia***
         10. Trees, pinnae 4–8 pairs .... ***Leucaena***

Acacia P. Miller

1. Plants unarmed, phyllodes present .... ***A. auriculiformis***
1. Plants armed, phyllodes absent:
   2. Straggling shrubs:
      3. Branchlets yellowish pubescent, thorns straight .... ***A. torta***
      3. Branchlets glabrous, thorns hooked .... ***A. caesia***
   2. Erect trees or shrubs:
      4. Flowers in globose heads:
         5. Heads in terminal extensive panicles .... ***A. leucophloea***
         5. Heads in axillary clusters:
            6. Leaves alternate, leaflets 16–20 pairs .... ***A. nilotica***
            6. Leaves densely clustered, leaflets 5–14 pairs:
               7. Pods flat, pulpy .... ***A. farnesiana***
               7. Pods linear, not pulpy .... ***A. roxburghii***
      4. Flowers in elongated spikes:
         8. Spines long, straight, hollow .... ***A. horrida***
         8. Spines short, hooked, solid:
            9. Flowers white .... ***A. chundra***
            9. Flowers yellow:
               10. Branchlets glabrous, pods lanceolate, grey, horned ... ***A. ferruginea***
               10. Branchlets dark-brown to purple, pods flat, thin, dark-brown .... ***A. catechu***

Acacia auriculiformis A. Cunn. ex Benth. in Lond. Journ. Bot. 1: 377. 1842.

Medium sized, unarmed trees up to 5 m tall; phyllodes coriaceous oblong-elliptic, falcate, 11–20 × 1.4–2 cm with parallel-nerves, tapering at the ends; flowers bright yellow, fragrant in cylindrical lax spikes; pods woody, brown, coiled at maturity, dehiscent; seeds more than 5, with yellow aril.

Cultivated as an avenue tree by the forest department on roadsides. Fl. & Fr.: Oct.–Aug. Eng.: Australian wattle. *Vern.:* Australian tumma. It is a good fire wood. Its sensitivity to water stress has become a matter of controversy. Sattenapalle, *VRK* 5876.

Acacia caesia (L.) Willd. Sp. Pl. 4: 1090. 1806; Gamble 428 (304). *Mimosa caesia* L. Sp. Pl. 522. 1753. *p.p. Acacia intsia* Willd. var. *caesia* (L.) Baker in Hook. f. Fl. Brit. India 2: 297. 1879.

Woody straggling shrubs up to 4 m tall; internodal thorns hooked; leaves up to 15 cm long, alternate, 2-pinnate, leaflets 10–15 pairs, oblong 1–1.5 × 0.5–0.8 cm, glabrescent beneath, obtuse or subacute, base truncate; flowers in creamish-white heads in terminal panicles; pods stipitate, 7–12 × 1.5–2 cm, glabrous, horned; seeds more than 10, brown.

Common in open and scrub forests. Fl. & Fr.: June–Dec. *Vern.*: Korinta, Yerra cheeki. A useful remedy for a type of cough named after the plant. Loyapalle, *VRK* 3989; Davupalli, *DMR* 18140.

Acacia catechu (L.f.) Willd. Sp. Pl. 4: 1079. 1806; FBI 2: 295; Gamble 428 (303). *Mimosa catechu* L.f. Suppl. Fl. 439. 1781.

Shrub like armed trees, 4–5 m tall; young branchlets dark brown to purple, 2 curved, stipular spines at axils; leaves 2-pinnate, pinnae 7–15 pairs, leaflets 20–40 pairs, elliptic-oblong, 3–6 × 0.5–1 mm, glabrous, sessile, subacute; flowers light greenish yellow in axillary spikes; pods flat, thin, dark-brown, shining; seeds 4–10.

Common in dry deciduous or in open forests. Fl. & Fr.: Aug.–Nov. Eng.: Red catechu, cutch tree. *Vern.*: Nalla sundra. The bark is bitter and acrid, used for cooling, astringent to the bowels, anthelmintic; antidysentric and antipyretic; it also cures itching, sore throat, bronchitis, indigestion, ulcers, boils, leucoderma etc. Bark yields the important brown 'Khathi' dye. The khari or kath extracted from the heart wood is widely used with betel. The wood is also used for construction of houses and agricultural implements. N. Varam, *VRK* 3523.

Acacia chundra (Roxb. ex Rottl.) Willd. Sp. Pl. 4: 1078. 1806; FBI 2: 295; Gamble 428 (303) (*sundra*). *Mimosa chundra* Roxb. ex Rottl. in Ges. Naturf. Freunde Berlin Neue Schriften 4: 207. 1803. *M. sundra* Roxb. Pl. Cor. t. 225. 1811 & Fl. Ind. 2: 562. 1832. *Acacia sundra* (Roxb.) DC. Prodr. 2: 458. 1825; FBI 2: 295. *p.p.*; Gamble 428 (303).

Armed trees, up to 8 m tall; leaves 2-pinnate, leaflets 5–25 pairs, elliptic-oblong, 2–3 × 0.8–1 mm, subacute, base oblique; flowers pale yellow or white in axillary spikes; pods flat, apically horned.

Common in forests. Fl. & Fr.: July–Dec. Kair. Red cutch. The hard wood is used for agricultural implements. Vinukonda RF, *VRK* 6846; Loyapalli, *SSR & KSM* 21648, Nagarjunakonda, *KTH* 9629 (CAL).

Acacia farnesiana (L.) Willd. Sp. Pl. 4: 1083. 1806; FBI 2: 292; Gamble 426 (301). *Mimosa farnesiana* L. Sp. Pl. 521. 1753.

Small armed trees, up to 4 m tall; branchlets reddish or brownish lenticellate, glabrous, leaves 2-pinnate, 2–6 cm long, pinnae 4–5 pairs, 1.5 cm long, leaflets 14–20 pairs, narrowly oblong, 2–4 × 0.5–1 mm, overlapping, glabrous; stipular thorns 1 cm long, straight; flowers yellow, very fragrant in axillary globose heads; pods brown, oblong, 5–7 × 1–1.5 cm, curved, glabrous, turgid, pulpy, apex hooked, seeds more than 10, 2-seriate, brown, irregularly globose.

Occasional in forests. Fl. & Fr.: Nov.–May. *Vern.*: Kasturi, Kampu tumma. Fruits used for cough, Nagarjuna Konda, *VRK* 6711.

Acacia ferruginea DC. Prodr. 2: 458. 1825; FBI 2: 295; Gamble 428 (303). *Mimosa ferruginea* Roxb. Fl. Ind. 2: 561. 1832, non Rottler, 1803.

Medium sized trees, up to 5 m tall; branchlets brownish or blackish glabrous; leaves alternate; leaflets 15–25 pairs, oblong, slightly falcate, 0.3–1 × 0.1–0.3 cm, glaucous, sparsely ciliate, subacute, base oblique, greyish when dry; flowers yellow, in axillary spikes; pods lanceolate, grey, glabrous, reticulately veined, apex obtuse with apicule, seeds 6.

Occasional in scrub forests. Fl. & Fr.: Mar.–Aug. *Vern.*: Inupa thumma, Sundra. The bark is bitter, acrid and anthelmintic. The extract of leaves and bark are used as astringent. Wood hard and very heavy used in cart wheels, posts, beams and agricultural implements. Reported in Guntur Forest Division Work Plan.

Acacia horrida (L.) Willd. Sp. Pl. 4: 1082. 1806. *Mimosa horrida* L. Sp. Pl. 521. 1753. *Acacia latronum* (L.f.) Willd. Sp. Pl. 4: 1077. 1806; FBI 2: 296; Gamble 427 (302). *Mimosa latronum* L.f. Suppl. Pl. 438. 1781.

Small or low armed trees; up to 5 m tall; crown spreading; branchlets black or brownish, warty; spines white, paired, longer one hollow and horn like; leaves 2-pinnate, up to 2 cm long, pinnae 2–4 pairs, 1 cm long, leaflets 10–12 pairs, linear-oblong, 1–3 × 0.8–1 mm, glabrous; stipular thorns white, hollow, 1–6 cm long; flowers creamish or yellow in axillary spikes; pods flat, 2.5–3.5 × 0.8–1.5 cm, falcate-oblong, turgid, curved; brown; seeds 4–8, ovoid, flattened.

Common in scrub jungles. Fl. & Fr.: May–Dec. *Vern:* Paki tumma. Javuku, *VRK* 3287; 3920; Loyapalle, *VRK* 3828; Vinukonda RF, *VRK* 6856; Mangalagiri, *SSR & KSM* 21615; Karempudi, *CAB* 4751 (MH).

Notes: Dorothy Hillcoat and J.P.M. Brenan (Kew. Bull. 1958: 39–40. 1958) have discussed at length the origin of this taxon which they have traced from Plukenat (Phytographie t. 121. f. 4. 1962)—the material for Linnaeus's *Mimosa horrida*, collected from Madras; *Mimosa latronum* L.f. then is a superfluous name for *M. horrida* L.

Acacia leucophloea (Roxb.) Willd. Sp. Pl. 4: 1083. 1806; FBI 2: 294; Gamble 427 (302). *Mimosa leucophloea* Roxb. Pl. Cor. t. 150. 1800.

Deciduous trees, up to 12 m tall; branchlets yellowish—violet; leaves 2-pinnate, up to 10 x 6 cm, pinnae 5–6 pairs, up to 3.5 cm long, leaflets 20–25 pairs, elliptic, 3–6 × 0.5–1 mm, glabrous-sparsely puberulous beneath, acute or rounded, margin ciliate, base oblique; spines opposite, 0.6–1 cm long; flowers yellow-creamish, globose heads in axillary panicles; pods flat, slightly velvety, linear, 5–10 × 0.4–1 cm, rusty tomentose; seeds 10–12, ovoid, dark brown, smooth and shining.

Common in forests. Fl. & Fr.: Sept.–Jan. Eng.: White kikar. *Vern.:* Thella tumma. The tree yields a gum which is used in indigenous medicine; it is demulcent and used in emulsifying agent. Paste of fresh leaves is applied on local swellings. Young fruits are used as vegetables. Bark yields a fibre, used for preparing fishing nets and cordage; bark mainly used in distillиries for clearing and flavouring liquors. Loyapalle, *VRK* 3972; Karempudi, *CAB* 4753 (MH).

Acacia nilotica (L.) Willd. ex Del. ssp. ***indica*** (Benth.) Brenan in Kew Bull. 12: 84. 1957. *Mimosa nilotica* L. Sp. Pl. 521. 1753. *Acacia arabica* (Lam.) Willd. var. *indica* Benth. in London J. Bot. 1: 500. 1842. *A. arabica* auct. non (Lam.) Willd. 1806; Gamble 425 (301). *Mimosa arabica* Lam. Encycl. 1: 19. 1783.

Armed trees, up to 6 m tall; branchlets grey-pubescent; leaves 2-pinnate, 4–5 cm long, pinnae 4–5 pairs, 1–3 cm long, leaflets oblong, 3–4 × 0.8–1 mm, apex rotund, entire, obtuse; flowers yellow, fragrant in axillary globose heads; pods moniliform, stipitate, apex horned, seeds 10–15, dark brown, compressed, shining.

Occasional in forests. Fl. & Fr.: May–Nov. Eng.: Black Babul. *Vern.:* Nalla tumma. It is a strong light demander and moderately drought resistant species. The pods make an excellent cattle fodder. Gum exudes largely from wounds in the bark mostly from Mar.–May. The bark and pods are used for tanning. A decoction of the bark used as substitute for soap. Bark in the form of powder is used along with pepper and salt in the treatment of tooth trouble. Koppanur, *VRK* 3270; Javuku, *VRK* 3862; Krishna river, *K.M. Sabastine* 9856 (MH).

Acacia roxburghii Wight & Arn. Prodr. 276. 1834; Gamble 427 (302). *A. planifrons* Wight & Arn. Prodr. 276. 1834; FBI 2: 293 *p.p.*

Small, thorny trees, up to 3 m tall; crown flat, umbrella like branches spreading; spines paired, unequal; 3–5 in a cluster, leaves 2-pinnate, pinnae 4–6 pairs, leaflets linear-oblong, 2–3 × 0.5–1 mm, pubescent, acute, entire, base sub-cordate, flowers yellow, scented; heads in axillary fascicles; pods linear, curved, to 8 × 1.5 cm, slightly turgid, glabrous; seeds 5–8.

Rare in forests. Fl. & Fr.: Nov.–Feb. Karempudi, *CAB* 4753 (MH).

Acacia torta (Roxb.) Craib. in Bull. Misc. Infor. 1915: 410. 1915; Gamble 428 (304). *Mimosa torta* Roxb. Fl. Ind. 2: 566. 1832. *Acacia caesia* auct. non Willd.; Wight & Arn. Prodr. Fl. Ind. Orient. 278. 1834.

Stragglers, up to 5 m tall; branchlets pubescent; internodal thorns straight; leaves alternate; pinnae 10–14 pairs; leaflets 25–40 pairs, elliptic, 4–8 × 1–2 mm, glabrous, base truncate, entire, subacute; flowers yellow, in terminal heads; pods flat; seeds up to 6.

Occasional in forests. Fl. & Fr.: Apr.–Aug. *Vern.:* Konda korinta. Kondavedu, *SSR* & KSM 21470, Piduguralla, *JSG* 18653 (MH).

Acacia mangium Willd. is widely cultivated as social forestry species in the district.

Albizia Durazzini

1. Pinnae 6–9 pairs, leaflets more than 20 pairs .... ***A. amara***
1. Pinnae 2–4 pairs, leaflets less than 20 pairs:
   2. Flowers sessile, pod brownish .... ***A. odoratissima***
   2. Flowers pedicelled, pod yellow .... ***A. lebbeck***

Albizia amara (Roxb.) Boivin in Encycl.19(2): 1838; FBI 2: 301; Gamble 432 (306). *Mimosa amara* Roxb. Pl. Cor. t. 122. 1799.

Deciduous unarmed trees, up to 6 m tall; crown spreading, branchlets yellowish or grey-pubescent; leaves 2-pinnate, pinnae 6–9 pairs, 2–3 cm long, glands in between 4 uppermost pinnae, leaflets 20–25 pairs, linear-elliptic, 4–5 × 0.5–1 mm; acute, entire, unequal, flowers creamish, heads in axillary racemes; pods flat, 6–15 × 1.5–2.5 cm, horned, indehiscent, margins undulated, seeds more than 10, ovoid to suborbicular.

Dominant in Macherla RF. It is used for shelter belts and wind breaks for afforestation. Fl. & Fr.: May–Nov. *Vern.:* Narlinga, konda chigara. The leaves provide green manure used as an emergency cattle food. The flowers are externally applied for skin diseases. Wood is used as an excellent fuel wood. Loyapalle, *VRK* 1818.

Albizia lebbeck (L.) Willd. Sp. Pl. 4: 1066. 1806; FBI 2: 298; Gamble 432 (306). *Mimosa lebbeck* L. Sp. Pl. 516. 1753.

Large deciduous unarmed trees, up to 15 m tall; leaves 2-pinnate, 15–20 cm long, pinnae 2–3 pairs, leaflets 4–8 pairs, elliptic-oblong, terminal pair of leaflets obovate, 3–5 × 1–2 cm, glabrescent, rounded or obtuse, entire, base obliquely obtuse, flowers pale yellow, scented, heads in axillary umbellate clusters; pods 10–20 × 3–5 cm, yellow with pale brown patches when old, wavy margin, flat, dry, attenuate at base, seeds 5–15, glabrous, shiny, pale brown.

Grown as avenue trees and occasional in forests. Fl. & Fr.: Apr.–Dec. Eng.: Lebbeck tree. *Vern.:* Dirisana, Sirisha. Leaves as green manure and wood used for agricultural purposes. The flowers on steam distillation give a colourless sweet smelling oil. Javuku, *VRK* 3950.

Albizia odoratissima (L.f.) Benth. in London J. Bot. 3: 88. 1844; FBI 2: 299; Gamble 431 (306). *Mimosa odoratissima* L.f. Suppl. Pl. 437. 1781.

Unarmed trees with extensive crown, up to 12 m tall; foliage drooping; leaves 2-pinnate, pinnae 3 pairs; leaflets 3–15 pairs, elliptic-oblong, 0.5–2.5 × 0.5–1 cm, glabrescent, obtuse, entire, obliquely rounded, usually drying grey, flowers creamish, fragrant, heads in umbellate corymbs; pods flat, oblong, compressed, 15–30 cm long, glabrous, finely reticulate veined, apex rounded, shortly stipitate.

Common in forests and rare in plains. Fl. & Fr.: Aug.–Jan. *Vern.:* Konda chigara. The twigs are lopped for fodder. The leaves boiled in ghee are used as cough remedy. The timber is widely employed for many purposes, particularly for construction of houses, agricultural implements, carts etc. The plant yields gum and dye (from bark). Vinukonda RF, *VRK* 6851.

Dichrostachys (A.p. deCandolle) Wight et Arnott *nom. cons.*

Dichrostachys cinerea (L.) Wight & Arn. Prodr. 271. 1834; FBI 2: 228; Gamble 419 (297). ssp. **cinerea** var. **cinerea**; Brenan & Brummitt in Bot. Soc. Brot. (Ser. 2) 39: 110. 1965. *Mimosa cinerea* L. Sp. Pl. 520. 1753.

Large, armed shrubs, up to 6 m tall; branchlets pubescent, spine tipped; leaves alternate or 2 to 3 in a cluster, 2-pinnate, pinnae 7–9 pairs, 1–1.5 cm long, leaflets 10–15 pairs, linear-oblong, 2–2.5 × 0.5–8 mm, pubescent; flowers; upper yellow, bisexual, fertile; lower pink, male and sterile, in axillary spikes; pods linear, flat, up to 6 × 0.6 cm, irregularly twisted, seeds 3–6, ovoid.

Common in forests. Fl. & Fr.: Apr.–Nov. *Vern.*: Veluthuru Chettu, Nela Jammi. Wood hard and tough, used as a good fuel. N. Varam, *VRK* 3955; Loyapalle, *VRK* 3987; *VRK* 3858; Nagarjunasagar, *KMS* 9763 (MH); Nagarjunakonda, *K. Thothathri* 9773 (CAL).

Entada Adanson *nom. cons.*

Entada pursaetha DC. Prodr. 2: 425. 1825. *Mimosa entada* L. Sp. Pl. 518. 1753. *Entada scandens* auct. non Benth. 1841; FBI 2: 86; Gamble 417 (296).

Giant, woody, unarmed climbers; leaves even pinnate; pinnae 2-pairs, leaflets 4-pairs, ovate-obovate, 2–5 × 1.5–2.5 cm, glabrous; flowers creamish, scented, in axillary spikes; pods elongate, flat, up to 100 × 12 cm, woody, with 1-seeded segments.

Occasional in forests. Fl. & Fr.: Feb.–Dec. *Vern.*: Gilla teega. Reported in Guntur Forest Division Work Plan.

Leucaena Bentham

Leucaena latisiliqua (L.) Gillis in Taxon 23: 190. 1974. *Mimosa latisiliqua* L. Sp. Pl. 519. 1753. *Leucaena leucocephala* (Lam.) de Wit in Taxon 10: 53. 1961. *Mimosa leucocephala* Lam. Encycl. 1: 12. 1783. *Leucaena glauca* auct. non Benth. 1842; FBI 2: 290; Gamble 419 (297).

Unarmed deciduous trees, up to 5 m tall; leaves 2-pinnate, pinnae 4–8 pairs, leaflets 15–20 pairs, linear-oblong, up to 1 × 0.2 cm, glabrous above, sparsely or finely pubescent beneath, acute, entire, base oblique; flowers whitish in dense globular heads; pods flat, thin, 8–15 × 1–1.5 cm, glabrous; seeds 10–15, polished, brown.

The subabul. Cultivated, introduced as a Social Forestry species in forest, wastelands, roadsides and fields. Fl. & Fr.: Oct.–Jan. It is recommended for plantation in a variety of soils although its best growth is obtained on deep and fertile soils. It is drought resistent but cannot tolerate prolonged drought. It is good coppicer. Eng.: White popinae. *Vern.:* Subabul. Wood is used as a firewood and leaves and fruits as fodder. Nallapadu, *VRK* 6761.

High density energy plantations in 15 hactares in NU campus planted under DNES funded work.

Mimosa Linnaeus

1. Spreading herbs, pinnae 2–3 pairs, sensitive .... ***M. pudica***
1. Straggling shrubs, pinnae 4–7 pairs, non sensitive:
   2. Pods prickly, sub sessile .... ***M. hamata***
   2. Pods glabrous, stipitate .... ***M. intsia***

Mimosa hamata Willd. Sp. Pl. 4: 1033. 1806; FBI 2: 291; Gamble 421 (298).

Armed scandant shrubs, up to 3 m tall; branchlets pubescent, with numerous slightly curved or straight prickles; leaves 2-pinnate, pinnae 4–7 pairs, leaflets 6–10 pairs, elliptic-oblong, 2–4 × 1–2 mm, pubescent, acute, mucronate, entire, oblique, flowers pink in axillary globose heads, heads solitary or in short terminal racemes, pods flat, curved, 4–8-jointed, prickly along both sutures; seeds brown, 4–8, ellipsoidal.

Occasional in wastelands and rare in scrub jungles. *Vern.:* Undrug Fl. & Fr.: Aug.–Jan. Vinukonda RF, *VRK* 6843.

Mimosa intsia L. Sp. Pl. 522. 1753. *Mimosa rubicaulis* Lam. Encycl. 1: 20. 1784; FBI 2: 291; Gamble 421 (299).

Armed straggling trees, up to 3 m tall, prickles short, hooked; leaves 2-pinnate, 9–13 cm long, pinnae 4–7 pairs, rachis without prickles, leaflets 10–15 pairs, elliptic-oblong, 0.5–1.5 × 0.2–0.4 cm, pubescent, obtuse, entire, base truncate; flowers pink, heads in a cluster on racemes; pods curved with prickles along margins, stipitate jointed, thin, strongly nerved, apices obtuse, horned; seeds 5–8, flat, beaked, dark brown.

Common in the scrub jungles. Fl. & Fr.: July–Dec. *Vern.:* Kandra tige, Kotha korinta. N. Varam, *VRK* 3235.

Mimosa pudica L. Sp. Pl. 518. 1753;FBI 2: 291; Gamble 421 (798).

Prickly prostrate herbs to undershrubs; leaves very sensitive, 4 cm long, pinnae 2–3 pairs, 4–6 cm, leaflets 15–20 pairs, elliptic-oblong, 4–5 × 1–1.5 mm, overlapping, glabrous above, clothed with appressed bristles beneath, acute, margin ciliate, base truncate-obtuse, stipules lanceolate; flowers pink, heads 2 or 3 in axillary clusters, pods flat, 1.5 × 0.5 cm, bristly along margins, slightly recurved, horned; seeds 3–5, compressed, ovoid.

Common weed on roadsides and in cultivated fields. Fl. & Fr.: Oct.–Feb. Eng.: Touch me not plant. *Vern.:* Athipathi. Sensitive to touch, the leaves folding up on contact. Nizampatnam, *VRK* 6790.

Neptunia Loureiro

Neptunia triquetra (Willd.) Benth. in J. Bot. (Hooker) 4: 355. 1842; FBI 2: 286; Gamble 416 (295). *Desmanthus triquetrus* Willd. Sp. Pl. 4: 1045. 1806.

Perennial, prostrate, glabrous herbs, woody at base; stem slender, compressed, angled; leaves 2–6 cm long, abruptly 2-pinnate; pinnae 1–3 pairs, 5 cm long, leaflets 10–20 pairs; sessile, elliptic-oblong, 3–5 × 1–2 mm, glabrous, ciliate, rounded, base oblique; flowers yellow in solitary, axillary small heads; pods 1–2 cm, flat, stipitate, oblong, compressed.

Occasional among grasses. Fl. & Fr.: May–Nov. It is known for headache. Phirangipuram, *CAB* 4610 (MH); Guntur, *G.V. Narayana* 16808 (MH).

Pithecellobium Martius *nom. cons.*

Pithecellobium dulce (Roxb.) Benth. in London J. Bot. 3: 199. 1844; FBI 2: 302; Gamble 434 (308). *Mimosa dulcis* Roxb. Pl. Cor. t. 99. 1798 & Fl. Ind. Orient. 2: 556. 1832.

Armed, large trees, up to 15 m tall; leaves 2-pinnate, pinnae 1 pair, leaflets oblong-oblanceolate, 2–4 × 0.5–1 cm, glabrous, base and apex obtuse, apiculate, entire; stipular spines erect, 1 cm long; flowers creamish; globose heads; in terminal or axillary panicles, pods spirally twisted, reddish-brown, 10–15 × 0.5–1 cm, circinate, torulose, dehiscent; seeds black, orbicular, 5–7, covered by reddish-white aril.

Cultivated in gardens as hedge and along the roadsides. It grows on a variety of soils. It can withstand any amount of prunning, lopping or browsing by animals. Native of tropical America. Fl. & Fr.: Oct.–May. *Vern.*: Seemachinta. Eng.: Manila tamarind. Wood strong and useful to general construction as packing cases, cart building, and agricultural implements; it is an excellent fuel wood. Pods (aril part) are edible. The bark contains yellow dye. V.P. South, *VRK* 3939.

Prosopis Linnaeus

1. Pods cylindrical, prickles small, leaflets 7–10 pairs .... *P. cineraria*
1. Pods compressed, thorns large, leaflets15–22 pairs .... *P. chilensis*

Prosopis chilensis (Molina) Stuntz in U.S.D.A. Bur. Pl. Industr. Invent. seeds 31: 85. 1914. *Ceratonia chilensis* Molina Sagg. Chil. 172. 1782. *Prosopis juliflora* (Sw.) DC. Prodr. 2: 447. 1825. *Mimosa juliflora* Sw. Prodr. 85. 1788.

Moderate sized drought resistant trees, with drooping branches and spreading crown, up to 7 m tall; leaves alternate or clustered, 2-pinnate, pinnae generally 1–2 pairs, 6 cm long, leaflets 15–20 pairs, obliquely oblong, 0.8–1.3 × 0.1–0.2 cm, glabrous, acute, entire, obtuse, spines straight, 1.5 cm long; flowers creamish or yellow in axillary hanging spikes; pods drupaceous, yellow, 15–20 × 0.8–1.2 cm, slightly curved, compressed, indehiscent; seeds brown, ovoid, 12–34, yellowish-brown, shiny.

Grows wild along rivers, streams and rare in forests. Native of Mexico. Introduced in India. It is suitable for afforestation for arid and semi-arid lands. Fl. & Fr.: Dec.–May. Eng.: Mosquito been. *Vern.:* Sarkaruthumma, Mulla tumma. It is immediate source of fuel for local people and for agriculural implements. It yields smooth, light yellowish brown gum which can be used as emulsifying agent.N. Varam., *VRK* 3595.

Prosopis cineraria (L.) Druce in Bot. Exch. Club. Soc. Brit. Isles 3: 422. 1914. *Mimosa cineraria* L. Sp. Pl. 517. 1753 ('*cinerea*'). *Prosopis spicigera* L. Mant. Pl. 68. 1767; FBI 2: 288; Gamble 421 (297).

Armed, evergreen trees, up to 10 m tall; prickles short; leaves pinnate, pinnae 2-pairs, leaflets 8–10 pairs, oblong, 0.5–1.2 × 0.1–0.3 cm, appressed hairy, rounded, entire, base oblique; flowers yellow in slender spikes; pods flat, thick; cylindric, straight or curved, filled with a dry sweetish pulp; seeds 10–15, oblong.

The Samee tree of Mahabharata. The plant worshipped by Hindus in Dasara festival. Rare in open forest. Fl. & Fr.: Apr.–Sept. *Vern.:* Jammi. The pods form a valuable food for cattle and goats. Wood hard, suitable for interior construction work such as windows, doors etc. and used as good fuel. The bark (which has sweetish taste) as well as the galls formed on the leaves are used for tanning. The tree exudes gum from the cut ends of branches. Koppanur, *VRK* 1292.

Samanea (A.P. de Candolle) Merril

Samanea saman (Jacq). Merr. in J. Wash. Acad. Sci. 6: 47. 1916. *Mimosa saman* Jacq. Fragm. 15. t. 9. 1800. *Pithecellobium saman* (Jacq.) Benth. in Hook. Lond. J. Bot. 3: 216. 1844. *Enterolobium saman* (Jacq.) Prain in J. Asiat. Soc. Bengal, Pt. 2. Nat. Hist. 66: 252. 1897; Gamble 435 (308).

Densely foliaceous, unarmed trees, up to 15 m tall; branchlets yellowish-tomentose; leaves alternate, 2-pinnate; pinnae 4 or 5 pairs; basal pair oblong-elliptic; upper ones obovate; 1–3.5 × 0.7–2 cm, glabrous above, pubescent below, entire, obtuse with minute apicule, base truncate; flowers orange-yellow, in axillary corymbose racemes; pods linear-oblong; slightly twisted; woody, septate between seeds; indehiscent, seeds 2-seriate, up to 20, brown, surrounded by brownish pulp.

The Rain Tree, the Umbrella Tree. Introduced, common avenue trees of plains. It can be grown on dry and barren soils, suitable for planting in saline and alkaline areas. It is a valuable Social forestry species. Fl. & Fr.: Mar.–Oct. The leaves and pods are esteemed as fodder for livestock. The pulp of the ripe pods is rich in fermentable sugars and may be employed for the production of alcohol. Wood is light and soft suitable for furniture or even as fuel. *Vern.*: Nidra ganneru. V.P. South, *VRK* 6998.

CRASSULACEAE

Kalanchoe Adanson

Kalanchoe pinnate (Lam.) Pers. Sys. Pl. 1: 446. 1805. *Cotyledon pinnata* Lam. Encycl. 2: 141. 1786. *Bryophyllum pinnatum* (Lam.) Oken. Allg. Naturgesch. 3: 1966. 1841. *B. pinnatum* (Lam.) Kurz in J. Asiat. Soc. Bengal Pt. 2, Nat. Hist. 40: 52. 1871; Gamble 451 (319). *B. calycinum*

Salisb. 2: 413. Parad. Lond. t. 3. 1805 & in Curtis. Bot. Mag. t. 1409. 1811; FBI 2: 413.

Semisucculent herbs, up to 1 m tall; leaves decussate, oblong or ovate-elliptic, 7.5–14 × 4–6 cm, thick coriaceous, glabrous, crenate, obtuse; flowers yellowish-green, in lax panicled cymes; follicles 1.5 cm; seeds linear, striate.

An introduced plant, now running wild. Cultivated as an ornamental and rare in forests. Reproduces vegetatively by producing bulbils in leaf crenatures. Fl. & Fr.: Jan.–Aug. *Vern.:* Asthibaksha, Simajamudu. Javuku, *VRK* 3902.

RHIZOPHORACEAE

1. Trees, up to 10 m tall, stilt roots prominent, calyx-lobes 4, hypocotyl >20 cm .... ***Rhizophora***
1. Shrubs (or small trees), up to 4 m tall, stilt roots absent, calyx lobes 5 or more:
   2. Leaves acute, dark-punctate below, bracteoles 0, stamens 18–20, ovary 4-celled ... . ***Bruguiera***
   2. Leaves obtuse, not punctate, bracteoles 4, stamens 10, ovary 3-celled .... ***Ceriops***

Bruguiera Lamarck

1. Flowers solitary, 3.5 cm across, sepals erect in fruit, petals reddish outside, cream inside .... ***B. gymnorrhiza***
1. Flowers in cymes, 1.5 cm across, sepals reflexed in fruit, petals greenish ... ***B. cylindrica***

Bruguiera cylindrica (L.) Blume, Enum. Pl. Javae 1: 93. 1827; quoad basionym; Gamble 459 (325). *Rhizophora cylindrica* L. Sp. Pl. 443. 1753.

Small trees, up to 5 m tall; leaves elliptic-oblanceolate, 5.5–10 x 2–3 cm, entire, acute or shortly acuminate, base cuneate; flowers greenish, in axillary, pedunculate cymes; hypocotyl 10 × 1 cm, straight, obscurely curved.

Common mangrove along with *Rhizophora* in mada forest. Fl. & Fr.: July–Oct. *Vern.: Vurada.* Nizampatnam port, *VRK* 6916; Nizampatnam, *DLR* 132 (NUH).

Bruguiera gymnorrhiza (L.) Savigny in Lam. Encycl. 4; 696. 1798; FBI 2: 437. *Rhizophora gymnorhiza* L. Sp. Pl. 443. 1753. *Bruguiera conjugata* Merr. in Philipp. J. Sci. 9: 118. 1914; Gamble 458 (324). *B. rheedii* Blume Enum. Pl. Javae 1: 92. 1827.

Evergreen mangrove tree, up to 5 m tall, with pneumatophores, leaves lanceolate or elliptic, 7.5–15 × 3–6 cm, glabrous, base cuneate, rarely obtuse, entire, acute; flowers orange-red, solitary, drooping; fruit oblong-obovoid; hypocotyl terete, elongate, slightly ribbed throughout on drying.

Common mangrove tree along the sea coast. Fl. & Fr.: Oct.–May. *Vern.:* Thuddu panna. Wood extremely hard, used for building and fuel. The radi-

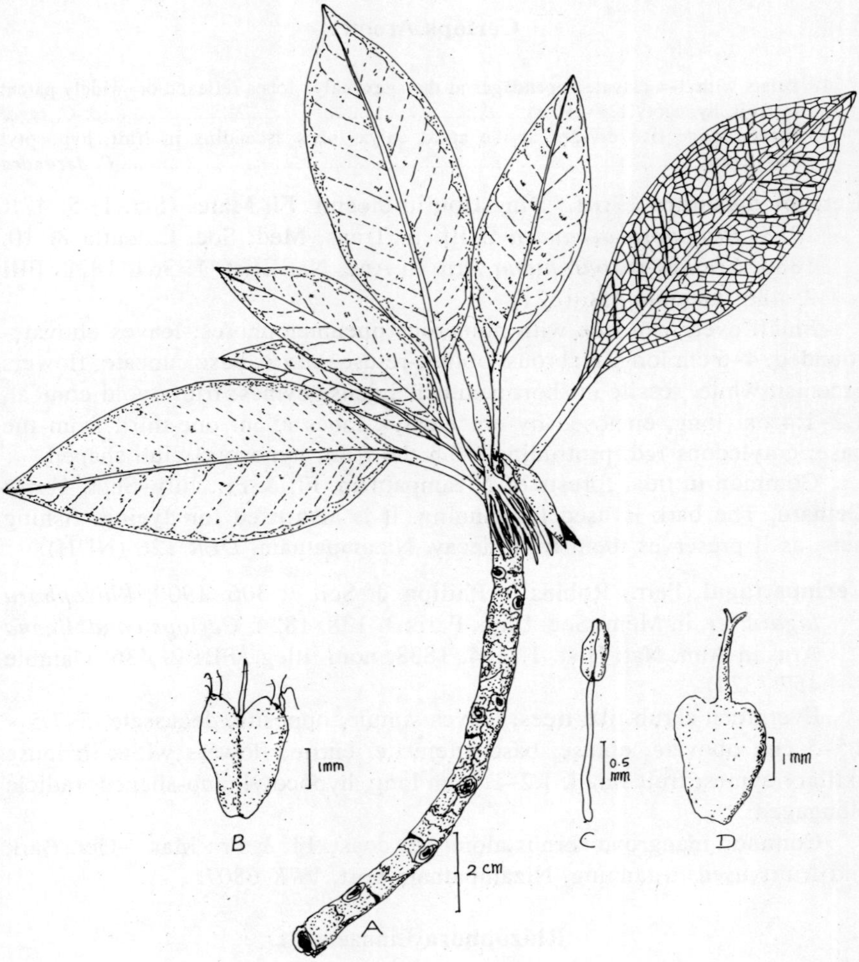

Fig. 10. *Bruguiera gymnorhiza* (L.)
A. Twig, B. Calyx, C. Stamen, D. Pistil.

cles form an article of diet for the local people. Kothapalem, *VRK* 5764; Nizampatnam, *DLR* 134 (NUH).

Ceriops Arnott

1. Petals with 3–4 clavate appendages at the apex, calyx lobes reflexed or widely patent in fruit, hypocotyl 20–25 cm .... ***C. tagal***
1. Petals fringe like divided at the apex, calyx lobes ascending in fruit, hypocotyl 10–12 cm .... ***C. decandra***

Ceriops decandra (Griff.) Ding Hou in Steenis. Fl. Males (Ser. 1) 5: 471. 1958. *Bruguiera decandra* Griff. in Trans. Med. Soc. Calcutta 8: 10. 1836. *Ceriops roxburghiana* Arn. in Ann. Nat. Hist. 1: 364. 1838; FBI 2: 436; Gamble 456 (321).

Small evergreen tree with numerous pneumatophores; leaves obovate-rounded, 4–6 cm long, glabrous, obtuse-retuse, entire, base cuneate; flowers greenish-white, sessile in short peduncled dense cymes; fruit ovoid-conical, 1.2–1.4 cm long, enclosed by erect calyx lobes about one-third from the base; cotyledons red, protruding from the fruit, hypocotyl club shaped.

Common in tidal forests of Nizampatnam. Fl. & Fr.: July–Sept. *Vern.:* Getharu. The bark is used for tanning. It is also used for dyeing, fishing nets, as it preserves them from decay. Nizampatnam, *DLR* 126 (NUH).

Ceriops tagal (Perr.) Robins in Philipp. J. Sci. 3: 306. 1908. *Rhizophora tagal* Perr. in Mem. Soc. Linn. Paris 3: 138. 1824. *Ceriops candolleana* Arn. in Ann. Nat. Hist. 1: 364. 1838; nom. illeg., FBI 2. 436; Gamble 457 (324).

Evergreen shrub like trees; leaves simple, opposite, decussate, 5–7.5 × 2.5–5 cm, obovate, obtuse, base attenuate, entire; flowers white in loose axillary cymes; fruit small, 1.2–1.4 cm long, hypocotyl club-shaped; radicle elongaged.

Common mangrove shrubs along sea coast. Fl. & Fr.: Mar.—Oct. Bark and fruits used in tanning. Nizampatnam port, *VRK* 6807.

Rhizophora Linnaeus

1. Leaves elliptic-lanceolate, flowers sessile in 2-flowered cymes, fruit 2.5 cm long, calyx lobes erect .... ***R. apiculata***
1. Leaves oblong-elliptic, flowers pedicellate in 3-flowered cymes, fruit 3–3.5 cm long, calyx lobes reflexed .... ***R. mucronata***

Rhizophora apiculata Blume Enum Pl. Javae 1: 91. 1827; Ding Hou in Steenis Fl. Males 1. 5: 452. 1958. *R. candelaria* DC. Prodr. 3: 32. 1829; Gamble 456 (323). *R. conjugata* auct. non L: FBI 2: 436.

Small evergreen trees, up to 5 m tall; stilt roots branched; leaves elliptic-oblong, 5–19 × 3–5 cm, dark green above, paler beneath, base cuneate,

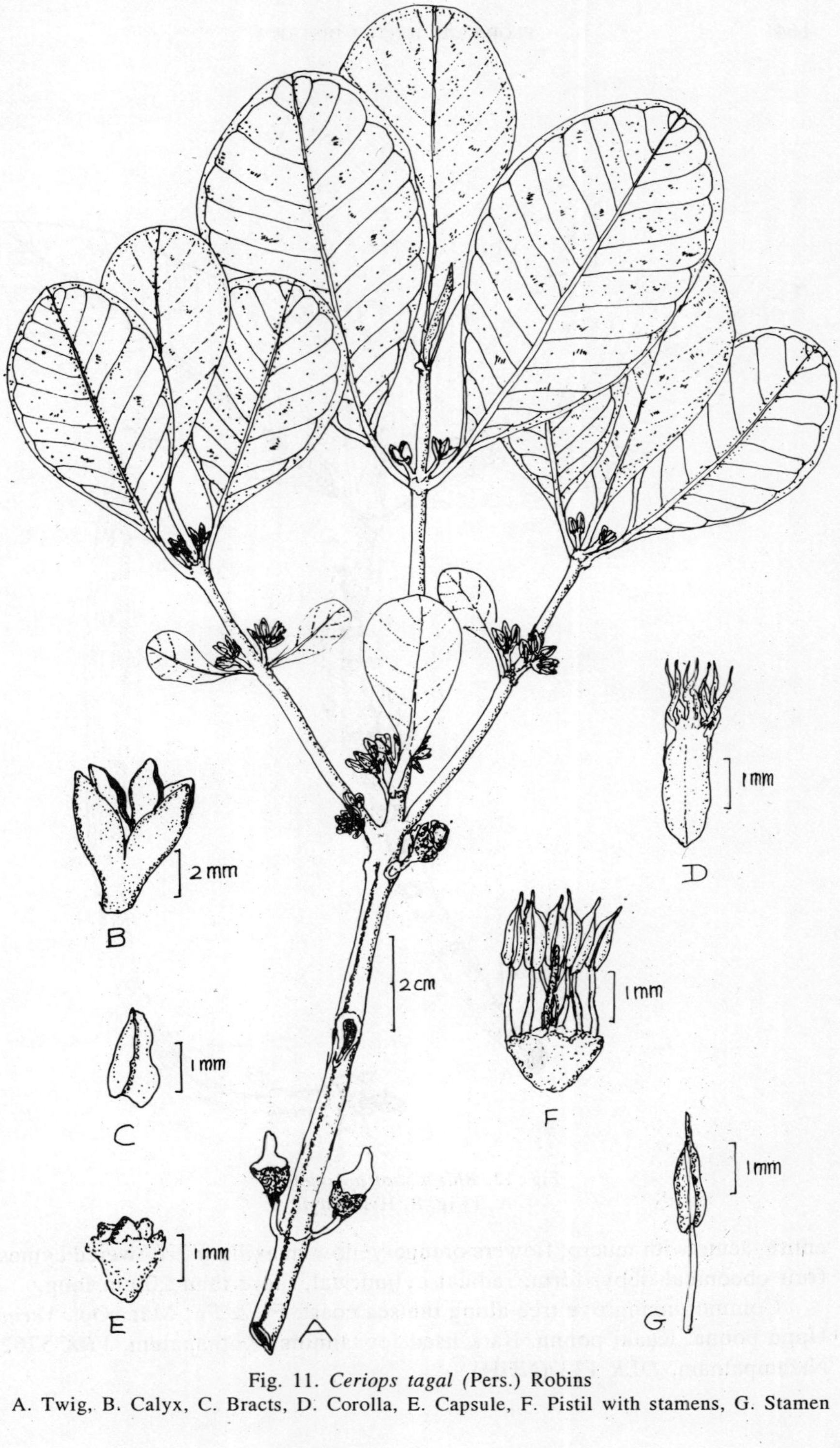

Fig. 11. *Ceriops tagal (*Pers.) Robins
A. Twig, B. Calyx, C. Bracts, D. Corolla, E. Capsule, F. Pistil with stamens, G. Stamen

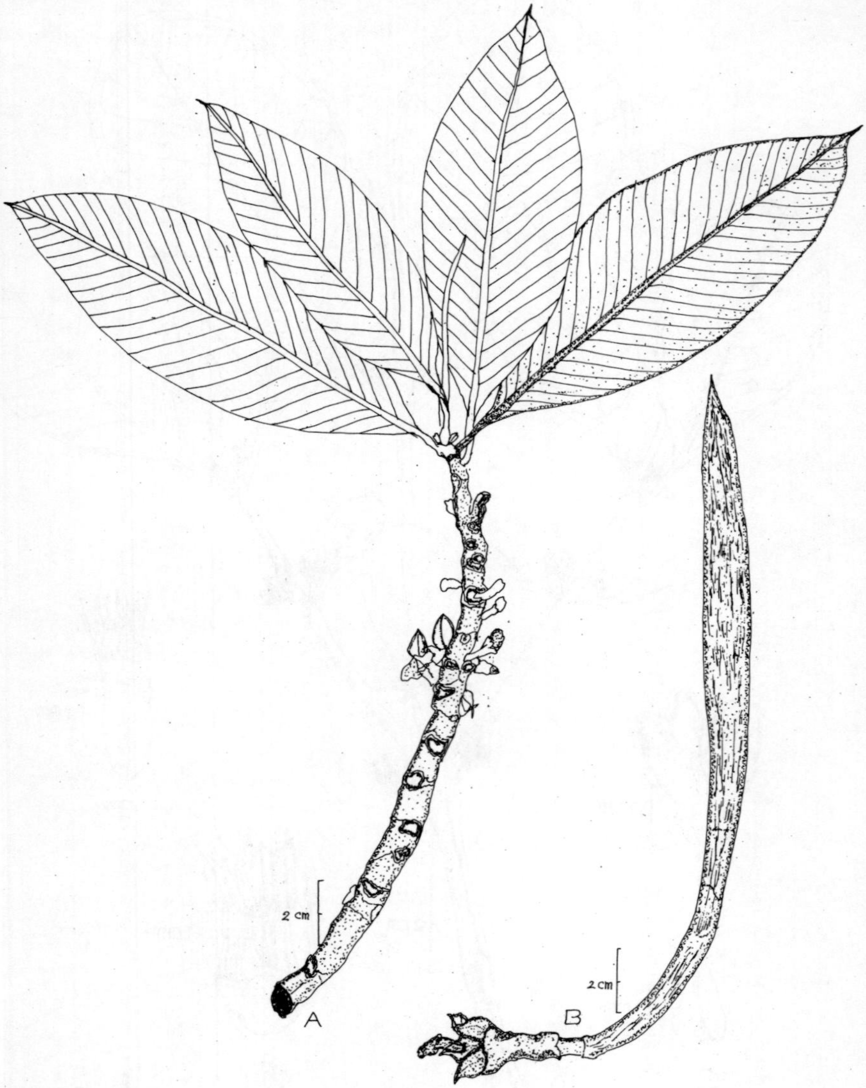

Fig. 12. *Rhizophora apiculata L.*
A. Twig, B. Hypocotyl

entire, acute with mucro; flowers orange-yellow in axillary 2-flowered cymes; fruit obconical-obpyriform, radical cylindrical, more than 20 cm long.

Common mangrove tree along the sea coast. Fl. & Fr.: Mar.–Oct. *Vern.:* Uppu ponna, Kaaki ponna. Bark used for tanning. Kothapalem, *VRK* 5762; Nizampatnam, *DLR* 133 (NUH).

Rhizophora mucronata Poir. in Lam. Tabl. Encycl. (text) 2: 517. 1794. t. 396. f. 2. 1797 & Encycl. 6; 189. 1804; FBI 2: 435; Gamble 456 (323).

Medium sized evergreen trees, up to 5 m tall, often buttressed with aerial stilt roots and pneumatophores, leaves oblong-elliptic, 7–15 × 4–7.5 cm, glabrous, brown or black dotted beneath and also pale punctualate, entire, mucronate, base cuneate; flowers crimson-red in axillary branched cymes; fruit ovoid, brown, crowned with persistent, reflexed sepals; hypocotyl cylindric, ca. 30 cm long, rugose.

Occasional along the back water mangroves of coast. Fl. & Fr.: June–Oct. Eng.: True mangrove, salt ponna, Demon candel. *Vern.:* Uppu ponna, Adavi ponna. Bark is powerful astringent and also gives a valuable tanning material. Wood very hard and heavy used for temporary bridges and heavy construction, also an excellent fuel. Fruit is sweet and edible, the juice is made into a light wine. Reported in Guntur Forest Division Work Plan.

Note: For impact of harbour construction on the denudation of mangrove flora at Nizampatnam, Guntur Dt., Please refer M.B.P. Rao *et al.* 1987. 'Ecological changes in tropical mangrove ecosystem due to human impact—Tropical Ecology 28: 232–238) and M.B.P. Rao and P.N. Rao in Water Sci. Tech. Rotterdam, 1984.

COMBRETACEAE

1. Trees or shrubs:
   2. Mangrove shrubs, petals 5. .... ***Lumnitzera***
   2. Trees, nonmangroves, petals 0:
      3. Flowers bisexual, in globose heads, leaves eglandular, fruits small, 2 winged .... ***Anogeissus***
      3. Flowers polygamous, in simple or branched spikes, leaves glandular, fruit entire-ridged or 3–5-winged .... ***Terminalia***
1. Stragglers, lianas or climbers:
   4. Petals 0, calyx accrescent, fruits ribbed .... ***Calycopteris***
   4. Petals 4 or 5, calyx not accrescent, fruits winged .... ***Combretum***

Anogeissus Wallich ex Guillemin & Perrottet

Anogeissus latifolia (Roxb. ex DC.) Wall. ex Guill. & Perr. Fl. Seneg. Tent. 7: 820. 1832; FBI 2: 450; Gamble 466 (330); Scot in Kew Bull. 33: 560. 1979. *Conocarpus latifolia* Roxb. ex DC. Prodr. 3: 16. 1828.

Deciduous trees, up to 10 m tall; bark grey, smooth, branchlets grey-pubescent; leaves elliptic or oblong-elliptic, 5–10 × 3–5.5 cm, glabrous, obtuse-rotund, entire, base cuneate or rounded, flowers pale green or yellow with red disc, axillary, aggregated in cymose heads; fruits compressed, 2-winged, crowned by persistent calyx tubes which are modified into stiff beak, seed solitary and minute.

Common in Vinukonda range. Fl. & Fr.: July–Dec. Eng.: Axle-wood tree. *Vern.:* Chiru manu, Velama. Wood very tough and strong, useful for various purposes. Timber makes an excellent charcoal and fuel. Leaves are rich in tannin and are collected for tanning and yields gum which is much used in calico-printing. It is an astringent, antidote against snake-bite and scorpion sting. Gum is edible. Vinukonda RF, *VRK* 6856; Jayantipuram, *VRK* 6960.

Calycopteris Lamark

Calycopteris floribunda Lam. Tabl. Encyl. 2: 485. t. 357. 1797; FBI 2: 449; Gamble 467 (331).

Densely tomentose woody stragglers, young branches slender, rusty puberulous; leaves elliptic-oblong, 5–10 × 2–3.5 cm, glabrous above, tomentose below, acuminate, entire, base obtuse; flowers sessile, yellowish-green in terminal and axillary panicles; fruits ovoid-ellipsoid, 5-ribbed, crowned by the persistent calyx-lobes.

Occasional on the outskirts of forests and along stream banks. Fl. & Fr.: Feb.–Aug. *Vern.:* Bontha. Ethipothala, *VRK* 3878.

Combretum Loefling *nom. cons.*

Combretum albidum G. Don in Trans. Linn. Soc. London. 15: 429. 1827. *C. ovalifolium* Roxb. Fl. Ind. 2: 226. 1832; FBI 2: 458; Gamble 469 (332).

Glabrescent woody stragglers; leaves decussate, elliptic or elliptic lanceolate, 8–10 × 2.5–3 cm, glabrous, obtusely acuminate, entire, base cuneate; flowers greenish yellow in axillary terminal panicles; fruits ovoid, up to 1.8 × 1 cm; 4 or 5-angular, 4–5-winged, greenish red, seed solitary.

Occasional along stream banks on trees. Fl. & Fr.: Sept.–Apr. *Vern.:* Shirtal baddu, Yada-tiga, Satrasala, *VRK* 6778; Vinukonda RF, *VRK* 6890; Jayanthipuram, *VRK* 6872.

Lumnitzera Willdenow

Lumnitzera racemosa Willd. Ges. Naturf. Freunde Berlin Neue Schriften 4: 187. 1803; FBI 2: 452; Gamble 468 (311).

Small mangrove trees, up to 5 m tall; leaves simple, sessile, fleshy, crowded at the ends of the branches, alternate, oblong or oblanceolate, obovate, 4.5–7.5 × 1.5–3 cm, thick-coriaceous, glabrous, base attenuate, entire, obtuse-emarginate; flowers white in axillary and terminal spikes; fruit ellipsoid, woody, longitudinally striate when dry, crowned by the persistent calyx-tube.

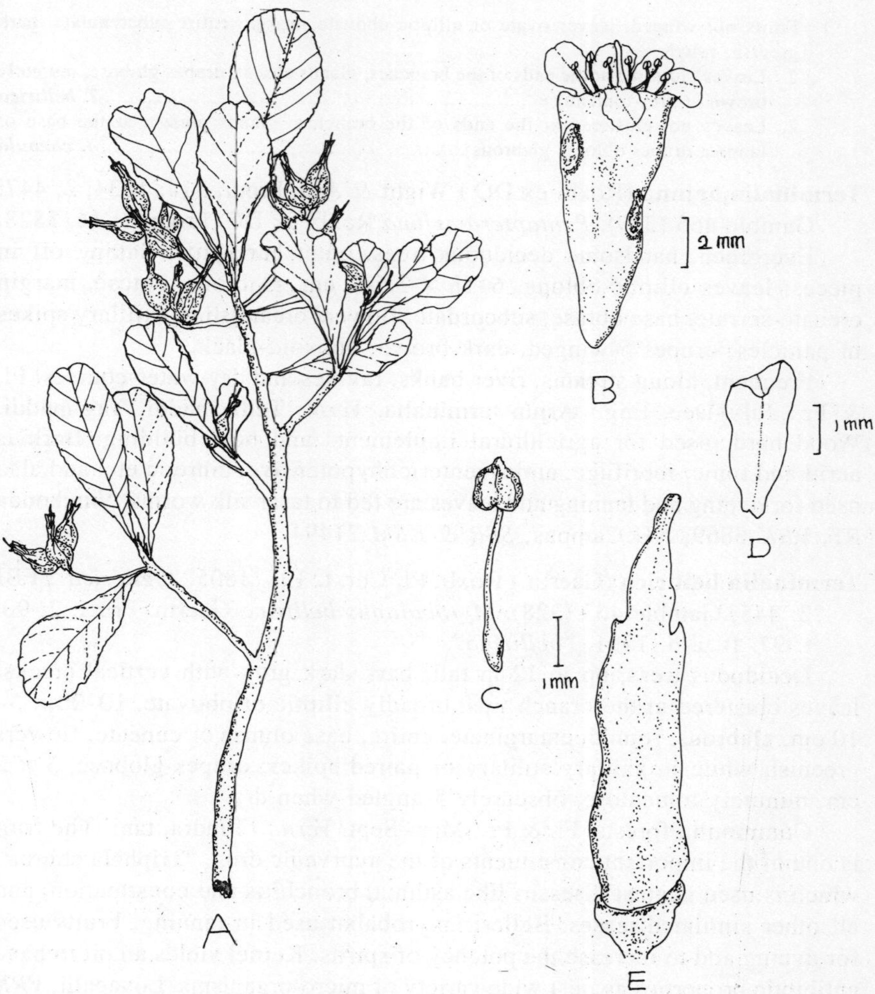

Fig. 13. *Lumnitzera racemosa Willd.*
A. Twig, B. Flower, C. Stamen, D. Petal, E. Pistil

Common in Kothapalem mangroves. Fl. & Fr.: Apr.–June. *Vern.*: Thanduga. The wood is strong and durable and is used for house posts and as fuel. Kothapalem, *VRK* 5767; Nizampatnam port, *VRK* 6810 & 6926. Nizampatnam, *DLR* 101 (NUH).

Terminalia Linnaeus

1. Fruits 5-winged, leaves oblong-oblanceolate, margin crenate-serrate, bark white and smooth .... *T. arjuna*

1. Fruits not winged, leaves ovate or elliptic-obovate, margin entire-subcrenulate, bark greyish, rough:
   2. Leaves clustered at the ends of the branches, glands absent, drupes globose, minutely brown or grey tomentose .... ***T. bellirica***
   2. Leaves not clustered at the ends of the branches, glands present at the base of lamina, drupes oblong, glabrous .... ***T. chebula***

Terminalia arjuna (Roxb. ex DC.) Wight & Arn. Prodr. 314: 1834; 2: 447; Gamble 465 (329). *Pentaptera arjuna* Roxb. ex DC. Prodr. 3: 14. 1828.

Evergreen, handsome deciduous trees, outer bark stiff, flaking off in pieces; leaves elliptic-oblong, 6–18 × 3.5–7 cm, glabrous, obtuse, margin crenate-serrate, base obtuse, subcordate, flowers creamish in axilllary spikes in panicles; drupes 5-winged, dark brown, obovoid-black.

Frequent, along streams, river banks, ravines and dry water-courses. Fl. & Fr.: July–Dec. Eng.: Arjun terminalia. *Vern.*: Tella maddi, Vag-maddi. Wood hard, used for agricultural implements and boat building. Bark is acrid and tonic, febrifuge, antidysenteric, hypotensive, diuretic etc. and also used for dyeing and tanning and leaves are fed to tasar silk worms. Vinukonda RF, *VRK* 6869; NU Campus, *SSR & KSM* 21494.

Terminalia bellirica (Gaertn.) Roxb. Pl. Cor. t. 198. 1805. "*bellerica*"; FBI 2: 445; Gamble 463 (328). *Myrobalanus bellirica* Gaertn. Fruct. 2: 90. t. 97. ff. a-d. 1791. "*bellirina*".

Deciduous trees, up to 12 m tall; bark dark grey with vertical cracks; leaves clustered at the branch tips, broadly elliptic or obovate, 10–20 × 5–10 cm, glabrous, rotund, emarginate, entire, base obtuse or cuneate; flowers greenish white in axillary solitary or paired spikes; drupes globose, 3 × 2 cm, minutely tomentose, obscurely 5-angled when dry.

Common in forests. Fl. & Fr.: Mar.–Sept. *Vern.:* 'Tandra, tani' The fruit is one of the important constituents of the ayurvedic drug, "Triphala churna" which is used against diseases like asthma, bronchitis and constipation, and all other similar diseases. Belleric myrobalan used in tanning. Fruits used for dyeing and to increase the potency of spirits. Kernel yields an oil to have antibiotic properties against wide variety of micro-organisms. Loyapalli, *VRK* 3978; Kakirala, *DMR* 18144.

Terminalia chebula Retz. Obs. Bot. 5: 31. 1789; FBI 2: 446. *p.p.;* Gamble 464 (328). *Myrobalanus chebula* (Retz.) Gaertn. Fruct. 2: 91. t. 97. 1791.

Deciduous trees, up to 10 m tall; branchlets rusty villous or glabrescent, leaves elliptic-obovate, 6–10 × 3–5 cm, densely woolly below, obtuse, entire; base rounded; glands 2 at the base of leaf; flowers creamish, in axillary spikes; drupes oblong, horned, 4 × 2.5 cm, glabrous, faintly angled.

Occasional in forest. Fl. & Fr.: July–Dec. *Vern.*: Karakkaya. Fruit an important source of tannin material and used externally for chronic ulcers, wounds etc., and in leather curing; dye is also extracted from fruits.

"Triphala," the traditional Indian medicine of Ayurvedic consisting of the fruits of *T. chebula, T. bellirica* and *Phyllanthus emblica* is widely used for the treatment of any sort of digestive or intestine disorders. Vulusulapenta, *DMR* 18195.

Terminalia catappa L. is planted for almond. *Quisqualis indica* L. (The Rangoon vine) is grown as garden favourite and is a bioherbicidal source to control *Cuscuta chinensis* Lamk.

MYRTACEAE

1. Fruits berries with 1-seed, mature leaves opposite .... *Syzygium*
1. Fruits capsular with many seeds, mature leaves alternate .... *Eucalyptus*

Eucalyptus L' Heritier

Eucalyptus camaldulensis Dehnh. Cat. Pl. Hort. Camald. ed. 2. 20. 1832.

Tall trees, up to 20 m tall, woody, bark papery; leaves linear-lanceolate, alternate, 10–25 × 1–2 cm, acuminate, glabrous; flowers yellowish-cream, in axillary clusters; capsule globose; seeds minute, many.

A Social forestry species for fuel, poles, minor building. One of the most abundantly planted tree by the forest department in wastelands and open forests. Fl. & Fr.: May–Oct. *Vern.*: Nilagiri chettu. Wood for pulp and agricultural and building purposes. Oil is used for headaches and body pains. N. Varam, *VRK* 1911.

Several hybrids are introduced by Forest Department for Social forestry and 'bioenergy'.

Syzygium J. Gaertner *nom. cons.*

Syzygium cumini (L.) Skeels in U.S.D.A. Bur. Pl. Industr. Bull. 248. 2. 1912. *Myrtus cumini* L. Sp. Pl. 471. 1753. *Eugenia jambolana* Lam. Encycl. 3: 198. 1789; FBI 2: 499. *Syzygium jambolanum* (Lam.) DC. Prodr. 3: 259. 1828; Gamble 481 (340).

Large evergreen trees, 12–15 m tall; bark smooth, grey; leaves opposite, elliptic or ovate-lanceolate, 9–15 × 4–5 cm, glabrous, acuminate, entire, base obtuse to narrowed; flowers creamish, sweet scented in axillary trichotomous cymes, berries ovoid or oblong; 1–2.5 × 1–1.5 cm, purplish-red and black when ripe, glabrous, seeds soilitary.

Occasional in river banks and forests both in the plains and hills, planted along the roads and in gardens. Fl. & Fr.: Apr.–July. Eng.: Java plum. *Vern.*: Neredu, Allaneredu. The ripe fruits are widely eaten and a preparation is given in stomachache, as a remedy for enlarged-spleen and in chronic diarrhoea. The bark and seed also used for diabetes. Bark is used in dyeing and tanning works and for colouring fishing nets; wood as timber, N. Varam, *VRK* 3989; Loyapalle, *VRK* 3838.

Callistemon citrinus (Curtis) Skeels (the bottle brush), is cultivated as an ornamental in gardens. *Psidium guajava* L. (The Guava) is cultivated for it edible fruits. *Eucalyptus* spp. are planted in gregarious patches under social forestry scheme.

LECYTHIDACEAE

Couroupita Aublet

Couroupita guianensis Aublet, Hist. Pl. Guiane 708. t. 282. 1775; Curtis, Bot. Mag. tt. 3158 & 3159. 1832.

Medium sized deciduous trees, up to 10 m tall; leaves simple, clustered at the ends of branchlets, alternate, 15–20 × 3–6 cm, obovate-oblong, base attenuate, entire, acute or obtuse, glabrous; flowers light yellow in racemes on trunk, staminal tube produced into a prominent hood; fruits pale reddish-brown, woody, pendulous, 15 cm dia, inner pulp ill-smelling.

The cannon ball tree. Introduced, cultivated in gardens and rare in forests. Fl. & Fr.: Through the year. *Vern.*: Nagalingam. V.P. South, *VRK* 6754.

BARRINGTONIACEAE

Barringtonia J.R. Forester et J.G.A. Forster

Barringtonia acutangula (L.) Gaertn. Fruct. Sem. Pl. 2: 97. t. 101. 1790; FBI 2: 508; Gamble 487 (344), ssp. **acutangula** Payens in Blumea 15: 228. 1967. *Eugenia acutangula* L. Sp. Pl. 471. 1753.

Trees, up to 10 m tall; trunk short; branchlets densely tomentose; leaves oblanceolate, 7–14 × 4–8 cm, glossy above, denticulate-crenulate obovate, 5–15 × 2–9 cm, obtuse-round, base attenuate, flowers scarlet, in axillary pendulous spicate racemes; berries ellipsoid to ovoid; seeds solitary, ovoid.

Occasional on the banks of rivers and tanks. Fl. & Fr.: June–Sept. Eng.: Indian-Oak. *Vern.*: Kurpa, Kanapa. It is an expectorant, laxative, emetic, and tonic and for diarrhoea and headache. Vinukonda RF, *KCJB* 17403 (MH).

MELASTOMATACEAE

1. Shrub or small tree, branchlets glabrous, leaves not 3-nerved, apex obtuse-retuse, flowers in umbels, fruits single-seeded .... ***Memecylon***
1. Herb, branchlets hirsute, leaves 3-nerved, apex acute, flowers in cymes, fruits many-seeded .... ***Osbeckia***

Memecylon Linnaeus

Memecylon umbellatum Burm. f. Fl. Ind. 87. 1768; Gamble 504 (300). *M. edule* Roxb. Pl. Cor. t. 82. 1795; FBI 2: 563; Gamble 504 (356).

Medium sized bush-like trees, up to 5 m tall; leaves oblong-lanceolate or elliptic, 4–6 × 1–2 cm, glabrous, acute, entire, base acute; flowers blue or bluish-pink in axillary umbels; berries globose, 5–6 × 3–4 cm, blue when ripe; seeds solitary.

Rare in forests. Fl. & Fr.: Apr.–Aug. *Vern.: Alli.* Fruits edible. Reported in Guntur Forest Division Work Plan.

Note: Vivekanandan (Flora of Tamilnadu 1: 160. 1983) treated *M. edule* and *M. umbellatum* as two distinct species.

Osbeckia Linnaeus

Osbeckia zeylanica L. f. Suppl. Pl. 215. 1781; FBI 2: 516; Gamble 494 (349).

Annual hirsute, erect herbs, up to 25 cm tall; stem sparsely appressed hairy, leaves elliptic-lanceolate, 1.5–3 × 0.4–0.8 cm, membranous, acute, base acute or rarely attenuate; flowers pinkish-purple, axillary, solitary; capsules oblong; seeds many.

Occasional in Vinukonda range. Fl. & Fr.: Sept.–Apr. Vinukonda RF, *VRK* 6930.

LYTHRACEAE

1. Erect herbs, flowers in cymes, fruit indehiscent .... *Ammannia*
1. Shrubs or trees; flowers in panicles, fruit dehiscent:
    2. Flowers zygomorphic, calyx-tube curved .... *Woodfordia*
    2. Flowers actinomorphic, calyx tube not curved.
        3. Stamens 8, flowers 4-merous, ovary 2–4-celled .... *Lawsonia*
        3. Stamens numerous, flowers 6-merous, ovary 3–6-celled .... *Lagerstroemia*

Ammannia Linnaeus

Ammannia baccifera L. Sp. Pl. 120. 1753. ssp. **aegyptica** (Willd.) Koehne Engl. Bot. Jahrb. Syst. 1: 260. 1880; FBI 2: 569; Gamble 510 (360). *A. salicifolia* sensu Clarke in Hook. f. Fl. Brit. India 2: 569. 1879. non Blume 1856.

Erect glabrous herbs, up to 50 cm tall; leaves opposite below, often alternate above, oblong-elliptic or oblanceolate, 4–5 × 0.5–0.7 cm, acute, entire, base cuneate; flowers white, in condensed axillary racemes or clusters; petals absent, capsules red when ripe, 1.5–2 mm, glabrous; seed subhemispheric, hollowed on the plane face.

Occasional in marshy places; often found in thick patches in drying up puddles and ponds and in paddy fields. Fl. & Fr.: Oct.–May. *Vern.*: Agnivendrapaku. Highly medicinal. Javuku, *VRK* 3879; Nagarjunakonda, *KTH* 9821 (CAL).

Lagerstroemia Linnaeus

Lagerstroemia indica L. Syst. Nat. ed. 10. 1076. 175; FBI 2: 575; Gamble 513 (362).

Large deciduous glabrous shrubs, up to 2 m tall; leaves elliptic-oblong, 3–5 × 2–3 cm; flowers pink in axillary racemes; capsules globoid, woody.

An ornamental plant grown in gardens for its beautiful flowers. Introduced from China. Fl. & Fr.: June–Oct. *Vern.*: Chinnagoranta. N. Varam, *VKR* 3969.

Lagerstroemia reginae Roxb. (=*L. reginae* Retz.). The pride of India, cultivated in gardens, compounds etc.

Lawsonia Linnaeus

Lawsonia inermis L. Sp. Pl. 349. 1753; Gamble 514 (363). *L. alba* Lam. Encycl. 3: 106. 1789; FBI 2: 573.

Erect shrubs, up to 4 m tall; branchlets end with spines; leaves obovate-elliptic, 1.2–2 × 0.5–0.7 cm, glabrous, acute or obtuse often mucronulate, entire, base attenuate; flowers yellow fragrant, in terminal panicles, capsules globose, up to 6 mm across, exceeding calyx tube, seeds trigonal.

The Henna Dye plant. Occasional shrub along streams. Commonly grown in private gardens. Fl. & Fr.: Sept.–May. *Vern.*: Gorintaku. Dye extracted from leaves used for nails and palms. Loyapalle, *VRK* 3992.

Woodfordia Salisbury

Woodfordia fruticosa (L.) Kurz in J. Asiat. Soc. Bengal 40: 56; 1871; Gamble 511 (361). *Lythrum fruiticosum* L. Sp. Pl. 641. 1762. *Woodfordia floribunda* Salisb. Parad. Lond. t. 42. 1806; FBI 2: 572.

Much-branched woody shrubs, up to 4 m tall, bark reddish-brown; leaves ovate-lanceolate, 6–10 × 2–2.5 cm, glabrescent, acuminate; flowers scarlet red, showy, in axillary fascicled cymes; capsules enclosed by persistent calyx lobes, irregularly dehiscent; seeds obovoid, brown, smooth.

Occasional in forests. Fl. & Fr.: Dec.–Apr. Eng.: Fire flame bush. *Vern.*: Jargi, Seringi. Flowers yield yellow or red dye, used for colouring silk and flowers are also used in medicine. Vinukonda RF, *VRK* 6883.

SONNERATIACEAE

Sonneratia Linnaeus f.

Sonneratia apetala Buch.—Ham. in M. Symes, Embassy Aya 3: 477. 1800; FBI 2: 579; Gamble 515 (363).

Evergreen mangrove trees, up to 6 m tall; pneumatophores numerous, conical; branchlets glabrous, slender, drooping; leaves decussate, oblong-obovate, 3.5–10 × 2–4 cm, coriaceous, glaucous, base attenuate, obtuse, entire; flowers cream coloured, solitary or 3–5 in terminal cluster; berries globose, 2 cm across, tipped with the style, supported by persistent calyx, seeds curved, angular, numerous.

Occasional mangrove tree in sea coast. Fl. & Fr.: Apr.–Sept. *Vern.*: Kalinga. Wood is used as fuel. Kothapalem, *VRK* 4769.

Punica granatum L. family Punicaceae, known as Pomegranate (Danimma), a native of Persia, cultivated for its edible fruits.

ONAGRACEAE

Ludwigia Linnaeus

Ludwigia perennis L. Sp. Pl. 119. 1753. *L. parviflora* Roxb. Fl. Ind. 1: 440. 1820; FBI 2: 588; Gamble 517 (365).

Annual erect, glabrescent herbs, up to 2 m tall; leaves narrowly elliptic-lanceolate, 2–5 × 0.5–0.7 cm, acuminate, entire, base attenuate; flowers yellow, axillary, solitary; capsules subterete, 1–1.5 × 0.3 cm, 4-ribbed; seeds many, multiseriate, ellipsoid-ovoid.

Occasional in wet lands. Fl. & Fr.: July–Dec. Ethipothala, *VRK* 5788; Krishna River, *VRK* 6736; Vinukonda RF, *VRK* 6859 ; NU campus, *SSR & KSM* 21495.

TURNERACEAE

Turnera Linnaeus

Turnera ulmifolia L. Sp. Pl. 271. 1753. L. var. **angustifolia** (Mill.) Willd. ex Urb. Monog. Turn. 141. 1883; Gamble 523 (369). *T. angustifolia* Mill. Gard. Dict. ed. B.n. 2. 1768.

Perennial herb, up to 1 m tall, branches densely pubescent; leaves lanceolate or ovate-lanceolate, 4–10 × 1–2 cm, more or less hairy on both sides, glandular beneath, acute to acuminate, margin coarsely dentate-serrate, base cuneate; petiole with 2, rather larger, apical lateral glands, flowers yellow in the leaf axils; capsule ovoid, 7.5 mm long, obtuse, hairy, 3-valved; seeds many, elliptic-oblong, brown, aril white.

Often found along road sides (Mohan Rao & Aasiya, 1995). Fl. & Fr.: Nearly throughout the year.

PASSIFLORACEAE

Passiflora Linnaeus

Passiflora foetida L. Sp. Pl. 499. 1753; Gamble 524 (370).

Hispid-hairy, tendrillar climbers; leaves 3-lobed, 3–7 × 2–5 cm, glandular-pubescent, acute, often serrate, base cordate; tendrils axillary; flowers white, axillary, solitary; berries sub-globose, 5 × 3 cm, subtended by persistent involucre; seeds ellipsoid.

Occasional on outskirts of forests on bushes and open hilly area. Tropical American species, introduced to India during 19th century (Srivastava, 1964). Fl. & Fr.: Feb.–Oct. Eng.: Stinking passion flower. *Vern.*: Juka Malli, Tella jumiki. Leaf extract used for itches and decoction for asthma. Javuku, *VRK* 3560.

CARICACEAE

Carica papaya L. Small tree. It is a native of West Indies, mostly cultivated for its edible fruits. Milky latex used for abortions by village people.

Trunk lofty, fluted, weak with crown of leaves; dioecious; seeded and seedless cultivars occur. Latex source of chewing gum and other purposes. Fruits contain 'papain' digestive enzyme; leaves in Africa for washing clothes; stem bark in ropes, strings, thread (Africa, W. Indies); fruits in Jams, Jellies, Candies, Toffees.

Nutritive value in 100 g of fruits

	Unripe	Ripe
Energy (Kilo cal.)	27	32
Calcium (mg)	28	17
Iron (mg)	0.9	0.5
Vit. A	0	666
Vit. C. (mg)	12	67

In one hactare 250–375 kg papain obtained.

CUCURBITACEAE

1. Tendrils simple:
   2. Bracts ciliate .... *Ctenolepis*
   2. Bracts glabrous:
      3. Male flowers solitary:
         4. Corolla lobes divided nearly to the base:
            5. Fruits oblong, calyx with often 2–3 scales at its base, male flowers with large enveloping bracts .... *Momordica*
            5. Fruits globose, calyx without scales, bracts absent .... *Citrullus*
         4. Corolla gamopetalous, bell-shaped .... *Coccinia*
      3. Male flowers in clusters or racemes:
         6. Seeds not turgid .... *Cucumis*
         6. Seeds turgid:
            7. Fruits flask-shaped, more than 2 cm across .... *Kedrostis*
            7 Fruits not flask-shaped, less than 2 cm across:
               8. Seeds pitted .... *Mukia*
               8. Seeds smooth:
                  9. Fruits globose .... *Melothria*
                  9. Fruits ellipsoid .... *Corallocarpus*
1. Tendrils branched:
   10. Leaves deeply lobed, flowers < 2.5 cm across, fruits globose .... *Diplocyclos*
   10. Leaves shallowly lobed or angular, flowers > 4 cm across, fruits oblong ...... *Luffa*

Citrullus H.A. Schrader *nom. cons.*

Citrullus colocynthis (L.) Schrad. in Linnaea 12: 414. 1838; FBI 2: 620; Gamble 536 (378); Chakrav. 20. 1982. *Cucumis colocynthis* L. Sp. Pl. 1011. 1753. *Colocynthis vulgaris* Schrad. Pl. Sem. Hort. Gotting. 2: 1883; Chakrav. 113. 1959.

Perennial trailing herbs; stem diffuse or creeping, slender, angled, hirsute; leaves narrowly triangular, deeply 3–5-lobed, each lobe deeply pinnatifid, rigid or scabrid, deltoid, tendril simple; flowers monoecieus, yellow or greenish-white, solitary, axillary; fruits globose, 3–4 cm long, striped with green and white bands when young, yellow when ripe, pulp spongy, very bitter, seeds ovoid, pale brown.

Rare in sandy wastelands and forests. Fl. & Fr.: Apr.–Nov. *Vern.*: Pitchi puccha. Fruits, seeds, roots used for jaundice and urinary diseases. N. Varam, *VRK* 3592; Nagarjunasagar, *KMS* 9715 (MH).

Coccinia R. Wight et Arnott

Coccinia grandis (L.) Voigt. Hort. Sub. Calc. 59. 1845. Chakrav. 24. 1982. *Bryonia grandis* L. Mant. Pl. 126. 1767. *Coccinia indica* Wight & Arn. Prodr. 347. 1834; Gamble 537 (379). *Cephalandra indica* (Wight & Arn.) Naud. in Ann. Sci. Nat. Bot. Ser. 5. 5: 16. 1866; FBI 2: 621. excl. syn.

Perennial, much branched scandent or prostrate vines, leaves 5-angled, or deeply 5-lobed, 5–8 × 5–7 cm, glabrous, green above, pale beneath, glandular below, obtuse or acute, more or less sinuate toothed, tendrils slender, striate, simple; flowers white, axillary, solitary; fruits ovoid-oblong, pulp juicy, seeds compressed, oblong-ovoid.

Common on bushes in forests. Cultivars as source of fruit vegetable. Fl. & Fr.: Throughout the year. Eng.: Kovai fruit. *Vern.*: Donda. It is used for curing bronchitis, catarrh, diabetes, skin affections and gonorrhoea. N. Varam, *VRK* 3502; Loyapalli, *VRK* 3852; Vinukonda, RF, *CAB* 4675 (MH); Nagarjuna Konda, *KTH* 9633 (CAL).

Corallocarpus Welwitch

Corallocarpus epigaeus (Rottl. et Willd.) Clarke in Hook. f. Fl. Brit. India 2: 628. 1879; Gamble 541 (382); Chakrav. 28. 1982. *Bryonia epigaea* Rottl. et Willd. Ges. Naturf Freunde Berlin Neue Schriften 4: 223. 1803.

Prostrate or climbing herbs; monoecious; tendrils simple; leaves broadly sub-orbicular, deeply 3–5-lobed, 1–5 × 2–4.5 cm, pubescent, lobes obtuse, irregularly dentate, leaf-base cordate; flowers greenish yellow, unisexual, axillary, male in racemes, female solitary; fruits ellipsoid, suddenly contracted into a slender beak, scarlet in the middle, base and beak green, seeds 6–10, ovoid or subglobose, in orange-coloured pulp.

Rare in forests. Fl. & Fr.: Aug.–Feb. *Vern.*: Naga donda. It is an alterative, and for chronic mucus enteritis, dysentery, snake-bite and syphilic rheumatism. Root contains a bitter principle like bryonin. Loyapalle, *VRK* 3853; Nagarjunakonda, *KTH* 9701 (CAL).

Ctenolepis J.D. Hooker

Ctenolepis garcinii (Burm. f.) Clarke in Hook. f. Fl. Brit. India 2: 629. 1879. *Sicyos garcinii* Burm. f. Fl. Indica 211. 1768. *Blastania garcinii* (Burm. f.) Cogn. in DC. Monogr. Phan. 3: 629. 1881; Gamble 540 (381); Chakav. 13. 1982.

Scabrid climbers, often forming mats; tendrils capillary; leaves palmately 5-lobed, 4–6 × 3–8 cm, villous, obtuse or acute, denticulate; flowers greenish, axillary, unisexual, male in racemes, female solitary; fruits ellipsoid, glabrous; seeds oblong, yellowish-grey.

Occasional in bushes. Fl. & Fr.: June–Dec. *Vern.*: Pamu Donda. Javuku, *VRK* 3540; Nagarjunakonda, *KTH* 9747 (CAL).

Cucumis Linnaeus

1. Perrennials; leaves deeply 5–7-lobed .... *C. callosus*
1. Annuals; leaves angular or lobed:
   2. Ovary softly hairy; young fruits not tuberculate .... *C. melo*
   2. Overy hispidulous, young fruits tuberculate .... *C. sativus*

Cucumis callosus (Rottl.) Cogn. in Engler Das Pflanzenr. 4. 275. 2: 129. 1924; Chakravarthy in Fasc. Fl. Ind. 1: 31. 1982. *Bryonia callosa* Rottl. in Ges. Naturf. Freunde Berlin Naue Schriften 4: 210. 1803. *Cucumis trigonus* Roxb. Fl. Ind. 2: 722. 1824; FBI 2: 619; Gamble 353 (378).

Perennial, monoecious, weak climbing herbs, branches grooved, hispid, almost prickly, tendrils simple; leaves 5–7-lobed, lobes round or ovate-oblong, often narrowed at base, apex round, dentate or lobulate; flowers yellow, male axillary, solitary, sometimes 2–3 in clusters, female solitary; fruit obovoid, striped white; seeds many, oblong, flattened, smooth.

Occasional. Fl. & Fr.: July–Feb. Nagarjunakonda, *KTH* 9730 (CAL).

Cucumis melo L. Sp. Pl. 1011. 1753. var. **melo**; FBI 2: 620; Gamble 535 (378); Chakrav. 32. 1982. *C. melo* L. var. *cultis* Kurz. in J. Asiat. Soc. Bengal 46: 102. 1877. *C. melo* var. *agrestis* (Naud.) Greb. in Kulturpflanze Beih 2: 424. 1959. *C. melo* Naud in Ann. Sci. Pl. 4: 614. 1805; Gamble 535 (378).

Prostrate, spreading vines; leaves orbicular or ovate, 3–5-lobed, 6–9 × 6–8 cm, hispid, lobes sub-angular, short, margin denticulate; flowers yellow, male flowers usually 5–6 in axillary fascicles, female flowers solitary; fruits spherical, ovoid, ellipsoid or oblong, yellow when ripe; seeds many, brown.

Cultivated for its fruits, occasional in hedges. Fl. & Fr.: Apr.–Sept. Eng.: The Melon. *Vern.*: Kharbuja. V.P. South, *VRK* 5812; Nagarjunakonda, *KTH* 9730 (CAL).

Fruits polymorphous, smooth, used as vegetable (raw) and as table fruits (when ripened). Cultivated on sandy beds of rivers.

Cucumis sativus L. Sp. Pl. 1012. 1753; FBI 2: 260; Gamble 535 (378); Chakrav. 36. 1982.

Scabrous, annual, trailing or climbing vines; tendrils simple, leaves broadly cordate-ovate, 7–10 × 7.5–10.5 cm, densely hairy, pedately 5–7 nerved, palmately 3–5-lobed; rough surface, denticulate; flowers yellow, unisexual, axillary, male in fascicles, female solitary; fruits ovate or oblong with tubular prickles when young, smooth when ripe; seeds numerous, white, compressed.

Cultivated in fields and rarely as an escape in hedges. Fl. & Fr.: Apr.–Sept. Eng.: Cucumber. *Vern.*: Dosa. Fruits edible and also used as vegetable. V.P. South, *VRK* 5830.

Diplocyclos (Endlicher) Post et Kuntze

Diplocyclos palmatus (L.) Jeffrey in Kew Bull. 15: 352. 1962; Chakrav. 48. 1982. *Bryonia palmata* L. Sp. Pl. 1012. 1753. excl. syn. *Bryonopsis laciniosa* auct. non (L.) Naud; Gamble 534 (377).

Glabrous vines; tendrils branched; leaves orbicular-ovate, palmately, deeply 5–7-lobed, 5–8 × 7–10 cm, scabrid above, glabrous below, lobes acuminate; flowers greenish-yellow, unisexual, male in axillary clusters, female solitary; fruits globose, brick red when ripe, smooth.

Occasional in forests on hedges. Fl. & Fr.: Sept.–Feb. *Vern.*: Linga donda. Vinukonda RF, *VRK* 6865.

Note: Jeffry (op. cit. pp. 347 and 352) has discussed at length the identity of the Indian plant hither to known as *Bryonia laciniosa* L. and points out that the American plant previously have the same specific epithet was under *Cayaponia* Manso, and that (p. 347), 'The Linnaen epithet has long been erroneously applied to wide spread plant of the old world tropics in the combination *Bryonopsis laciniosa* (L.) Naud., for which speices the correct name appears to be *Diplocyclos palmatus*...'

Kedrostis Medikus

Kedrostis foetidissima (Jacq.) Cogn. in DC. Monog. Phan. 3: 634. 1881. *Trichosanthes foetidissima* Jacq. Collect. 2: 341. 1789. *Kedrostis rostrata* (Rottl.) Cogn. in DC. Monog. Phan. 3: 636. 1881; Gamble 540 (381). *Bryonia rostrata* Rottl. in Ges. Naturf. Freunde Berlin Neue Schriften ser. 4. 213. 1803. *Rhynchocarpa foetida* sensu Clarke in FBI 2: 627. 1879.

Prostrate or climbing herbs; monoecious, branchlets sparsely pilose; leaves broadly ovate-cordate, faintly 3–5-angled, 4.5–5.5 × 4–6 cm; chartaceous, softly pubescent, irregularly distantly dentate, acute, cordate at base, tendrils simple, filiform, flowers monoecious, male flowers yellow in racemes; female flowers solitary; fruits flask-shaped up to 2.5 cm long, deep red when ripe, tapering into a long narrow beak, seeds 6–10, obovoid, turgid, granular, pubescent.

Common among roadside hedges and rare in forests. Fl. & Fr.: Aug.–Feb. *Vern.*: Kunkuma donda. Macherla RF, *VRK* 5885.

Luffa P. Miller

1. Seeds winged, stamens 5 .... ***L. cylindrica***
1. Seeds not winged, stamens 3:
    2. Roots not tuberous .... ***L. acutangula***
    2. Roots tuberous .... ***L. tuberosa***

Luffa acutangula (L.) Roxb. Fl. Ind. 3: 713. 1832; FBI 2: 615; Gamble 533 (377). var. **acutangula**; Chakrav. 69. 1982. *Cucumis acutangulus* L. Sp. Pl. 1011. 1753.

Extensively trailing herbs; tendrils branched; leaves palmately 5–7-angled, 10–25 × 12–18 cm, scabrous, base cordate; flowers yellow, unisexual, axillary, male in fascicles, female solitary; fruits 10-angled, oblong, 10–35 cm long.

Cultivated in gardens and rarely as an escape. Fl. & Fr.: Sept.–Apr. *Vern.:* Beera. Fruits as vegetable. Krishna river, *KMS* 9808 (MH).

Luffa cylindrica (L.) M. Roem. Syn. Monog. 2: 63. 1846. *Momordica cylindrica* L. Sp. Pl. 1009. 1753. *Luffa aegyptica* Mill. Gard. Dict. ed. 8. 1768; FBI 2: 614; Gamble 533 (376).

Climbing vines; tendrils branched; leaves 3–5-lobed; 8–12 × 7–10 cm, gland dotted, scabrid; flowers yellow, unisexual, axillary, male in racemes, female solitary; fruits oblong, striped, glabrous; seeds many, ovoid.

Grows wild in hedges in Macherla range. Fl. & Fr.: Aug.–Mar. *Vern.*: Adavibeera. Ethipothala, *VRK* 5801.

Luffa tuberosa Roxb. Fl. Ind. 3: 717. 1832; Chakrav. 74. 1982. *Momordica tuberosa* Cogn. in DC. Manogr. Phan. 3: 454. 1881; Gamble 533 (376). *M. cymbalaria* Fenzl. ex Naud. in Am. Sci. Nat. Ser. 4. 12: 134. 1959; FBI 2: 618.

Perennial climbing or creeping herbs; roots tuberous; tendril simple; leaves orbicular-reniform, 3–5-angled or slightly lobed, 2–6 × 2–5 cm, puberulous, acute, dentate, base cordate; flowers white, unisexual, axillary, male in fascicles, female solitary; fruits ribbed, pyriform, 2–4 × 0.4–0.8 cm.

Common weed of black-cotton soils. Fl. & Fr.: Aug.—Jan. *Vern.*: Adavi kakara. Dachepally, *KCJB* 17466 (MH).

Note: The species has been removed to *Momordica* under *M. cymbalaria* Fenzl. by C.B. Clarke in Hooker's Flora of British India (2: 618. 1879). Cogniaux as referred above named it as *M. tuberosa* (Roxb.) Cogn., based on Roxburgh's *Luffa tuberosa.* The fruit is specific character in *Luffa* and there is no reason to shift this species to *Momordica* which has either muriculate or echinate fruit but never angular. In his original description, Roxburgh stated that the fruit is exactly like *L. amara* Roxb. (*L. acutangula* var. *amara*) but without the stopple. Moreover the leaves of all the species of *Momordica* contain true cystoliths on the lower surface. Cystoliths are absent in this species. Foliaceous bracts which are a common feature in *Momordica* are also absent in this species. So *Luffa tuberosa* Roxb. is retained against *Momordica cymbalaria* Fenzl.

Melothria Linnaeus

Melothria leiosperma (Wight & Arn.) Cogn. in DC. Monogr. Phan. 3: 622. 1881; Gamble 539 (380); Chakrav 140. 1959. *Bryonia leiosperma* Wight & Arn. in Ann. Mag. Nat. Hist. 8: 268. 1842; FBI 2: 623.

Annual monoecious climbing herbs; stems hispid, furrowed, slightly branched, villous; tendrils simple; leaves broadly ovate-cordate, deeply 5-lobed, 2–3 × 2–4 cm, bright green above, paler beneath, scabrid, acute, broadly dentate; flowers yellow, axillary, solitary; fruit globose, hairy, red when ripe; seed ellipsoid, turgid, conspicuous margined, smooth on the faces.

Occasional climber on hedges and forests. Fl. & Fr.: Throughout the year. Sattenapally RF, *VRK* 5884; Vinukonda, *VRK* 6946.

Momordica Linnaeus

Momordica charantia L. Sp. Pl. 1009. 1753; FBI 2: 616; Gamble 532 (375).

Annual trailing or climbing herbs, much branched, and grooved; tendrils simple; leaves deeply 5–7-lobed; 5–6 × 4–5 cm, sparsely villous above and dense below, dentate-lobulate; flowers yellow, unisexual, axillary, male in clusters, female solitary; fruit oblong, 6–15 × 1–3 cm, orange when ripe; seeds ovoid, compressed.

Cultivated in fields, rarely as an escape in hedges. Fl. & Fr.: Oct.–May. *Vern.*: Kakara. Eng.: Bitter gourd. Fruits as vegetable and also for medicinal purpose. Ethipothala, *VRK* 3922; Sattenepally, *VRK* 5839.

Mukia Arnott

Mukia maderaspatana (L.) Roem. Syn. Monog. 247. 1846. *Cucumis maderaspatanus* L. Sp. Pl. 1012. 1753. *Mukia scabrella* (L.f.) Arn. in J. Bot. 3: 276. 1841; FBI 2: 623. excl. syn. *Melothria maderaspatana* (L.) Cogn. in DC. Mong. Phan. 3: 623. 1981; Gamble 539 (381); Chakrav. 83. 1982.

Annual, scabrous climbing vines; leaves ovate-deltoid, angular or 3–5-lobed, 2–5 × 2–4 cm, denticulate, base cordate, tendrils simple; flowers yellow, unisexual, male in clusters, female solitary; fruits globose, small, bright red when ripe, seeds pitted.

Occasional climbers on bushes on the outskirts, of forests and waste lands. Fl. & Fr.: Oct.–Mar. *Vern.*: Pottibudama, Narapatta teega. Javuku, *VRK* 3549; N.U. Campus, *SSR* & KSM 21489.

Benincasa hispida Cogn., the 'Ash or white pumpkin' *Lagenaria siceraria* (Molina) Standl. the 'Bottle gourd,' Petiole glandular at apex.

Fruit of various shapes and sizes with hard, durable, smooth shell. *Citrullus lanatus* (Thunb.) Matsumura & Nakai. The 'Water melon.' Berry fleshy, oblong, large, are cultivated as vegetable crops.

Trichosanthes cucumeriana var. *anguina* (L.) Haines, the snake gourd, Lingapotla is a valuable medicinal species—cultivated NU campus Staff Quarters area. Fruits are used as vegetables.

CACTACEAE

1. Stems flattened, jointed, spines 5–7 per areole, flowers to 5 cm across .... *Opuntia*
1. Stems columnar, winged, spines ca 15, flowers 15 cm across .... *Cereus*

Cereus P. Miller

Cereus pterogonus Lemoire. Cact. Gen. Sp. Nov. 59. 1839; Curtis, Bot. Mag. t. 5360. 1863; Burkill in Rec. Bot. Surv. India 4: 294. 1911.

Erect, armed, giant shrubs, up to 5 m tall; latex milky; stems columnar, 3–9-winged or angled; wings to 4 cm across; areoles echinate; leaves 0; flowers greenish-white, lateral, sessile, solitary.

Introduced, wild in wastelands and along hedges. Fl. & Fr.: Mar.–July. Vinukonda RF, *VRK* 6980.

Opuntia P. Miller

Opuntia stricta (Haw.) Syn. Pl. Succ. 191. 1812. var. ***dillenii*** (Ker-Gawl.) Benson Cact. Succ. Jour. Amer. 41; 126. 1969. *Cactus strictus* Haw. Misc. Nat. 188. 1803. *Opuntia dillenii* (Ker-Gawl.) Haw. Suppl. Pl. succ. 79. 1819; FBI 2: 657; Gamble 548 (387). *Cactus dillenii* Ker-Gawl. Reg. t. 255. 1818.

Thorny bushes, up to 2.5 m tall; stems jointed, modiified into fleshy cladode like plates, 15–25 × 10–20 cm, leaves modified into spines up to 3 cm long, hard, 3–7 per areole, yellow; flowers yellow; berries obovoid, purple, fleshy, areolar; seeds many.

Common in the forests, roadsides and hedges. Native of Mexico, introduced into India during 18th century. Fl. & Fr.: Thoughout the year. Eng.: Prickly pear. *Vern.*: Nagajamudu. Ripe fruits are eaten raw. Javuku, *VRK* 3917; Nagarjunakonda, *VRK* 6720.

AIZOACEAE

1. Stamens more than 30, ovary 3-celled, ovules 2 per cell on axile placentae; mangroves .... ***Sesuvium***
1. Stamens less than 20, ovary 1- or 2-celled, ovules 1 or 2 (few) on basal placentae; non-mangroves:
    2. Flowers solitary, capsules enclosed in the calyx tube .... ***Trianthema***
    2. Flowers in axillary clusters, capsules not enclosed in the calyx tube .... ***Zaleya***

Sesuvium Linnaeus

Sesuvium portulacastrum (L.) L. Syst. Nat. (ed. 10) 1058. 1759; FBI 2: 659; Gamble 550 (388). *Portulaca portulacastrum* L. Sp. Pl. 446. 1753. *Sesuvium repens* Willd. Enum. Hort. Berol. 521. 1809.

Spreading succulent prostrate or sub-erect herbs, rooted at nodes; leaves linear or spathulate-oblong, 1.3–3.5 × 0.5–0.7 cm, fleshy, base attenuate; entire, obtuse; flowers axillary, solitary; sepals purple or red; capsule oblong, enclosed within the persistent calyx, circumscissile; seeds many, dark, smooth, reniform.

Common succulent mangrove weed in dense masses on the sandy sea shores and saline areas. Fl. & Fr.: Dec.–May. *Vern.*: Thikka kura. Nizampatnam port, *VRK* 6796.

Trianthema Linnaeus

1. Leaves narrowly elliptic or linear, flowers fascicled, free from the base of the petiole, stamens 5, seeds with concentric smooth lines .... *T. triquetra*
1. Leaves ovate or broadly oblong, flowers solitary, sheathed by the base of the petiole, stamens 10–20, seeds 4–9 with concentric muricate lines .... *T. portulacastrum*

Trianthema portulacastrum L. Sp. Pl. 223. 1753; Gamble 550 (389). *T. monogyna* L. Mant. Pl. 69. 1767; FBI 2: 660.

Annual prostrate or ascending diffuse glabrous herbs, up to 30 cm tall; branchlets glandular; leaves unequally paired, bigger ones orbicular-obovate, smaller ones oblong, 0.5–3.5 × 0.5–3 cm, subsucculent, membranous, apex obtuse, rounded or retuse, base obtuse or attenuate, margin entire or undulate, flowers pinkish or yellowish-white, axillary, solitary; capsules turbinate, breaking transversely into an upper coriaceous lid and a lower membranous cup; seeds 2 in the operculum and 7 in the lower cup, black.

Abundant weed in the cultivated fields and in wastelands and rare in forests. Fl. & Fr.: Sept.–May. Eng: Horse purselane. *Vern.*: Atakamamidi, Pappu Aku, Galijeru. Leaves as vegetable. It aids in urination and is useful in dropsy. The alkaloid also reduces swellings in the body brought about by liver or kidney disorders. V.P. South, *VRK* 5783. Four ecotypes were reported by B.V.N. Reddy, 1983–Ph.D., Nagarjuna University.

Trianthema triquetra Rottl. ex Willd. in Ges. Naturf. Freunde Neue Schriften 4: 181. 1803; Gamble 551 (389). *T. crystallina* auct. (non Vahl 1790); FBI 2: 660.

Annual prostrate glabrous herbs; branchlets radiating from root stock, densely matty, leaves very small, succulent, narrowly lanceolate or elliptic, 2–6 × 1.5–2 mm, unequal, glabrous, acute or obtuse, base subacute; flowers pink in pairs or in threes or rarely solitary; capsules globoid with 1 seed in operculum part and 1 in the basal part, seeds black, with concentric smooth line.

Common in moist wastelands, often saltish localities and rare in forests. Fl. & Fr.: June–Oct. *Vern.*: Kukka-palakoora. Nizampatnam, *VRK* 6922, N.U. campus, *PNR, SSR & KSM* 21487.

Note: The plant has been confused with *T. crystallina* Vahl in most of the Indian Floras except Cooke (1; 554, 1903). The Indian plant is not *T. crystallina* Vahl, as the latter possesses oval or lanceolate leaves, up to 2 cm in length, often unequal sides (fide Backer in Fl. Males 1, 4(3): 273, 1951).

Zaleya Linnaeus

Zaleya decandra (L.) Burm. f. Fl. Ind. 110. t. 31. 1768. *Trianthema decandra* L. Mant. Pl. 70. 1767; FBI 2: 661; Gamble 651 (389).

Prostrate glabrescent herbs, much branched, striate, leaves ovate-elliptic, sub-fleshy, 2.2–3. × 0.8–1.6 cm, opposite pairs somewhat unequal, entire, obtuse, usually apiculate at apex; flowers pink inside, greenish outside, in axillary subumbellate clusters; capsules subterete, beak two-lobed; seeds 4, 2 in base and 2 in beak, black, orbicular-reniform, striate.

Occasional weed in cultivated fields and moist waste places in forests. Fl. & Fr.: Sept.–Jan. *Vern.*: Kamerigadda. Root juice is used for jaundice. V.P. South, *VRK* 9837 (MH); Nagarjunakonda, *KTH* 9719 (CAL).

MOLLUGINACEAE

1. Leaves radical .... ***Mollugo (nudicaulis)***
1. Leaves cauline:
    2. Leaves elliptic-lanceolate, linear or linear-oblanceolate .... ***Mollugo p.p.***
    2. Leaves obovate, oblong or obovate-suborbicular:
        3. Flowers pink, staminodes present, petaloid .... ***Corbichonia***
        3. Flowers greenish or white, staminodes absent:
            4. Fruits syncarpous, dehiscent, seeds numerous per cell .... ***Glinus***
            4. Fruits apocarpous, indehiscent, seeds 1 per cell .... ***Gisekia***

Corbichonia Scopoli

Corbichonia decumbens (Forssk.) Excell in J. Bot. 73: 80. 1935. *Orygia decumbens* Forssk. Fl. Aeg.—Arab. 103. 1775; FBI 2: 664; Gamble 551 (389).

Prostrate herbs; branchlets glabrous; leaves decussate and alternate, pairs unequal, obovate-suborbicular, 1–2 × 0.8–1.5 cm, fleshy, glabrous, entire, rotund-obtuse, base cuneate or attenuate; flowers pink, in leaf-opposed or terminal cymes or lax racemes; capsules subglobose, loculicidal; seeds reniform, minutely strophiolate.

Occasional in river banks. Fl. & Fr.: Sept.–Feb. Nagarjunakonda, *KTH* 9703 (CAL).

Gisekia Linnaeus

Gisekia pharnaceoides L. Mant. Pl. 562. 1771; FBI 2: 664; Gamble 553 (391); Jeffrey in Kew Bull. 14: 235. 1960.

Diffusely branched prostrate herbs; leaves oblong or obovate-oblong, elliptic-lanceolate, 1–2 × 0.4–0.7 cm, sub-fleshy, obtuse, entire; base cuneate; flowers small, yellowish, in axillary or terminal cymes, stamens pink; capsules 5-celled, apocarpous, indehiscent; seeds 5-granular, solitary, with scattered white glandular prominences.

Common in sandy soils. Fl. & Fr.: Throughout the year. *Vern.*: Isukadantikura. V.P. South, *VRK* 5782; 6839; Nagarjunakonda, *KTH* 9704 (CAL).

Glinus Linnaeus

Note: In most of our floras this genus is placed under the genus *Mollugo* L. Fenzel in *Ann. Wien* 1: 346–353 et 372, 1836 separated from *Mollugo* those plants with seeds having a thread from the raphe twining round the seed and placed them under *Glinus*. Bentham and Hooker, f. in *Gen. Pl.* 1: 857. 1867 stated that this difference was insufficient to separate plants which are closely allied in all other respects. Subsequent authors followed their opinion. Merrill & Perry and Baker have reviewed the separation of *Glinus* from *Mollugo*.

1. Branchlets thickly stellate hairy, sepals acute, densely stellate hairy .... *G. lotoides*
1. Branchlets glabrous, sepals obtuse, glabrous .... *G. oppositifolius*

Glinus lotoides L. Sp. Pl. 463. 1753. *Mollugo lotoides* (L.) Kuntze Rev. Gen. Pl. 214. 1891; Gamble 552 (390). *M. hirta* Thunb. Prodr. Fl. Cap. 24. 1794; FBI 2: 662. *Glinus dictamnoides* L. Mant. Pl. 243. 1771; FBI 2: 662.

Densely tomentose, spreading herbs, leaves orbicular or obovate, 1–2 × 0.5–1.5 cm, stellate tomentose, obtuse, entire, base cuneate, flowers greenish in lateral clusters; capsules globose, loculicidal, 5-valved; seeds many, brown, granulate, subreniform, strophiole membranous, with a filiform appendage.

Occasional in dried tank beds and streams, Fl. & Fr.: July–Apr. N. Varam, *VRK* 3594; Krishna river, *KMS* 9815 (MH).

Glinus oppositifolius (L.) A. DC. in Bull. Herb. Bois. (Ser. 2) 1: 552. 1901. *Mollugo oppositifolia* L. Sp. Pl. 89. 1753; Gamble 552 (390). *M. spergula* L. Syst. Nat. (ed. 10) 881. 1789; FBI 2: 662.

Slender decumbent herbs; leaves lanceolate, elliptic or obovate, 1–3 × 0.3–1.5 cm, glabrous or thinly hairy, apex rounded, obtuse, acute or minutely cuspidate, subacute, fascicled at nodes; flowers white in axillary clusters; capsules 3-locular; seeds mumerous, black, finely granulate.

Occasional in plains in moist waste lands and in rocky shady places. Fl. & Fr.: July–Apr. *Vern.*: Chayuntarashi. Leaves are eaten as vegetable. The plants are valued for medicinal properties. Nizampatnam, *VRK* 6797; Nagarjunasagar, *KMS* 9837 (MH).

Mollugo Linnaeus

1. Leaves radical .... ***M. nudicaulis***
1. Leaves cauline:
    2. Flowers in axillary umbellate cymes, stem reddish, leaves narrow-linear .... ***M. cerviana***
    2. Flowers in dichasial cymes; leaves other than linear or acicular, stem greenish:
        3. Inflorescence rachis straight; pedicels curved, seeds prominently tuberculate .... ***M. pentaphylla***
        3. Inflorescence rachis zig-zag; pedicels straight; seeds smooth or obscurely granular .... ***M. disticha***

Mollugo cerviana (L.) Ser. in DC. Prodr. 1: 392. 1824; FBI 2: 663; Gamble 553 (390). *Pharnaceum cerviana* L. Sp. Pl. 272. 1753.

Erect slender glabrous herbs, up to 12 cm long; stems many, almost filiform, radical leaves rosulate, narrow, 0.5–1 cm long; cauline leaves 5–7 cm, in a cluster, narrowly linear or acicular, 1–1.5 cm long, chartaceous, glaucous below, entire, obtuse-apiculate, base attenuate; flowers numerous on long filiform stiff pedicels, in terminal umbellate cymes; capsule ellipsoid; seeds many, ovoid-sub-reniform, minutely granular.

Occasional on sandy soils as a weed. Fl. & Fr.: Jan.–Mar. Nagarjunakonda. *KTH* 9838 (CAL).

Note: Easily distinguished by its needle like leaves and slender branches.

Mollugo disticha Ser. in DC. Prodr. 1: 392. 1824; FBI 2: 663; Gamble 553 (390).

Erect annual herbs; branchlets slightly pubescent; leaves 3–5 in fascicles at nodes, obovate-oblanceolate or elliptic, 0.5–1.5 × 0.2–0.4 cm, pubescent, cuneate at base, entire, apex acute, flowers white, shortly pedicellate in leaf-opposed and/or terminal cymes; capsule ovoid, seeds many, pale pink, obscurely granuler.

Common in Nagarjuna University campus. Fl. & Fr.: June–Oct.

Mollugo nudicaulis Lam. Encyl. 4: 234. 1797; FBI 2: 664; Gamble 553 (390).

Annual erect herbs, up to 20 cm tall; leaves all radical, elliptic-oblong, 2–4 × 0.8–2 cm, glabrous, obtuse, entire, base attenuate, forming a rosette on the ground; flowers white in trichotomous cymes; capsules ellipsoid, brown; seeds many, black, granular.

Occasional in forest undergrowth. Fl. & Fr.: Through the year. *Vern.*: Pedda paripata. Koppanur, *VRK* 3267; Vijayapuri, *KMS* 9844 (MH); Nagarjunakonda, *KTH* 9609 (CAL).

Mollugo pentaphylla L. Sp. Pl. 89. 1753; Gamble 553 (390). *M. stricta* sensu Clarke in Hook. f. Fl. Brit. India 2: 663. 1879. *p.p.*; non L. 1762.

Annual glabrous erect or spreading herbs with dichotomous branches; leaves linear-oblanceolate, 2–3 × 0.2–0.4 cm, obtuse, entire, base cuneate, 2–8 at each node; flowers white in leaf-opposed or terminal elongated racemes; capsules oblong; seeds many, dark-brown, granular.

Common weed in wastelands and fields. Fl. & Fr.: Throughout the year. *Vern.*: Chinavaragogu. Whole plant is used to cure post delivery pains. Sattenapalle, *VRK* 5849; Vulasalapadu, *DMR* 21139, Loyapalli, *SSR & KSM* 21628.

APIACEAE (*nom alter* UMBELLIFERAE)

Centella Linnaeus

Centella asiatica (L.) Urban in Mart. F. Bras. 11: 287. t. 78. f. 1. 1879; Gamble 1: 556 (392). *Hydrocotyle asiatica* L. Sp. Pl. 234. 1753; FBI 2: 669.

Herbs with large creeping stems rooting at nodes; leaves orbicular-reniform, 1.2–3.7 cm across, glabrous, base cordate, margin shallowly crenate, coarsely dentate or subentire; flowers 3–5 or 6, red in an umbel, subsessile, sepals absent; fruit orbicular to ellipsoid, 2–3 mm across, mericarps with 5 primary ridges; seeds much compressed.

Occasional in wet places. Fl. & Fr.: Throughout the year. *Vern.*: Sarswathi aku. Whole plant made into pieces is cooked as vegetable. It is also used as an antidote to snake-bite. The dried leaf powder, mixed with honey and ghee, is given orally early in the morning for improving memory power. Davupalli, *SSR & KSM* 21430.

Coriandrum satium L. and *Foeniculum vulgare* Miller are cultivated as spice crops; often met with as escapes in rural outskirts. *Daucus carota* L. (Vern. Carrot) and *Trachyspermum ammi* (L.) Spr. are cultivated for edible roots and seeds respectively.

ALANGIACEAE

Alangium Lamarck *nom cons.*

Alangium salvifolium (L.f.) Wang. in Engl. Pfl. Fam. 41: 9. 1910. ssp. **salvifolium**; Gamble 572 (404). *Grewia salvifolia* L.f. Suppl. Pl. 409. 1781. *Alangium lamarckii* Thw. Enum. Pl. Zeyl. 133. 1859; FBI 2: 741.

Deciduous bushy trees, up to 10 m tall; branchlets tomentose, spine tipped; leaves oblong-lanceolate, ovate or elliptic, 3.5–10 × 1–2.5 cm, glabrous above, puberulous below, attenuate or acute, entire, base obtuse; flowers white, scented, in axillary cymes or clusters; berries globose, 2 × 1.5 cm, blackish purple when ripe, succulent with bony endocarp; seed solitary, ovoid.

Common in forests and plains. Fl. & Fr.: Feb.–July. *Vern.*: Vooduga. Wood as Timber. Fruits are edible. They are antiphlegmatic, laxative, tonic, refrigerent. Root bark used as a purgative and for skin diseases, wood hard, useful for ornamental work and also as a good fuel. Davupalli, *SSR & KSM* 21405.

RUBIACEAE

1. Herbs:
    2. Prostrate herbs, leaves distichous .... ***Dentella***
    2. Erect or diffuse herbs, leaves decussate to whorled:
        3. Leaf-base acute to decurrent, seeds more than 40, pitted or reticulate:
            4. Leaves whorled, flowers in a lax branched raceme .... ***Kohautia***
            4. Leaves decussate, flowers 1 or 2 in the leaf axils or in sub-terminal umbels .... ***Hedyotis***
        3. Leaf base truncate to rounded, seeds less than 20, with ventral cavity .... ***Spermacoce***
1. Trees, shrubs or lianas:
    5. Flowers aggregated in globose heads:
        6. Leaves elliptic-oblong to obovate, corolla spreading, > 1.5 cm across, ovule solitary in each carpel, fruit a syncarpium .... ***Morinda***
        6. Leaves cordate or oblong-orbicular, corolla not spreading, < 1 cm across, ovules many in each carpel, cluster of capsules:
            7. Leaf apex obtuse, peduncle terminal, solitary, seeds flat, winged ... ***Mitragyna***
            7. Leaf apex acuminate, peduncle axillary, 1–3 together, seeds elongate, not winged .... ***Haldinia***
    5. Flowers in fascicles, cymes or panicles, rarely solitary:
        8. Leaves ovate, flowers 7-merous, planted .... ***Guettarda***
        8. Leaves elliptic-oblong to orbicular, flowers 4- or 5-merous:

9. Flowers 4-merous:
  10. Stigma 2-fid, style less than twice as long as tube .... ***Ixora***
  10. Stigma fusiform, style twice as long as tube .... ***Pavetta***
9. Flowers 5-merous:
  11. Cymes terminal .... ***Tarenna***
  11. Cymes axillary or leaf-opposed:
    12. Armed shrubs or small trees:
      13. Berries more than 2 cm across, seeds numerous in fleshy pulp .... ***Catunaregum***
      13. Berries less than 1 cm across, seeds few, not embedded in pulp ... ***Benkara***
    12. Unarmed trees or shrubs (except *Canthium parviflorum*)
      14. Leaves rotund-orbicular, berries up to 2 cm across .... ***Deccania***
      14. Leaves other than rotund-orbicular, berries more than 2 cm across:
        15. One calyx-lobe extended into a large petal like structure ... ***Mussaenda***
        15. Calyx-lobe not extended:
          16. Stipules interpetiolar, flowers 4-merous, fruit a drupe, with 2 pyrenes .... ***Canthium***
          16. Stipules intrapetiolar, flowers 5-merous, fruit a berry with woody rind .... ***Gardenia***

Benkara Tirvengadum

Benkara malabarica (Lam.) Tirveng. in Taxon 32: 440. 1983. *Randia malabarica* Lam. Encycl. 3: 25. 1789; FBI 2: 111; Gamble 616 (435).

Armed shrubs, up to 6 m tall; branches grey, divaricate; spines short; leaves coriaceous, decussate or clustered, obovate-oblanceolate, 3–6 × 1–2 cm, shiny, entire, obtuse, base cuneate; flowers white, in dense, sessile axillary corymbose cymes; berries scarlet, globose; seeds conical or cubical.

Common in scrub forests. Fl. & Fr.: Feb.–Aug.. *Vern.*: Peddalli. Papaiahpalem, *VRK* 5848; Kondavedu Fort, *C.A. Barber* 4682 (MH).

Note: Tirvengadum (1983) re-established the genus *Benkara* for the taxon segregated from *Randia.* He also discussed the reasons for re-establishing the genus and also rejected the name *Griffithia.*

Canthium Lamarck

1. Armed woody shrubs, leaves under 7 cm long .... ***C. parviflorum***
1. Unarmed trees, leaves over 7 cm long .... ***C. dicoccum***

Canthium dicoccum (Gaertn.) Teijsm. & Binn. Cat. Hort. Bog. 113. 1866. *Psydrax dicoccus* Gaertn. Fruct. Sem. Pl. 1: 125. t. 26. f. 2. 1788. *Canthium didymum* Roxb. Fl. Ind. 1: 535. 1824; FBI 3: 132. *Plectronia didyma* (Gaertn.) Kurz Forest. Fl. Burma 2: 35. 1877; Gamble 624 (440).

Densely foliaceous, unarmed medium sized trees, up to 6 m tall; young branches 4-angled, leaves elliptic-lanceolate, 8–10 × 5–7 cm, glabrous, acute to acuminate, base acute-attenuate. Flowers creamish-white in axillary pedunculate umbels; drupes ellipsoid-globose, black; seed solitary.

Occasional in deciduous forests. Fl. & Fr.: Aug.–Apr. *Vern.*: Nalla Balusu, Nakkari, Niralli. Wood is used for making combs and for agricultural implements. Mellavagu, *VRK* 5707.

Canthium parviflorum Lam. Encycl. 1: 602. 1785; FBI 3: 136. *Plectronia parviflora* Bedd. For. Man. 134–35. 1874; Gamble 625 (441).

Armed woody shrubs, up to 4 m tall; bark grey with longitudinal flakes; spines 2–3 cm long; opposite-decussate, supra-axillary; leaves elliptic-ovate to obovate, 1.5–3 × 0.5–2 cm, glabrous, acute, entire, base attenuate; flowers small, greenish-white in axillary cymes; drupes obcordate, 1 × 0.8 cm, furrowed, brown when ripe, seeds solitary.

Common in the forests on rocky hills. Fl. & Fr.: Apr.–Sept. *Vern.*: Chinna Balusu. Fruits edible, Leaf juice taken internally to cure fever. Root juice taken internally to kill intestinal worms (anthelmintic). Loyapalle, *VRK* 3980; Bollapalle, *CAB* 17428 (MH).

Catunaregum Wolf

Note: Tiruvengadum (1978a) re-established the genus *Catunaregum* for the old world *Randia sensu lato*. He describes that *Randia* is an exclusively American genus characterised by its one-celled ovary with parietal placentation. He transferred the South Indian *Xeromphis* (*Randia*) to *Catunaregum*.

Catunaregum spinosa (Thunb.) Tirveng. in Bull. Mus. Nat. Hist. Paris 3. Ser. 521. Bot. 35: 13. 1978. *Gardenia spinosa* Thunb. Diss. Gard. 7: 16. Pl. 2. f. 4. 1780. *Xeromphis spinosa* (Thunb.) Keay in Bull. Jard. Bot. Etat. Brux. 28: 37–39. 1958. *Randia dumetorum* (Retz.) Poir. var. *floribunda* (DC.) Gamble Fl. Pres. Madras 616. 1921. 434. *R. floribunda* DC. Prodr. 41: 386. 1830. *R. longispina* Wight & Arn. Prodr. 398. 1834; Gamble 616 (434). *R. brandisii* Gamble, Fl. Pres. Madras 616. 1921 & 434. 1957.

Medium sized tree-like armed shrubs, up to 2.5 m tall; spines axillary, opposite-decussate, 1–2 cm long; leaves obovate, to oblanceolate, 2–3 × 1–2 cm, pubescent, obtuse or shortly acuminate, base narrowed into a slender petiole, flowers white or yellow, in axillary clusters; berries globose, yellow when ripe, 3 cm across, crowned by the calyx.

Common in the forests. Fl. & Fr.: Feb.–Sept. *Vern.*: Manga. The fruit paste is used to cure dandruff and to prevent falling of hairs. Dried fruit powder is used as poison and also used as detergent to wash clothes. The fruit is given in chronic cases of diarrhoea and dysentery. Loyapalle, *VRK* 3977; Mellavagu, *VRK* 5716; V.P. South, *VRK* 6827; Bollapalle, *VRK* 5742.

Deccania Tirvengadum

Deccania pubescens (Roth) Tirveng. in Nordic J. Bot. 3: 456. 1983. var. **pubescens.** *Gardenia pubescens* Roth Nov. Pl. Sp. 151. 1821. *Randia candolleana* Wight & Arn. Prodr. 39. 1834; FBI 3: 113. *p.p.;* var. *corymbosa* sensu Gamble 616 (435).

Unarmed small trees, up to 4 m tall; leaves orbicular, 1.5–4 × 1–1.5 cm, glabrescent above, tomentose below, obtuse; flowers white or light yellow, solitary at the spines; berries globose with a crown of calyx lobes, seeds smooth.

Common in the forests. Fl. & Fr.: Nov.–June. *Vern*.: Kondamanga. Fruits are used as fish poison. Javuku, *VRK* 3874.

Dentella Forster & Forster f.

Dentella repens (L.) Forst. & Forst. f. Char. Gen. 26. t. 13. 1776; FBI 3. 42; Gamble 590 (416). *Oldenlandia repens* L. Mant. 1: 40. 1767.

Annual, slender, prostrate, hairy herbs; leaves minute sub-spathulate, entire, acute, glabrous; flowers white, solitary in the forks of stem; capsules globose, indehiscent, 3 mm, densely villous with coarse hairs; seeds angular, punctate.

Weed in moist localities especially in cultivated fields. Fl. & Fr.: June–Nov. Krishna river, *VRK* 6740; Nizampatnam, *VRK* 6798.

Gardenia J. Ellis *nom. cons.*

1. Calyx lobes triangular, acute, stamens included .... ***G. gummifera***
1. Calyx lobes lanceolate, acuminate, stamens excluded:
   2. Leaves sessile or sub-sessile; calyx and corolla lobes 7–9, berries globose .... ***G. latifolia***
   2. Leaves petiolate; calyx and corolla lobes 5, berries ellipsoid .... ***G. resinifera***

Gardenia gummifera L. Suppl. Pl. 164. 1781; FBI 3: 116; Gamble 618 (436).

A small tree, up to 3 m tall; leaves obovate-oblong, 3–6 × 1–3 cm, glabrescent, base rounded to subcordate, margin entire, apex obtuse, subsessile, flowers yellowish white or white, solitary, axillary; berry oblong or ellipsoid, 2–3.5 cm; seeds rugose.

In deciduous forests. Fl. & Fr.: Jan.–Sept. *Vern.*: Manchibikki. Gum from the stem dissolved in water and given in constipation and also given to kill intestinal worms (anthelminitc). Davupalli, *SSR & KSM* 21415.

Gardenia latifolia Ait. Hort. Kew. 1: 294. 1789; FBI 3: 116; Gamble 619 (436).

Densely foliaceous deciduous trees, up to 10 m tall; leaves whorled in threes, obovate-orbicular, 12–18 × 6–12 cm, glabrous, obtuse, entire, base rounded to attenuate; flowers white, solitary, axillary; berries globose with a crown of calyx, 3–5 × 3 cm; seeds many, rugose, flattened.

Common in forests. Fl. & Fr.: Apr.–Dec. Eng.: Indian Box wood. *Vern.*: Pedda bikki. The growing tips exude a yellow resin which is made into a paste and used as fixative to fix wooden handles of sickles, knives and arrows. Wood strong, hard and heavy used for making combs and turnery articles, light furniture, toys etc. Fruits edible. Vinukonda, RF, *VRK* 6886, 6887.

Gardenia resinifera Roth, Nov. Pl. Sp. 150. 1821. *G. lucida* Roxb. Fl. Ind. 2: 553. 1824; FBI 3: 115; Gamble 618 (436).

Unarmed trees, up to 10 m tall; leaves elliptic-ovate or oblong 6–16 × 3–5 cm, glabrous above, bullate in the nerve axils, entire, obtuse-acute, base attenuate-rounded; flowers white, axillary, solitary; berries ellipsoid, with a crown of calyx; seeds rugose.

Occasional in forests. Fl. & Fr.: Jan.–Apr. *Vern.*: Bikki, Manoribikki. Source of a medicinal gum, microbicide round worms destroyed. Wood useful for turned articles. Kondavedu Fort, *CAB* 17495 (MH).

Gardenia jasminoides Ellis. Cultivated in gardens for its white fragrant flowers. *Vern.*: Garudavardhanam.

Guettarda Linnaeus

Guettarda speciosa L. Sp. Pl. 991. 1753; FBI 3: 126.

Evergreen trees, up to 10 m tall; branchlets stout; leaves decussate, ovate, 8–14 x 5–8 cm, puberulous along nerves above, pubescent beneath; entire, obtuse-shortly acute, base rounded; flowers white in axillary cymes; drupes woody, globose.

Occasional in forests and planted along the coast. *Vern.*: Pannir chettu. Guntur, (*Uhl*, 1929). (A.C College compound, Guntur).

Haldinia Ridsdale

Haldinia cordifolia (Roxb.) Ridsd. in Blumea 24: 361. 1978. *Nauclea cordifolia* Roxb. Pl. Cor. t. 53. 1796. *Adina cordifolia* (Roxb.) Hook. f. ex Brandis Forest Fl. N.W. India t. 33. 1874; FBI 3: 24; Gamble 584 (412).

Very large deciduous trees, trunk often with buttresses, tender parts pubescent; leaves thick-coriaceous, cordate-orbicular, 10–17 × 11–20 cm, dark green, puberulous along nerves above, densely pubescent and pale beneath, abruptly acuminate, entire, base cordate, stipules inter-petiolar, foliaceous, caducous; flowers capitate, yellow, axillary; capsule of 2 dehiscent cocci; seeds elongate.

Common in forests. Fl. & Fr.: Mar.–Aug.. *Vern.*: Bandaru. Pasupa ganapa, Rudraganapa. The roots are used as an astringent in dysentry. The bark is acrid and bitter used in biliousness; it is also regarded as febrifuge and antiseptic. Leaves are used as fodder. Wood moderately hard, durable under water. Vinukonda RF, *VRK* 6847.

Note: This species has so far been treated in floras under *Adina cordifolia* (Roxb.) Hook. f. ex Brandis. But Ridsdale (l.c.) in his revision of tribe *Naucleae* s.s. feels that the Indian taxon does not belong to the genus *Adina* Salisb., which does not occur in India and has therefore, raised the monotypic new genus *Haldinia* Ridsd. to accommodate it. He differentiates the two as follows:

1. Terminal vegetative bud loosely surrounded by stipules, stipules more or less free in bud, deeply bifid, somewhat persistent, ovules up to 4 per locule .... ***Adina***
1. Terminal vegetative bud flattened, stipules depressed in bud, entire, pubescent, keeled .... ***Haldinia***

Hedyotis Linnaeus

Note: In the present study the concept of W.H. Lewis (1961) and Backer & Bakhuizen. f. (1965) is followed. They treated the genus *Oldenlandia* L. under *Hedyotis* L.

1. Peduncles from the upper axils and in terminal leafless cymose panicles ..... ***H. affinis***
1. Peduncles not in terminal leafless cymose panicles:
   2. Peduncles mostly from the upper axils, several flowered, sometimes cymose .... ***H. puberula***
   2. Peduncles axillary from the lower or all the axils:
      3. Diffuse or sub-erect herbs, pedicels 2.5 to 12 mm long, usually paired on slender peduncle .... ***H. corymbosa***
      3. Erect, corymbosely branched herbs, pedicels 0.6–2 cm long, solitary, axillary .... ***H. herbacea***

Hedyotis affinis Roem. & Schult. in L. Syst. Veg. 3: 194. *Oldenlandia affinis* (Roem. & Schult) DC. Prodr. 4: 428. 1830; *Hedyotis dichotoma* Heyne ex Roth Nov. Pl. Sp. 93. 1821. *Oldenlandia dichotoma* (Roth) Hook. f. in Fl. Brit. India 3: 67. 1880; Gamble 601 (424).

Slender, erect herbs, up to 30 cm tall; leaves decussate, linear, 2–5 × 0.3–0.5 cm, glabrous, revolute, acuminate, base decurrent; subsessile; flowers pinkish white, in terminal lax, dichotomous panicles; capsule loculicidal; seeds reticulate.

Occasional on hill slopes. Fl. & Fr.: Oct.–Mar. Nagarjunakonda, *KTH* 9789 (CAL).

Hedyotis corymbosa (L.) Lam. Encycl. 1: 272. 1972. *Oldenlandia corymbosa* L. Sp. Pl. 119. 1753; FBI 3: 64; Gamble 600 (423).

Small prostrate or ascending herbs, up to 30 cm tall; leaves sessile, linear-lanceolate, 1–3 × 0.2–0.3 cm, acute to acuminate, base decurrent; flowers white with blue tinge in axillary peduncled corymbose fascicles; capsules globose, loculicidal, 3 × 2 mm, seeds reticulate.

Occasional weed of wastelands and cultivated fields. Fl. & Fr.: Oct.–Mar. V.P. South, *VRK* 5816.

Hedyotis herbacea L. Sp. Pl. 102. 1753. *Oldenlandia heynii* R. Br. ex G. Don Gen. Syst. 3: 531. 1834; FBl 3: 65. *O. herbacea* (L.) Roxb. Fl. Ind. 1: 445. 1820; Gamble 601 (424).

Erect, much corymbosely branched glabrous herb, up to 40 cm tall; stems 4-angled, leaves linear-lanceolate, 2–4 × 0.3–0.5 cm, glabrous or nearly so, acute, entire, decurrent at base; flowers white, solitary or 2 in each axil; capsule globose, very slightly didymous, 2.5 mm across, top ridged and projecting above the hypanthium; seeds black or brown, ellipsoid, minute, 3-angled.

Rare in plains and hills in the forests (Mohan Rao & Aasiya, 1995). Fl. & Fr.: June–Dec.

Hedyotis puberula (G. Don) Arn. Pugill. Pl. Ind. r. 342. 1836. *Oldenlandia puberula* G. Don. Gen. Syst. 3: 530. 1834. *Oldenlandia umbellata* L. Sp. Pl. 119. 1753; FBI 3: 66; Gamble 601 (424). *O. wightii* Hook. f. Fl. Brit. India 3: 66. 1880; Gamble 601 (424).

Spreading herbs with ascending branches, up to 30 cm tall; stems 6–8-ridged or lined, hispidulous; leaves in clusters, sessile, linear-lanceolate, 1–2.5 × 0.2–0.3 cm, glabrescent, margin undulate, acute to apiculate, base decurrent; flowers white or lilac in axillary or terminal pedunculate umbels; capsules loculicidal, globose, often didymous, covered by calyx lobes, glabrescent; seeds reticulate, obscurely angled.

Common weed in wastelands, roadsides and occasional in forests. Fl. & Fr.: Throughout the year. *Vern.*: Chiriveru, Mellavagu, *VSR* 5727; Ethipothala, *VRK* 5793; Settenapalle, V*RK* 5859.

Ixora Linnaeus

1. Fruits berries, calyx lobes reduced to teeth and shorter than tube .... ***I. finlaysoniana***
1. Fruits drupes, calyx lobes longer than tube:
   2. Leaves oblanceolate .... ***I. pavetta***
   2. Leaves linear-oblong .... ***I. arborea***

Ixora arborea Roxb. ex J.E. Smith in Rees. Cyclop 19: 5. 1811; *I. parviflora* Vahl Symb. Bot. 3: 11. t. 52. 1794. non Lam. 1791; FBI 3: 142; Gamble 631 (445).

Small trees, up to 4 m tall; leaves linear-oblong or obovate-oblong, 6–15 × 3–6 cm, glabrous, entire, acute to obtuse, base rounded; flowers white, odorous, in terminal, corymbose cymes; drupes didymous, red; seeds brown, smooth.

Common in forests. Fl. & Fr.: Feb.–May. *Vern.: Gorivi chettu.* It is used for coloured urine, general debility, anaemia and whooping cough. Vinukonda RF, *VRK* 6891.

Ixora finlaysoniana Wall. ex G. Don Gen. Syst. 3: 572. 1834; Gamble 629 (444).

Small trees, up to 8 m tall; branchlets glabrous, brownish-white; leaves coriaceous, elliptic-obovate, 5–15 × 4–5 cm, glabrous, entire, acute, base acute-cuneate; flowers white, fragrant; in terminal corymbs; berries globose.

Occasional in forests. Also cultivated in gardens. V.P. South, *VRK* 6726.

Ixora pavetta Andr. Bor. Rep. t. 78. 1799. *I. parviflora* Vahl Symb. 3: 11. t. 52. 1794. non. Lam. 1791; FBI 3: 142; Gamble 631 (445).

Small evergreen trees, up to 4 m tall; leaves oblanceolate, 6–15 × 1.5–6 cm, shining, glabrous, acute, entire, base rounded, or subcordate; flowers white in terminal corymbose cymes; drupes globose or ellipsoid.

Common in forests. Fl. & Fr.: June–Nov. *Vern.*: Gorivi. Loyapalle, *VRK* 3986.

I. coccinea L. with red umbels and *I. singaporensis* (Nursery nance) with deep orange umbels are cultivated as ornamentals.

Kohautia Chamisso et Schlechtendal

Note: Kohautia Cham. & Schlecht. (in Linnaea 4: 156. 1829) was treated as a section of genus *Hedyotis* L. by Wight & Arn. in Prodr. 417. 1834. Hook. f. (in Benth. & Hook. Gen. Pl. 2: 58, 1873) recognised it as a subgenus of the genus *Oldenlandia* L. Bremekamp (in Verb. Kon. Ned. Akad. Wet. 11, 48(2); 56, 1952) has revalidated as a genus stating "the genus *Kohautia* Cham. & Schlecht. is well characterised by the structure of its flowers. The latter are always monomorphic, with the anthers, as well as the stigma included and the stigmata at a lower level than the anthers or occasionally just touching them."

Kohautia aspera (Heyne ex Roth) Bremek. Verh. Kon. Ned. Akad. Wetensch; Afd. Natuurk., Tweede Sect-Ser. 2. 48 (2): 113. 1952. *Oldenlandia aspera* (Roth) DC. Prodr. 4: 428. 1830; Gamble 602 (424). *Hedyotis aspera* Heyne ex Roth Nov. Pl. Sp. 94. 1821; FBI 3: 68.

Erect herbs, up to 75 cm tall; leaves whorled, linear, 2–5 × 0.3–0.5 cm, glabrous, chartaceous, 1-nerved, lateral nerves obscure, attenuate, revolute, acute, base attenuate; flowers white, small in lax branched racemes; capsules loculicidal; seeds reticulate.

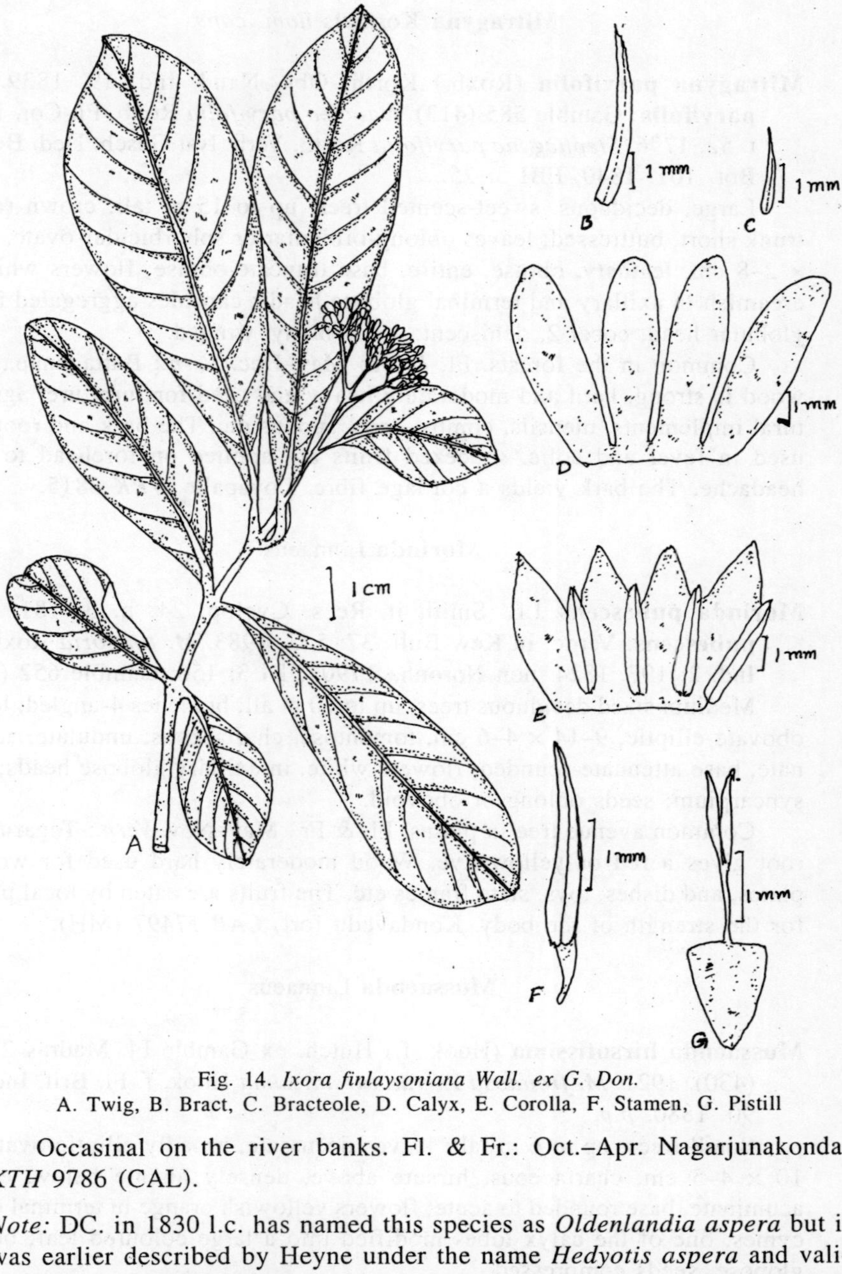

Fig. 14. *Ixora finlaysoniana Wall. ex G. Don.*
A. Twig, B. Bract, C. Bracteole, D. Calyx, E. Corolla, F. Stamen, G. Pistill

Occasinal on the river banks. Fl. & Fr.: Oct.–Apr. Nagarjunakonda, *KTH* 9786 (CAL).

Note: DC. in 1830 l.c. has named this species as *Oldenlandia aspera* but it was earlier described by Heyne under the name *Hedyotis aspera* and validated by Roth in 1821 (l.c). While transferring it to the revalidated genus *Kohautia* Bremekamp in 1952 (l.c) has given the present combination based on the above.

Mitragyna Korthals *nom. cons.*

Mitragyna parvifolia (Roxb.) Korth. Obs. Naud. Ind. 19. 1839. var. **parvifolia**; Gamble 585 (413). *Nauclea parvifolia* Roxb. Pl. Cor. 1. 40. t. 52. 1796. *Stephegyne parvifolia* Korth. Verh. Nat Gesch. Ned. Bezitt., Bot. 161. 1840; FBI 3: 25.

Large, deciduous, sweet-scented trees, up to 15 m tall; crown round, trunk short buttressed; leaves oblong-orbicular or suborbicular-ovate, 4–13 × 2–8 cm, leathery, obtuse, entire, base truncate-obtuse; flowers white or creamish in axillary and terminal globose heads; capsules aggregated into a globular head, cocci 2, dehiscent; seeds many, winged.

Common in the forests. Fl. & Fr.: May–Dec. *Vern*.: Battaganapa. The wood is strong, hard and moderately heavy and used for furniture, agricultural implements, utensils, combs, cups, spoons etc. The bark and roots are used in fever and colic, squeezed fruits are applied on forehead to treat headache. The bark yields a cordage fibre. Loyapalle, *VRK* 3815.

Morinda Linnaeus

Morinda pubescens J.E. Smith in Rees. Cyclop. 24: n. 3. 1813. var. **pubescens**; Verdc. in Kew Bull. 37: 543. 1983. *M. tinctoria* Roxb. Fl. Ind. 2: 197. 1824. non Noronha 1790; FBI 3: 156; Gamble 652 (459).

Medium sized deciduous trees, up to 10 m all; branches 4-angled, leaves obovate-elliptic, 9–14 × 4–6 cm, tomentose, chartaceous; undulate, acuminate, base attenuate-rounded; flowers white, in axillary globose heads; fruit syncarpium; seeds oblong or obovoid.

Common avenue tree in plains. Fl. & Fr.: May–Nov. *Vern*.: Togaru. The root gives a red or yellow dye. Wood moderately hard used for wooden plates, and dishes, toys, slate frames etc. The fruits are eaten by local people for the strength of the body. Kondavedu fort, *CAB* 17497 (MH).

Mussaenda Linnaeus

Mussaenda hirsutissima (Hook. f.) Hutch. ex Gamble Fl. Madras 2: 610 (430). 1921. *M. frondosa* L. var. *hirsutissima* Hook. f. Fl. Brit. India 3: 90. 1880. *p.p.*

Small trees, up to 6 m tall; leaves decussate, broadly elliptic-ovate, 7–10 × 4–5 cm, chartaceous, hirsute above, densely hirsute below, entire, acuminate, base rounded to acute; flowers yellowish orange in terminal dense cymes; one of the calyx lobes modified into a large coloured leaf; berries globose; seeds compressed.

Planted as avenue tree. Fl. & Fr.: Apr.–Sept. Guntur, (*Uhl,* 1929).

Pavetta Linnaeus

Pavetta indica L. Sp. Pl. 110. 1753; FBI 3: 150; Gamble 633 (446).

Small trees, up to 5 m tall; leaves elliptic-lanceolate, 10–15 × 4–5 cm, glabrous, acute, base obtuse; flowers white or greenish white in terminal corymbs, drupes globose, black.

Common in Loyapalle section. Fl. & Fr.: June–Nov. *Vern.*: Papati chettu. It is used as a diuretic and purgative and for dropsy, haemorrohoidal pain, jaundice, rheumatism, ulcer and visceral obstructions. Loyapalle, *VRK* 3836; Kondavedu Fort, *CAB* 4664 (MH).

Spermacoce Linnaeus

1. Stems erect, with short internodes .... ***S. pusilla***
1. Stems procumbent with long internodes .... ***S. articularis***

Spermacoce articularis L.f. Suppl. Pl. 119. 1781. *Borreria articularis* (L.f.) F.N. Will. in Bull. Herb. Boisser. ser. 2. 5: 956. 1905. *Spermacoce hispida* L. Sp. Pl. 102. 1753; FBI 3: 200. *Borreria hispida* (L.) Schum. in Engler & Prantl Pfl. Fam. 4(4): 144. 1891; non Spruce ex Schum. 1888; Gamble 654 (461).

Annual procumbent hispid herbs; branches quadrangular; leaves elliptic-lanceolate or ovate-oblong, 1.5–3 × 0.5–1 cm, leathery, more or less scabrid on both sides, obtuse or rounded; flowers pale violet in axillary whorls; capsules 5 × 3 mm, oblong, hispid above, glabrous below, fine leathery septum in between two unicarps; seeds 2, ovoid, 2 × 1 mm, dark brown.

Common weed in open forests. Fl. & Fr.: July–Dec. *Vern.*: Madanakada Aku. Koppanur, *VRK* 3268; Mellavagu, *VRK* 5719; V.P. Singh, *VRK* 5817; Jayanthipurm, *VRK* 6975, 6937.

Spermacoce pusilla Wall. in. Roxb. Fl. Ind. 1: 379.1820, *Borreria pusilla* (Wall) DC. Prodr. 4: 453. *Borreria stricta* (L.f.) Schum. in Engler & Prantl Pfl. Fam. 4(4): 143. 1891; Gamble 654 (461). (non *B. stricta*; Mey. 1818). *Spermacoce stricta* L.f. Suppl. 120. 1781; FBI 3: 200 (non *B. stricta* Mey. 1818).

Annual, erect, small herbs, up to 40 cm tall, stems obtusely 4-angled with scabrous hairs; leaves sessile, linear or ovate-oblong, 2–5 × 0.5–0.8 cm, scabrous, acute to acuminate; stipules filiform-bristly; flowers white in dense, axillary and terminal clusters; capsules 3 x 1 mm, obovoid, glabrescent; seeds 2, 2 × 0.5 mm, brown, grooved ventrally.

Common in forests. Fl. & Fr.: Sept.–Dec. *Vern.*: Patcha nuvi. N. Varam, *VRK* 3521; Jayanthipuram, *VRK* 6976.

Tarenna J. Gaertner

Tarenna asiatica (L.) Kuntze ex Schum. in Bot. Tidsskr. 24. 332. 1902. *Rondeletia asiatica* L. Sp. Pl. 172. 1753. *Chomelia asiatica* (L.) Kuntze Rev. Gen. Pl. 1: 278. 1891; Gamble 613 (432). *Webera corymbosa* Willd. Sp. Pl. 1. 1224. 1798; FBI 3: 102.

Small trees, up to 5 m tall; branchlets glabrous, leaves oblong to oblanceolate, 8–12 × 2–4 cm, glabrous, shining, turning black on drying, acute, entire, base truncate to attenuate; flowers creamish, fragrant, in terminal corymbose cymes; berries black, globose with a crown of calyx lobes; seeds 3–4 in each cell.

Occasional in the forests. Fl. & Fr.: Apr.–Nov. *Vern.*: Komi chettu. Ripe fruits are eaten raw. The resin exuding leaf buds are made into a paste and used as fixative to fix wooden handles of sickles, knives and arrows. Loyapalle, *VRK* 3848.

ASTERACEAE (*nom. alter.* COMPOSITAE)

1. Achenes armed:
    2. Heads homogamous in axillary or terminal clusters .... *Xanthium*
    2. Heads heterogamous, axillary, solitary .... *Acanthospermum*
1. Achenes not armed:
    3. Flowers in axillary heads, sessile; marshy succulent herbs .... *Caesulia*
    3. Flowers not in axillary sessile heads; not succulent:
        4. Involucral bracts spiny:
            5. Leaves glabrous, pinnatified, spiny; ovary linear, 6 mm long, pubescent .. .. *Echinops*
            5. Leaves velvety, simple, entire, not spiny, ovary obovoid, 2.5 mm long, densely villous .... *Dicoma*
        4. Involucral bracts not spiny:
            6. Plants white woolly .... *Gnaphalium*
            6. Plants not woolly:
                7. Prostrate or procumbent herbs:
                    8. Leaves radical, rosette .... *Launaea*
                    8. Leaves cauline:
                        9. Leaves alternate, decumbent:
                            10. Leaves pinnatisect .... *Glossocardia*
                            10. Leaves entire or dentate .... *Sphaeranthus*
                        9. Leaves opposite, procumbent:
                            11. Heads axillary .... *Eclipta*
                            11. Heads terminal .... *Tridax*
                7. Erect herbs:
                    12. Leaves opposite:
                        13. Heads homogamous:
                            14. Leaves pubescent; pappus silky white, copious, heads 1-5-flowered .... *Lagascea*
                            14. Leaves glabrescent, pappus absent, heads more than 15-flowered .... *Ageratum*
                        13. Heads heterogamous:

15. Heads subsessile, solitary .... *Synedrella*
15. Heads stalked, in panicles .... *Bidens*
12. Leaves alternate:
16. Heads with ligulate florets:
17. Leaves radical and cauline .... *Emilia*
17. Leaves cauline:
18. Leaves pinnatifid .... *Parthenium*
18. Leaves entire:
19. Florets white, achenes not ciliated .... *Blainvillea*
19. Florets yellow, achenes ciliate or hairy .... *Pulicaria*
16. Heads with all tubular florets:
20. Anther-bases tailed:
21. Leaves amplexicaul .... *Pentanema*
21. Leaves petiolate .... *Blumea*
20. Anther-bases not tailed:
22. Bracts spinous, pappus silvery brown .... *Oligochaeta*
22. Bracts not spinous, pappus silky white:
23. Heads solitary, flowers yellow .... *Grangea*
23. Heads in corymbs, flowers pink or white .... *Vernonia*

Acanthospermum Schrank

Acanthospermum hispidum DC. Prodr. 5: 522. 1836; Gamble 704 (495).

Annual erect hispid herbs, up to 60 cm tall; leaves subsessile, opposite, obovate, oblanceolate or spathulate, 3–6 × 1.5–3.5 cm, scabrous, acute, coarsely serrate, base cuneate; flowers yellow in heterogamous heads in the forks of the dichotomous branches; involucral bracts 2-seriate, outer lanceolate, inner connate; achenes compressed, softly prickly with 2 long straight apical awns, pappus absent.

Common weed of wastelands and cultivated fields. Native of Brazil, introduced into India. Fl. & Fr.: Aug.–Jan. *Vern.*: Seemapalleru. V.P. South, *VRK* 3542.

Ageratum Linnaeus

Ageratum conyzoides L. Sp. Pl. 839. 1753; FBI 3: 243; Gamble 677 (476).

Annual erect pubescent herbs, up to 80 cm tall; leaves opposite, 6–10 x 4–6 cm, broadly ovate, hairy, acute, crenate, base truncate; flowers pale blue or white in terminal homogamous heads in corymbs; achenes 5-angled, black, glabrous.

Occasional in moist places. Native of South America. Fl. & Fr.: Throughout the year. Javuku, *VRK* 3915; Koppanur, *VRK* 3895.

A South American species, now well naturalized throughout India.

Bidens Linnaeus

Bidens bipinnata L. Sp. Pl. 832. 1753. *B. pilosa* L. var. *bipinnata* (L.) Hook. f. Fl. Brit. Ind. 3: 309. 1881; Gamble 710 (499).

Annual erect herb to 75 cm tall; branchlets striate; leves opposite, lower ones bipinnately compound, upper most ones pinnate, apex acute, serrate, bark obtuse, lateral segments 2–4 × 1–1.5 cm, terminal, 5–7 × 1–2 cm, heads heterogamous in terminal divaricate branches, yellow; achenes 4-ribbed.

Rare in open forests of Davupalli.

Blainvillea Cassini

Blainvillea acmella (L.) Phillips. in Blumea 6: 350. 1950. *Verbesina acmella* L. Sp. Pl. 901. 1753. *Eclipta latifolia* L. f. Suppl. Pl. 378. 1781. *Blainvillea latifolia* (L.f.) DC. in Wight Contrib. Bot. India. 17: 1834; FBI 3: 305. *B. rhomboidea* Cass. in Cuvier. Dict. Sci. Nat. 29: 494. 1823; Gamble 706 (496).

Erect herbs, up to 1.5 m tall; branches striate; scabrid; leaves opposite, ovate, 3–8 × 2–4.5 cm, hairy on both sides, acute to acuminate, serrate, base cuneate; flowers yellow or white, heads peduncled, solitary, axillary, terminal or at the forks of dichotomy; achenes obovate, angled, pubescent.

Occasional, both in plains and forests. Fl. & Fr.: Aug.–Apr. V.P. south, *VRK* 5809; Kondavedu Fort, *CAB* 4646 (MH).

Note: Although *Blainvillea rhomboidea* Cass. (1823 l.c.) appears to be prior combination for the plant, but *Verbesina acmella* Linn. (1753 l.c.) being the basionym gets priority over it; hence the combination *Blainvillea acmella* (L.) Philipson (1950 lc.) is used here.

Blumea A.P. de Candolle *nom. cons.*

1. Heads yellow or white:
   2. Upper leaves opposite, lower leaves alternate, heads yellow .... ***B. bifoliata***
   2. All leaves alternate, heads white .... ***B. eriantha***
1. Heads rose, pink or purple:
   3. Leaf-base acute, capitula more than 5 .... ***B. mollis***
   3. Leaf-base semi-amplexicaul, capitula 1 or 2 .... ***B. obliqua***

Blumea bifoliata (L.) DC. in Wight Contrib. Bot. Ind. 14. 1834; FBI 3: 261; Gamble 686 (482); Randeria in Blumea 10: 285. 1960. *Conyza bifoliata* L. Sp. Pl. 1207. 1753.

Much branched, erect, villous herbs, up to 40 cm tall; leaves lower ones spathulate or oblanceolate, upper ones obovate or elliptic, 2–7 × 1–3 cm, hairy, acute, dentate, base in lower ones attenuate, upper ones semi-amplexicaul, flowers yellow, heads solitary in the terminal axils; achenes narrowly oblong, angles hairy; pappus white, uniseriate.

Occasional in forests. Fl. & Fr.: Feb.–May. Koppanur, *VRK* 3263.

Blumea eriantha DC. in Wight Contrib. Bot. Ind. 15. 1834; FBI 3: 266; Gamble 686 (482).

Slender erect herbs, up to 50 cm tall; leaves obovate or oblanceolate, 2–11 × 1–3 cm, membranous, densely silky, acute, base attenuate, margin spinulose, dentate, sessile; flowers yellow, heads on long slender peduncles of dichotomous cymes; achenes oblong to elliptic-oblong, hairy.

Occasional weed along the streams. Fl. & Fr.: Jan.–May. *Vern.*: Chepputhattaku. Decoction of the root is given internally to treat rabies. Koppanur, *VRK* 3870; Krishna river, *VRK* 6739.

Blumea mollis (D. Don) Merr. in Philipp. J. Sci. (Bot.) 5: 395. 1910; Randeria in Blumea 10: 261. 1960. *Erigeron molle* D. Don Prodr. Fl. Nepal. 172. 1825. *Blumea wightiana* DC. in Wight Contrib. Bot. Ind. 14. 1834; Gamble 686 (483). *B. neilgherriensis* Hook. f. Fl. Brit. India 3: 261. 1881; Gamble 686 (483).

Annual, erect villous herbs, up to 60 cm tall; stem grooved, villous, leves elliptic-obovate or lanceolate, 3–10 × 2.5–4 cm, thinly pubescent, serrate, usually scented; flowers yellow, heads axillary and terminal, solitary; achenes hairy, minutely angled; pappus pink.

Occasional weed of wastelands and forests. Fl. & Fr.: Oct.–Mar. Krishna river, *VRK* 6732; Anupu, *VRK* 3589.

Blumea obliqua (L.) Druce in Rep. Bot. Soc. Exch. Club. Brit. Isles 4: 609. 1917. *Erigeron obliquum* L. Mant. Pl. 573. 1771. *Blumea amplectans* DC. in Wight Contrib. Bot. Ind. 13. 1834; FBI 3: 260; Gamble 686 (482).

Erect annual herbs, up to 1 m tall; stems covered by woolly sticky glandular hairs; leaves simple, elliptic-oblong to obovate, 5–6 × 0.5–3 cm, dentate-serrate, base attenuate, acute, aromatic, villous-tomentose; flowers yellow in terminal heads; achenes oblong and ribbed.

Common weed of waste places and clultivated fields. Fl. & Fr.: July–Dec. Macherla RF, *VRK* 3263; Sattenapally, *VRK* 5867; Satrasala, *VRK* 6782.

Caesulia Roxburgh

Caesulia axillaris Roxb. Cor. Pl. t. 93. 1798; FBI 3: 291; Gamble 704 (494).

Annual prostrate or erect herbs, up to 30 cm tall; leaves alternate, linear or linear oblong, narrowly lanceolate, 8–13 × 0.5–1.5 cm, glabrous, acute, remotely denticulate, sheathing base embracing the cluster of heads; flowers white in axillary homogamous heads; achenes flat, winged, brown, smooth, glabrous.

Occasional weed of marshy places. Fl. & Fr.: Oct.–Apr. B.K. Palem, Macherla RF, *VRK* 3277.

Dicoma Cassini

Dicoma tomentosa Cass. Bull. Sci. Soc. Philom Paris 1818; 12. 1818; FBI 3: 387; Gamble 728 (512). *D. lanuginosa* DC. in Wight, Contrib. Bot. India 26. 1834.

Rigid herbs, up to 75 cm tall; branchlets densely covered with cottony wool; leaves alternate, clustered, obovate or linear-lanceolate, 3–5 × 0.3–0.6 cm, thick, pilose above, white-woolly below, dentate, apiculate, base cuneate; capitulum solitary, axillary, homogamous, disciform, florets whitish, bisexual; achenes obovate, slightly enlarged, with densely silky hairs.

Occasional in plains. Fl. & Fr.: Dec.–Mar. Nagarjunakonda, *KTH* 9836 (CAL).

Echinops Linnaeus

Echinops echinatus Roxb. Fl. Ind. 3: 447. 1832; FBI 3: 358; Gamble 724 (509).

Erect white woolly, much branched, armed, woody herbs, up to 60 cm tall; leaves alternate, oblong-lanceolate, pinnatifid, 8–12 × 4–5 cm, (including spine tips), pilose above, white-woolly below, lobes ending in sharp spines; flowers greenish white or pale yellow in terminal, solitary, homogamous, globose, compound heads; outer involucre spiny; achenes elongate, angled, villous.

Common weed of waste lands and open forests. Fl. & Fr.: Oct.–Apr. *Vern.*: Gunta-kala garaku. It is used as an alterative, aphrodisiac and diuretic and for cough, dyspepsia, hysteria, ophthalmic affections and scrofula. The paste of roots is applied to cure wounds. Koppanur, *VRK* 1294.

Eclipta Linnaeus *nom. cons.*

Eclipta prostrata (L.) L. Mant. Pl. 286. 1771. *Verbesina prostrata* L. Sp. Pl. 902. 1753. *Eclipta alba* (L.) Hassk. Pl. Jav. Rar. 528. 1848; FBI 3: 304; Gamble 705 (496). *Verbesina alba* L. Sp. Pl. 902. 1753.

Annual erect or prostrate herbs with ascending branches, up to 50 cm tall; leaves oblong-lanceolate, 2–6 × 1–3 cm, hirsute, serrate-dentate, apex acute, base obtuse; flowers white in axillary or terminal heterogamous heads; achenes oblong, hairy, triquetrous.

Common weed in moist localities. Plant extract ingredient of hair oils, and herbal hair dye. Fl. & Fr.: Nov.–May. *Vern.*: Bhringa Raj, Guntagalagara. V.P. South, *VRK* 3544; Krishna river, *VRK* 6731; Nizampatnam, *VRK* 6792.

Note: Verbasina prostrata, *V. alba* and *V. pseudoacmella* were the three combinations made by Linnaeus (in his Species Plantarum, 1753, page 902) for three different taxa, which in course of time have been worked out to be conspecific. The problem then as to which of the three should be selected as the basionym. Linnaeus himself in 1771 (loc. cit.) reduced the three into two species merging *V. alba* and *V. prostrata* into one–*Eclipta prostrata*. In the light of Art. 57 of the code (Int. code of Bot. Nom. 1978) the priority is to be given to the combination chosen by the author who fuses the two taxa and as such *Eclipta prostrata* (L.) L. is the valid name for this taxon and the combination *E. alba* made by Hassk. (1848 loc. cit.) on the basis of *Verbasina alba* L. (1753) does not stand since Linnaeus himself in 1771 (loc. cit.) has preferred the other epithet.

Emilia Cassini

Emilia sonchifolia (L.) DC. in Wight Contrib. Bot. Ind. 24. 1834; FBI 3: 336; Gamble 716 (503). *Cacalia sonchifolia* L. Sp. Pl. 335. 1753.

Annual erect herb to 50 cm tall; leaves both radical and cauline, 6–9.5 × 1.6–2.5 cm, lower ones lyrately lobed, spathulate, base auricled, margin dentate, apex obtuse, sessile; heads pink in loose terminal corymbs, homogamous; achenes angled, 5-ribbed.

Occasional, in moist places. Flowers with a peak during Dec.–Apr. Kondavedu fort, *SSR & KSM* 21465.

Glossocardia Cassini

Glossocardia bosvallea (L.f.) DC. in Wight Contrib. Bot. Ind. 19. 1834; Gamble 709 (499). *Verbesina bosvallea* L.f. Suppl. 379. 1781. *Glossocardia linearifolia* Cass. in Dict. Sci. Nat. 19: 62. 1821; FBI 3: 308.

Annual aromatic, prostrate or ascending herbs, up to 35 cm tall; leaves bipinnatisect, pinnules filiform, 1–2 × 0.1–0.4 cm, puberulous, base decurrent, apex apiculate; flowers yellow in terminal and axillary heterogamous heads; achenes compressed, black, ciliate along angles.

Occasional weed of cultivated fields, wastelands and roadsides. Fl. & Fr.: Aug.–Oct. *Vern.*: Thuraka peesu. V.P. South, *VRK* 5808; Jayanthipuram, *VRK* 6964; Kondavedu Fort, *CAB* 4649 (MH).

Gnaphalium Linnaeus

Gnaphalium polycaulon Pers. Syn. Pl. 2: 421. 1807. *G. indicum auct. non* L.; FBI 3: 289; Gamble 698 (491).

Annual white woolly slender herbs, up to 15 cm tall; branches decumbent, diffuse; leaves 0.5–2.5 × 0.2–0.5 cm, obovate, white woolly, apiculate.

entire, base cuneate; heads white in terminal spikes, heterogamous, silvery woolly, achenes oblong, minutely scaly, pappus hairy.

Common in moist places especially on the banks of ponds and lakes. Fl. & Fr.: Sept.–Mar., Vinukonda RF, *VRK* 6871.

Grangea Adanson

Grangea maderaspatana (L.) Poir. in Lam. Encycl. (Suppl. 2): 825. 1811; FBI 3: 247; Gamble 680 (478). *Artemesia maderaspatana* L. Sp. Pl. 849. 1753.

Prostrate spreading radially pubescent herbs, to 30 cm long; branches striate, hirtellous-pubescent; leaves 2–7.5 cm long, sinuately pinnatifid, lobes opposite, oblong, 0.5–1 × 0.3–0.6 cm, apiculate, entire; flowers yellow, red with age, heads axillary and terminal, solitary, globose; achenes obovate, truncate, glandular.

Occasional weed in wastelands and cultivated fields. Fl. & Fr.: July–Mar. *Vern.*: Savi, Masipatri, Nemali-padham. Vinukonda RF, *VRK* 6878; Nizampatnam, *VRK* 6799.

Lagascea Cavanilles

Lagascea mollis Cav. Anales Cienc Nat. 6: 332. t. 44. 1803; FBI 3: 302; Gamble 703 (494).

Annual, erect, white silky pubescent herbs, up to 50 cm tall; stems grooved; leaves opposite and alternate, elliptic-ovate, rhomboid, 3–5 × 1–3 cm, white pubescent, acuminate, base acute, capitula 1-flowered, aggregated, in terminal long peduncled compound inflorescence, subtended by leafy bracts; achenes compressed, densely pubescent, pappus white.

Common in moist places and roadsides and occasional in open forests. Native of Mexico. Fl. & Fr.: Sept.–May. *Vern.*: Nallagada. The juice of the fresh leaves used as a styptic. Koppanur, *VRK* 3251.

Launaea Cassini

Launaea sarmentosa (Willd.) Sch.-Bip. ex O. Kuntze Rev. Gen. Pl. 1: 350. 1891. *Prenanthes sarmentosa* Willd. Phyt. 10: 35. t. 6. f. 2. 1794 et Sp. Pl. 3: 1540. 1803. *Launaea pinnatifida* Cass. in Ann. Sc. Nat. (Ser. 1) 23: 85. 1831. FBI 3: 416; Gamble 733 (515).

Trailing herbs with milky sap, rooting at nodes; leaves radical and cauline, fleshy, 3–4 × 5–6 cm, oblong, pinnatifid or sinuate-lobed, dentate, acute, glabrescent; flowers pale yellow in terminal heads; heads solitary or in clusters of 1–2; achenes pale, pappus silky, obtusely 4-gonous.

A common weed along the sandy sea coast. It is one of the good sand-binding species. Fl. & Fr.: Aug.–Mar. Nizampatnam coast, *VRK* 6806.

Oligochaeta Koch

Oligochaeta ramosa (Roxb.) Wagentiz Veroff. Geobot. Inst. ETH Stiftung Rubel Zurich 37: 323. 1962. *Carduus ramosus* Roxb. Fl. Ind. 3: 407. 1832. *Voluterella divaricata* Benth & Hook. f. Gen. Pl. 2: 476. 1873, *p.p.*, FBI 3: 383; Gamble 727 (511).

Erect, dichotomously branched herbs, up to 60 cm tall; stems stiff, white tomentose, rough; leaves alternate, oblanceolate, up to 3 × 0.6 cm, deeply lobed, lobes with small white spines, roughly pubescent, base decurient; flowers violet-white in terminal homogamous solitary heads; involucre bracts spinescent at tip; achenes elongated, oblong, ribbed, brown with silvery pappus.

Occasional weed in wastelands and roadsides. Fl. & Fr.: Nov.–Apr. Nagarjunakonda, *VRK* 6703.

Parthenium Linnaeus

Parthenium hysterophorus L. Sp. Pl. 988. 1753; Rao in J. Bombay Nat. Hist. Soc. 54: 219. 1956.

Annual, erect herbs, up to 1.5 m tall; stems angular; leaves alternate, 4–10 x 1–3 cm, pinnatifid, pinnules opposite, hairy; flowers white in terminal heads, panicled, achenes compressed.

Abundant in open forests and roadsides especially near human habitations and fast spreading all over. Native of tropical and subtropical America. It was first reported in India by Rao in 1956. Fl. & Fr.: Throughout the year. *Vern.*: Vayyaribhama; Pitchi Machipatri. N. Varam, *VRK* 3233.

The carrot weed or congress weed. Surface hairs possess 'Parthenin'—a sesquiterpene lactone; an irritant of skin & eyes, pollen allergic. World's worst weed. Towers from Canada, Dr, P.V. Subba Rao from IISc, Bangalore contributed to its biology and control. Several biotypes are possible. Defies all control measures. Safe recommendation is uprooting the plants before flowering, near human dwellings and playgrounds.

Pentanema Cassini

Pentanema indicum (L.) Ling in Acta Phyt. Sin. 10: 179. 1965. *Inula indica* L. Sp. Pl. ed. 2. 1236. 1762. *Vicoa auriculata* Cass. in Ann. Sci. Nat. (Paris) Ser. 1. 17: 418. 1829; FBI 3: 297. *Vicoa indica* (L.) DC. in Wight Contrib. Bot. Ind. 10. 1834; Gamble 701 (493).

Annual, erect, ridged herbs, up to 80 cm tall; stem pale brown, grooved; leaves alternate, oblong-lanceolate, 3–10 × 0.5–2 cm, minutely scabrous, minutely serrulate, apex acute, base amplexicaul, auriculate, flowers bright yellow in terminal, solitary, heterogamous heads; achenes obovate, villous.

Occasional weed in forests and plains. Fl. & Fr.: Nov.–Apr. Javuku, *VRK* 3925; Vijayapuri, *KMS* 9852 (MH).

Pulicaria Gaertner

Pulicaria wightiana (DC.) Clarke Comp. Ind. 118. 1876; Gamble 702 (493). *Poloa wightiana* DC. Arch. Bot. (Paris) 2: 515. 1833.

Annual erect, hairy or glabrescent herbs, up to 50 cm tall; leaves alternate, oblong-lanceolate or linear-ovate, amplexicaul at base, 2.5–6 × 0.4–1 cm, sessile, glabrescent to hairy, serrate-dentate; flowers yellow in terminal, solitary heads; achenes oblong, ribbed, nearly glabrescent.

Occasional weed in waste places and in cultivated crops. Fl. & Fr.: Sept.–Feb. Mellavagu, *VRK* 5710; V.P. South, *VRK* 5804.

Sphaeranthus Linnaeus

Sphaeranthus indicus L. Sp. Pl. 927. 1753; FBI 3; 275. *p.p.,* Gamble 692 (487).

Prostrate or ascending, aromatic, glandular herbs, up to 30 cm tall; branches with dentate wings; leaves oblong-spathulate, 3–7 × 0.5–3 cm, hairy, coarsely serrate-dentate, somewhat spiny, apex obtuse, subsessile, flowers purple to pink, in globose, terminal heterogamous heads, with stout peduncle, achenes smooth or angled, glabrous or villous.

Common weed in dried tanks and moist waste places. Fl. & Fr.; Dec.–May. *Vern.*: Bodataramu; valuable medicinal species as microbicide and blood purifier. Leaves are used to cure elephantiasis. Nallapadu, *VRK* 6766.

Synedrella J. Gaertner *nom. cons.*

Synedrella nodiflora (L.) Gaertn. Fruct. 2: 456. t. 171. 1791; FBI 3: 308; Gamble 708 (498). *Verbesina nodiflora* L. Cent. Pl. 1: 28. 1755.

Annual erect herb, up to 70 cm tall; branches obtusely angled, grooved, sparsely pilose when young; leaves ovate-lanceolate, 5–11 × 3–5 cm, scabrous above, densely so below, base rounded to attenuate, margin crenate-serrate, apex acute; flowers yellow in axillary, solitary heads; achenes of ray florets obovate to elliptic, winged, that of bisexual florets obovate-cuneate, 3-ribbed.

Common in waste lands in Loyapalli. Fl. & Fr.: Sept.–Apr.

Tridax Linnaeus

Tridax procumbens L. Sp. Pl. 900. 1753; FBI 3: 331; Gamble 711 (500).

Annual procumbent hirsute herbs, up to 30 cm tall; leaves elliptic, 2–3 x 1–1.5 cm, strigose on both sides, deeply dentate, acute at base; flowers yellowish-white in terminal heterogamous heads; achenes compressed, obconical, ribbed with silky white pappus.

Common weed in open forests near fields. A native of Central America. Fl. & Fr.: Throughout the year. *Vern.*: Gayapaku. The powder of the leaves is topically applied on cuts and wounds as an antiseptic. Leaf juice is dropped into the eyes for treatment of inflammation. N. Varam, *VRK* 3231; Loyapalle, *VRK* 3849.

Vernonia Schreb. *nom. cons.*

1. Phyllaries sparsely pubescent, apex glabrescent, outer pappus > 0.5 mm ... *V. albicans*
1. Phyllaries densely pubescent throughtout, outer pappus < 0.3 mm .... *V. cinerea*

Vernonia albicans DC. in Wight Contrib. Bot. Ind. 6: 1834; Gamble 676 (475). *V. cinerea* sensu FBI 3: 233 *p.p.*

White tomentose, erect herbs, up to 50 cm tall; branches striate, leaves alternate, ovate-lanceolate, 3–6 × 0.8–2 cm, pubescent above, tomentose below, obtuse, serrate, base acute-cuneate; flowers purple in axillary and terminal panicled heads; achenes villous, angled.

Occasional in moist waste places. Fl. & Fr.: Nov.–Jan. Sattenapally, *VRK* 5846.

Vernonia cinerea (L.) Less. in Linnaea 4: 291. 1829; FBI 3: 233; Gamble 676 (475). *Conyza cinerea* L. Sp. Pl. 862. 1753.

Annual erect herbs, up to 60 cm tall; stems grooved, hairy; leaves alternate, ovate, broadly elliptic or lanceolate, 2–5 × 1–3 cm, glabrescent above, pubescent below, undulately toothed, obtuse, serrate; flowers pink in terminal panicled heads; achenes angled, silky pappus.

Common in roadsides and occasional in forests. Fl. & Fr.: Nov.–Jan. Eng.: Ash coloured feabone. Tel.: Saha devi. Whole plant extract is given against urinary infection and abdominal pain. Fresh juice of the leaves used against eczema and ringworm, Koppanur, *VRK* 3281; Kondavedu fort, *CAB* 4650 (MH).

Xanthium Linnaeus

Xanthium strumarium L. Sp. Pl. 987. 1753. *p.p.;* FBI 3: 303. *p.p.;* Gamble 703 (494).

Scabrous undershrubs, up to 1.5 m tall; stem grooved, hispid-pubescent; leaves subcordate, or acute 3-lobed, 5–8 × 5–9 cm, glandular, rough, margin coarsely serrate-crenate, irregularly toothed; flowers yellowish green in axillary or terminal monoecious heads; achenes enlarged in prickly glandular envelop; pappus absent.

Common in moist open forests and roadsides. Native of Europe (Ridley 1930) but Srivastava (1964) considers it as a native of South America. Fl. & Fr.: Throughout the year. *Vern.*: Marula matangi. Juice of leaves is applied to cure wounds and cuts. Leaf paste with turmeric applied externally to treat skin diseases. N. Varam, *VRK* 3230.

Artemesia nilagirica (Clarke) Pampan., *Cosmos caudatus* Kunth. *Dahlia imperialis* Roezel ex E. Ort., *Chrysanthemum indicum* L., *Tagetes patula* L., *Vernonia elaeagnifolia* DC., *Solidago canadensis* L. Golden rod, *Rudbeckia bicolor*, *Cosmos bipinnatus*, *C. sulphureus* Cav., *Aster, Eupatorium, Ageratum, Zinnia* sp. (multicoloured), are some of the cultivated garden favourites in the district. *Helianthus annuus* L. is cultivated for the oil.

PLUMBAGINACEAE

Plumbago Linnaeus

Plumbago zeylanica L. Sp. Pl. 151. 1753; FBI 3: 480; Gamble 2: 744 (524).

Undershrub; stems spreading, terete, striate, glabrous; leaves elliptic ovate, 3–7.5 × 2–3.7 cm, glabrous, somewhat glaucous beneath, apex acutely apiculate, entire, base shortly and abruptly attenuated into a short petiole; flowers white, in elongate, terminal, often panicled racemes; calyx glandular, narrowly tubular; capsule oblong, acute with five furrows.

Occasional in open forests among bushes. Fl. & Fr.: Sept.–Apr. *Vern.*: Tella chitramulamu. Leaf paste mildly heated and massaged for chest pain, muscle pain and sprains; also applied on wounds. Roots of this paste and applied externally to treat elephantiasis.

MYRSINACEAE

Aegiceras J. Gaertner

Aegiceras corniculatus (L.) Blanco, Fl. Filip. 1: 79. 1837; Gamble 757 (532); *Rhizophora corniculata* L. Amoen. Acad. 4: 123. 1759. *Aegiceras majus* Gaertn. Fruct. Sem. Pl. 1: 216. t. 46. f. 1. 1788.

Small trees up to 5 m tall; leaves alternate, opposite at apices, obovate or oblong-obovate, 4–6 × 2–4 cm, coriaceous, base cuneate, entire, obtuse to emarginate, glabrous; flowers white in axillary and terminal umbels; fruit falcate, 5–7 cm long, cylindrical, curved, sharply pointed, seeds elongate.

Common mangrove tree along the sea coast at Kothapalem and Nizampatnam. Fl. & Fr.: Mar.–July. *Vern.*: Dudumara, Guggilam. Bark is used for tanning. The pretty white flowers are made into garlands. Kothapalem, *VRK* 5766; Nizampatnam, *VRK* 6812; Nizampatnam *DLR* 104 (NUH).

SAPOTACEAE

1. Corolla lobes without petaloid appendages, calyx lobes 4 .... ***Madhuca***
1. Corolla lobes with petaloid appendages, calyx lobes 6–8:
    2. Leaf-apex acute to acuminate, calyx-lobes and stamens 8 each .... ***Mimusops***
    2. Leaf-apex obtuse to emarginate, calyx-lobes and stamens 6 each .... ***Manilkara***

Madhuca Linnaeus

Maduca indica J. Gmelin Syst. Nat. 2(1): 799. 1791. *M. longifolia* (J. Koeing) Macbr. var. *latifolia* A. Chev. Rev. Int. Bot. Appl. Agric. Trop. 23: 149. 1943. *Bassia latifolia* Roxb. Pl. Cor. t. 19. 1795 & Fl. Ind. 2: 526. 1832; FBI 3: 544; Gamble 763 (536).

Deciduous tree, up to 15 m tall, trunk short; branchlets glabrous, spreading; leaves clustered near the ends of branches, thick, elliptic or elliptic-oblong, 8–15 × 2–5 cm, chartaceous, glabrescent, apex acute or shortly acuminate, margin entire, base rounded or acute; flowers creamish, dense in fascicles on drooping densely tomentose pedicels; berry ovoid-oblong, 4 × 2 cm, fleshy, greenish; seeds 1–4, brown, smooth.

Rare in deciduous forests. It is a drought resistant hardy species, needs bright sunshine and coppices well. Fl. & Fr.: Feb.–July. Eng.: Indian butter tree. *Vern.*: Ippa, Ippi: Corolla of flowers contain much sugar and are an important article of food and for the distillation of spirit. The fruits have large fleshy cotyledons which give valuable edible oil. Wood very hard, heavy to very heavy used for all kinds of building purposes. Kakirala, *DMR* 21129.

Manilkara Adanson *nom. cons.*

Manilkara hexandra (Roxb.) Dubard in Ann. Mus. Col. Mars. (Ser. 3) 23: 9. f. 2. 1915; Roy in Blumea 7: 408. 1953. *Mimusops hexandra* Roxb. Pl. Cor. t. 15. 1795; FBI 3: 549; Gamble 766 (538).

Evergreen trees, up to 8 m tall, forming a dense crown, leaves clustered at the ends of branches, elliptic-obovate, 5–9 × 2–5 cm, shining above, emarginate, entire,base obtuse or acute; flowers white, solitary or in fascicles; berries ellipsoid, reddish-yellow, smooth; seeds ovoid, reddish brown, 1–1.5 cm long, smooth.

Occasional in dry evergreen forests. Fl. & Fr.: July–Jan. It favours bright sunshine. Eng.: Milk fruit tree. *Vern.*: Puttapala, Pala Nemmi, Manchi pala.

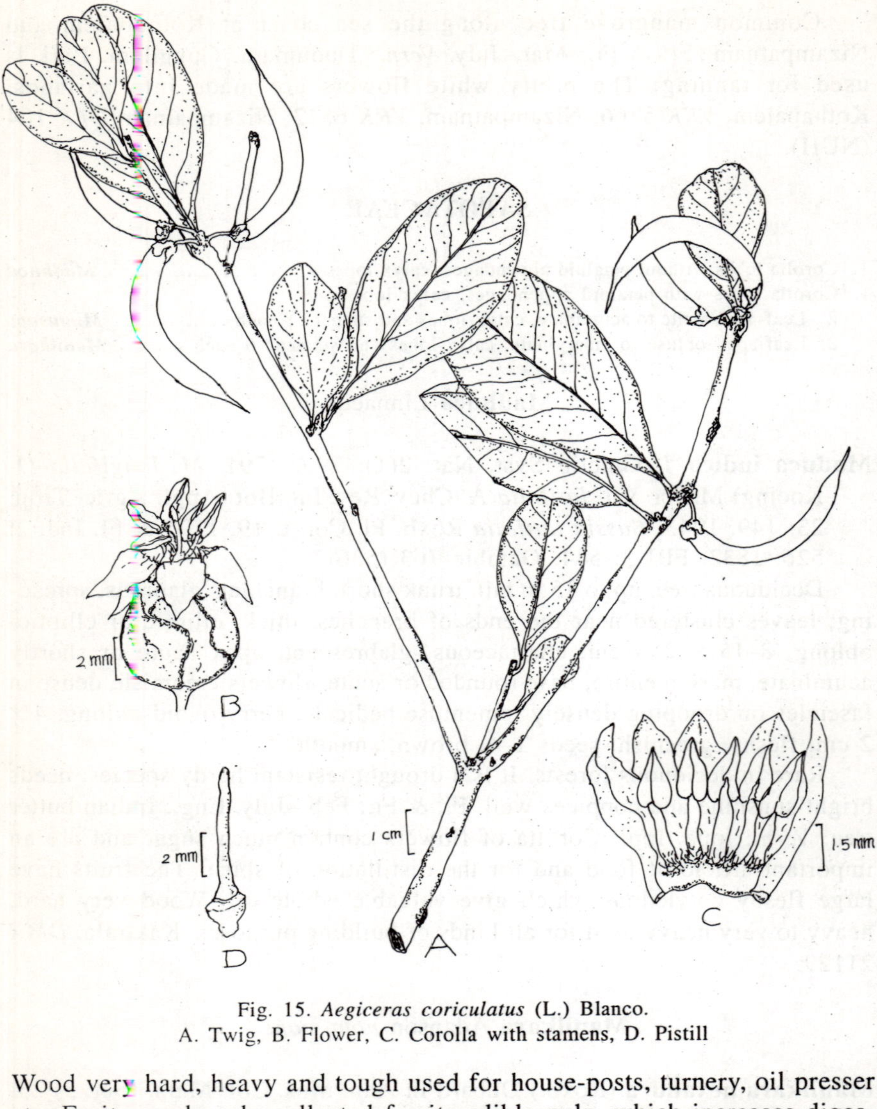

Fig. 15. *Aegiceras coriculatus* (L.) Blanco.
A. Twig, B. Flower, C. Corolla with stamens, D. Pistill

Wood very hard, heavy and tough used for house-posts, turnery, oil presser etc. Fruits are largely collected for its edible pulp, which increases digestion. Branches used in religious ceremonies. Latex applied on teeth and gums against toothache. Mellavagu, *VRK* 576.

Mimusops Linnaeus

Mimusops elengi L. Sp. Pl. 349. 1753; FBI 3: 548; Gamble 765 (538); Royen in Bumea 6: 594. 1952.

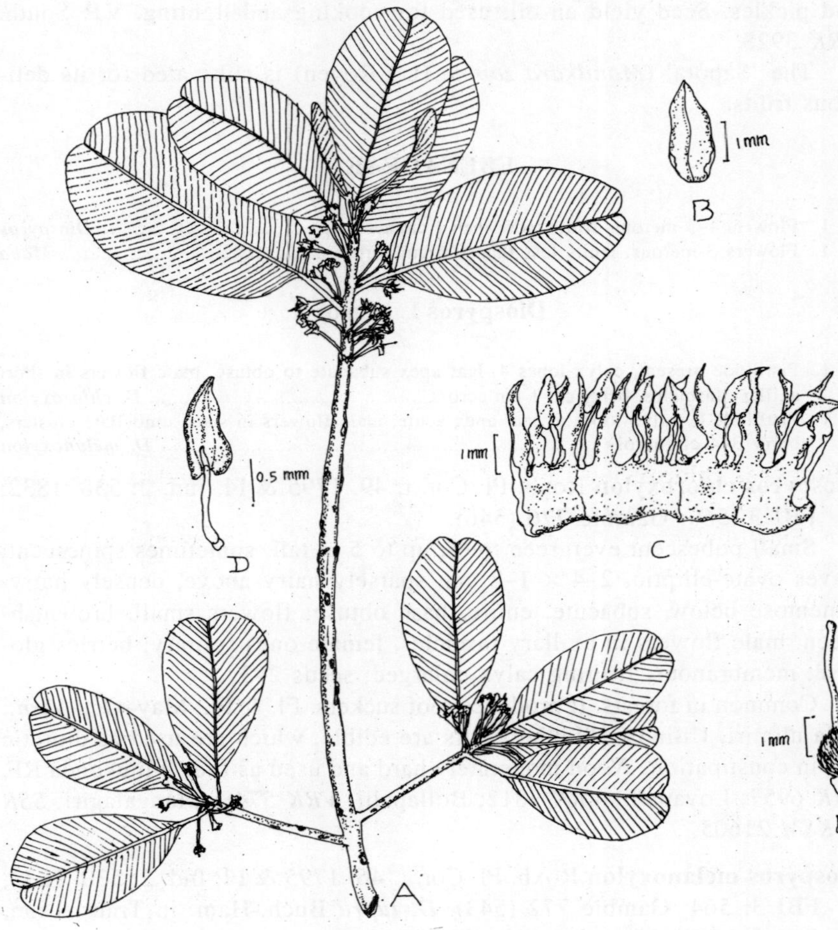

Fig. 16. *Manilkara hexandra* (Roxb.) Dubard.
A. Twig, B. Calyx lobe, C. Corolla with stamens, D. Stamen, E. Pistil.

Evergreen handsome trees, up to 10 m tall; leaves variable, elliptic-oblong to lanceolate, 5–12 × 2–5 cm, coriaceous, acute-acuminate, entire, base truncate to acute; flowers brownish-white, scented, solitary or a few to upper axils; berries orange-yellow, fleshy, 1.5 cm across, ovoid; seed solitary, brownish-black, shining, compressed, smooth with hard testa and soft whitish albumin.

Common in plains and rare in forests. Fl. & Fr.: Feb.–Sept. Eng.: Mograh. *Vern.*: Pagada, Vakulamu. Eng.: Bullet wood. Young twigs are used for cleaning teeth. The fragrant flowers are used for making garlands and a perfume is distilled from them. Fruits edible, sometimes used for making preservaties and pickles. Seed yield an oil, used for cooking and lighting. V.P. South, *VRK* 3928.

The 'Sapota' (*Manilkara zapota* (L.) Royen) is cultivated for its delicious fruits.

EBENACEAE

1. Flowers 4–5-merous, fruits more than 1 cm across .... ***Diospyros***
1. Flowers 3-merous, fruits less than 0.6 cm across.. ***Maba***

Diospyros Linnaeus

1. Pistillode present, calyx-lobes 4, leaf apex subacute to obtuse, male flowers in short axillary fascicles, berry ca. 1 cm across .... ***D. chloroxylon***
1. Pistillode 0, calyx-lobes 5, leaf apex acute, male flowers in short umbellate clusters, berry 3–4 cm across .... ***D. melanoxylon***

Diospyros chloroxylon Roxb. Pl. Cor. t. 49. 1795 & Fl. Ind. 2: 538. 1832; FBI 3: 560; Gamble 776 (546).

Small pubescent evergreen trees, up to 5 m tall, sometimes spinescent; leaves ovate-elliptic, 2–4 × 1–2 cm, sparsely hairy above, densely hairy-tomentose below, subacute, entire, base obtuse; flowers small, brownish-green, male flowers in axillary fascicles, female ones solitary; berries globose, membranous, fruiting calyx enlarged; seeds 2–3.

Common in forests. It produces root suckers. Fl. & Fr.: May–Oct. *Vern.*: Nela ulimiri, Ullinda, Ellinda. Fruits are edible, which are used as diuretic and in constipation. Wood moderately hard and used as fuel. Vinukonda RF, *VRK* 6957; Loyapalli, *VRK* 3812; Bollapalli, *VRK* 5748; Mangalagiri, *SSR & KSM* 21605.

Diospyros melanoxylon Roxb. Pl. Cor. t. 46. 1795 & Fl. Ind. 2: 530. 1832; FBI 3: 564; Gamble 772 (543). *D. tupru* Buch.-Ham. in Trans. Linn. Soc. London 15: 111. 1827; FBI 3: 563.

Small trees, up to 6 m tall, with soft hairy branches; leaves broadly elliptic-oblong, 6–15 × 5–9 cm, glabrous above, thinly pubescent below, obtuse, entire, base rounded; flowers unisexual, male white, in axillary umbellate cymes, female solitary; berries globose, 3–4.5 cm across, yellow when ripe, smooth, supported by thick calyx; seeds 2–8, oblong, compressed.

Common in forests. It is good coppicer. Fl. & Fr.: Nov.–June. Eng.: Ebony, Tendu. *Vern.*: Tumkaku, Beedi aku. The leaves are extensively used as wrapper for local cigarettes known as bidies. The leaves are diuretic, carminative, laxative and styptic. The bark is astringent and its decoction is used in diarrhoea and dypspepsia. Fruits are edible, astringent. Wood hard, used for posts, rafters etc. It is good fuel wood. Davupalli, *SSR & KSM* 21404.

Maba J.R. Forester et J.G.A. Forster

Maba buxifolia (Rottb.) A.L. Juss. Ann. Mus. Natl. Hist. Nat. 5: 418. 1804; FBI 3: 551. *p.p.;* Gamble 768 (539). *Pisonia buxifolia* Rottb. Nyesaml. Kongel Danskeuiden S.K. Salsk. Skr. 2: 536. t. y. f. 2. 1783.

Woody shrubs, up to 4 m tall; branches radiating, stiff, elongated; leaves obovate, 1–3 × 1–1.5 cm, coriaceous, obtuse to emarginate, entire, base cuneate; flowers yellowish-white in axillary fascicles, 5 mm across, crowned by calyx, berries orange when ripe; seed one.

Occasional in forests. Fl. & Fr.: Feb.–May. *Vern.*: Uti. The fruits are pulpy and edible. Wood hard, heavy and durable, used for boat anchors, handles and sheaths of weapons and rafters. Reported in Guntur Forest Division Work Plan.

OLEACEAE

Jasminum Linnaeus

Jasminum auriculatum Vahl Symb. Bot. 3:1. 1794; FBI 3: 600; Gamble 790 (555).

Pubescent vines; leaves 3-foliolate, rarely simple, lateral leaflets reduced or absent, terminal leaflet ovate-orbicular, 1.5–5 × 1–2 cm, pubescent, acute to obtuse, entire, base rounded; flowers white, scented in compound, many-flowered, pubescent, lax pedunculate, terminal cymes; berries globose, black, clustered, 7 mm across.

Common as a twinner in the forests. Fl. & Fr.: Jan.–June. *Vern.*: Adavi Jaji. Adavi Teega—Malli, Sanna Jaji. Tender shoots and leaves are consumed as vegetable after cooking. The leaf juice mixed with honey put in ear to cure pus in the ear. Leaf paste used for wounds and herps. N Varam, *VRK* 3232; Loyapalle, *VRK* 3973; Bollapalle, *JSG* 18233 (MH); Nakkirekal, *CAB* 17471 (MH).

The Arabian Jasmine includes Double flowered. *Jasminum arborescens* Roxb, known as malle, *J. multiflorum* (Burm. f.) Andr. known as Virajaji, *J. officinale* L. known as Sanna Jaji. Roots are said to be antidote for Cobra-venom. *Jasminum angustifolium* Vahl, *Jasminum grandiflorum* L., *Jasminum sambac* (L.). Aiton are cultivated commonly in gardens and back yards for their scented flowers.

NYCTANTHACEAE

Note: The genus *Nyctanthes* has been included in Oleaceae by Bentham and Hooker, while Airy Shaw included it in Verbenaceae. But authors have raised it to a separate family Nyctanthaceae.

Nyctanthes Linnaeus

Nyctanthes arbor-tristis L. Sp. Pl. 6. 1753; FBI 3: 603; Gamble 792 (556).

Erect evergreen shrubs to small trees, up to 4 m tall; leaves elliptic-ovate, 4–6 × 2–3.5 cm, glabrous above, pubescent below, acute to acuminate, entire, base obtuse; flowers white with orange streaks, very fragrant in terminal and axillary cymes; capsules obovoid, compressed, 2-valved, glabrous; epicarp thick, seeds 2.

Occasional in forests and cultivated in gardens and temples. Fl. & Fr.: Feb.–Oct. Eng.: Tree of sorrow. *Vern.*: Parijathamu, Krishti. Wood is mainly useful for fuel. The rough leaves are used in smoothening surface of the famous Kondapalle toys. The decoction of the leaf is given to treat fever. Vinukonda RF, *VRK* 6849.

SALVADORACEAE

Salvadora Linnaeus

Salvadora persica L. Sp. Pl. 122. 1753; FBI 3: 619; Gamble 799 (562).

Medium sized evergreen trees, up to 10 m tall; bark grey, deeply cracked; branchlets drooping, pale yellowish; leaves elliptic-oblong, or ovate-lanceolate, 3–6 × 1.5–3 cm, glabrous to glaucous, thick-chartaceous, obtuse, entire, base rounded to acute; flowers yellowish-green, in axillary and terminal copious panicles; drupes globular, red when ripe, smooth, juicy; seeds solitary, 2 mm across.

Rare in forests. Fl. & Fr.: Dec.–Apr. *Vern.*: Chekkar chettu. Pedda gogu. Eng.: Tooth brush tree. Twigs are used for tooth cleaning. Tree is reported to be suitable for growing in shelter-belts and as wind-breaks. Fruits are edible and leaves are used as camel fodder. Decoction of leaves is used in asthma and cough, and poultice is applied to painful tumors and piles. Javuku, *VRK* 3912.

APOCYNACEAE

1. Plants armed, fruits berries .... ***Carissa***
1. Plants unarmed, fruits other than berries:
    2. Lianas or stragglers—fruits follicles:
        3. Anthers exerted .... ***Vallaris***
        3. Anthers included .... ***Ichnocarpus***
    2. Trees, shrubs or erect or trailing herbs:
        4. Corolla-throat with corona, anthers connate around stigma, trees .... ***Wrightia***
        4. Corolla throat without corona, anthers free from stigma:
            5. Leaves verticillate or alternate, corolla funnel-shaped:
                6. Corolla yellow, fruits indehiscent drupes: .... ***Cascabela***
                6. Corolla other than yellow, fruits dehiscent follicles:

7. Deciduous trees, leaves alternate, seeds without coma .... ***Plumeria***
7. Evergreen shrubs, leaves verticillate, seeds with coma .... ***Nerium***
5. Leaves decussate, corolla salver-form:
8. Shrubs, seeds with a tuft of coma, wild .... ***Holarrhena***
8. Herbs or under shrubs, seeds without coma, introduced .... ***Catharanthus***

Carissa Linnaeus *nom. cons.*

Carissa spinarum L. Mant. 559. 1771; FBI 3: 631; Gamble 805 (565).

Bushy evergreen armed shrubs, up to 3 m tall; branches zig-zag; thorns divaricate, straight, up to 3 cm long; leaves elliptic-ovate, 3–3.5 × 1–2 cm, young leaves densely pubescent; flowers white, scented in terminal and axillary corymbose cymes; berries globose, purplish black when ripe.

Runs wild in scrub jungles. Fl. & Fr.: May–Dec. *Vern.*: Kalivi. Fruits edible. N. Varam, *VRK* 3962.

Cascabela Rafinesque

Cascabela thevetia (L.) Lippold in Feddes Report. 91: 52. 1980. *Cerbera thevetia* L. Sp. Pl. 209. 1753. *Thevetia nerifolia* Juss. ex Steud. Nom. Bot (ed. 2) 2: 680. 1841; Gamble 821 (577).

Evergreen small trees, up to 4 m tall; latex milky; leaves linear, 10–15 × 0.6–1 cm, glabrous, acuminate, entire, base narrowed; flowers yellow, fragrant, in terminal cymes; drupes angular, poisonous.

Cultivated in gardens, rarely as an escape. Native of tropical America. Fl. & Fr.: Throughout the year. *Vern.*: Pachaganneru. Fruit is used to get abortion and to treat fevers. Javuku, *VRK* 3300.

Catharanthus G. Don

1. Annual erect herbs, corolla < 1 cm across, leaves elliptic-lanceolate, apex acute to acuminate .... ***C. pusillus***
1. Perennial undershrubs, corolla > 3 cm across, leaves elliptic-obovate, apex obtusely apiculate .... ***C. roseus***

Catharanthus pusillus (Murray) G. Don Gen. Syst. 4: 95. 1837. *Vinca pusilla* Murray, Novi Comment-Soc. Regiae Sci. Gott. 3: 66. t. 2. f. 1. 1773; FBI 3: 640. *Lochnera pusilla* (Murray) Schum. in Engler & Prantl Pfl. Fam 4(2): 145. 1895; Gamble 809 (568).

Annual erect herbs, up to 50 cm tall; branchlets terete below, quadrangular above, glabrous; latex milky; leaves elliptic-lanceolate, 2–6 × 1–1.5 cm, glabrous, entire, acute to acuminate, base acute to attenuate; flowers white, in terminal or axillary corymbs; mericarps glabrous.

Occasional weed of cultivated fields and wastelands. Fl. & Fr.: July–Dec. *Vern.*: Yerrimirapa. The fresh leaves are pounded and made into a fine paste is applied on externally to cuts and wounds for healing and to relieve pain. Nagarjunakonda, *KTH* 9838 (CAL).

Catharanthus roseus (L.) G. Don Gen. Hist. 4: 95. 1837. *Vinca rosea* L. Syst. Nat. (ed. 10).944. 1759; FBI 3: 640. *Lochnera rosea* (L.) Reinchenb. Consp. Regni Veg. 134. 1828; Gamble 809 (569).

Perennial undershrubs, up to 75 cm tall; leaves elliptic-oblong or obovate, 2.5–6 x 1–3 cm, pubescent when young, glabrous with age, apiculate, entire, base cuneate; flowers white in axillary cymes; follicles up to 4 × 0.2 cm, pubescent; seeds brownish-black.

Cultivated in gardens, temples, also run wild. Fl. & Fr.: Throughout the year. Eng.: The red periwinkle. *Vern.*: Billaganneru. Roots attributed anticardiac depressant properties and source of antitumour alkaloids vinblastine, vincristine and others. Hence subjected to overexploitation. V.P. South, *VRK* 3935.

Holarrhena R. Brown

Holarrhena pubescens (Buch.-Ham.) Wall. ex G. Don Gen. Hist. 4: 78. 1837. *Echites pubescens* Buch.-Ham. in Trans. Linn. Soc. London 13: 521. 1821. *Holarrhena antidysenterica* Wall. ex DC. Prodr. 8: 413. 1844; FBI 3: 644; Gamble 811 (570).

Deciduous small trees, unarmed, up to 5 m tall; bark thick, brown, peeling off in irregular flakes, leaves elliptic-oblong, 8–20 × 5–10 cm, shortly acuminate, entire, base rounded; flowers white, with yellow corona in terminal corymbose cymes; follicles paired, smooth, terete, slender, curved to 20 cm long, seeds oblong-linear.

Occasional in forests. Fl. & Fr.: Mar.–Oct. *Vern.:* Kodisapala, Kolamukki, Pulvareni. Leaves and roots, source of a herbal drug specific for dysentary. It is recommended that the specific epithet *antidysenterica* should be a *nom. cons*. Known as Kutaja in Sanskrit, the active principle is *Conerrine*—which is antidysenteric (Satyavati, G.V. *et al.* 1987, Medicinal Plants of India, ICMR Vol. 2). The stem bark powder is given orally for the treatment of asthma and cough. Vinukonda RF, *VRK* 6885.

Recommends conservation of specific epithet *antidysenterica* which is self explanatory.

Ichnocarpus R. Brown *nom. cons.*

Ichnocarpus frutescens (L.) R.Br. Mem. Wern. Nat. Hist. Soc. 1: 62. 1809 & Ait. f. in Hort. Kew. ed. 2. 2:69. 1811; FBI 3: 669. Gamble 822 (577). *Apocynum frutescens* L. Sp. Pl. 213. 1753.

Evergreen, much branched, straggling shrubs; branchlets rusty pubescent; leaves elliptic-lanceolate, 5–9 × 2–4 cm, glabrous above, tomentose below; acute; flowers creamish, scented in axillary and terminal paniculate cymes; follicles slender; seeds linear.

Common in scrub jungles. Fl. & Fr.: Aug.—Feb. *Vern.*: Nalla teega. Stems are used for making ropes. Roots are used in medicine as substitute for *Hemidesmus indicus*. Kondavedu Fort, *SSR & KSM* 21658, Kondavedu Fort, *CAB* 4693 (MH).

Note: In our floras it has been described under *Ichnocarpus frutescens* of R. Brown (1809, l.c.). Hook. f. in FBI (l.c.) attributed its authority to Brown while citing Aiton's 2 ed. of Part 2 of *Hortus Kewensis* (1811), which is not correct. Its earliest report was made by Linnaeus in "Species Plantarum" in 1753 (l.c) under *Apocynum frutescens*. This prior most and validly published name is the basis of the combination *Ichnocarpus frutescens* (L.) Ait. in Ait. f. (l.c).

Nerium Linnaeus

Nerium indicum Mill. Gard. Dict. ed. 8 n, 2. 1768. *N. odorum* Soland. in Ait. Hort. Kew. ed. 1. 1: 297. 1787; Gamble 821 (577). *N. oleander* L. Sp. Pl. 209. 1753.

Large evergreen, xerophytic shrubs, up to 3 m tall; leaves linear-lanceolate, 10–18 × 1–2.5 cm, glabrous, thick green when aged; arranged in whorls or in clusers; flowers brick red, rose or white, fragrant, in terminal polychasial cymes.

The Indian Oleander. Cultivated as an ornamental. Fl. & Fr.: Throughout the year. *Vern.*: Ganneru. Several cultivars are available with flowers of various shades, double flowered and variegated leaves, Guntur (*Uhl*, 1929).

Note: *N. indicum* and *N. oleander* are commonly treated as separate species. However some are inclined to consider them conspecific (Cf. Gurant, Fosberg, Smithsonian Contrib. Bot. 17: 53. 1974).

Plumeria Linnaeus

Plumeria rubra L. Sp. Pl. 209. 1753. *P. acutifolia* Poir. in Lam. Encycl. (Suppl. 2): 667. 1812; FBI 3: 641; Gamble 821 (577).

Deciduous trees, up to 10 m tall; leves alternate, broadly obovate-spathulate, 20–30 × 7–10 cm, subcoriaceous, glabrous to pubescent beneath, acuminate, base cuneate to attenuate; flowers red or white cream, with yellow centre, fragrant, large in terminal corymbose cymes; follicles paired; seeds flat, winged.

The Pagoda tree. Native of tropical America. Fl. & Fr.: Mar.–Aug.. Eng.: Temple tree. *Vern.*: Deva ganneru, Deva Kanchan. Decoction of bark is used as purgative, it is also used for venereal diseases. Latex from plant is used for making ointments for itch, rheumatic pains and gum troubles, wood is used for making musical instruments. Guntur, (*Uhl*, 1989).

Rauwolfia serpentina nurseries are vigorous maintained by Agri-Gold near Chinakakani farm on the way to Mangalagiri.

Vallaris Burman

Vallaris solanacea (Roth) Kuntze. Rev. Gen. Pl. 2: 447. 1891; Gamble 815 (573). *Peltanthera solanacea* Roth, Nov. Pl. Sp. 132. 1821. *Vallaris heynei* Spreng. Syst. Veg. 1: 635. FBI 3: 650.

Twining shrub, bark yellowish-grey, lenticellate; tender parts puberulous; leaves decussate, oblong or elliptic, 4–10 × 2–3.5 cm, membranous, glabrous, shortly acuminate, entire, acute to obtuse, some what decurrent, lateral nerves ca. 6 pairs; flowers white in subcorymbose cymes; follicles terete, connate, slowly splitting later, gradually tapering, to 15 × 3.7 cm; seeds oblong, yellowish grey, beaked at one end, with a tuft of silky white coma.

Occasional. Ornamental as bowers in parks of Guntur. Fl. & Fr.: Apr.–Jan. *Vern.*: Nityamalle, Madhu malati. Nizampatnam, *VRK* 6813.

Wrightia R. Brown

1. Leaves hispid below, corona-scales fimbriate, follicles smooth, connate only at apex .... *W. tinctoria*
1. Leaves tomentose below, corona scales oblong, follicles pustulate, connate throughout .... *W. arborea*

Wrightia arborea (Dennst.) Mabb. in Taxon 26: 533. 1977. *Periploca arborea* Dennst. Schlussel Hortus Malab. 13, 23 & 25. 1818 & in Fortsetz. Allg. Tentsch. Gart. Mag. 3: 32, 41 & 83. 1818. *Wrightia tomentosa* Roem. & Schult. in L. Syst. Veg. 4: 414. 1819; FBI 3: 654.

Unarmed trees, up to 10 m tall; branchlets tomentose; leaves elliptic-obovate, 6–15 × 3–6 cm, thinly pubescent along nerves above, tomentose below, apex acuminate, entire, base acute; flowers white with corona, in short dense, erect, terminal corymbose cymes; follicles paired, beaked, connate throughout, rough with white tubercles; seeds slender, attenuated at the apex.

Occasional in forests. Fl. & Fr.: Apr.–Dec. *Vern.*: Tella pala, Tedla pala, Pala bariki. The leaves are eaten as pot herb; the tender fruit is also edible. The fruit contain floss which is used for stuffing. Wood very light to light and moderately hard used for various purposes. The fresh seeds have a high percentage of fertility. Vinukonda RF, *VRK* 6850.

Wrightia tinctoria R. Br. Mem. Wern. Soc. 1: 74, 1811; FBI 3: 653; Gamble 815 (573).

Densely foliaceous, deciduous trees, up to 6 m tall; latex milky; leaves opposite, elliptic-oblong, or oblong-lanceolate, 13–15 × 3–4 cm, glabrescent, base truncate; entire, apex bluntly acuminate; flowers white, scented in terminal dichotomous cymes; follicles paired, basally and apically jointed, cylindric, 25–35 × 0.2–0.5 cm; seeds pointed at the apex with a deciduous coma often more than 2–3.5 cm long at the base.

Common in wastelands and roadsides and rare in forests. Fl. & Fr.: Apr.–Oct. *Vern.*: Ankudu, Aku pala, Tedda pala, Reppala. The flowers are occasionally used as a vegetable. The leaves are the source of blue dye indigo called Mysore Pala Indigo, which is used as a substitute for indigo. Leaves are used as wrapper for bidies. Wood commonly used for toys, carving of frames, spoons with carved handles, combs, screws. Leaf paste is applied on externally to treat scorpion bite. V.P. South, *VRK* 3533; Kondavedu Fort, *CAB* 17488 (MH).

Allamanda cathartica L. and *Cascabela thevetia* (L.) Lippold two South American species are cultivated for their golden yellow flowers.

Another important but common cultivated shrubs are white simple and double flowered cultivars of *Tabernmontana coronaria* grown in gardens and temples. Propagation by cuttings and suckers.

That the simple and double flowered varieties are likely to be of a higher taxonomic status, an issue which needs synthetic evidence.

Alstonia scholaris R. Br. Trees cultivated in NU campus Boys Hostel. 'Edakulapala' used in Green belt programme at several places.

ASCLEPIADACEAE

Note: A.A. Bullock (1959) has separated the genera *Hemidesmus* R.Br., *Cryptolepis* R.Br. and *Cryptostegia* R.Br. from Asclepiadaceae and treated them under Periplocaceae based in stamen morphology. However, in the present study the above three genera are retained under Asclepiadaceae.

1. Erect herbs or shrubs:
    2. Mature stem succulent, angular, leafless, pollinia pendulous, corona double, follicles paired .... ***Caralluma***
    2. Mature stem woody, terete, leafy, pollinia erect, corona single, follicles united throughout .... ***Calotropis***
1. Climbers, straggling vines or lianas:
    3. Leaves reduced, branchlets jointed, fleshy, green .... ***Sarcostemma***
    3. Leaves well developed, branchlets not jointed woody or wiry:
        4. Leaves cordiform, as long as or longer than broad:
            5. Corolla base inflated, lobes oblong, leafbase deeply cordate, peduncle to 12 cm long, corona double .... ***Pergularia***
            5. Corolla base cylindric, lobes ovate, leaf-base truncate to rounded, peduncle to 2.5 cm long, corona single .... ***Wattakaka***
        4. Leaves other than cordiform, twice, or more as long as broad:
            6. Corolla tube inflated basally, ovules in 2 vertical rows .... ***Ceropegia***
            6. Corolla tube not inflated, ovules in many rows:
                7. Inflorescence an axillary umbel:
                    8. Latex watery, leaf-apex apiculate, corolla rotate:

9. Calyx glandular, corolla valvate, dry leaves chartaceous .... ***Pentatropis***
9. Calyx not glandular, corolla twisted, dry leaves subcoriaceous.............. ***Tylophora***
8. Latex milky (watery in *Leptadenia*), leaf-apex acute, corolla .... companulate:
10. Calyx glandular, corona single .... ***Gymnema***
10. Calyx not glandular, corona double .... ***Leptadenia***
7. Inflorescence other than umbel:
11. Follicle > 3 cm across; corona double .... ***Decalepis***
11. Follicle < 2 cm across (except *Cryptostegia*), corona single (double in *Oxystelma*):
12. Leaves oblong to linear-lanceolate, apex gradually acute, anthers adnate to stigma:
13. Cymes or fascicles shorter than leaves; corona single .... ***Hemidesmus***
13. Racemes longer than leaves, corona double .... ***Oxystelma***
12. Leaves elliptic to obovate, apex abruptly acute, anthers free, distinct:
14. Corolla purple, cymes terminal .... ***Cryptostegia***
14. Corolla white or yellowish, umbellate racemes axillary .... ***Cryptolepis***

Calotropis R. Brown

1. Flowers pale mauve .... ***C. gigantea***
1. Flowers purplish mauve .... ***C. procera***

Calotropis gigantea (L.) R.Br. in Ait. f. Hort. Kew. ed. 2. 2: 78. 1811; FBI 4: 17; Gamble 832 (585). *Asclepias gigantea* L. Sp. Pl. 214. 1753.

Soft, white woolly shrubs, up to 2 m tall; latex milky, leaves subsessile, decussate, elliptic-oblong or obovate, 9–15 × 4–6 cm, obtuse, rounded or acute, entire, base auriculate; flowers pale mauve, in terminal umbellate cymes; follicles oblong, inflated; seeds oblong with silky white coma.

Common weed of wastelands and fields. Fl. & Fr.: Throughout the year. *Vern.*: Tella Jilledu. Latex pasted on stomach with turmeric for controlling the stomach ache. Leaves are used for treating eye troubles and root bark in dystentry. Root bark paste is said to be applied for elephantiasis. It is an expectorant, emetic, diaphoretic, irritant, and purgative and for dysentery, intermittent fever, cold, cough, asthma and indigestion. V.P. South, *VRK* 3926.

Calotropis procera (Ait.) R. Br. in Ait. f. Hort. Kew. ed. 2: 2: 78. 1811; FBI 4: 18; Gamble 832 (585). *Asclepias procera* Ait. Hort. Kew. ed. 1. 1: 305. 1789.

Erect milky shrubs, up to 2 m tall; leaves opposite decussate, obovate, 9–15 × 5–10 cm, acute, entire, glaucous, base auriculate; flowers purplish mauve in axillary and terminal umbellate cymes; follicles recurved; seeds oblong with silky white coma.

Common weed in open forests. Fl. & Fr.: Throughout the year. *Vern.*: Rakta Jilledu. Latex pasted on stomach with tumeric for controlling the stomach ache. N. Varam, *VRK* 3505.

Caralluma R. Brown

Caralluma adscendens (Roxb.) Haw. Syn. Pl. Succ. 47. 1812; FBI 4: 76; Gamble 861 (605). *Stapelia adscendens* Roxb. Pl. Cor. t. 30. 1795.

Erect, succulent herbs; stem and branches angled; latex watery; leaves 0; flowers dark purple, solitary or in pairs, lateral on elongate very slender branches; follicles paired, slender, 10–12.5 cm long, apically beaked; seeds flattened, broadly margined.

Common in dry open forests on rocky soils. Fl. & Fr.; July–Sept. *Vern.*: Kundelu kommulu. The juice of the stem is used to treat indigestion. V.P. South, *VRK* 6826; Kondavedu Fort, *CAB* 4657 (MH).

Ceropegia Linnaeus

Note: The genus has evolved a trap mechanism for pollination by small dipterans. The mouth of the tubular corolla is typically equipped with long, often coloured, hairs which move with the slightest breeze, and these combined with the corolla color hairs serve to attract the dipterans to the flowers. The neck of the corolla has retrorse hairs or is quite smooth, preventing the insects from easily escaping the flower after entering. The insects are trapped in the inflated base of the corolla, which is often lighted by translucent areas on the sides (probably keeping insects active), and pollinia become attached to them to be transferred to other flowers upon escape. Escape is faciitated during the last stage of anthesis by the corolla turning downward and the hairs in the neck becoming flaccid.

1. Leaves deciduous, branchlets succulent; corolla brownish-purple, beak longer than the corolla tube, pollinia oblong .... ***C. juncea***
1. Leaves persistent, branchlets not succulent; corolla greenish-yellow, beak shorter than corolla tube; pollinia spherical .... ***C. candelabrum***

Ceropegia candelabrum L. Sp. Pl. 211. 1753; FBI 4: 70; Gamble 857 (603). *C. tuberosa* Roxb. Pl. cor. t. 9. 1795; FBI 4: 70; Gamble 857 (603).

Twiners; leaves ovate-oblong or elliptic to obovate, 2–3 × 0.8–1 cm; chartaceous, glabrous, entire, apex acutey apiculate, base acute to rounded; flowers greenish-yellow, in axillary umbellate cymes; follicles paired; 10–14 × 0.5–0.7 cm, cylindric, tapering at apex; seeds oblong.

Occasional in forests. Fl. & Fr.: July–Dec. *Vern.*: Bakkali manda. Kondavedu Fort. *CAB* 4662 (MH).

Ceropegia juncea Roxb. Pl. Cor. t. 10. 1795; FBI 4: 68. Gamble 856 (602).

Fleshy, twining herb, root tuberous; leaves lanceolate, ca. 1 × 0.2 cm, deciduous; flowers greenish, speckled and veined with brownish purple, solitary or in cymes; follicle to 7 mm long.

Rare in plains found in Nagarjunanagar (Mohan Rao & Aasiya, 1995).

Cryptolepis R. Brown

Cryptolepis buchanani Roem. & Schult. Syst. Veg. 4: 409. 1819; FBI 4: 5; Gambe 826 (580).

Perennial straggling shrubs; latex milky, leaves elliptic-oblong, 4.5–13 × 2.5–7.5 cm, glabrous, with an abrupt long apiculus, entire, base acute; lateral nerves 10–15 pairs, horizontally parallel, forming an intramarginal ring; flowers white or yellow in axillary umbellate cymes; follicles paired, straight, up to 6 x 1 cm; seeds obovoid.

Common in forests. Fl. & Fr.: May–Dec. *Vern.*: Adavi Pala teega, Gunji. The plant is used in a preparation given to children as a cure for ricketsia. Stem fibre used for making ropes and cordage. Dried root powder is used for cooling. Reported in Guntur Forest Division Work Plan.

Cryoptostegia R. Brown

Cryptostegia grandiflora R. Br. Bot. Reg. t. 435. 1820; FBI 4: 6; Gamble 862 (606).

Large straggling shrubs, latex milky; leaves elliptic-oblong, or ovate-elliptic, 6–10 × 3–5 cm, glabrous, glaucous above, pale below, sub acute, entire, base obtuse to acute, flowers purplish or rosy in terminal cymes; follicles ovoid–oblong, paired, 3-winged, tapering towards apex.

Occasional along stream banks and bushes in forests. It is a native of Tropical Africa naturalised in India. Fl. & Fr.: May–Dec. *Vern.:* Rabbaru chettu. Guntur (*Uhl*, 1929).

Decalepis R. Wight & Arnott

Decalepis hamiltonii Wight & Arn. in Wight Contrib. Bot. India 64. 1834; FBI 4: 11; Gamble 828 (582).

Deciduous straggler to liane to 15 m; branchlets terete with swollen, winged nodes; leaves ovate or elliptic-oblong, 2–6 × 1–4 cm, glabrous, subcoriaceous, apex subacute to obtuse, margin entire to undulate, base attenuate; flower cream coloured in trichotomous cymes; follicle oblong or lanceolate, cylindric, 5 x 3 cm, epicarp wrinkled; seeds ovate.

Occasional, on the slopes, among rocks, by stream banks etc. Fl: May–July. Fruits: Jan. onwards. *Vern.*: Nannari, Barresugandhi. Aromatic roots are used as spice and condiment, also root juice as a soft drink, considered to be appetiser and blood purifier.

Gymnema R. Brown

Gymnema sylvestre (Retz.) R. Br. ex Schult. in Roem. & Schult. in L. Syst. Veg. 6: 57. 1819; FBI 4: 29; Gamble 839 (590). *Periploca sylvestris* Retz. Obs. Bot. 2: 15. 1781.

Woody straggling pubescent shrubs, latex milky, young branches pubescent; leaves elliptic-ovate, 2–3 × 1–2 cm, pubescent along nerves below, acute, entire, base obtuse; flowers yellow in axillary pairs; follicles oblong, 7 × 0.5 cm, narrow tipped.

Common forest twiner. Fl. & Fr.: Sept.–Mar. *Vern.*: Podapatri Aku. Leaves ascribed oral antihyperglyaecmic property, recommended in local medicine as specific to non insulin dependant diabetes along with a few other ingredients. There is the other view that it numbs taste buds but has no hypoglycaemic role. Root paste is taken internally relieves fever and excess of heat. Loyapalle, *VRK* 3823; Bollapalle, *VRK* 5740; NU Campus, *SSR* 21630.

Chewing the leaves of this plant makes one temporarily unable to taste sugar and consequently it has been fancied by some as a treatment for diabetes.

Hemidesmus R. Brown

1. Leaves linear, yellow-pubescent, follicles terete .... ***H. indicus*** var. ***indicus***
1. Leaves elliptic-oblong, tomentose, follicles incurved .... ***H. indicus*** var. ***pubescens***

Hemidesmus indicus (L.) R. Br. in Ait. f. Hort. Kew. 2: 75. 1811. var. **indicus**; FBI 4: 5; Gamble 825 (580). *Periploca indica* L. Sp. Pl. 211. 1753.

Woody straggling shrubs; latex milky, leaves opposite, decussate, linear, 5–7 × 2–2.5 cm, yellow-pubescent, acute, entire, base obtuse; flowers yellow, in axillary fascicles; follicles terete; seeds oblong with pale brown coma.

The Indian Sarasapasilla. Occasional in forests. Fl. & Fr.: Sept.–Mar. *Vern.*: Sugandhi pala. This is commonly substituted by *Decalepis* in herbal medicine. Mellavagu, *VRK* 5737.

Hemidesmus indicus (L.) R. Br. var. **pubescens** (Wight & Arn.) Hook. f. Fl. Brit. India 4: 5. 1883; Gamble 825 (580). *H. pubescens* Wight & Arn. in Wight Contrib. Bot. India 63. 1834.

Slender twining shrubs; leaves elliptic to broadly oblong, 4–7 × 2–3 cm, softly pubescent above, velvety-tomentose below, base broadly acute to rounded, acute; flowers purple in axillary cymes; follicles 14 × 0.5 cm, tip incurved.

Rare on hedges. Fl. & Fr.: June–Dec. V.P. South, *VRK* 3538.

Leptadenia R. Brown

Leptadenia reticulata (Retz.) Wight & Arn. in Wight Contrib. Bot. India 47. 1834; FBI 4: 63; Gamble 850 (598); Bullock in Kew Bull. 10: 291. 3. 1955. *Cynanchum reticulatum* Retz. Obs. Bot. 2: 15. 1781.

Diffuse straggling shrubs; bark corky, thick; latex not milky; young branchlets hoary pubescent; leaves elliptic-oblong or lanceolate, 4–8 × 2–6 cm, grey pubescent, acute, entire, base obtuse, truncate or rounded; flowers green in axillary or terminal umbellate-cymes; follicles paired, cylindrical with a curved beak, bluntly acute at both ends, 6 x 2 cm, seeds oblong with pale brown coma, winged.

Common in forests, wasteplaces and roadsides. Fl. & Fr.: May–Dec. *Vern.*: Mukkuthummudu tiga. Increases lactation in cattle. It yields fibre from wood. Leaf paste applied for skin diseases. Latex inhaled for cold. Alcoholic extracts of roots and leaves show antibacterial activity. N. Varam, *VRK* 3970; Loyapalli to Sirigiripadu road, *SSR & KSM* 21638.

Oxystelma R. Brown

Oxystelma esculentum (L.f.) R. Br. ex Schult. in L. Syst. Veg. 6: 89. 1820; FBI 4: 17; Gamble 883 (596). *Periploca esculenta* L.f. Suppl. Pl. 168. 1781.

Glabrous straggler or twining herbs; latex milky, leaves linear-lanceolate, 5–8 × 0.5–1 cm, glabrous, mucronate, base rounded; flowers purplish or pinkish-violet in axillary long peduncled racemes; follicles ovoid-lanceolate, 5 × 2 cm, inflated, mucronate, seeds ovate with silky coma.

Occasional among bushes along canal banks. Fl. & Fr.: Mar.–Oct. *Vern.*: Palakura. The latex of the plant is used as an antiseptic on cuts and wounds. Javuku, *VRK* 3562; 3875.

Pentatropis Wight & Arnott

Pentatropis capensis (L.f.) Bull. in Kew Bull. 9: 284. 1955. *Cynanchum capense* L.f. Suppl. Pl. 169. 1781. *Asclepias microphylla* Heyne ex Roth Nov. Pl. Sp. 177. 1821. *Pentatropis microphylla* (Heyne ex Roth) Wight Contrib. Bot. Ind. 52. 1834; 4: 20; Gamble 835 (587).

Slender twining shrubs or straggler; leaves ovate, 1.5–3 × 1.5–2.5 cm, glabrescent below, obtuse, mucronate, entire, base cordate; flowers greenish purple, in axillary umbels; follicle 5 × 1 cm, basally inflated, tapering above, seeds ovate, flattened, winged with white silky coma.

Common among hedges in plains, scrub jungles on thickets, NU campus, Guntur channel tank bund areas. Occasional in forests. Fl. & Fr.: May–

Nov. *Vern.*: Poola pala. Plant is considered cooling agent. Papaiah palem, *VRK* 5837; Macherla RF, *VRK* 5889; NU campus, *SSR* 21682.

Pergularia Linnaeus

Note: N.E. Brown in Kew. Bull. 323–25,1907 has shown that R. Brown founded the genus *Daemia* upon the same plant on which Linnaeus had based his description of *Pergularia*. Hence *Daemia* R.Br. should be reduced to *Pergularia*.

Pergularia daemia (Forssk.) Chiov. Result. Sc. Miss. Stefan.Paoli Somal. Ital. 1: 115. 1976. *Asclepias daemia* Forssk. Fl. Aeg.–Arab. 51. 1775. *Cynanchum extensum* Jacq. Misc. 2: 353. 1782. *Daemia extensa* (Jacq.) R. Br. in Ait. f. Hort. Kew. 2: 76. 1811; FBI 4: 20. *Pergularia extensa* (Jacq.) N.E. Br. in Harvey & Sonder. Fl. Cap. 4: 758. 1908; Gamble 836 (588).

Perennial woody hispid climbers, latex milky; leaves cordate, 2–5 × 2–4 cm, glabrescent, acute to acuminate, entire, base cordate; flowers greenish, foetid in axillary, umbelliform cymes; follicles paired, basally swollen, lanceolate, up to 6 × 1 cm, softly echinate; seeds obovate with pale brown coma.

Common among bushes in forests. Fl. & Fr.: Sept.–Mar. *Vern.*: Dushtapu teega. Extract of the leaves is used to cure scabies. Plant extract is used for uterine and menstrual disorders. Koppanur, *VRK* 3240; Nagarjunasagar, *KMS* 9716 (MH).

Sarcostemma R. Brown

Sarcostemma acidum (Roxb.) Voigt Hort. Sub. Calc. 542. 1845. *Asclepias acida* Roxb. Fl. Ind. 2: 31. 1832. *Sarcostemma brevistigma* Wight & Arn. in Wight Contrib. Bot. Ind. 59. 1834; FBI 4: 26; Gamble 838 (589).

Leafless, much branched climbing shrubs, branches pendulous; latex milky; stems jointed, succulent, green; flowers white, scented, in terminal umbels; follicles ovoid-lanceolate.

Common in the bushes along roadsides. Fl. & Fr.: July–Jan. *Vern.*: Sannajamudu. Stems medicinally used. N. Varam, *VRK* 3960; Davupalli, *DMR & SSR* 18158; Kondavedu Fort, *CAB* 4640 (MH).

Tylophora R. Brown

1. Leaves glabrescent above, pubescent below, seeds without coma .... *T. indica*
1. Leaves glabrous on both sides, seeds with coma .... *T. macrantha*

Tylophora indica (Burm. f.) Merr. in Philipp. J. Sci. Bot. 19: 373. 1921. *Cynanchum indicum* Burm. f. Fl. Ind. 70. 1768. *Tylophora asthamatica* (L.f.) Wight & Arn. in Wight Contrib. Bot. Ind. 51. 1834; FBI 4: 44; Gamble 843 (593). *Asclepias asthmatica* L.f. Suppl. Pl. 171. 1781.

Hispid climbing shrubs; latex watery; leaves coriaceous, ovate-elliptic, oblong-lanceolate, 5–8 × 2–5 cm, fleshy, glabrescent above, pubescent below, acute to obtusely mucronate, base truncate or cordate; flowers purplish or brownish-violet, in axillary umbels; follicles linear-cylindric, acute at both ends, 7 × 1.5 cm; seeds ovate, flat.

Occasional in forests. Fl. & Fr.: May–Oct. *Vern.*: Mekameyani aku; Kukkapala tiga. Leaves with black pepper made into paste and taken with water as an emetic in snake bites, poison will come out with vomiting. Also used for treating asthma, bronchitis, whooping cough, diarrhoea and dysentary. Kothapalem, *VRK* 5773.

The specific epithet *asthmatica* is better retained as the a *nom. cons.* as leaves of the species help in reducing bronchitis and asthma where the specific epithet implied an exclusive property it should be retained; also ref. Hariharan. In defence, of Taxonamists, Curr. Sci. 1998.

Tylophora macrantha Hook. f. Fl. Brit. India 4: 40. 1883; Gamble 843 (592).

Slender climbing shrubs; leaves oblong-lanceolate, 7–8 × 2.5–3 cm, glabrous, entire, acuminate; flowers small, in axillary umbellate cymes; follicle smooth, acuminate, up to 8 cm long, seeds ovoid, flattened with silky coma.

Occasional twiner on hedges in the forests. Fl. & Fr.: July–Dec. *Vern.*: Pelusaku. Loyapalle, *VRK* 3830.

Wattakaka Hasskarl

Note: Some authors synonymize *Wattakaka* under the African genus *Dregea*, others regard it as a subgenus of *Dregea*. We follow Huber (in Abeywikrama, Fl. Ceylon 1 (D: 50. 1973) in maintaining it as a distinct genus.

Wattakaka volubilis (L.f.) Stapf in Curtis Bot. Mag. Su 8976. 1922. *Asclepias volubilis* L.f. Suppl. Pl. 170. 1781. *Dregea volubilis* (L.f.) Benth. ex Hook. f. in Fl. Brit. India 4: 46. 1893. *Marsdenia volubilis* (L.f.) T. Cook Fl. Bombay 2: 166 (230) 1904; Gamble 846 (595).

Woody pubescent straggling shrubs; bark somewhat corky; branchlets lenticellate; latex milky; leaves broadly ovate, 7–15 × 5–8 cm, sparsely hairy above, pubescent below, apex shortly acuminate, entire, base rounded; penni nerved, lowest 1 or 2 pairs arising from base; flowers greenish in terminal drooping umbels; follicles paired, oblong, to 10 × 2.5 cm, tapering; seeds obovoid, flattened, 7 × 4 mm with soft, dull white coma.

Common in Loyapalle section. Fl. & Fr.: Mar.–Dec. *Vern.*: Bandiguri-ginga, Doodi-pala. The fresh dry root paste is used as an antidote for snake bite. Loyapalle, *VRK* 3834; Bollapalle, *VRK* 5756; NU capus, *SSR & KSM* 21683, Vinukonda RF, *CAB* 17396 (MH).

LOGANIACEAE

Strychnos Linnaeus

1. Bark blackish, cracked; leaf petiole 7.5–12.5 mm long, corolla lobes longer than tube, berries up to 1.5 cm across, deep blue when ripe .... ***S. potatorum***
1. Bark grey, smooth, leaf petiole under 5 mm long, corolla lobes shorter than tube, berries more than 2 cm across, orange when ripe .... ***S. nux-vomica***

Strychnos nux-vomica L. Sp. Pl. 189. 1753; FBI 4: 90; Gamble 868 (610).

Deciduous trees, up to 10 m tall; leaves coriaceous, orbicular, 6–8 × 4–5.5 cm, shining, obtuse to acuminate, entire, base obtuse, 5-nerved, transverse nerves irregular; flowers greenish-yellow in terminal compound cymes; berries globose, orange when ripe, seeds many, discoid, much compressed, ashy grey, with silky hair radiating from the centre.

Common in Loyapalle and V.P. South sections. Fl. & Fr.: Feb.–June. Eng.: Snake-wood tree. *Vern.*: Musti, Vishamushti. Leaves, fruits and bark medicinal. Strychnine and brucine obtained from the seed are extremely bitter and most poisonous bark extract of late used in antihyperglycaemic formulations, used as an insecticide to kill vermine infection. Wood is largely employed for axe-handles, plough, cart wheels etc. Loyapalle, *VRK* 3811.

Shoots, bark indiscretely harvested for illegal commerce. To be conserved on top priority.

Strychnos potatorum L.f. Suppl. Pl. 148. 1781; FBI 4: 90; Gamble 868 (610).

Deciduous trees, up to 10 m tall; branches fluted; leaves coriaceous, elliptic, 5–7 × 2–4 cm, glabrous, shining, bright green, apex acute-acuminate, entire, base rounded, 3–5-nerved, transverse nerves paralled, conspicuous; flowers white in cymes on old wood; berries globose, deep blue when ripe; seeds 1 or 2.

Occasional in the forests. Fl. & Fr.: July–Dec. *Vern.*: Chilla. Indupu. Eng.: Cleaning nut tree. Fruit pulp is edible. The dried stem bark powder is topically applied on cuts and wounds as an antiseptic. Seeds well known for precipitating pollutants of water. Needs further study about scope as a tailor-made tool in reducing water pollution. Javuku, *VRK* 3906; Loyapalle, *VRK* 3841; V.P. South, *VRK* 6831.

GENTIANACEAE

1. Ovary 2-celled; stamens connivent around style, exerted, flowers blue or lilac ... *Exacum*
1. Ovary 1-celled; stamens away from style, included, flowers other than blue:
    2. Corolla regular, flowers in fascicles .... *Enicostemma*
    2. Corolla irregular, flowers in cymes:
        3. Flowers yellow,sessile, stigma sub-entire .... *Hoppea*
        3. Flowers pink or white, stalked stigma deeply bilobed .... *Canscora*

Canscora Lamarck

1. Stems obtusely quadrangular or obscurely winged .... *C. diffusa*
1. Stems strongly 4-winged .... *C. decurrens*

Canscora decurrens Dalz. in Hook. J. Bot. Kew Misc. 2: 136. 1850; FBI 4: 103; Gamble 878 (617).

Annual erect herbs, up to 50 cm tall; stems 4-winged; branches opposite and alternate; leaves sessile or subsessile, lower ones oblong, 1.5–4 × 0.5–2 cm, upper ones ovate to lanceolate, much smaller than lower ones, passing into bracts, glabrous, acute or apiculate; flowers pink or white in terminal dichotomous or trichotomously branched cymes; capsules oblong, seeds many, minute.

Occasional along the streams in forests. Fl. & Fr.: Dec.–May. N. Varam, *VRK* 3597; Javuku, *VRK* 3871; Macherla, *VRK* 5879; Jayanthi puram, *VRK* 6950.

Canscora diffusa (Vahl) R. Br. Prodr. 451. in Obs. 1810. *Gentiana diffusa* Vahl Symb. Bot. 3. 47. 1794; FBI 4: 103; Gamble 876 (617).

Erect herb, to 40 cm tall; stem obtusely quadrangular, obscurely winged; leaves membranous, 3-nerved, reticulately veined, the upper broadly ovate, 1.2–2.1 × 1–1.2 cm, acute, apiculate, entire, base rounded, sessile, the lower elliptic, to 5.8 × 2.2 cm, acute, entire, base narrowed, petiolate; flowers pink in lax diffuse, terminal and axillary paniculate cymes, capsule narrowly oblong, membranous.

Occasional, in moist places. Fl. & Fr.: Aug.–Mar. Botlaka bavi, *DMR* 21104.

Enicostemma Blume *nom. cons.*

Enicostemma axillare (Lam.) Raynal in Adansonia 9: 75. 1969. ssp. **axillare**. *Gentiana axillaris* Lam. Tabl. Encycl. 1 (2): 487. 1793. non Raf, 1828, nec Reichb. 1828. *Enicostemma littorale* auct. non Blume 1826; FBI 4: 101; Gamble 875 (615).

Erect herbs, up to 50 cm tall; stems angular; leaves lanceolate-oblong, 4–9 × 1–1.5 cm, glabrous, obtuse or acute, base cuneate; flowers white, in axillary clusters; capsules septicidal, 3.5 x 3 mm, seeds minute, reticulate, circular.

Common in moist places gravelly waste lands, river banks and forests. Fl. & Fr.: June–Nov. *Vern.*: Nelagulimidi. Plant powder antidiabetic and antidote to snake bite. Macherla RF, *VRK* 5882; Krishna river, *VRK* 6742; Mellavagu, *VRK* 5728; Konduvedu, *SSR & KSM* 21466.

Exacum Linnaeus

Exacum pedunculatum L. Sp. Pl. 111. 1753; FBI 34: 97; Gamble 874 (614).

Annual erect herbs, up to 30 cm tall; stems 4-angled; leaves sessile, decussate, elliptic or lanceolate, 2–5 × 0.8–1.5 cm, glabrescent, acute, entire,base usually tapering, 3 or 5-nerved; flowers pale violet or pink with yellow stamens, in axillary and dichasial cymes; capsules globose; seeds circular, minute.

Occasional in forests. Fl. & Fr.: Nov.–Mar. *Vern.*: Nalagorimidi, Rasuka. The paste mixed with *Allium sativum* and 3 black pepper seeds is used as an antidote for snake bite. Javuku, *VRK* 3569.

Hoppea Willdenow

Hoppea dichotoma Willd. Ges. Naturf. Freunde Berlin Neue Schriften 3: 435. 1801; FBI 4: 100; Gamble 877 (616).

Erect, profusely branched herbs, up to 10 cm; branches 4-angled, winged; leaves elliptic-ovate, 6–8 × 2.5–4 mm, base cuneate, apiculate; flowers pale yellow, in apical, dichasial cymes, and in the forks of branches, capsule globose, 4 mm across.

Common weed in moist places especially on the canal bunds. Fl. & Fr.: Jan.–Mar. Krishna river bed, *VRK* 6755.

MENYANTHACEAE

Nymphoides J. Hill

Nymphoides hydrophylla (Lour.) Kuntze Rev. Gen. Pl. 429. 1891. *Menyanthes hydrophylla* Lour. Fl. Cochinch. 1: 129. 1790. *Limnanthemum cristatum* (Roxb.) Griseb. Gen. Sp. Gent. 342. 1839; FBI 4: 131; Gamble 883 (620).

Aquatic free floating herbs; rhizome short, erect, with petiole like branches; branchlets long petioled, up to 75 cm long; leaves orbicular, 3–9 × 3–10 cm, obtuse, entire, base deeply cordate, flowers white, in clusters at distal node of stem; capsules broadly ovoid; seeds tuberculate.

Occasional in the ponds of Guntur district. Fl. & Fr.: May–Oct. *Vern.*: Kukka-Tamara. Nallapadu, *VRK* 6774; Vinukonda RF, *VRK* 6884.

HYDROPHYLLACEAE

Hydrolea Linnaeus

Hydrolea zeylanica (L.) Vahl Symb. Bot. 2: 46. 1791; FBI 4: 133; Gamble 2: 884 (621). *Nama zeylanica* L. Sp. Pl. 226. 1753.

Erect or diffuse, often subsucculent herb, 20–60 cm, often rooting at nodes; branchlets pubescent when young; leaves linear, lanceolate or oblanceolate, 3 × 1 cm, glabrous above, glaucescent below, acute or acuminate, entire, truncate; flowers blue in lax, axillary or terminal racemes; capsule ovoid to ellipsoid.

Occasional in wet lands of Nagarjuna nagar grass lands (Mohan Rao and Aasiya, 1995), river and canal banks, fallow fields and moist places. Fl. & Fr.: Nov.–Apr.

BORAGINACEAE

1. Flowers aggregated in 1-sided spikes or racemes, ovary not lobed .... *Heliotropium*
1. Flowers solitary, ovary 4-lobed:
   2. Erect herbs, corolla blue or pinkish white, stamens connivent, style undivided .... *Trichodesma*
   2. Prostrate herbs, corolla cream, stamens free, style arms 2 .... *Coldenia*

Coldenia Linnaeus

Coldenia procumbens L. Sp. Pl. 125. 1753; FBI 4: 144; Gamble 892 (627); Kazmi in J. Arnold Arbor. 51: 148. 1970.

Deep rooted, prostrate, scabrid herbs, stems stout, shaggy with white hairs, lying flat on the ground; leaves hispid, elliptic-ovate or obovate-oblong, 1–2 × 0.5–1 cm, pilose, obtuse-emarginate, serrate-crenate, base oblique; flowers yellow or white, axillary, solitary, subsessile; drupes pyramidal, 4-lobed with a short central double beak, pubescent, seed solitary in each lobe.

Common weed in dried tanks and ponds. Fl. & Fr.: Feb.–Aug.. *Vern.*: Hamsapadu. The fresh leaf paste is applied externally to treat rheumatic swellings and whole plant paste is used to cure skin diseases like white patches and allergy. N. Varam, *VRK* 3591.

Heliotropium Linnaeus

1. Erect subshrubs, leaves more than 7 cm long, nutlets connate in pairs .... *H. indicum*

1. Erect spreading or procumbent herbs, leaves less than 5 cm long, nutlets free:
   2. Inflorescencence a raceme, flowers shortly stalked, stamens apically connivent, anthers ovate-cordate:
      3. Plants spreading or procumbent, leaves elliptic-oblong, margin flat ... ***H. scabrum***
      3. Plants erect, leaves linear-lanceolate, margin revolute:
         4. Branchlets glabrescent to sparsely scabrous, bracts linear ..... ***H. zeylanicum***
         4. Branchlets softly sericeous, bracts lanceolate .... ***H. bracteatum***
   2. Inflorescence a spike, flowers subsessile, stamens free, anthers oblong or linear-lanceolate:
      5. Leaves glabrous, nutlets glabrous .... ***H. curassavicum***
      5. Leaves white pubescent, nutlets tuberculate or hairy:
         6. Perennial herbs, leaves covered by white strigose hairs .... ***H. strigosum***
         6. Annual herbs, leaves white pubescent:
            7. Leaves up to 6 cm long, lanceolate .... ***H. subulatum***
            7. Leaves up to 2 cm long, elliptic-ovate or obovate-spathulate:
               8. Stigma sessile, fruiting calyx-lobes spreading, leaf base cuneate, apex obtusely apiculate; .... ***H. ovalifolium***
               8. Stigma raised, calyx-lobes enveloping the drupe, leaf base attenuate, apex subacute .... ***H. supinum***

Heliotropium bracteatum R. Br. Prodr. 493. 1810; FBI 4: 151; Gamble 2: 897.

Erect herb, up to 50 cm tall; branchlets softly sericeous; leaves linear-lanceolate, to 4.2 × 0.2 cm, chartaceous, appressed sericeous, apex acute, margin revolute, base cuneate, petiole short, flowers white in subterminal racemes, nutlets rounded, smooth, 1 mm.

Common, in dry rocky plains and fields. Fl. & Fr.: July–Dec. Macherla, *VRK* 3517.

Heliotropium curassavicum L. Sp. Pl. 130. 1753; Gamble 896 (630).

Annual erect or ascending herbs, up to 20 cm tall; leaves linear-lanceolate, 2–3 × 0.2–0.3 cm, glabrous, acute to obtuse, entire, base cuneate; flowers white in terminal spikes; fruits pyramidal with 4 nutlets, nutlets glabrous.

Occurs along the sandy coast of Nizampatnam. Fl. & Fr.: Dec.–May. *Vern.*: Nela-Golividi. Powdered roots are applied to sores and wounds. Nizampatnam, *VRK* 6795, 6920; Kothapalem, *VRK* 5759.

Heliotropium indicum L. Sp. Pl. 130. 1753; FBI 4: 152; Gamble 896 (630).

Erect herbs, with woody base, up to 50 cm tall; young branches hirsute; leaves broadly elliptic-ovate, 5–8 × 2–4 cm, chartaceous, strigose, hairy on nerves, acute, base truncate to attenuate, petiole long; flowers white in terminal or leaf-opposed, helicoid spikes; nutlets connate in pairs, 3 mm, angled, beaked above.

Occasional in wastelands, way sides of plains. Fl. & Fr.: Throughout the year. Leaf paste applied for scorpian-sting and also on ulcerous wounds. Nallapadu, *VRK* 6773; V.P. South, *VRK* 5811.

Heliotropium ovalifolium Forssk. Fl. Aeg.—Arab. 38. 1775; FBI 4: 150; Gamble 896 (629); Kazmi in J. Arnold Arbor. 51: 178. 1970.

Annual prostrate or ascending white-villous herbs, up to 30 cm tall; leaves elliptic-obovate or spathulate, 1–2 × 0.5–0.8 cm, white woolly tomentose, obtuse apiculate, base cuneate; flowers white in axillary or terminal spikes; nutlets 1 mm, hairy; seed solitary.

Common on roadsides in coastal belt. Fl. & Fr.: Apr.–Dec. Nagarjunasagar, *KMS* 9810 (CAL).

Heliotropium scabrum Retz. Obs. Bot. 2: 8. 1781; FBI 4: 152; Gamble 897 (630). *H. marifolium* Retz. Obs. Bot. 2: 8. 1781; FBI 4: 152; Kazmi in J. Arnold Arbor. 51: 154. 1970.

Spreading or procumbent herbs; branches twiggy, radiating from root stock, appressed hirsute; leaves elliptic-oblong, 0.8–1.8 × 0.3–0.5 cm, strigosely hirsute, acute, entire; flowers white, in subterminal short racemes; nutlets subglobose, appressed-hirsute, to 1.3 mm.

Occasional, in Nagarjunanagar grass lands (Mohan Rao & Aasiya, 1995). Nambar fields, on sandy lands, waste places, roadisdes etc. Fl. & Fr.: Throughout the year.

Heliotropium strigosum Willd. Sp. Pl. 1: 743. 1797; FBI 4: 151. *H. brevifolium* Wall. ex Roxb. Fl. Ind. 2: 2. 1824; Gamble 895 (630).

Small much branched often procumbent herbs, up to 30 cm tall; leaves narrow, strigose, linear, acute, up to 3 cm long; flowers white in axillary and terminal spicate racemes; fruits with 4-nutlets, broad, depressed, conical at the apex.

Common in forest undergrowth. Fl. & Fr.: May–Nov. Plant juice is applied to sore eyes, gum boils and sores. Plant is laxative and diuretic. N. Varam, *VRK* 3506.

Heliotropium subulatum Hochst ex DC. Prodr. 9: 528 & 551. 1845. *H. zeylanicum* Clarke in Hook. f. Fl. Brit. India 4: 148. 1883.

Annual erect, hairy herbs, up to 50 cm tall; leaves 1.5–6.3 × 0.3–1.5 cm, lanceolate, densely strigose, entire, acute, base cuneate; flowers sessile in 25–45 cm long, terminal scabrid, scorpioid cymes; nutlets 4, conical, glabrous, brown, minutely tuberculate.

Occasional weed in wastelands. Fl. & Fr.: Throughout the year. Nagarjunakonda, *KTH* 9842 (CAL).

Heliotropium supinum L. Sp. Pl. 130. 1753; FBI 4: 149; Gamble 896 (629); Kazmi in J. Arnold Arbor. 51: 179. 1970.

Prostrate, villous herbs; leaves elliptic-obovate; 1.5–2 × 0.8–1.5 cm, softly white woolly, obtuse, entire, base attenuate; flowers white in terminal scorpioid spikes; nutlets closed by persistent woolly calyx.

Common in Kaza, Kanteru and China Kakani as post harvest weed. Occasional in the forests. Fl. & Fr.: Dec.–May. N. Varam, *VRK* 3586.

Heliotropium zeylanicum (Burm. f.) Lam. Encycl. 3: 94. 1789; Gamble 896 (630). ssp. **zeylanicum**; Kazmi in J. Arnold Arbor. 51: 156. 1970. *H. paniculatum* R.Br. Prodr. 494. 1810. FBI 4: 151; *H. curassavicum* L. var. *zeylanicum* Burm. f. Fl. Ind. 41. t. 16. f. 2. 1768.

Erect herbs, up to 60 cm tall; branchlets glabrescent to sparsely scabrous; leaves linear-lanceolate, 1–2 × 0.3–0.4 cm, chartaceous, minutely scabrous above, appressed-strigose below, revolute, acute, base cuneate; flowers white in axillary or terminal racemes; nutlets hairy.

Common roadside weed near Kaza, Mangalagiri, Tadepalli. Fl. & Fr.: Oct.–Feb. Madinepadu, *CAB* 17457 (MH).

Trichodesma R. Brown

1. Leaves petiolate, calyx-lobes rounded, corolla lobes eglandular .... *T. zeylanicum*
1. Leaves sessile, calyx-lobes cordate, corolla lobes glandular spotted .... *T. indicum*

Trichodesma indicum (L.) Lehmann Pl. Fam. Asperif. 193. 1818 (*indica*); FBI 4: 153; Gamble 899 (631); Kazmi in J. Arnold Arbor. 52: 516. 1971. *Borago indica* L. Sp. Pl. 137. 1753.

Annual erect villous herbs, up to 75 cm tall; leaves sessile, elliptic-oblong, 3–6.5 × 1–2 cm, glabrescent above, hirsute-pubescent below, margins and veins hairy, obtuse, or subacute, base auriculate; flowers pale violet blue, axillary, solitary; calyx lobes villous; fruit pyramidal, subtruncate at the apex, tipped with the persistent style, mutlets ovoid, 0.4 cm long, smooth.

Occasional weed of the moist places and cultivated crops. Fl. & Fr.: Sept.–Mar. *Vern.*: Guvoa chettu. Mellavagu, *VRK* 5729.

Trichodesma zeylanicum (Burm. f.) R. Br. Prodr. 496. 1810; FBI 4: 154; Gamble 899 (632). *Borago zeylanica* Burm. f. Fl. Ind. 41. 1768.

Erect hispid herbs, up to 1 m tall; leaves elliptic-lanceolate, 3–7 × 0.5–1 cm, densely clothed above with short stiff hairs from bulbous white bases, finely pubescent and with a few pustular hairs beneath, acute, entire, base truncate; flowers pale blue, axillary, solitary, upper ones in leafy racemes; nutlets long, broadly ovoid, smooth.

Rare in forests. Common post harvest weed in Kunchenapalli, Duggirala. Fl. & Fr.: Nov.–May. Nagarjunasagar, *KMS* 9724 (MH).

CORDIACEAE

1. Trees, flowers in cymes, leaves alternate or opposite:
   2. Calyx-lobes connate, accrescent in fruit, drupes pulpy .... *Cordia*
   2. Calyx-lobes divided, not accrescent in fruit, drupes not pulpy .... *Ehretia*

1. Subshrubs, flowers solitary or in clusters, leaves fascicled:
   3. Flowers pink, leaves scabrous, riparian .... *Rotula*
   3. Flowers cream, leaves white-glandular, not riparian .... *Carmona*

Carmona Cavanilles

Carmona retusa (Vahl) Masam. in Trans. Nat. Hist. Soc. 30: 61. 1940. *Cordia retusa* Vahl Symb. Bot. 2: 42. 1791. *Ehretia microphylla* Lam. Tabl. Encycl. 1: 425. 1792; Gamble 891 (626). *E. buxifolia* Roxb. Pl. Cor. t. 57; 1795; FBI 4: 144.

Moderate shrubs, up to 1.5 m tall; leaves alternate or clustered at nodes, obovate, 1.5–2.5 × 0.5–0.7 cm, white glandular above, sparsely scabrous along nerves below, apex obtuse, bluntly 3–7-toothed, entire, base cuneate; flowers white or pale yellow, axillary, solitary or in clusters; drupes globose, of 4 pyrenes, 4 mm across.

Rare in forests. Fl. & Fr.: Oct.–Mar. Kondavedu, *SSR* & KSM 21651; Kondavedu Fort, *CAB* 4637 (MH).

Cordia Linnaeus

1. Fruiting calyx not ribbed .... *C. dichotoma*
1. Fruiting calyx ribbed:
   2. Filaments glabrous, leaves scabrous above .... *C. monoica*
   2. Filaments hairy at the base, leaves glabrous above .... *C. evolutior*

Cordia dichotoma Forst. f. Prodr. 18. 1786. *C. obliqua* Willd. Phytogr. 4. t. 4. 1794; FBI 4: 137. 1883; Gamble 887 (623).

Deciduous tree, up to 10 m tall; branches drooping; bark thick, greyish brown with shallow furrows and wrinkles; leaves elliptic-oblong or orbicular 4–11 × 3–8 cm, glabrous, sometimes hairy in the nerve axils, obtuse or acute, entire or remotely dentate or sinuate, base acute or cuneate, 3-nerved; flowers white, fragrant in corymbose, glabrous panicles; drupe pinkish, ovoid or oblong tipped by short spine, shiny, pulp scanty, very viscid; seed one, large.

Common, in open lands and forests. Fl. & Fr.: Feb.–July. *Vern*.: Bankira, Banka Nakkara. Fruits edible. Bark is astringent, wood is used for agricultural implements. The young tender leaves used as vegetable. Kakirala-Bellamkonda, *DMR* 21130, Davupalli, *SSR & KSM* 21410.

Cordia evolutior (C.B. Clarke) Gamble Fl. Madras 888. 1923 & 624. 1957. *C. fulvosa* Wight var. *evolutior* C.B. Clarke in Hook. f. FBI 4: 140. 1883.

Medium to small tree, crown low spreading; young branches clothed with short fulvous pubescence; leaves ovate, ovate-oblong or suborbicular, 2.5–8 × 3.2–6.5 cm, glabrous above, fulvous tomentose beneath, subacute, entire, rounded or acute; flowers white, 1–2 cm long, in short fulvous-tomentose panicles; drupe yellow, ovoid-conical, apiculate, persistent calyx cup shaped.

Rare, in forests. Fl. & Fr.: Jan.–May. Davupalli, *SSR* & *KSM* 21425.

Cordia monoica Roxb. Pl. Cor. t. 58. 1796; FBI 4: 137; Gamble 888 (624).

Small trees, up to 5 m tall; leaves alternate, ovate-elliptic, 5–9 × 2.5–5 cm, scabrous and spotted with whitish points above, pubescent below, obtuse, or subacute, distantly crenate, base rounded; flowers yellow in dense cymes; drupes ellipsoid, 1 cm long, calyx lobes erect.

Occasional in forests. Fl. & Fr.: July–Dec. *Vern.*: Pachabodupu. Ripe fruits are eaten raw. Ethipothala, *VRK* 5796; Jayanthipuram, *VRK* 6974; Bollapalle, *CAB* 17419 (MH); Guntur, *JSG* 33785 (MH).

Ehretia P. Browne

Ehretia pubescens Benth. in Royle Ill. Bot. Himal. 306. 1836; Gamble 891 (626). *E. leavis* Roxb. var. *pubescens* (Benth.) Clarke in Hook. f. Fl. Brit. India 4: 142. 1883.

Medium sized trees, up to 8 m tall; leaves elliptic-oblong, 3–6 × 2–4 cm, scabrous-pubescent, acute, entire, base obtuse to truncate; flowers white in terminal corymbose cymes; drupes globose, red when ripe.

Occasional in forests. Fl. & Fr.: July–Dec. Kondavedu, *CAB* 17487 (MH); Vinukonda; *CAB* 17393; Nakarekkal, *CAB* 17483 (MH).

Rotula Loureira

Rotula acquatica Lour. Fl. Cochinch. 1: 121: 1790; Gamble 893 (627). *Rhabdia lycioides* Mart. & Zucc. Nov. Gen. Sp. 2: 136. t. 195. 1827; FBl 4: 145.

Erect, stiff, subshrubs, branches virgate, with numerous short lateral arrested branchlets, up to 75 cm tall; leaves alternate, oblong to obovate-spathulate, 1.5–2.5 × 0.5–1 cm, thick, glabrescent, base cuneate, margin ciliate, apex obtuse or apiculate; flowers pink in terminal or axillary clusters; drupes of 4 pyrenes, 5 mm across, orange when ripe.

Occasional in Krishna river bed. Fl. & Fr.: Dec.–Mar. Krishna river, *VRK* 6747.

CONVOLVULACEAE

1. Leafaless parasitic twiners, branchlets yellowish .... *Cuscuta*
1. Leafy autotrophs:
    2. Small spreading herbs:
        3. Leaves ovate, flowers blue, sytles 2, free or basally united .... *Evolvulus*

3. Leaves not ovate, flowers whitish yellow, style 1, entire .... *Merremia*
2. Climbing or erect shrubs or herbs:
4. Fruits indehiscent, woody:
5. Corolla salver-shaped, stamens attached near the middle of the tube, fruit nearly dry.. *Rivea*
5. Corolla funnel-shaped, stamens attached below the middle of the tube, fruit fleshy or leathery .... *Argyreia*
4. Fruits capsular:
6. Pollen spinulose .... *Ipomoea*
6. Pollen smooth:
7. Styles 1, stigma bilobed:
8. Outer sepals much longer and broader than the inner ones .... *Jacquemontia*
8. Outer sepals not as above .... *Convolvulus*
7. Styles 2, free, stigma capitate .... *Cressa*

Argyreia Loureiro

1. Leaves glabrous on the upper surface, sepals ovate or orbicular .... *A. nervosa*
1. Leaves hairy on the upper surface, sepals orbicular or obovate .... *A. cymosa*

Argyreia cymosa (Roxb.) Sweet. Hort. Brit. (ed. 2) 373. 1830; FBI 4: 190; Gamble 908 (637). *Lettsonia cymosa* Roxb. Fl. Ind. 2: 82. 1824.

Stragglers on thickets; stem woody, terete, pubescent; leaves deltoid to cordiform, 6–8 × 4–6 cm, chartaceous, thinly pubescent on both sides, entire, acute or obtuse, base truncate or cordate; flowers pinkish in axillary, corymbose cymes; fruit globose, 1.7 × 1.4 cm, glabrous; seeds 2, 3 or 4, ovate to elliptic, black.

Common on coast line. Fl. Fr.: Aug.–Jan. Guntur, (*Uhl*, 1929).

Argyreia nervosa (Burm. f.) Boj. Hort. Maurit. 224. 1837. *Convolvulus nervosus* Burm. f. Fl. Ind. 48. 1768. *Argyreia speciosa* Sweet Hort. Brit. 289. 1827; FBI 4: 185; Gamble 907 (637).

A very large perennial climber; stem woody, terete, hollow; young branches densely tomentose, whitish or fulvous; leaves large, broadly ovate-cordate, 10–30 × 9–25 cm, apex obtuse or cordate, entire, base cordate, glabrous above, densely fulvous or white tomentose beneath; flowers rose-purple, axillary, 1-few-flowered subcapitate cymes; fruit indehiscent, subglobose, 2 cm across, brown when ripe; seeds 2–4, ovate to orbicular.

Occasional, in dry deciduous forests. Fl. & Fr.: Nov.–Feb. *Vern.*: Samudrapala, Chandrapala. It is reported to be bitter, laxative and useful in urinary, heart and skin diseases. The roots have aphrodisiac, antiseptic and emmollient properties. The plant is highly fibrous. It is cultivated in NU Botanical gardens and opposite Mangalagiri Post Office.

It is popularly known by the name Baby Wood rose, Woolly Morning glory and Elephant creeper. The attractive leaves deep green above, silvery white beneath and reddish pink flowers made this an ornamental in demand.

Convolvulus Linnaeus

Convolvulus arvensis L. Sp. Pl. 153. 1753; FBI 4: 219; Gamble 925 (650).

Annual vine, stem trailing or twining, terete or slightly angular, glabrous or sparsely pubescent, leaves ovate-oblong, 1–5 × 0.4–1.5 cm, apically acute to obtuse, mucronulate, base hastate, glabrous or slightly pubescent, flowers axillary, solitary or 2–3-flowered cymes; capsule ovoid-globose, 5–8 mm long; seeds 4, dark brown or black, 3-4 mm long.

In disturbed habitat, on hedges, thickets, waste places and proximity of marshes. (Mohan Rao and Aasiya, 1995) Eng.: Deer foot.

Cressa Linnaeus

Cressa cretica L. Sp. Pl. 223. 1753; FBI 4: 225; Gamble 922 (647).

Erect herbs, up to 20 cm; basally woody, much branched above, leaves, ovate, 5–8 × 2–3 cm, pubescent on both sides, entire, acute, base truncate; flowers white, subsessile, solitary, clustered in upper axils; capsules ellipsoid-globose, 3–4 × 2.5–3 mm, crowned with long hairs, seeds glabrous, 1–4, shining, dark brown.

Common in salt marshes and interior as post harvest weed in Kakani, Kanteru, Kaza, Repalle. Fl. & Fr.: Mar.–Oct. *Vern.*: Uppusenaga. Unconfirmed reports on its antitumour activity, also on record as 'Sanjeevani' plant. The weed is common in Nagarjunanagar. The ecotypic differentiation and productivity parameters were studied by Ramesh Babu, T. 1984 using local species populations, also Rao, P.N. and Ramesh Babu, 1983. National Symp. Adv. Fr. Pl. Sci. 222–223, Jodhpur. Nizampatnam, *VRK* 6929.

Cuscuta Linnaeus

1. Slender twiners; flowers in fascicles; styles filiform, stigmas capitellate; capsule dry .... *C. chinensis*
1. Stout twiners, flowers in racemes; styles stout, stigmas acute; capsules succulent .... *C. reflexa*

Cuscuta chinensis Lam. Encycl. 2: 229. 1786; FBI 4: 226; Gamble 931 (654).

Stem twining, slender, yellow, glabrous, terete; flowers sessile or shortly pedicellate, bracts fleshy, glabrous; corolla yellowish; fruit capsular, depressed-globose, 2.5 × 3 mm, circumscissile; seeds 4, smooth, glabrous, brown, minutely muricate.

Common along waysides, waste places and cultivated lands on a variety of hosts. Fl. & Fr.: Almost throughout the year.

This species is a native of China, now distributed throughout India. The plant is purgative and is used internally to cure protracted fevers, acid induration of the liver.

Holoparasite on pulse crops in Nambur, Kaza of Guntur Dt. Virulent in Jan.–Mar. Causes serious losses and the pulse crop to be abandoned. It is common on roadisde weed hosts especially in cleared localities. It spreads on a clump of hosts at a time without host specificity. The hosts are hence designated as 'host complex'. Prolific flowering, and fruiting, Viviparous, vegegative propagation also occurs by 'segments of shoots.

Atleast 200 hosts of Dicots & Monocots recorded in Nambur and Kaza. The herbarium is in NU Botany Department. For details, refer P.N. Rao and K.K. Neelai, 1997 Tech Rep. to M.O.E. & F, New Delhi; & K.K. Neelai (1998) Ph.D. NU; P.N. Rao (1994). China dodder menace. Int. Par. Weed Symp. Jakarta.

Cuscuta reflexa Roxb. Pl. Cor. t. 104. 1799 & Fl. Ind. 1: 446. 1832; FBI 4: 225; Gamble 931 (654).

Leafless, twining, parasitic vines; branchlets much branched, glabrous, yellowish green, flowers creamish, in subsessile racemes; capsules succulent, globose, 5 mm across; seeds 4 or less, to 3.5 × 2.5 mm.

Occasional parasite on trees. Fl. & Fr.: Dec.–Apr. Reported in Guntur Forest Division Work Plan.

Evolvulus Linnaeus

Evolvulus alsinoides (L.) L. Sp. Pl. ed. 2. 2: 392. 1762; FBI 4: 220; Gamble 923 (648). *Convolvulus alsinoides* L. Sp. Pl. 157. 1753.

Spreading, much-branched, hairy herbs with woody root stocks; leaves elliptic or oblong-lanceolate, 1–2 × 0.4–0.6 cm, thinly sericeous on both sides, obtuse or shortly acute with apicule, entire, base narrowed; flowers blue, axillary, solitary; capsules depressed-globose, 5 mm across; seeds glabrous.

Common in open forests. Fl. & Fr.: Throughout the year. *Vern.*: Vishnukranta. Plant infusion mixed with that of *Merremia gangetica* (L.) Cuf. is a valuable herbal drug. N. Varam, *VRK* 3519; Mellavagu, *VRK* 5739; Vijayapuri, *KMS* 9833 (MH).

Ipomoea Linnaeus

1. Aquatic herbs .... ***I. aquatica***
1. Terrestrial herbs or shrubs:
    2. Erect shrubs .... ***I. carnea***
    2. Prostrate or twining shrubs or herbs:
        3. Prostrate herbs-leaves 2-lobed .... ***I. pes-caprae***
        3. Twining herbs or shrubs:
            4. Corolla salver-form, lobes triangular:
                5. Branchlets pubescent, flowers red, seeds not woolly, sepals distinctly awned below the apex, petals or scarlet, seeds pubescent ..... ***I. hederifolia***

5. Branchlets glabrous, flowers pink-purple, seeds woolly, sepals not awned, petals pink-purple, seeds woolly .... *I. mauritiana*
4. Corolla funnel-form, lobes distinct (except in *I. sepiaria*):
6. Leaves divided to the base (or below the middle):
7. Leaves 5-lobed, capsules ovoid, papery .... *I. coptica*
7. Leaves more than 5-lobed, capsules globose, not papery:
8. Leaves sericeous, flowers few in dense heads, calyx-lobes sericeous, unequal, petals pink .... *I. pes-tigridis*
8. Leaves glabrous, flowers solitary or 1–3 on axillary peduncles, calyx lobes equal, petals purple or purple blue .... *I. cairica*
6. Leaves entire, or divided up to the middle:
9. Flowers subsessile, aggregated in capitate cluster .... *I. eriocarpa*
9. Flowers stalked, solitary or in cymes/panicles:
10. Calyx-lobes linear-lanceolate, acuminate, more than 1 cm long .... *I. nil*
10. Calyx-lobes not linear-lanceolate, apex obtuse-apiculate, less than 1 cm long:
11. Leafbase truncate, flowers in panicles, stout climber ...... *I. staphylina*
11. Leafbase deeply cordate, flowers in umbellate cymes, weak climbers:
12. Peduncle stout, thickened towards the apex, basal lobes triangular-acute, corolla salver-form, cream with purple throat, seeds short tomentose .... *I. sepiaria*
12. Peduncles filiform, basal lobes rounded, obtuse, corolla funnel-form with tinged pink, seeds puberulent .... *I. obscura*

Ipomoea aquatica Forssk. Fl. Aeg.-Arab. 44. 1775; FBI 4: 210. *I. reptans* Poir. in Lamk. Encycl. Suppl. 3: 460. 1814; Gamble 916 (643).

Aquatic twiners in moist localities and ponds, rooting at nodes, branches hollow, striate; leaves variable in shape, ovate, ovate-oblong, ovate-lanceolate or linear, 6–8 × 2–4 cm, glabrous, acute or obtuse with apicule, entire, base cordate; flowers purple or rarely white, axillary, solitary; capsules ovoid, apiculate.

Common in ponds in the district. A problematic weed choking water tanks and irrigation channels. Fl. & Fr.: Throughout the year. *Vern.*: Thootikura. Young roots and leaves are nutritive and palatable. It is used as food for fish and pig. Javuku, *VRK* 3866; Macherla RF, *VRK* 3574. Achampeta, *VRK* 5868, 5872.

Ipomoea cairica (L.) Sweet Hort. Brit. 287. 1827; Gamble 919 (645). *Convolvulus cairicus* L. Syst. ed. 10. 922. 1759. *Ipomoea palmata* Forssk. Fl. Aeg.-Arab. 43. 1775; FBI 4: 214.

Glabrous slender twiners; leaves palmately dissected, 3–7 lobed, oblanceolate, glabrous, acute or obtuse, mucronate, base cordate; flowers violet, in axillary cymes; capsules globose; seeds pubescent.

Occasional in forests. Often cultivated in gardens. It is popularly known as Railway creeper. *Vern.:* Kakara thooti. This plant has numerous large showy flowers with purple colour on both sides and deep puple throat. It is suitable for fences, seeds are used as purgative. V.P. South, *VRK* 6725.

Ipomoea carnea Jacq. Enum. Syst. Pl. 13. 1760; Gamble 919 (645), ssp. **fistulosa** (Mart. ex Choisy) D. Austin in Taxon 26: 237. 1977. *I. fistuolsa* Mart. ex Choisy in DC. Prodr. 9: 349. 1845.

Large diffuse leafy shrubs, up to 2 m tall, branchlets hollow, glaucous; leaves broadly ovate, deltoid to hastate, 8–10 × 3–5 cm, chartaceous, gradually acuminate, entire, base cordate to truncate; flowers pink or rose in dichotomous, axillary and subterminal cymes; capsules globose; seeds hairy.

Common on roadsides and open forests and river banks, planted shrub by household fences, canal banks etc. Native of tropical America. Fl. & Fr.: Throughout the year. *Vern.*: Thuta kada, Rubber chettu. Juice of the stem is used as an antiseptic on cuts and wounds. Javuku, *VRK* 3282.

Ipomoea coptica (L.) Roem. & Schult. Syst. Veg. 4: 208. 1819. *Convolvulus copticus* L. Mant. Pl. 559. 1771. *Ipomoea dissecta* Willd. Phytographia 5. t. 2. f. 3. 1794; FBI 4: 200; Gamble 918 (645).

Annual herb, stem twining or prostrate, slender, terete; leaves pedately 5-lobed, mid lobe long, side lobes small, lobes ovate, oblong-obovate or oblanceolate, 2 × 0.5 cm, glabrous, acute, serrate, cuneate; stipular leaves 5-fid; flowers whitish-cream, in 1–3-flowered axillary cymes; bracts palmately laciniate; capsule ovoid, glabrous, fruiting calyx slightly enlarged; seeds 4, densely pubescent.

Rare, on coastal sands, semi-arid, sandy or rocky soils in the plains and hills. (Mohan Rao and Aasiya, 1995). Fl. & Fr.: Sept.–Apr.

Ipomoea eriocarpa R. Br. Prodr. 484. 1810; FBI 4: 204. *Convolvulus hispidus* Vahl Symb. Bot. 3. 29. 1794. *Ipomoea hispida* (Vahl) Roem. & Shult. Syst. Veg. 4: 238. 1819; Gamble 915 (643).

Slender, very twining herb; leaves narrowly lanceolate, to deltoid or ovate, 5–7.9 × 2–2.8 cm, chartacous, hairy, acute to acuminate, with apicule, margin hairy, base cordate; flowers small, pink in sessile or sub-sessile, capitate, axillary hirsute cymes; capsule globose, ca. 6 mm diam., hairy, 4-seeded; seeds grey.

Occasional, in waste lands and open places. Fl. & Fr.: Aug.–Feb. *Vern.*: Purti teega. Vijayapuri South, *VRK* 5818; Loyapalli, *SSR* & *KSM* 21662.

Ipomoea hederifolia L. Syst. Nat. ed. 10. 925. 1759. *Quamoclit phoenicea* (Roxb.) Choisy Mem. Soc. Phys. Geneva 6: 433. 1834; Gamble 919 (645). *Ipomoea coccinea* auct. FBI 4: 199. 1883.

Annual twining herbs; leaves broadly ovate, entire, or coarsely dentate, 10–20 × 10–18 cm, glabrous, acuminate with apicule, base cordate; flowers red in axillary cymes; capsules globose, 6–7 mm long, glabrous; seeds 4, pubescent.

Occasional in plains on hedges and rare in forests. Fl. & Fr.: Sept.–Mar. *Vern.*: Kasiratnam. The whole plant is used to cure the influence of evil spirits. Nizampatnam, *VRK* 6789; Sattenapalle, *VRK* 5842.

Native of tropical America, now widely naturalized in the tropics. It is popularly known as Star Ipomoea.

Ipomoea mauritiana Jacq. Collect. 4: 216. 1791. *Convolvulus paniculatus* L. Sp. Pl. 156. 1753. *Ipomoea paniculata* (L.) R. Br. Prodr. 486. 1810. *I. digitata* auct. plur non L. 1759; FBI 4: 202; Gamble 918 (644).

Large perennial climbers; leaves lobed nearly to the base; lobes lanceolate or elliptic, glabrous, acute, base cordate; flowers pink-purple, in cymose clusters; capsules ovoid; seeds woolly.

Occasional in forests. Fl. & Fr.: Throughout the year. *Vern.*: Nelagummudu, Penneru. Medicinal part of this plant is the large tuberous root with yellowish brown coat. It is brittle, mucilaginous and bitter in taste and is considered aphrodisiac, demulcent, lactogogue and cholagogue. Kondavedu Fort, *C.B. Barber* 4656 (MH).

Needs steps for protection and also conservation—recommended for micropropagation and molecular farming.

Ipomoea nil (L.) Roth Catal. Bot. 1: 36. 1797. *Convolvulus nil* L. Sp. Pl. (ed. 2) 1: 219. 1972. *Ipomoea hederacea* auct. non. (L.) Jacq. FBI 4: 199; Gamble 917 (644).

Villous twining shrubs; leaves cordate, 3-lobed, 4–5 × 4–6 cm, hairy on both sides, acute to shortly acuminate, base cordate; flowers blue in axillary long peduncled umbellate cymes; capsules 3-celled, ovoid, glabrous.

Common in forestes. Fl. & Fr.: Sept.–Jan. Koppanur, *VRK* 3272; Papaiahpalem, *VRK* 5852; Kondavedu Fort, *CAB* 4655 (MH); Loyapalli, *SSR & KSM* 21641.

Ipomoea obscura (L.) Ker.-Gawl. Bot. Reg. t. 239. 1817; FBI 4: 207; Gamble 916 (643). *Convolvulus obscura* L. Sp. Pl. ed. 2. 220. 1762.

Slender annual twining herbs, leaves suborbicular or ovate, 2–5 × 2–4.5 cm, glabrescent above, villous below, acuminate, base cordate, 7-nerved; flowers pale yellow with purplish base, small, axillary, solitary or in sub-umbellate cymes; capsules globose, apically beaked, 7–11 mm long, seeds glabrescent, 4.

Occasional in plains on hedges, roadsides and rare in forests. Fl. & Fr.: Feb.–Apr. *Vern.*: Nalla kokkiti. Juice of root is given orally as abortificient. Macherla RF, *VRK* 5900; Satrasala, *VRK* 6786; Loyapalli, *SSR & KSM* 21674.

Ipomoea pes-caprae (L.) R. Br. in Tuckey. Narr. Exp. Cango. App. 477. 1818; Gamble 917 (644). *Convolvulus pes-capre* L. Sp. Pl. 159. 1753. *Ipomoea biloba* Forssk. Fl. Aeg.-Arab. 44. 1775; FBI 4: 212.

Prostrate and long creeping herb from a perennial elongate root; sap milky; leaves thick, 2-lobed, base truncate, or rounded, 2–4 × 4–15 cm, glabrescent, emarginate, entire, parallel nerved from base; flowers bright pink, axillary, solitary; capsules globose, 1.2–1.8 cm long; seeds brown, velvety.

Common on sandy sea coast. It is a good sand binder. Fl. & Fr.: Oct.–Mar. *Vern.*: Bala banthi Tige, Beda Teega, Chevulapilli Tige. Fl. & Fr.: Oct.–Mar. Nizampatnam, *VRK* 6802.

Ipomoea pes-tigridis L. Sp. Pl. 162. 1753; FBI 4: 204; Gamble 918 (644).

Twining herbs; branches villous hairy; leaves palmately 5–7-lobed, lobes oblanceolate or obovate, 3–5 × 1–2 cm, acute or acuminate, contracted at base, appressed or patent hairs on either side; flowers pale pink in axillary long peduncled capitate clusters; capsules globose, papery, glabrous; seeds 4, pubescent.

Occasional in forests and on hedges. Fl. & Fr.: Sept.–Mar. *Vern.:* Meka Madugu, Puritikada. It is used as cattle-feed both in green and as hay. Mellavagu, *VRK* 5717; Kondavedu Fort, *CAB* 4677 (MH).

Ipomoea sepiaria Koen. ex Roxb. Fl. Ind. 2: 90. 1824; FBI 4; 209; Gamble 916 (643).

Hispid, perennial twining herbs; branchlets pilose; leaves deeply cordate, slightly 3-lobed, 4–6 × 4.5–7 cm, pubescent, acute to acuminate, or shortly caudate, base cordate; flowers white or light purple in axillary subumbellate cymes; capsules globose, 6–7 mm long, 2–4-seeded; seeds black, tomentose.

Occasional weed on natural road side, and hedges in forests. Fl. & Fr.: July–Oct. It is diuretic, deobstruent and antidote to arsenic poisioning. Macherla RF, *VRK* 5888; Kondavedu Fort, *CAB* 4654 (MH).

Ipomoea staphylina Roem. & Schult. in L. Syst. Veg. 4: 249. 1819; FBI 4: 210; Gamble 917 (643).

Stout straggling shrubs, leaves broadly elliptic-ovate, 6–15 × 4–10 cm, glabrescent, greenish above, pale beneath, acute, entire, base sub-cordate to truncate, flowers pink in lax panicled cymes; capsules ovoid.

Abundant in forests. Fl. & Fr.: Oct.–Apr. *Vern.*: Theendra tiga. The fresh leaf smeared with castor oil and burned, applied on fresh burned wounds for healing and to relieve pain. Loyapalle, *VRK* 3821.

Jacquemontia J.D. Choisy

Jacquemontia paniculata (Burm. f.) Hallier f. Bot. Jahrb. Syst. 16: 541. 1893; Gamble 926 (650). *Ipomoea paniculata* Burm. f. Fl. Ind. 50. t. 21. f. 3. 1768. *Convolvulus parviflorus* Vahl Symb. Bot. 3: 29. 1794. non Desr. 1791; FBI 4: 220.

Twining herbs, more or less pubescent or puberulous; leaves simple, ovate, 5–9 × 4–6 cm, chartaceous, entire, apex acuminate or acute and apiculate, pubescent, base shallowly cordate or truncate; flowers white, small in axillary corymbose cymes; capsules subglobose, 5 mm across, glabrous; seeds trigonous, margins slightly winged.

Common twiner in the forests. Fl. & Fr.: Dec.–Apr. Javuku, *VRK* 3880; Koppanur, *VRK* 3900; Vinukonda RF, *VRK* 6866.

Merremia Dennsted ex Endlicher *nom. cons.*

1. Leaves palmately compound .... ***M. aegyptia***
1. Leaves simple, entire or lobed:
    2. Peduncle very short or nearly absent .... ***M. gangetica***
    2. Peduncles longer;
        3. Capsule circumscissile, 1.5 cm diam., the upper part of the epicarp separateing from the lower part, stems narrowly winged .... ***M. turpethum***
        3. Capsule not circumscissile, ca. 6 mm diam., opening by 4-valves or irregularly:
            4. Capsule coarsely wrinkled, branchlets tuberculate, leaves ovate, base cordate or subcordate .... ***M. hederacea***
            4. Capsule smooth, branchlets smooth, leaves linear-oblong, oblanceolate, base hastate and often with several spinulose teeth .... ***M. tridentata***

Merremia aegyptia (L.) Urban. Symp. Antill. 4. 505. 1910; Gamble 928 (652). *Ipomoea aegyptia* L. Sp. Pl. 162. 1753. *I. pentaphylla* Jacq. Collectanea 2: 297. 1789; FBI 4: 202.

Vines; branchlets sericeous; leaves palmately compound; leaflets 5, obovate, 7–12 × 2.5–5 cm, chartaceous, patently hirsute, entire, shortly acuminate, subsessile, base acute; flowers white, large in long peduncled dichasial cymes; capsules covered by silky calyx lobes; seeds 4, glabrous.

Occasional vines on the thickets. Fl. & Fr.: Aug.–Dec. *Vern.*: Elika jemudu. Seeds are considered edible. Sattenapally, *VRK* 5856; Macherla RF, *VRK* 5893; Nagarjunakonda, *KTH* 9824 (CAL); Nallapadu, *VRK* 6789.

The species with beautiful white flowers and green leaves are attractive in early summer mornings. It is an elegant creeper for fences and garden walls. Fruits are used for day flower arrangements

Merremia gangetica (L.) Cuf. in Bull. Jard. Bot. Etat 31 (Suppl.): 743. 1961. *Convolvulus gangeticus* L. Cent. Pl. 2; 9. 1756. *Merremia emarginata* (Burm. f.) Hallier f. in Bot. Jahrb. Syst. 16: 552. 1893; Gamble 928 (652). *Evolvulus emarginatus* Burm. f. Fl. Ind. 77. t. 30. f. 1. 1768. *Ipomoea reniformis* (Roxb.) Choisy in Mem. Soc. Phys. Geneve 6: 446. 1834; FBI 4: 206.

Hispid creeping herbs; rooting at the nodes, glabrous or glabrescent, leaves usually reniform, or broadly ovate, 1–3.5 × 0.8–4 cm; glabrous, obtuse, or slightly emarginate, crenate-dentate, base cordate; flowers small, yellow in axillary, solitary or 2–3-flowered cymes; capsules globose, 3.5 mm diam., seeds black, often only 2, glabrous.

Abundant weed of cultivated fields and moist waste lands. Fl. & Fr.: Sept.–Feb. As per Ayurvedic literature, included among anthelmentics, also used to overcome urinary afflictions, worms, fever and gastric problems, leaf juice in migraine and snake bites, infusion of the leaves in 'burning motions', the decoction for cough, rheumatism and neuralgia. Mellavagu, *VRK* 5732; Krishna river, *VRK* 6746.

Merremia hederacea (Burm. f.) Hallier f. in Bot. Jahrb. Syst. 18: 118. 1894. *Evolvulus hederaceus* Burm. f. Fl. India 77. t. 30. f. 2. 1768. *Ipomoea chryseides* Ker.-Gawl. Bot. Reg. 4: t. 270. 1818; FBI 4: 206. *Merremia chyrseides* Hallier f. in Bot. Jahrb-Syst. 16: 522. 1893, Gamble 929 (652).

Annual herb, stem twining or prostrate, rooting at nodes, terete, minutely tuberculate; leaves ovate or cordate or shallowly 3-lobed, 3–5 × 3–4 cm, glabrous, acuminate, entire, cordate or truncate, mid rib and lateral veins raised beneath; flowers yellow, 1-many-flowered dichasial or monochasial cymes, peduncle up to 8 cm long; capsule globose somewhat 4-angled; seeds 4, reddish brown pubescent.

Rare, in moist places, on the hills (Mohan Rao and Aasiya, 1995). Fl. & Fr.: Sept.–Feb.

Merremia tridentata (L.) Hallier f. in Bot. Jahrb. Syst. 16. 552. 1893; Gamble 928 (652). *Convolvulus tridentatus* L. Sp. Pl. 157. 1753. *Ipomoea tridentata* (L.) Roth in Roem. Arch. Bot. (Leipzig) 1. 2: 38. 1798; FBI 4: 205.

Prostrate or shortly twining glabrous herbs; leaves narrow-oblong or linear, 1–3 × 0.5–1.2 cm, glabrescent, muriculate, proximally toothed, distally entire, base shortly dentate; flowers yellow, axillary, solitary or in cymes; capsules globose, 6.5 mm dia., seeds glabrous, 2–4.

Common weed of open forests. Fl. & Fr.: Sept.–Mar. *Vern.*: Surapu teega. Entire plant made into powder, taken in internally to cure piles and night blindness. Entire plant paste used externally in bone fracture for bone setting by tribal people. Javuku, *VRK* 3576; V.P. South, *VRK* 5787; Bapatla, *Diwakar* 94203 (MH).

Merremia turpethum (L.) G.L. Shah & R.G. Bhat. in J. Bombay Nat. Hist. Soc. 74. 567. 1978. *Convolvulus turpethum* L. Sp. Pl. 155. 1753. *Operculina turpethum* (L.) A. Silva Manso, Enum. Subst. Braz. 16. 49. 1936; Gamble 929 (653). *Ipomoea turpethum* (L.) R. Br. Prodr. 485. 1810; FBI 4: 212.

Twiners with winged branchlets, leaves simple, broadly ovate to cordate, 8–12 × 7–13 cm, puberulous, base truncate to subcordate, entire, apex obtuse with apicule; penninerved, lower 2 pairs arising from base, flowers white in lax corymbose cymes; capsules 2.7 × 2.3 cm, enclosed in woody calyx-lobes; seeds solitary.

Occasional along the streams. Fl. & Fr.: Dec.–Apr. The plant is useful in treating dropsy, eye diseases, oedema, hepatic, haemophilic disorders, jaundice, enlarged spleen etc. Traditionally, it is administered against dropsy due to heart, kidney or liver diseases. Javuku, *VRK* 3886; Guntur, *JSG* 84023 (MH).

This plant with large showy cream flowers is best grown over sun-shades, bowers, parapets and arches in the gardens. Fruit of the plant important component in dry flower arrangements.

Rivea Choisy

1. Peduncles usually 1-flowered cymes; corolla glabrous outside; outer 3-sepals acute, inner obtuse .... ***R. hypocrateriformis***
1. Peduncles usually 3 or more-flowered cymes; corolla white-silky, all sepals acuminate .... ***R. ornata***

Rivea hypocrateriformis (Desv.) Choisy in Mem. Soc. Phys. Geneve 6: 408; 1834; FBI 4: 184; Gamble 903 (635). *Convolvulus hypocrateriformis* Desv. in Lam. Encycl. 3: 561. 1792.

Large woody straggling shrubs, sap milky; leaves orbicular, 3–5 × 2.5–6 cm, glabrous above, pubescent below, apex obtuse, apiculate, entire, base cordate; flowers white, axillary, solitary; capsules conical, 1.5 × 1 cm; seeds glabrous, 4 or a few enclosed by enlarged calyx lobes.

Common in forests. Fl. & Fr.: June–Dec. *Vern.*: Boddi Teega, Junguru. Juice of whole plant is used to treat fever. Leaves and young shoots are edible. Javuku, *VRK* 3565, 3570; Vinukonda RF, *CAB* 17397 (MH); Mangalagiri, *SSR & KSM* 21638.

Rivea ornata Choisy in Mem. Soc. Phys. Geneve 6: 409. 1834; FBI 4: 183; Gamble 2: 904 (635).

Perennial climbing or straggling shrubs; stem woody at base, densely silky pubescent when young; leaves broadly ovate to reniform, 3–6 × 3–7 cm, glabrous above, white silky-pubescent beneath, obtuse to acuminate, entire, cordate, midrib raised beneath; Flowers white, fragant, in 3- or more flowered axillary or terminal cymes, peduncle to 1 cm long; capsule glabrous, yellowish brown, fruiting calyx accrescent; seeds usually 2, glabrous to puberulous, black.

In dry situations, along roadisdes and waste lands and deciduous forests. (Mohan Rao and Aasiya, 1995). Fl. & Fr.: Aug.—Mar. *Vern.*: Bodditiga. Plant is considered cooling agent and tonic, in preparation of ointments for piles and phthiriasis.

Ipomoea batatas Lam. (Sweet potato) is cultivated as a tuber crop.

SOLANACEAE

1. Fruits capsules, dehiscent .... ***Datura***
1. Fruits berries, indehiscent:
   2. Calyx inflated into bladder-like, enclosing the berry:

3. Flowers solitary, corolla broadly campanulate, lobes plicate in bud, leaves glabrous to strigose pubescent, berry yellow when ripe .... ***Physalis***
3. Flowers in fascicles, corolla narrowly campanulate, lobes not plicate leaves hoary tomentose, berry orange when ripe .... ***Withania***

2. Calyx not inflated:
4. Anthers connate, longer than filaments, flowers violet .... ***Solanum***
4. Anthers free, as long as or shorter than filamens, flowers white .... ***Capsicum***

Capsicum Linnaeus

Capsicum frutescens L. Sp. Pl. 189. 1753; FBI 4: 239; Gamble 950 (661); Heiser & Pickersgill in Taxon 18: 280. 1969.

Perennial shrubs, up to 2 m tall; leaves elliptic or elliptic-ovate, 4–9 × 2–4 cm, glabrous, acute or shortly obtusely acuminate, entire, base acute to obtuse; flowers greenish white, terminal, 2–5 in bunches; berries elongate, 0.8–3 × 0.1–0.2 cm, red when ripe; seeds circular, compressed.

The Chilli plant. Mostly cultivated, occasionally found as an escape near habitations. Native of tropical America. Fl. & Fr.: Throughout the year. *Vern.*: Konda mirapa. Fruit used as spice. V.P. South, *VRK* 3538.

Guntur District is well known for Chilli cultivation and export. 'C'-vitamin source, several cultivars designed including male sterile lines. It is also popularised 'A chilly a Day keeps the doctor away.'

Datura Linnaeus

1. Capsules armed with stout, short spines .... ***D. metel***
1. Capsules armed with weak, long spines .... ***D. innoxia***

Datura innoxia Mill. Gard. Dict. n. 5. 1768; Sant. in J. Bombay Nat. Hist. Soc. 47. 659. 1948. *D. metel* auct. non. L.; Sims in Curtis. Bot. Mag. t. 1440. 1812; FBI 4: 243; Gamble 941 (660).

Annual stout shrub with grey tomentum, up to 1 m tall; leaves elliptic-ovate to broad ovate, 8–10 × 5–8 cm, entire or shallowly lobed, glabrescent, acute, base oblique; flowers white, solitary; capsule globose, covered with sharp, weak spines.

Occasional in wastelands and open forests. Native of America (Mexico). Fl. & Fr.: Oct.–Mar. *Vern.*: Ummetha. Leaves, fruit and seeds medicinal. It is used as an anodyne, antiseptic, sedative and cerebral depressant and for fever, catarrh, diarrhoea and skin troubles. It is a commercial source of scopalamine. V.P. South, *VRK* 3938.

Daura metel L. Sp. Pl. 179. 1753. *D. fastuosa* auct. non L. 1759; FBI 4; 242; Gamble 949 (660).

Annual tomentose shrub like herbs, with dichotomous branches, up to 1 m tall; leaves ovate, 8–11 × 5–6 cm, acute, entire, rarely angled, tomentose, base unequal; flowers white, axillary or terminal, solitary, capsules globose with hard deltoid spines.

Common weed in waste places and cultivated fields. Fl. & Fr.: May–Jan. *Vern.:* Ummetha. It is used as an anodyne, antispasmodic, mydriatic, narcotic and catarrh. insanity, fever. diarrhoea, cerebral affections, asthma and skin trouble. Guntur, *JSG* 83101 (MH).

The identity, nomenclature and evolution of different spieces of *Datura* needs a thorough examination in view of reports of a natural interspecific hybrid *D. quercifolia* H.B.K. which is also likely to occur in Guntur district.

Physalis Linnaeus

1. Fruiting calyx less than 2 cm long, berries less than 0.8 cm across .... ***P. minima***
1. Fruiting calyx more than 3 cm long, berries more than 1 cm across .... ***P. peruviana***

Physalis minima L. Sp. Pl. 183. 1753; FBI 4: 238; Gamble 939 (659).

Erect glabrescent herbs, up to 60 cm tall; leaves elliptic-ovate, 5–9 × 3.5–5 cm, sparsely strigose; acute, margin shallowly toothed, base cuneate; flowers yellow, axillary, solitary, berries globose, entirely enveloped in the enlarged membranous calyx; seeds discoid, finely muriculate, orange-yellow.

Occasional in wastelands and roadsides. Fl. & Fr.: June–Dec. Budama, Budda. Fruits are edible. Achampeta, *VRK* 5860; Nagarjuna University campus, *VRK* 6987.

Physalis peruviana L. Sp. Pl. (ed. 2) 1670. 1763; FBI 4: 238; Gamble 939 (659).

Unarmed subshrubs, up to 1 m tall; leaves elliptic-ovate, 6–10 × 4–6 cm, chartaceous, sparsely strigose, coarsely toothed, acute, base obliquely rounded; flowers yellow, solitary, axillary; berries globose, yellow, overtopped by inflated calyx; seeds compressed.

Occaisonal weed in wastelands. Fl. & Fr.: July–Feb. Fruits are edible. Guntur (*Uhl*, 1929).

Several varieties are cultivated by Dr. G.R. Rao in NU campus during 1996–97. These include seed samples from USA and Mexico. The genus *Physalis* is a cytological complex with enormous biodiversity.

Solanum Linnaeus

1. Plants armed:
   2. Climbing shrubs .... ***S. trilobatum***
   2. Erect or diffuse herbs or shrubs:
      3. Diffuse subshrubs, berries yellow with green blotches .... ***S. surattense***
      3. Erect shrubs, berries uniformly yellow:
         4. Berry 7 mm diam. .... ***S. anguivi***
         4. Berry 1 to 2.5 cm diam. .... ***S. mèlongena*** var ***incanum***

1. Plants unarmed:
   5. Climbers, leaf-blade pinnatifid .... ***S. seaforthianum***
   5. Erect herbs or shrubs, leaf blade entire:
      6. Annuals, leaf margin bluntly toothed, ripe berries black .... ***S. nigrum***
      6. Perennials, leaf margin entire, ripe berries red or yellow .... ***S. pubescens***

Solanum anguivi Lam. Tabl. Encycl. 2: 23. 1794; Hepper in Bot. J. Linn. Soc. 76: 290. 1978. *S. indicum* auct. non L. FBI 4: 234; Gamble 938 (658).

Armed, yellowish villous undershrubs, up to 1 m tall; very prickly; leaves angulate, 4–6 × 2–4 cm, stellate-pubescent, prickles yellow, along the nerves, apex acute, margin lobed, base cordate, cuneate or truncate; flowers purple or purplish-violet, in racemose extra-axillary cymes; berry oblong-globose, dark yellow when ripe; seeds minutely pitted.

Occasional in the forests. Fl. & Fr.: Sept.–Mar. *Vern.*: Vankudu. Fruits are made into pieces and are cooked as vegetable and also given to cure fever. Loyapalle, *VRK* 3855; Nizampatnam, *VRK* 6801; Jayanthipuram, *VRK* 6954.

Solanum melongena L. var. **incanum** (L.) O. Kuntze, Rev. Gen. Pl. 188. 1753. *S. coagulans* Forssk. Fl. Aegypt.—Arab. 47. 1775; FBI 4: 236.

A branched prickly shrub, up to 1 m tall; prickles stout, very sharp, recurved, yellow; leaves ovate-elliptic, 3–8 × 2–4 cm, stellate hairy with velvety above, woolly beneath, apex subacute, sinuately lobed, base truncate to subcordate, sparsely prickly along the mid rib on both sides; berry 1.5–2.2 cm long, ovoid or subglobose, yellow; seeds pale yellow, minutely pitted.

Common on outskirts of forests and along roadsides. Fl. & Fr.: Jan.–May. *Vern.*: Adavi vankaya. Davupalli, *DMR & SSR* 18164.

Solanum nigrum L. Sp. Pl. 186. 1753; FBI 4: 229; Gamble 936 (657).

Annual erect much divaricately branched, unarmed herbs, up to 1.5 m tall; stem striate, glabrous; leaves elliptic-ovate, 6–10 × 3–5 cm, glabrous, acute, entire or bluntly toothed, base rounded or acute; flowers white, small in extra axillary, sub-umbellate, 3–8-flowered cymes; berries juicy, globose, 7 mm across, smooth, black when ripe, seeds yellow, minutely pitted.

Occasional in moist places and rare in forests. Fl. & Fr.: Throughout the year. *Vern.*: Kamanchi. Fruit decoction is used for gargling to cure throat infection. NU Campus, *SSR & KSM* 21686.

S. nigrum is a species complex. Naturally occurring *S. nigrum* in India is 6n, and a Hexaplod with 2n = 72. Diploid, tetraploid, natural and artificial are also experimentally raised. The artificial tetraploid has scarlet red fruits while others have black/purple fruits.

Solanum pubescens Willd. Phytographia 5. t. 3. 1794; FBI 4: 230; Gamble 936 (657).

Erect, fulvous-pubescent, unarmed shrubs, up to 1.5 m tall; leaves elliptic-ovate to ovate, 4–8 x 2–6 cm, hairy, subacute, entire, or more or less sinuately lobed, base rounded or subcordate; flowers purplish in axillary racemes; berries orange, globose; seeds minute, muriculate.

Common on roadsides and forest undergrowth. Fl. & Fr.: May–Nov. *Vern.*: Vushti. The seeds of the plant are used in rheumatism. Juice of fruit is given to cure indigestion; fruits as vegetable. N. Varam, *VRK* 3262; Nagarjunasagar, *KMS* 9729 (MH).

Solanum seaforthianum Andrews Bot. Repos 8. t. 504. 1808.

Glabrous, unarmed climbers; leaves pinnately lobed; lobes ovate-obovate, or elliptic-oblong, 2–5 × 1–2 cm, glabrous, decurrent, acute, base oblique; flowers violet in leaf opposed or terminal panicles; berries globose, red when ripe.

Native of South America. Introduced and runs wild. Fl. & Fr.: Oct.–Apr. Guntur, (*Uhl.* 1929).

Solanum surattense Burm. f. Fl. Ind. 57. 1768. (excl. syn.); Sant. in Rec. Bull. Bot. Sur. India. 18: 154. 1960. *S. xanthocarpum* Schrad & Wendl. Sert Hannov. 1: 8. t. 2. 1795; FBI 4: 236; Gamble 938 (658).

A very prickly perennial, diffuse herbs, somewhat woody at the base; leaves lanceolate, dissected, 5–8 × 3–5 cm, prickles yellow, straight, up to 2 cm long, numerous along the nerves, acute, base attenuate; flowers purple, in extra axillary 4–5-flowered cymes; berries globose, 1.5 cm across, yellow when ripe, seeds smooth.

Common weed in open forests and roadsides. Fl. & Fr.: Throughout the year. *Vern.*: Ramamulaga. Nela mulaka, Mullavanga. Root is an expectorant, forming an ingredient Ayurvedic medicine Dasamula. It is used in cough, asthma and pains in chest. Juice of fruits is used in sore throat and juice of leaves mixed with black pepper is recommended in rheumatism. Fruits yield glyco-alkaloids which can be source material for cortisone and sex hormone preparations. N. Varam, *VRK* 3585; Nagarjunasagar, *KMS* 9748 (MH).

Note: In most floras this plant has been referred as *Solanum xanthocarpum* of Schrad. and Wendl. (1795). Burm. f. *S. surattense* (1768), is an earlier one but its description is too meagre to identity a plant.

Solanum trilobatum L. Sp. Pl. 188. 1753; FBI 4: 236; Gamble 938 (659).

Armed trailing or sub-scandant undershrubs; branches divaricate; leaves ovate-angular, sinuate, 3–7.5 × 2–6 cm, chartaceous, 3–5-nerved, prickly on both sides, base cordate-truncate, obtuse; flowers violet-purple in extra axillary racemose cymes; berry globose, green with white spots, scarlet when ripe, seeds circular slightly pitted.

Occasional weed on hedges in roadsides. Fl. & Fr.: Dec.–Mar. *Vern.*: Mundla Musta. Its leaves are reported to be cooked and eaten as vegetable. Nizampatnam, *VRK* 6787.

Withania Pauquy *nom. cons.*

Withania somnifera (L.) Dunal in DC. Prodr. 13: 453. 1852; FBI 4: 239; Gamble 940 (660). *Physalis somnifera* L. Sp. Pl. 182. 1753.

Erect branching undershrub, up to 1.5 m tall, stellate tomentose; leaves ovate, 3–8 × 2–6 cm, pubescent, acute, entire, shortly cuneate, flowers yellowish, fascicled in the leaf-axils; berry globose, enclosed in the inflated, accrescent calyx, 5-angled, red when ripe.

Occasional, in waste places and hill slopes. Fl. & Fr.: June–Jan. *Vern.*: Aswagandha, noticed plant during 1976–80 along Kommanur and Guntur channel banks. Now fast disappearing due to over exploitation; under pot culture in NUBG.

Ashwagandha—recommended for conservation by micropropagation—therapeutic properties under investigation (p. 922, Curr. Sci., 70. 1997).

The Day Queen (*Cestrum diurnum* L.) and Night Queen (*Cestrum nocturnum* L.) are grown as ornaments. The tomato (*Lycopersicon esculentum* Miller), Brinjal (*Solanum melongena* L.) are cultivated as vegetable crops. The tobacco (*Nicotiana tabacum* L.) is extensively cultivated as tobacco crop in the district. Guntur is one of the world exporters of tobacco. Several strains occur based on Nicotine and tar content. The Petunia (*Petunia hybrida*) is under extensive cultivation.

SCROPHULARIACEAE

1. Leaves both alternate and decussate in the same plant .... ***Striga***
1. Leaves decussate or whorled (alternate in *Verbascum*):
    2. Aquatic; submerged leaves capillaceo-multifid .... ***Limnophila***
    2. Combination of characters not as above:
        3. Corolla rotate:
            4. Corolla yellow, lobes 5, anthers transverse, filaments bearded ...... ***Verbascum***
            4. Corolla white, lobes 4, anthers erect, anther cells erect .... ***Scoparia***
        3. Corolla tublar, 2-lipped:
            5. Leaf-blade pinnatifid .... ***Sutera***
            5. Leaf-blade undivided:
                6. Leaf margin entire:
                    7. Erect or ascending herbs, staminodes present, leaves 3-nerved from base .... ***Lindernia p.p.***
                    7. Prostrate herbs, staminodes 0, leaves penninerved .... ***Bacopa***
                6. Leaf margin serrate-crenate:
                    8. Stamens 2 .... ***Lindernia p.p.***
                    8. Stamens 4:
                        9. Corolla glabrous within, anthers connate .... ***Lindernia p.p.***
                        9. Corolla villous within, anthers free:
                            10. Leaf-base auriculate, calyx deeply 5-lobed, stigma 2-lobed, anthers unequal .... ***Stemodia***
                            10. Leaf-base other than auriculate, calyx spathaceous, stigma simple, anthers equal .... ***Centranthera***

Bacopa Aublet *nom. cons.*

Bacopa monnieri (L.) Wettst. in Engler & Prantl Pfl. Fam. 4 (3b): 77. 1891 (*'monniera'*). *Lysimachia monnieri* L. in Torner, Cent. 2. Pl. 9. 1756. *Monniera cuneifolia* Michaux Fl. Bor. Am. 2: 22. 1803 (*'monniera'*); Gamble 953 (669). *Herpestis monnieri* (L.) Kunth, Nov. Gen. Sp. 2: 366. 1818. *nom. illeg.;* FBI 4; 272.

Prostrate, sub-succulent herbs with spreading branches, rooting at nodes; leaves decussate, narrowly obovate-oblong or spathulate, 1–2 × 0.2–0.6 cm, glabrous, obtuse, entire; flowers purple, axillary, solitary; capsule oblong-globose, 4 × 2 mm, included in the calyx, seeds oblong.

Common in marshy places along the streams. Fl. & Fr.: Throughout the year. *Vern.:* Brahmi, active compound—Bacosides. Improves memory (vide Sukh Dev 1997, Curr. Sci., Vol. 73, No. 11). Component of the dry Memory + along with Mandukaparni and Mahabringraj. Entire plant constitutes drug which is used in insanity, epilepsy and as potent diuretic cordiotonic and nerve tonic. Koppanur, *VRK* 3889; Javuku, *VRK* 3867; Nagarjunasagar, *KMS* 97.17 (MH); Nagarjunakonda, *KTH* 9709 (CAL).

Centranthera R. Brown

Centranthera indica (L.) Gamble Fl. Madras 2: 971 (683). 1924. *Rhinanthus indica* L. Sp. Pl. 603. 1753. *Centranthera hispida* R. Br. Prodr. 438. 1810; Gamble 971 (683).

Erect scabrid herbs up to 20 cm tall; stem 4-angled; leaves entire-toothed, oblong-oblanceolate, to 3 × 0.5 cm, hispid and scabird, acute or obtuse, flowers subsessile, solitary, axillary; capsule oblong, 5–7 mm long, enclosed in calyx.

Occasional weed in grass-lands and open forests. Fl. & Fr.: Sept.–Mar. V.P. South, *VRK* 6757.

Limnophila R. Brown *nom. cons.*

Limnophila indica (L.) Druce Bot. exch. Club. Soc. Brit. Isles 3: 420. 1914; Philcox in Kew Bull. 24: 115. 1970. *Hottonia indica* L. Syst. Nat. (ed.10) 919. 1759. *Limnophila gratioloides* R.Br. Prodr. 442. 1810; FBI 4: 271; Gamble 951 (668).

Prostrate herbs, puberulous and glandular-pubescent or glabrous; leaves decussate, linear-lanceolate, 1–2.5 × 0.2–0.5 cm, obtuse, serrulate, multifid; flowers pale pink, axillary, solitary; capsules globose, 3 mm across.

Occasional in moist places and in streams and ponds in forests. Fl. & Fr.: July–Dec. Leaves are eaten as vegetable, and herb is considered antiseptic, and with coconut oil used in elephantiasis. Jayanthipuram, *VRK* 6932; Vinukonda, *VRK* 6881.

Lindernia Allioni

Note: The generic alignments within the group *Lindernia* has long been a matter of discussion and controversy. Linnaeus treated *Lindernia* and *Vandellia*, both with four perfect stamens as two distinct genera, though Bentham and many others preferred to unit them. *Ilysanthes* and *Bonnaya* are characterised by two perfect stamens and two staminodes but differ in their relative position. In *Ilysanthes*, the posterior pair is fertile and the anterior staminodal, whereas in *Bonnaya* they are in reverse order. Bentham (1876), Hooker (1884) and many others have preferred to keep them under *Ilysanthes*. Haines (1922) reduced the four genera into two but, in a different manner. He included *Bonnaya* in *Vandelia* and *Ilysanthes* in *Lindernia*, thus mixed up taxa with four and two perfect stamens. He recognised these groups mainly based on major vein configurations in their leaves, *Vandelia* with penninerved leaves and *Lindernia* with basally veined leaves.

Neither staminal nor venation characters have provided adequate and reliable taxonomic criteria for circumscribing the different genera within this group. Pennell (1935, 1943) combined all these four taxa under the single generic name *Lindernia*, a natural assemblage characterised by the remarkably uniform corolla, cursiously recurving anterior filaments and by similar septicidal dehiscence of capsule. Indian species of this genus have been revised by Sivarajan and Mathew (1983) who recognised 22 species.

1. Prostrate herbs .... ***L. antipoda***
1. Erect herbs:
    2. Leaves entire, capsules oblong-globose .... ***L. parviflora***
    2. Leaves dentate, capsules linear-lanceolate .... ***L. ciliata***

Lindernia antipoda (L.) Alston in Trimen, Fl. Ceylon 6 (Suppl.); 214. 1931; Philcox in Kew Bull. 22: 57. 1968. *Ruellia antipoda* L. Sp. Pl. 635. 1753. *Ilysanthes veronicifolia* (Retz.) Urban, Ber. Deustsch. Bot. Ges. 2: 436. 1884; Gamble 962 (675). *Bonnaya veronicifolia* (Retz.) Spreng. in L. Syst. Veg. (ed. 16) 1: 41. 1824; FBI 4: 285.

Prostrate herbs; branches square, 10–25 cm long; leaves elliptic-obovate, 1.5-2.5 × 0.5–1 cm, glabrous, apex obtuse, margin shallowly serrate, cuneate; flowers bluish, axillary, solitary or terminal racemes; capsules linear-lanceolate, exceeding the calyx, 1.2–1.5 cm long.

Common wetland weed. Fl. & Fr.: Aug.–Mar. Ethipothala falls, *VRK* 5789.

Lindernia ciliata (Colsm.) Pennell in Brittonia 2: 182. 1936; Philcox in Kew Bull. 22: 51. 1968. *Gratiola ciliata* Colsm. Prodr. Des. Gratiol. 14. 1793. *Ilysanthes serrata* (Roxb.) Urban Ber. Deutsch Bot. Ges. 2: 436.

1884; Gamble 962 (675). *Bonnaya brachiata* Link & Otto Icon. Pl. Select: 26. t. 11. 1820; FBI 4: 284.

Annual erect herbs, up to 25 cm tall, branches 4-angled; leaves elliptic-obovate, 1–3 × 0.5–1 cm, obtuse, sharply and closely aristate serrate, base cuneate; flowers white with pink throat, in terminal racemes; capsules linear-lanceolate, up to 1.5 cm long, exceeding the calyx lobes; seeds usually truncate at both ends.

Common weed in moist places and open forests. Fl. & Fr.: July–Dec. Macherla RF, *VRK* 5891.

Lindernia parviflora (Roxb.) Haines Bot. Bihar, Orissa 2: 635 (665) 1922. *Gratiola parviflora* Roxb. Pl. Cor. t. 203. 1811. *Ilysanthes parviflora* (Roxb.) Benth. in DC. Prodr. 10: 419. 1846; FBI 4: 283; Gamble 961 (675).

Aquatic, erect, slender glabrous, diffusely branched herbs, up to 25 cm tall; leaves oblong-elliptic, 1–2 × 0.1–0.3 cm, glabrous, acute, entire, base rounded, 3–5-nerved from the base; flowers violet-white or pinkish-white on axillary slender pedicels; capsules oblong-globose, 0.6 × 0.3 cm, exceeding the calyx.

Common in perennial streams and marshy areas. Fl. & Fr.: Sept.–Mar. Javuku, *VRK* 3292.

Scoparia Linnaeus

Scoparia dulcis L. Sp. Pl. 116. 1753; FBI 4: 289; Gamble 964 (678).

Erect, much branched glabrous herb or undershrub up to 60 cm; stems 4–6 angled; leaves 3-nately whorled, upper often opposite, rhomboid, elliptic, ovate or lanceolate, 1.2–3.7 × 0.6–1.8 cm, glandular, acute, serrate, tapering; flowers white, 1 or 2 in axils of leaves; capsule subglobose, 3.5 × 3 mm, septicidal; seeds truncate or 4-angled, reticulate.

Occasional weed in waste lands, clearings, fallow fields etc. (Mohan Rao and Aasiya, 1995). Eng.: Sweet broom weed. *Vern.:* Potti pulli. Fl. & Fr.: Throughout the year.

An infusion of leaves is used in fever, cough and bronchitis, and as a gargle for toothache. Leaves and stems are used for diabetes. These are recent enquiries for illicit export of this weed to USA due to a undetermined medicinal value.

A native of tropical America, now Pantropic.

Stemodia Linnaeus

Stemodia viscosa Roxb. Pl. Cor. t. 163. 1802; FBI 2: 265; Gamble 949 (666).

Annual glandular-hairy, ascending herbs, up to 50 cm tall; leaves decussate or ternate, elliptic-oblong, 3–7 × 1–1.5 cm, serrate, acute, base aciculate; flowers violet in axillary leafy racemes; capsules oblong-globose, 4 mm, equalling calyx, tipped by withered style, seeds oblong, smooth.

Occasional in plains on river banks. Fl. & Fr.: Aug.–Dec. *Vern.:* Bodasaramama. Nagarjunakonda, *KTH* 9847 (CAL).

Striga Loureiro

1. Calyx 5-ribbed; corolla pink to purple; stem purple or brownish .... ***S. gesneriodes***
1. Calyx 15-ribbed, corolla white or yellow; stem greenish:
   2. Lateral ribs of calyx lobes converging at apex, calyx-lobes lanceolate, corolla exclusively white, >1.5 cm across .... ***S. angustifola***
   2. Lateral ribs of calyx lobes ending just above sinuses, calyx lobes linear; corolla white or yellow, < 1 cm across .... ***S. asiatica***

Striga angustifolia (D. Don) C.J. Saldanha in Bull. Bot. Surv. India 5: 70. 1963. *Buchnera angustifolia* D. Don Prodr. Fl. Nepal. 91. 1825. *Striga euphrasiodes* Benth. in Hook. Comp. Bot. Mag. 1: 364. 1836; FBI 4: 299.

Erect herb, 10–35 cm; stems quandrangular, scabrid; leaves linear or narrowly oblong to pinnatifid, 1–5 × 0.2–0.5 cm, margins thickened; flowers white, 1.2–1.8 cm long; capsule obovoid or ovoid, 6 × 5 mm, seeds longitudinally ribbed.

Occasional in grassy lands or in scrub jungles. Fl. & Fr.: Throughout the year. Eng.: Witch weed. *Vern.*: Nala badamika. Opposite NU campus, Nagarjunanagar and open grazed grass lands; ecology and biology-cotents of Ph.D. Thesis, NU by B.V.N. Reddy, 1953. New hosts of this hemiparasite reported by P.N. Rao, B.V.N. Reddy and B.V. Raghavaswamy, 1986–Adansonia in 2: 207–212. Together deposited in NUH. Species determinant is Dr. Chrio Parker of Weed Research Organization, UK; Davupalli, *SSR* & *KSM* 21673.

Striga asiatica (L.) Kuntze Rev. Gen. Pl. 2: 466. 1891. *Buchnera asiatica* L. Sp. Pl. 630. 1753. *Striga lutea* Lour. Fl. Cochinch. 22. 1790; FBI 4: 299; Gamble 968 (680).

Erect slender villous herbs, up to 40 cm tall; leaves linear-oblong, 2.5 × 0.2 cm, hispid; flowers white or yellow in spikes; capsules oblong, 4 × 2 mm, seeds oblong.

Parasite on millets and other grasses. Fl. & Fr.: Sept.–Dec. Loyapalle, *VRK* 3995; V.P. South, *VRK* 3534; Phirangipuram, *CAB* 4612 (MH).

Striga gesnerioides (Willd.) Vatke Oest. Bot. Z. 25: 11. 1875. *Buchnera gesnerioides* Willd. Sp. Pl. 3: 388. 1800. *Striga orobanchoides* (Endl.) Benth. Comp. Bot. Mag. 1: 361. 1836; FBI 4: 299; Gamble 967 (680).

Erect, root parasitic herbs on *Lepidagathis cristata*, up to 25 cm tall, black when dry; stem reddish brown, many from a tuberous root, leaves scaly, linear, 2–5 × 0.3–0.4 cm, dark reddish purple, acute, base truncate, flowers pink, sessile, numerous in terminal spikes; capsules obovoid-globose, up to 4 mm, tipped with the base of the style, seeds oblong.

Occasional in open forests, parasite on roots of *Lepidagathis cristata* Fl. & Fr.: Sept.–Dec. Javuku, *VRK* 3568; V.P. South, *VRK* 5834; Bollapalle, *CAB* 17420 (MH).

Sutera Roth

Sutera dissecta (Del.) Walp. Repot. 3: 271. 1844; Druce in Kew Bull. 1940. 63–64. *Capraria dissecta* Del. Fl. Egypt. 95. t. 32. f. 2. 1812. *Sutera glandulosa* Roth Nov. Pl. Sp. 291. 1821; FBI 4: 258; Gamble 946 (664).

Erect herbs, with spreading glandular-viscid branches; leaves 0.8–2 cm, ovate, pinnatifid, segments toothed and cut; flowers white or bluish, axillary, solitary; capsules ovoid-oblong; 5 mm long, seeds minute, spherical or oblong.

Occasional. Fl. & Fr.: Aug.–Dec. Krishna river, *KMS* 9800 (MH).

Verbascum Linnaeus

Verbascum chinense (L.) Sant. Fl. Purandhar 90, 1957. *Scrophularia chinensis* L. Mant. 250. 1771. *Celsia coromandeliana* Vahl Symb. Bot. 3: 79. 1794; FBI 4; 251; Gamble 945 (663).

Erect, pubescent herbs, up to 1 m tall; radical leaves lyrate, leaves alternate, oblanceolate, 3–6 × 1–2 cm, pubescent, crenate-serrate; flowers yellow in terminal racemes; capsules globose, 6–8 mm long, glandular, seeds rugose.

Common in moist waste places and forests. Fl. & Fr.: Nov.–May. Guntur (*Uhl,* 1929).

OROBANCHACEAE

Orobanche Linnaeus

Orobanche cernua Loefl. It. Hisp. 152. 1758; FBI 4: 325; Gamble 978 (687).

Annual, erect, leafless root parasitic herbs, up to 40 cm tall; scales ovate-lanceolate; flowers creamish, sessile in cylindric dense spikes; capsules ellipsoid, 3-valved, seeds numerous, blackish-brown.

Complete root parasite on *Nicotiana tabacum* which is grown all over the district especially on black cotton soils. Fl. & Fr.: Oct.–Jan. Guntur, *KCJB* 83102 (MH); Tadikonda *KCJB* 86607 (MH).

BIGNONIACEAE

1. Leaves 2–3-pinnate, large tree .... ***Millingtonia***
1. Leaves 1-pinnate, small to medium-sized trees:
    2. Flowers in pendulous racemes, capsules indehiscent .... ***Kigelia***
    2. Flowers in erect corymbs or racemes, capsules dehiscent:
        3. Calyx spathaceous .... ***Dolichandrone***
        3. Calyx tubular or campanulate:
            4. Leaflets sharply serrate, capsules up to 20 cm long .... ***Tecoma***
            4. Leaflets entire to minutely toothed, capsules over 30 cm long .... ***Stereospermum***

Dolichandrone (Fenzl.) Seemann *nom. cons.*

Dolichandrone falcata (Wall. ex DC.) Seem. Fl. Vill. Nov. 151. 1880; FBI 4: 380; Gamble 996 (700). *Spathodea falcata* Wall. ex DC. Prodr. 9: 206. 1845.

Densely foliaceous deciduous trees, up to 5 m tall; young branches tomentose; leaves 1-pinnate, rachises tomentose; leaflets 5–7, elliptic to orbicular, or obovate, 2–3.5 × 1.8–2.5 cm, chartaceous, thinly pubescent, apex rounded with short acute tip, entire, acute; flowers white, scented in axillary few-flowered corymbose racemes; capsules compressed, falcate, 20–30 × 1.5–1.8 cm; seeds numerous, rectangular, greyish.

Occasional in forests. Fl. & Fr.: July–Mar. *Vern.:* Chittu-Niruddi, Chittoddi, wood is hard and used for building and agricultural purposes. Loyapalle, *VRK* 3974; Javuku, *VRK* 3864.

Kigelia A.P. de Candolle

Kigelia africana (Lam.) Benth. in Hook. Niger Fl. 463. 1849. *Bignonia africana* Lam. Encycl. 1: 424. 1785. *Kigelia pinnata* DC. Prodr. 9: 247. 1845; Gamble 1000 (703).

Large deciduous trees, up to 10 m tall, trunk short; leaflets 5–7, coriaceous, elliptic-oblong, 4–11 × 2.5–5 cm, glabrous, obtuse, rarely emarginate, entire, base oblique; flowers dark-purple or bright pink with yellow streaks, foetid in long-peduncled pendulous racemes; capsules gourd-like, 30–40 × 10–15 cm, indehiscent, woody; seeds many, embedded in pulp.

The Indian Sausage Tree. Occasional on roadsides as avenue trees. Fl. & Fr.: Apr.–Sept. *Vern.*: Yenugula vatavruksham. It is used as a purgative and for rheumatic pains, dysentery, syphilitic sores and venereal diseases. V.P. South, *VRK* 6821.

Millingtonia Linnaeus

Millingtonia hortensis L.f. Suppl. Pl. 291. 1781; FBI 4: 377; Gamble 994 (699). *Bignonia suberosa* Roxb. Pl. Cor. t. 214. 1811.

Evergreen trees, up to 15 m tall; leaves 2-pinnate; leaflets elliptic-ovate, 3–5 × 2–4 cm, chartaceous, impressed above, raised below, entire to coarsely toothed, acuminate, base often oblique, rarely rounded; flowers fragrant, white, in long corymbose panicles; capsule terete; linear, 30 × 2 cm, woody; seeds numerous.

The Indian coral tree. Avenue tree, planted. Fl. & Fr.: Feb.–Aug. *Vern.*: Punnagamalle; beautiful, evergreen, easily propagated by root suckers. Reported in Guntur Forest Division Work Plan. Self sown populations in NU campus.

Stereospermum Chamisso

Stereospermum personatum (Hassk.) Chatterjee in Bull. Bot. Soc. Bengal 2: 70. 1948. *Dipterosperma personatum* Hassk. Flora 25. Pt. 2, Biebl. 28. 1842. *Stereospermum tetragonum* DC. Prodr. 9: 210. 1845; Gamble 998 (701). *S. chelonoides* auct. non (L.f.) DC.; FBI 4: 382.

Densely foliaceous, deciduous tree, up to 15 m tall; leaves 1-pinnate, 15–20 cm long, leaflets 9–11, coriaceous, elliptic, 7.5–10 × 2–3 cm, glabrous, caudate-acuminate, entire to minutely toothed, acute; flowers yellow, veined with red in large glabrous panicles; capsule slender, terete or 4-gonous, 30–40 cm long, epicarp thick with speckles; seeds trigonous, compressed.

Occasional, in dry deciduous forests. Fl.: Apr.–June; Fr.: May (following year). *Vern.*: Pulkevelama, Pisul, Tagada. The root bark and the flowers along with sugar are used in the preparation of cool drink. The juice of the leaves boiled with oil used in diseases of ear and teeth and in rheumatism. The leaves are lopped for fodder. Wood is used for canoes, and tea-boxes, match boxes, splints, furniture, fancy work, plywood and floor-boards. Wood is reported to yield excellent charcoal. Leaves are lopped for fodder. Davupalli, *SSR & KSM* 21426.

Tecoma A.L. Jussieu

Tecoma stans (L.) Kunth in H.B.K. Nov. Gen. Pl. 3: 144. 1819. *Bignonia stans* L. Sp. Pl. (ed. 2) 871. 1763. *Stenolobium stans* (L.) Seem. Ann. Mag. Nat. Hist. (ser. 3) 10: 30. 1862; Gamble 1000 (703).

Large shrubs, to small trees, up to 6 m tall; bark greyish, rough; leaves 1-pinnate; leaflets ovate or lanceolate, 3–8 × 2–4 cm, glabrous, acute-acuminate, serrate, base oblique; flowers yellow in terminal racemes; capsules oblong, 10–15 × 0.5–0.7 cm.

Cultivated in gardens, rarely as an escape. Native of tropical America. Fl. & Fr.: Throughout the year. *Vern.*: Suvarna Ganneru. The root is reported to be a powerful diuretic, vermifuge and tonic. The wood is highly prized for furniture, carving and agricultural implements. Guntur (*Uhl*, 1929).

Bignonia spp, *Jacaranda mimosifolia* D. Don, *Tecomaria caopensis* (Thunb.) Lindley, *Tabeubia rosea* (Bertol.) DC., *Adenocalyma alliacea, Pyrostegia venusta* (Ker.-Gawl.) Miers are the cultivated ones.

PEDALIACEAE

1. Leaves palminerved, flowers in racemes, stamens 2.—corolla purple with yellow throat .... ***Martynia***
1. Leaves penninerved, flowers solitary, stamens 4:
   2. Drupes indehiscent, 4-spinous, corolla, yellow, subequally 5-lobed, stem rounded .... ***Pedalium***
   2. Capsules dehiscent, unarmed, corolla pink or purple or white, irregular, stems obtusely quadrangular .... ***Sesamum***

Martynia Linnaeus

Martynia annua L. Sp. Pl. 618. 1753; Gamble 1003 (705). *M. diandra* Gloxin. Obs. Bot. 14. t. 1. 1785; FBI 4: 386.

Annual, glandular-hairy, sticky undershrubs, up to 1.5 m tall; leaves opposite, broad ovate, 10–15 × 9–15 cm, palmately nerved from base; flowers purple or pink with yellow throat in axillary racemes; drupes ovoid with two prominent hooks, seeds oblong.

Common in moist places. Native of tropical and sub-tropical America. Fl. & Fr.: Aug.–Dec. *Vern.*: Telukondikaya. Seed paste applied on affected areas for scabies. N. Varam, *VRK* 3504.

Pedalium D. Royen ex Linnaeus

Pedalium murex L. Syst. Nat. (ed. 10) 1123. 1759; FBI 4: 386; Gamble 1001 (704).

Annual fleshy, foetid, much branched spreading herbs, with minutely glistening glands on the young shoots; leaves subopposite or alternate, succulent, ovate, 3–5 × 2–3 cm, glabrous, repend-dentate, obtuse; flowers yellow, axillary, solitary; drupes indehiscent, 1.8 × 0.7 cm, spiny along the angles.

Common in waste places, dried ponds and roadsides. Fl. & Fr.: Apr.–Sept. Eng.: Devil's claw. *Vern.*: Enugu palleru. It is used as an alexiteric and for epilepsy, sore throat, inflammation and tuberculous glands of the neck. V.P. South, *VRK* 5827.

Sesamum Linnaeus

1. Leaves entire, flowers white, seeds glabrous .... *S. orientale*
1. Leaves pinnatifid, flowers purple, seeds winged .... *S. alatum*

Sesamum alatum Thonn. in Schum. & Thonn. Beskr. Guin. Pl. 284. 1827; Ramanujam & Joshi in Journ. Bombay Nat. Hist. Soc. 52: 657. 1954; Bruce in Fl. Trop. Afr. (Pedaliaceae) 17. *Sesamum ekambaramii* Naidu in Journ. Bombay Nat. Hist. Soc. 51: 697. 1953.

Erect branched herbs to shrubs, up to 2 m tall; leaves palmately compound, simple, opposite at the base, sub-opposite above; palmately dissected; flowers purple, axillary, solitary on short peduncles; capsules oblong-obovate, narrow at base with beak; seeds numerous, black, winged.

Rare in waste places and cultivated fields. Fl. & Fr.: Oct.–Dec. V.P. South, *VRK* 5803. Loyapalle, *VRK* 3993; Loyapalli, *SSR* & *KSM* 21635.

Note: B. Appala Naidu, Department of Botany, Agricultural College, Bapatla has described a new species of *Sesamum* as *S. ekambaramii* Naidu from Stuartpuram near Bapatla, Andhra Pradesh. But S. Ramanjuam and A.B. Joshi of IARI, New Delhi have corrected it as *S. alatum* Thonn.

Sesamum orientale L. Sp. Pl. 634. 1753. *S. indicum* L. Sp. Pl. 634. 1753; FBI 4: 387; Gamble 1002 (704).

Erect hairy tall herbs, 1–2 m tall; lower leaves opposite, lobed or deeply divided, upper ones undivided, linear-oblong or elliptic-lanceolate, 8–10 × 2–4 cm, chartaceous; flower pink or rosy, axillary, solitary; capsule erect oblong, 2.5 × 0.8 cm, valvular, bluntly 4-gonous; seeds black or brown, smooth, glabrous, ca. 2 mm long.

Commonly cultivated, rarely as an escape. Fl. & Fr.: July–Dec. *Vern.*: Nuvuulu. Oil mixed with *Vitex negundo* roots, taken internally cures leprosy. Seed is used in confectionary and as topping for bread and rolls, also used in Hindu religious ceremonies. The seed oil (gingelly oil) is edible and mainly used as an ingredient of confectionary and for making margarine. Also used in manufacture of soaps, cosmetics, perfumes, insecticides and pharmaceutical products. Oil-cake is used as cattle feed. Kondavedu Fort, *CAB* 17485 (MH); Loyapalli, *SSR & KSM* 21635.

Note: According to Backer and Backhuizen the first author to unite these equally priorable names is Graham (1839) and he adopted the name *S. orientale*. De Candolle in his Prodromous 9: 250. 1845 reduced *S. orientale* to *S. indicum* which has been followed by many workers which is contrary to the code.

ACANTHACEAE

1. Corolla-tube cartilaginous, upper lip of corolla reduced, fertile anther-cell 1:
    2. Armed erect shrubs, mangroves .... ***Acanthus***
    2. Unarmed herbs, inland .... ***Blepharis***
1. Corolla tube thin, fleshy, upper lip of corolla well developed, fertile anther-cells 2 (1 cell sterile in *Crossandra*):
    3. Corolla tube as long as the lips, lobes 2+3:
        4. Scapigerous herbs .... ***Elytraria***
        4. Non scapigerous herbs:
            5. Stamens 4, didynamous, flowers either in condensed spikes or in axillary whorls:
                6. Branchlets diffuse, prickly with white deflexed bristles .... ***Gantelbua***
                6. Branchlets not prickly with bristles:
                    7. Flowers pale blue, bracteoles 0 .... ***Hemigraphis***
                    7. Flowers white or purple, bracteoles present:
                        8. Stems not thorny, flowers in condensed spikes, capsules 2- or 4-seeded, bracts and bracteoles sharply mucronate .... ***Lepidagathis***
                        8. Stem thorny at nodes, flowers in axillary whorls, capsules 8- or more-seeded, bracts and bracteoles not mucronate .... ***Hygrophila***
            5. Stamens 2, equal, flowers in cymes, elongated spikes or panicles:
                9. Inflorescence a dense spike:
                    10. Calyx segments 5 .... ***Justicia***
                    10. Calyx segments 4:
                        11. Bracts and bracteoles prominently scarious margined, flowers secund .... ***Rungia***
                        11. Bracts and bracteoles not or faintly scarious margined; flowers not secund .... ***Rostellularia***
                9. Inflorescence a cyme or panicle:
                    12. Ovules 2 or rarely 1 in each cell, seeds papillose, stamens free, corolla tube twised .... ***Peristrophe***
                    12. Ovules 4–8 in each cell, seeds glabrous, stamens connivent; corolla tube not twisted:
                        13. Capsule 4-seeded, panicles exclusively axillary .... ***Indoneesiella***
                        13. Capsule more than 8-seeded, panicles both axillary and terminal .... ***Andrographis***
    3. Corolla tube more than twice as long as lobes, lobes 4+1:
        14. Corolla limb 1-sided, anthers wih 1 fertile cell, bracts 4-farious ... ***Crossandra***
        14. Corolla limb equally spreading, anthers with both fertile cells, bracts not 4-farious:
            15. Fertile stamens 4, corolla tube ventricose:
                16. Bracts 0, bracteoles leafy, capsules basally beaked, leafbase truncate or rounded to acute, roots not tuberous .... ***Dipteracanthus***
                16. Bracts present, bracteoles not leafy, capsules not basally beaked, leaf-base cuneate to attenuate, roots tuberous .... ***Ruellia***
            15. Fertile stamens 2, corolla tube cylindric:
                17. Corolla-lobes equally spreading:
                    18. Calyx segments 4, unequal stamens inserted within the tube .... ***Barleria***
                    18. Calyx segments 5, equal, valvate, stamens inserted at the throat .... ***Eranthemum***
                17. Corolla-lobes distinctly 2-lipped:

19. Flowers in panicles, stamens attached at the throat of the tube, corolla white, bracts shorter than calyx, capsule 4-seeded, leaves pubescent .... ***Rhinacanthus***
19. Flowers in spikes, stamens attached at the base of upper lip, corolla bluish-green, bracts longer than calyx, capsules 2-seeded, leaves glabrous .... ***Ecbolium***

Acanthus Linnaeus

Acanthus ilicifolius L. Sp. Pl. 639. 1753; FBI 4: 481; Gamble 1014 (712).

Armed subshrubs, up to 1.5 m tall; leaves oblong or elliptic usually pinnatifid, 8–12 × 4–6 cm, coriaceous, base attenuate, apex acuminate, with sharp spinous teeth on the margin terminating the lateral nerves and the mid rib; flowers bluish purple in terminal spikes; capsules brown, ellipsoid-oblong, 4 × 2 cm, with a short, conical point, smooth and shining, seeds 4, orbicular 1.5 cm across, softly wrinkled.

Common mangrove along navigating canals of Kothapalem coast. Fl. & Fr.: Aug.–Dec. *Vern.*: Aldhi. Kothapalem, *VRK* 5761; Nizampatnam, *DLR* 143 (NUH).

Andrographis Wallich ex Nees

Andrographis paniculata (Burm. f.) Wall. ex Nees in Wall. Pl. As. Rar. 3: 116. 1832; FBI 4: 501; Gamble 1048 (734); Sant. in Bot. Mem. Univ. Bombay 2: 50. 1951. *Justicia paniculata* Burm. f. Fl. Ind. 9. 1768. *Andrographis subspathulata* Clarke in Hook. f. Fl. Brit. India 4: 502. 1884.

Erect herbs, up to 1.5 m tall, sharply quandrangular, leaves lanceolate, 4–6 × 1–1.5 cm, glabrous, acute, entire, base acute to obtuse; flowers white, with maroon streaks on upper lip, in lax spreading axillary and terminal unilateral racemes; capsules elliptic or linear-oblong, acute at both ends; seeds numerous, rugosely pitted, yellowish-brown.

Common weed of open forests. Fl. & Fr.: July–Dec. *Vern.*: Nelavemu. It is used as an anthelmintic, alterative, anodyne, astringent and febrifuge and for cholera, bronchitis, dysentery, dyspepsia, diabetes, malaria, influenza, itches, piles, jaundice and debility. Nelavemu used in liver troubles. Modern confirmatory evidence obtained-active principles. Andrographiolide, picrosides I & II isolated useful in liver cirrhosis (vide Sukdev 1997, Vol. 70, Curr. Sci.). Illegal trade resulting in reduced popula-tions to be conserved by micropropagation. During 1976–80, large populations; occurred in Nagarjunanagar but as on date (during 1995–98), the species disappeared due to unauthorised clearing reaction a local threatened status. Jayanthipuram, *VRK* 6903; Davupalli, *SSR & KSM* 21422.

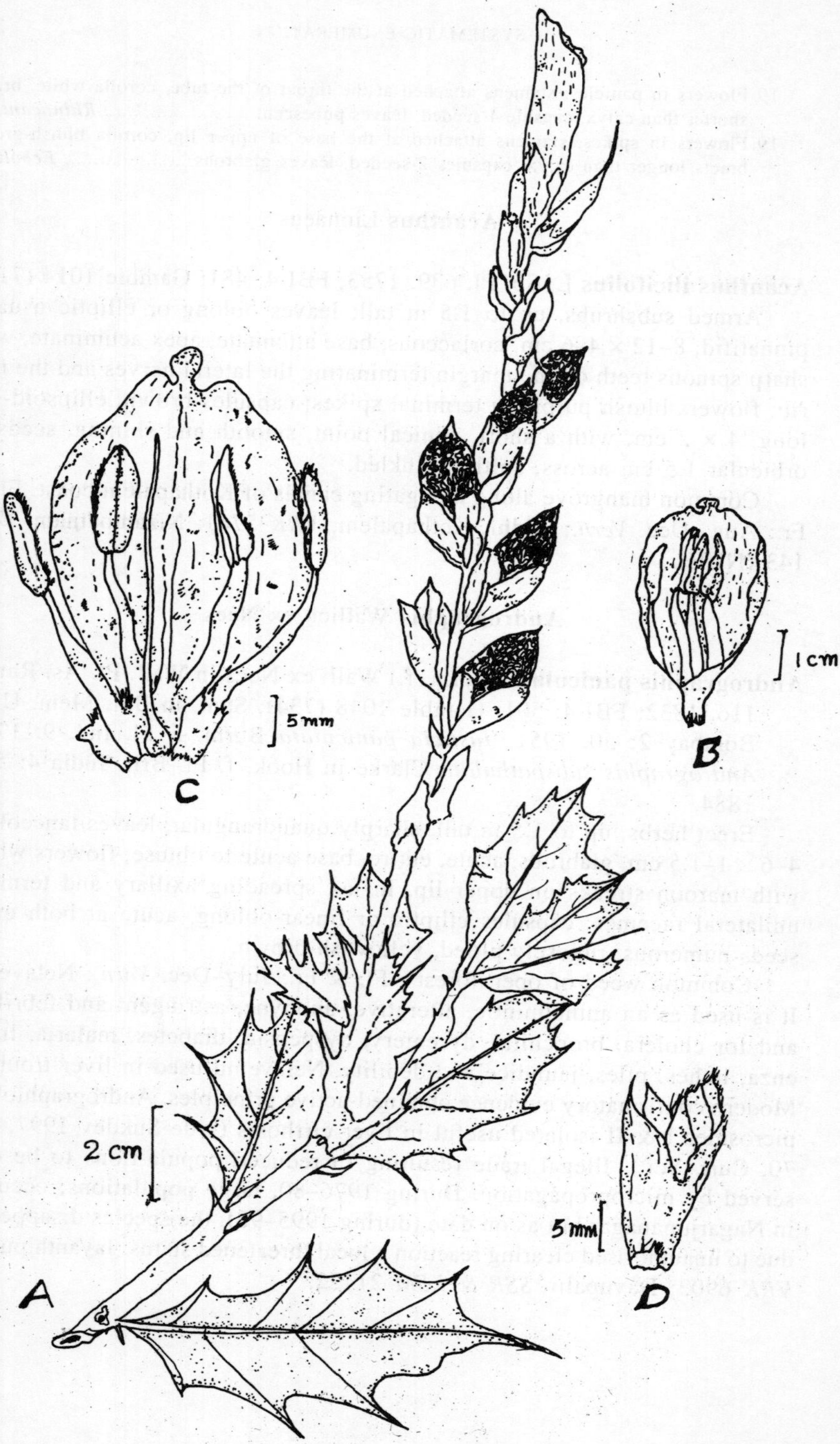
C
5 mm
B
1 cm
2 cm
5 mm
A
D

Barleria Linnaeus

1. Plants armed:
   2. Flowers pink to violet, seeds 4, capsule ellipsoid .... ***B. noctiflora***
   2. Flowers yellow, seeds 2, capsule ovoid:
      3. Leaves linear-lanceolate, chartaceous, calyx lobes acute, corolla yellowish, stamens 5 .... ***B. cuspidata***
      3. Leaves ovate, coriaceous, calyx lobes acuminate, corolla golden yellow to orange, stamens 4 .... ***B. prionitis***
1. Plants unarmed:
   4. Corolla tube ca. 8 cm long, plants stellate pubescent .... ***B. longiflora***
   4. Corolla tube ca. 4 cm long, plants not stellate pubescent:
      5. Flowers in secund spikes, leaves up to 20 cm long .... ***B. strigosa***
      5. Flowers not in secund spikes, leaves up to 13 cm long:
         6. Outer calyx lobes ovate, entire, leaves tomentose .... ***B. tomentosa***
         6. Outer calyx lobes ovate-lanceolate, spinous-ciliate along margin, leaves puberulous .... ***B. cristata***

Barleria cristata L. Sp. Pl. 636. 1753; FBI 4: 488; Gamble 1060 (743); Sant. in Bot. Mem. Univ. Bombay 2: 59. 1951.

Erect undershrubs, up to 1 m tall; leaves elliptic-lanceolate, 4–10 × 1.5–3.5 cm, pubescent, acute or acuminate; entire, base tapering; flowers purplish-pink in axillary and terminal short ovate spikes; outer calyx lobes with spiny margins; capsules oblong, 1.5 × 0.4 cm, acute at both ends; seeds 4, hairy, orbicular, compressed.

Occasional in open forests. Fl. & Fr.: Sept.–Feb. *Vern.*: Gobbi, Niru Goranta. It is an expectorant, antidote to snake bite and for inflammation. Apigenin, naringenin and epigeninglucoronide isolated from flowers. Macherla RF, *VRK* 3298; V.P. South, *VRK* 6728; Vinukonda RF, *VRK* 6855, 6888.

Barleria cuspidata Heyne ex Nees in Wall. Pl. As. Rar. 3: 93. 1832; FBI 4: 483; Gamble 1059 (741).

Armed, rigid erect subshrubs, up to 50 cm tall; stems and branches obscurely 4-gonous, spines white, 4; leaves linear-lanceolate, 3–5 × 0.3–0.7 cm, thinly scabrous, base cuneate, apex spiny mucronate; flowers yellow, solitary, axillary; capsules ovoid; seeds 2, hairy.

Common in forest undergrowth. Fl. & Fr.: Dec.–Mar. Loyapalle, *VRK* 3827.

Barleria longifolia L. f. Suppl. Pl. 290. 1781; FBI 4: 485; Gamble 1059 (742); Sant. in Bot. Mem. Univ. Bombay 2: 64. 1951.

Pubescent, unarmed shrubs, up to 2 m tall; branchlets tomentose; leaves ovate, 3–5 × 2–4 cm, sparsely tomentose above, densely below, acute, ciliate, base truncate-rounded oblique; flowers whitish yellow or white in ax-

Fig. 17. *Acanthus illicifolius* L.
A. Twig, B. Flower, C. Flower split open, D. Calyx

illary racemes; corolla tube long (ca. 8 cm), capsules elliptic-obong, up to 2 × 0.5 cm, black.

Occasional in scrub jungles and rocky places and on hillocks; Mangalagiri. Fl. & Fr.: Sept.–Feb. *Vern.*: Pinna-gorinta. Javuku, *VRK* 3909; Kondavedu Fort, *CAB* 4607 (MH).

Barleria noctiflora L. f. Suppl. Pl. 290. 1781; FBI 4: 484, Gamble 1059 (742).

Shrub, to 80 cm; branchlets white-pilose, prickly, spines sometimes forked; leaves clustered, oblong or obovate to spathulate, 1.2–2.2 × 0.4–0.6 cm, thick, sparsely pubescent, obtuse or acute, spine tipped, entire, cuneate; flowers white, fragrant, solitary, axillary, sessile, calyx-lobes spinous-toothed, white scarious; capsule ellipsoid, 1.5 × 0.6 cm; seeds 4, appressed hairy.

Occasional on waste lands. Fls.: opening at dusk with a peak during Dec.–Feb. Fr: Jan.–Mar. Davupalli, *SSR* & *KSM* 21677; Mallavagu, *SKW* 3685 (BLAT).

Barleria prionitis L. Sp. Pl. 636. 1753; FBI 4: 482; Gamble 1058 (741); Sant. in Bot. Mem. Univ. Bombay 2: 57. 1956; ssp. **prionitis**; Wood *et al.* in Kew Bull. 38: 436. 1983.

Bushy, armed undershrubs, up to 1.5 m tall; spines white, 2–4 cm, in the axils of leaves; leaves elliptic-obovate, 5–10 × 2–5 cm, glabrescent, acute, obtuse, mucronate,base attenuate; flowers orange-yellow, axillary, solitary or in cymose clusters; capsules ovoid, beaked, up to 1 × 0.5 cm, black.

Rare in open forests. Fl. & Fr.: Sept.–Feb. *Vern.*: Mulla Gorinta. It is used as a febifuge and used for anasarca, boils, catarrah affections, glandular swellings and toothache. Decoction of leaves is given internally to control irritation and also has moderate diuretic activity. Javuku, *VRK* 3555; Kondavedu Fort, *CAB* 4639 (MH), Nagarjunakonda, *KTH* 9732 (CAL).

Barleria strigosa Willd. Sp. Pl. 3: 379. 1800; FBI 4: 498; Gamble 1061 (743).

Shrub, up to 1.5 m tall; strigose; leaves ovate, 10–20 × 5–12 cm, glabrous, acuminate, entire, base long decurrent; flowers blue in dense unilateral spikes at the ends of branchlets; capsule 4-seeded; seeds silky.

Occasional (Mohan Rao and Aasiya, 1995). Fl. & Fr.: Aug.–Mar. *Vern.*: Nilambaram. It is used as an anti-inflammatory, spasmodic and for cough.

Barleria tomentosa Roth, Nov. Pl. Sp. 314. 1821; FBI 4: 485; Gamble 1059 (742).

Unarmed shrubs; branchlets tawny-pubescent; leaves ovate, 5 × 3 cm, thinly tomentose below, shortly acuminate at both ends; flowers pink, subsessile, axillary; bracts oblong-spathulate; capsules 4-seeded, seeds appressed-hairy.

Occasional in scrub forests. Fl. & Fr.: Aug.–Dec. Nagarjunakonda, *KTH* 9775 (CAL).

Blepharis A.L. Jussieu

Leaves petioled, thick, narrowly oblong, flowers violet to purple, corolla lip up to 2.5 cm long .... ***B. molluginifolia***

1. Leaves subsessile, membranous, elliptic-ovate, flowers white with pink nerves, corolla lip up to 6 mm long .... ***B. maderaspatensis***

Blepharis maderaspatensis (L.) Heyne ex Roth Nov. Pl.Sp. 320. 1821; Sant. in Bot. Mem. Univ. Bombay 2: 15. 1951. *Acanthus maderaspatensis* L. Sp. Pl. 639. 1753. *Blepharis boerhavifolia* Pers. Syn. Pl. 2: 180. 1806. FBI 4: 478; Gamble 1013 (712).

Erect much branched herbs; leaves whorled, elliptic-ovate or oblanceolate, 5–8 × 1–1.5 cm, pubescent, acute, ciliate; distantly toothed, sessile; flowers white with pale pink nerves; axillary, solitary or 2–3 in groups with bristly bracteoles; capsules oblong, 7 × 3 cm, brown; seeds 2, flattened, echlinate.

Common in plains and occasional in forests. Fl. & Fr.: Sept.–Feb. Loyapalle, *VRK* 3842; Krishna river bank, *VRK* 6735; Achampet, *VRK* 5866; Papaiah palem, *VRK* 5845; V.P. South, *VRK* 5781; Nagarjunakonda, *KTH* 9745 (CAL); Kondavedu Fort, *CAB* 4688 (MH).

Blepharis molluginifolia Pers. Syn. Pl. 2: 180. 1806; FBI 4: 479; Gamble 1013 (712).

Prostrate herbs, rooting at nodes; leaves narrowly oblong to oblanceolate or ovate, 1.5–4 × 0.5–1 cm, strigose, entire, obtuse at both ends; flowers violet to purple in axillary cymes; capsules 5 × 4 mm; seeds 4, flattened, orbicular, echinate on retinacula.

Common in plains. Fl. & Fr.: Sept.–Feb. Nagarjunakonda, *KTH* 9795 (CAL); Davupalli, *SSR* & *KSM* 21499.

Crossandra R.A. Salisbury

Crossandra infundibuliformis (L.) Nees in Wall. Pl. As. Rar. 3: 98. 1832. *Justicia infundibuliformis* L. Syst. Nat. (ed. 10) 850. 1758. *Crossandra undulaefolia* Salisb. in W. Hooker. Parad. Lond. t. 12. 1805; FBI 4: 492; Gamble 1055 (730).

Erect subshrubs, up to 1 m tall; leaves crowded near the inflorescence, elliptic to oblong, 5–10 × 2–4 cm, acute, entire; flowers golden yellow in terminal condensed spikes; capsules oblong, 1 × 0.5 cm; seeds orbicular, 4, 3 mm across.

Occasional in Macherla RF, as an escape. Fl. & Fr.: Dec.–Mar. *Vern.*: Kanakambaralu. Loyapalle, *VRK* 3840.

Several cultivars in various colours available in floriculture.

Dipteracanthus C.G.D. Nees

1. Erect herbs, leaves appressed hirsute, dull green when dry; corolla < 1.5 cm across, capsules glabrous .... ***D. patulus***
1. Prostrate herbs, leaves strigose, black when dry, corolla > 2.5 cm across, capsules pubescent .... ***D. prostratus***

Dipteracanthus patulus (Jacq.) Nees in Wall. Pl. As. Rar. 3: 82. 1832; Sant. in Bot. Mem. Univ. Bombay 2: 24. 1951. *Ruellia patula* Jacq. Misc. Bot. 2: 358. 1781; FBI 4: 412; Gamble 1071 (714).

Pubescent, erect, much branched undershrubs, up to 50 cm tall; leaves elliptic-ovate, 1–3 × 0.7–2 cm, pubescent, obtuse, base rounded to acute, crenulate; flowers pale purple or violet, axillary, solitary or 2–3 together; capsules elliptic, up to 1.5 × 0.4 cm, basally beaked, glabrous, seeds many, orbicular, flattened, thin, pubescent along margins.

Occasional in cultivated fields and open forests. Fl. & Fr.: July–Apr. V.P. South, *VRK* 6721; B.K. Palem, *VRK* 3244; Bollapalle, *VRK* 5744; Vinukonda RF, *VRK* 6857, 6896; Loyapalli, *SSR* & *KSM* 21644; .Nagarjunakonda, *KTH* 9672 (CAL).

Dipteracanthus prostratus (Poir.) Nees in Wall. Pl. As. Rar. 3: 81. 1832. *Ruellia prostrata* Poir. in Lam. Encycl. 6: 349. 1804; FBI 4: 411; Gamble 1917 (714).

Annual prostrate herbs; branchlets angled, often rooting at the nodes; leaves ovate-elliptic to deltoid, 2–6 × 1–3 cm, slightly strigose, subacute to obtuse, base truncate to acute; flowers violet, axillary, solitary; capsules thinly pubescent; seeds many, 4 mm across, flattened, thin, bearded along margins.

Occasional near forest border. Fl. & Fr.: Throughout the year. Nagarjunakonda, *KTH* 9610 (CAL); Loyapalli, *SSR* & *KSM* 21645.

Ecbolium S. Kurz

Ecbolium viride (Forssk.) Alston in Trimen Fl. Ceylon 6 (Suppl): 229. 1931. *Justicia viridis* Forssk. Fl. Aeg.—Arab. 5. 1775. *Ecbolium linnaeanum* Kurz in J. Asiat. Soc. Bengal, Pt. 2, Nat. Hist. 40: 75. 1871; FBI 4: 544; Gamble 1074 (752). *Justicia ecbolium* L. Sp. Pl. 15. 1753. *p.p.*

Glabrous shrubs, 1–2 m tall; leaves ovate to elliptic-ovate, 9–11 × 4–6 cm, coriaceous, obtuse to acute, entire, pubescent, base attenuate; flowers violet-blue in axillary spikes; capsules ovoid, compressed; seeds 2, tuberculate.

Occasional in forests. Fl. & Fr.: Sept.–Apr. Nakka-toka. Loyapalli, *SSR* & *KSM* 21640.Kondavedu Fort, *CAB* 17496 (MH).

Elytraria Michaux *nom. cons.*

Elytraria acaulis (L. f.) Lindau in Engl. & Prantl. Nat. Pfl. Fam. 1: 304. 1897; Gamble 1009 (709); Bremek. in Reinwardtia 3: 250. 1955. *Justicia acaulis* L. f. Suppl. Pl. 84. 1781. *Elytraria crenata* Vahl Enum. Pl. 1: 106. 1804; FBI 4: 394.

Annual, scapigerous herbs, stems reduced none; leaves radical, in close spirals, obovate to oblanceolate, 7–14 × 3–6 cm, glabrescent above, villous along the nerves below, obtuse, crenate; flowers white in axillary, slender spikes; capsules oblong, 0.8 × 0.2 cm, seeds many, flattened, reticulate; retinaculae absent.

Common in rocky waste places and forests. Fl. & Fr.: Throughout the year. *Vern.*: Cheppuattaaku. Fresh leaf paste is used for wounds and diseases of nails. It is used for veneral diseases and cough. Papaiah palem, *VRK* 5853; Davupalli, *SSR & KSM* 21687; Kondavedu Fort, *CAB* 4713 (MH); Vinukonda, *JSG* 17400 (MH).

Eranthemum Linnaeus

1. Bracts white with green nerves .... ***E. nervosum***
1. Bracts green, rarely whitish when old .... ***E. purpurascens***

Eranthemum nervosum (Vahl) R. Br. ex Roem. ex Schult. Syst. Veg. 1: 174. 1817; Gamble 1024 (719). *Justicia nervosa* Vahl, Enum. Pl. 1: 164. 1804. *Daedalcanthus nervosus* (Vahl) T. And. in J. Linn. Soc. London 9: 487. 1867; FBI 4: 418.

A pretty shrub, up to 2 m tall; lower larger, ovate, glabrous, acuminate at both ends; upper leaves many, smaller; flowers blue, in short, bracteate spikes in large terminal panicles, bracts elliptic-ovate; capsule cylindric.

Occasional, in plains of Nagarjuna University campus (Mohan Rao and Aasiya, 1995). Fl. & Fr.: Oct.–Mar.

Eranthemum purpurascens Nees in Wall. Pl. As. Rar. 3: 106. 1832; Gamble 1025 (719). *Daedalcanthus purpurascens* (Nees) T. And. in J. Linn. Soc. London 9: 488. 1867; FBI 4: 420. *Justicia pulchella* Roxb. Pl. Cor. t. 177. 1805. non *Eranthemum pulchellum* T. And. 1800.

Small erect herbs to shrubs, up to 1 m tall; leaves ovate to ovate-lanceolate, 4–18 × 2.5–8 cm, glabrescent, rounded at base and then decurrent, apex acuminate, upper subsessile; flowers pale blue in densely aggregate spikes, bracts ovate-lanceolate; capsule clavate-oblong, pointed, glabrous, up to 1 cm long; seeds 4.

Occasional in Nagarjuna University campus (Mohan Rao and Aasiya, 1995). Fl. & Fr.: Nov.–Feb.

Gantelbua Bremekamp

Gantelbua urens (Heyne ex Roth) Bremek. Mat. Mon. Strob. 148. 1944. *Ruellia urens* Heyne ex Roth Nov. Pl. Sp. 302, 1821. *Hemigraphis dura* T. And. in J. Linn. Soc. 9: 461. 1867 (excl. syn. *R. crispa* Nees non L.); FBI 4: 422; Gamble 1018 (715).

Procumbent, stiff hairy herbs; leaves oblong-obovate, 2–5 × 0.8–1.5 cm, obtuse, subcrenate, base attenuate; flowers blue in terminal or pseudo-axillary dense heads; capsules linear-oblong; grooved; seeds 6–8, hairy.

Occasional in forests; common in Nagarjunanagar. Fl. & Fr.: Sept.–Apr. Guntur, *JSG* 18607 (MH).

Note: The genus *Gantelbua* was separated from *Hemigraphis* on account of the peculiar structure of the inflorescence, but it differs in having deeply divided calyx and by the small size of the areole. The name is a latinization of the vernacular name of the type species which according to Cooke, p. 357 is *Gantelbu.*

Hemigraphis Nees

Hemigraphis latebrosa (Heyne ex Roth) Nees in DC. Prodr. 11: 723. 1847; FBI 4: 423; Gamble 1018 (715). *Ruellia latebrosa* Heyne ex Roth Nov. Pl. Sp. 307. 1821.

Procumbent pubescent herbs; stems faintly angled, leaves ovate, 4–9 × 1.5–6 cm, hairy, acute to acuminate, crenate-serrate, base cuneate to attenuate; flowers blue or purple, with yellow tube, in axillary capitate heads; capsules oblong; seeds 6.

Occasional in forests. Fl. & Fr.: Sept.–Mar. Davupalli, *DMR* 21106.

Hygrophila R. Brown emend Heine

Hygrophila auriculata (Schum.) Heine in Kew Bull. 16: 172. 1962. *Barleria auriculata* Schum. in Schum. & Thonn. Beskr. Buin. Pl. 285. 1827. *B. longifolia* L. Cent. Pl. 2: 22. 1756 (non *Hygrophila longifolia* Nees 1847). *Asteracantha longifolia* (L.) Nees in Wall. Pl. As. Rar. 3: 90. 1832; Gamble 1015 (712); Sant. in Rec. Bull. Bot. Sur. India 16: 173. 1960. *Hygrophila spinosa* T. And. in Thwaites Enum. Pl. Zeyl. 225. 1860; FBI 4: 408.

Erect, marshy, armed, strigose-hispid, stout herbs, up to 75 cm tall; leaves whorled, lanceolate, 5–10 × 0.5–1.5 cm, scabrous, spines 6 at each node, yellow, sharp, straight; flowers pink or bluish purple in axillary whorls; capsules linear-oblong.

Occasional in marshy places, streams and canals, Fl. & Fr.: Oct.–Apr. *Vern.*: Mudigubbaku. Nallapadu, *VRK* 6765.

Indoneesiella Sreemadhavan

1. Leaves elliptic to oblong, racemes branched, exceeding the leaves .... *I. longipedunculata*
1. Leaves oblong to oblanceolate, racemes scarcely branched, not exceeding the leaves .... *I. echioides*

Indoeesiella echioides (L.) Sreem. in Phytologia 15: 27. 1968. *Justicia echioides* L. Sp. Pl. 16 1753. *Andrographis echioides* (L.) Nees in Wall. Pl. As. Rar. 3: 117. 1832. *p.p.* FBI 4: 505; Gamble 1051 (736).

Annual erect hairy herbs, up to 50 cm tall; leaves oblanceolate or elliptic-ovate, 4–7 × 2–3 cm, decussate, acute, entire, sessile; flowers white with purple tinge in axillary racemes; capsules elliptic-lanceolate, up to 1 × 0.3 cm, glabrescent; seeds 4, rugose, pitted.

Occasional weed in waste places and forests. Fl. & Fr.: Aug.–Dec. Nagarjunasagar right canal, *VRK* 3538; Anupu, *VRK* 3516;Mangalagiri, *SSR* & *KSM* 21616; Bollapalle, *CAB* 17408 (MH); Karempudi, *CAB* 17446 (MH).

Indoneesiella longipedunculata (Sreem.) Sreem. Phytologia 16: 466. 1968. *Neesiella longipedunculata* Sreem. Phytologia 15: 271. 1967.

Erect herbs, up to 50 cm tall; branchlets glandular-hispid; leaves elliptic to oblong, 4–5 × 1–2 cm, glabrous, subacute to apiculate, base acute to auriculate; flowers white with pink tinge in terminal racemes; capsule 1.2 × 0.4 cm, glandular-ciliate.

Occasional in plains. Fl. & Fr.: Throughout the year. Nagarjunasagar, *KMS* 9743 (MH).

Justicia Linnaeus

1. Anthers cells with a white basal spur like appendage, bracts small:
   2. Bracts green, orbicular-spathulate, longer than calyx-lobes, capsule glabrous except at tip .... *J. glauca*
   2. Bracts white with green nerves, narrow, as long as or shorter than calyx-lobes, capsule pubescent .... *J. trinervia*
1. Anthers cells not appendaged spur-like base, bracts large .... *J. adhatoda*

Justicia adhatoda L. Sp. Pl. 15. 1753; Stearn in J. Arn. Arb. 52. 1971. *Adhatoda zeylanica* Medic. in Commentat. Acad. Elect. Sci. Theod.–Palat. 6: 393. 1790. *A. vasica* Nees in Wall. Pl. As. Rar. 3: 103. 1832; FBI 4: 540; Gamble 1082 (758).

Erect subshrubs, up to 1.5 m tall; stem with yellowish bark; leaves oblanceolate, 10–18 × 3–6 cm, thin, coriaceous, base acute to attenuate, margin entire, apex acuminate, minutely puberulous when young; flowers creamish-white in dense axillary spikes; capsules beaked, clavate; seeds 4, orbicular, rugose.

Common in Loyapalle section. Fl. & Fr.: Nov.–Mar. *Vern*.: Addasaram. The leaves and the roots are considered an efficaceous remedy for all sorts of coughs and colds. Plant is useful in soil reclamination programme. Also grown for sea side landscaping. Loyapalli, *VRK* 3826.

Justicia glauca Rottl. Ges. Naturf. Freunde Berlin Neue Schriften 4: 219. 1803; FBI 4: 529; Gamble 1079 (755).

Woody undershrubs, up to 50 cm tall; branchlets terete, leaves elliptic-ovate, 1–2 × 0.5–1.5 cm, glabrescent, subacute, entire, base obtuse to rounded, flowers purple or pale pink in axillary and terminal spikes; capsules compressed, apically pointed, up to 1.4 × 0.5 cm, brown.

Common on roadsides and open forests. Fl. & Fr.: July–Dec. N. Varam, *VRK* 3513; Mangalagiri, *SSR* & *KSM* 21614; Bollapalle, *CAB* 17414 (MH); Nagarjunasagar, *KMS* 9721 (MH); Nagarjunakonda, *KTH* 9608 (CAL).

Justicia trinervia Vahl Enum. 1: 156. 1834; FBI 4: 526; Gamble 1079 (755).

Low procumbent herbs, branching from woody root stock; leaves upper sessile, lower shortly petiolate, elliptic or oblong or obovate-oblong, obtuse; flowers in continuous or in lax glabrous spikes; bract and bracteoles white with green nerves, corolla whitish-pink.

Occasional in moist shady forests and open wastelands. Fl. & Fr.: Sept.–Dec. Sattenapally, *VRK* 5871.

Lepidagathis Willdenow

1. Bracts and calyx-lobes long-spinous, leaves linear-lanceolate, 3–6 cm long .... ***L. cristata***
1. Bracts and calyx-lobes ciliate, with long white hairs, leaves ovate-oblong, to 3 cm long .... ***L. subarmata***

Lepidagathis cristata Willd. Sp. Pl. 3: 400. 1800; FBI 4: 516; Gamble 1066 (747); Sant. in Bot. Mem. Univ. Bombay 2: 71. 1951.

Perennial procumbent herbs; branches numerous, spreading on all sides; leaves linear-lanceolate, 3–6 × 0.5–1 cm, pubescent, acute at both ends, margin entire to serrulate, flowers creamish with pink tinge in condensed, ovoid spikes crowded at the base of the stems; capsules oblong, 6 × 2 mm, compressed; seeds 2.

Common in rocky places in open forests and roadsides. Nagarjunanagar and Nambur Railway Station. Fl. & Fr.: Sept.–Mar. *Vern.*: Nakkapentika chettu. Entire plant is shade dried, burned and the ash is mixed with cheese or neem oil or coconut oil or castor oil, applied externally to cure various skin diseases like eczema, warts, drandruff for children and adults. Javuku, *VRK* 3580.

Lepidagathis subarmata (Clarke) Gamble Fl. Pres. Madras 1066, 1924 & 747. 1957 (repr. ed.), *L. mitis* Dalz. var. *subarmata* Clarke in Hook. f. Fl. Brit. India 4: 516. 1885.

Stiff undershrubs; branches diffusely spreading; subquadrangular, leaves sessile, linear-oblong, 1–3 × 0.5–1.5 cm, glabrous, entire, hairy on nerves below, subacute, entire, base narrowed; flowers pale pink with maroon spots near the throat, in globose heads; capsules long ovoid; seeds 2, ovoid-oblong.

Occasional in plains. Fl. & Fr.: Nov.–Mar. Mangalagiri, *SSR & KSM* 21427; Karempudi, *CAB* 4756 (MH).

Peristrophe C.G.D. Nees

Peristrophe paniculata (Forssk.) Brummit in Kew Bull. 38: 451. 1983. *Perianthera paniculata* Forssk. Fl. Aeg. Arab. 7. 1775. *Peristrophe bicalyculata* (Retz.) Nees in Wall. Pl. As. Rar. 3: 113. 1982; FBI 4: 554; Gamble 1084 (759).

Erect much branched herbs, up to 75 cm tall, hairy, stems 4–6-angled; leaves elliptic-ovate to lanceolate, 4–8 × 1–3 cm, pubescent below, acuminate, entire, base narrowed to obtuse; flowers pink in terminal trichotomously branched lax panicles; capsules ellipsoid, up to 0.3 x 0.8 cm, seeds 4, orbicular, papillose and slightly rugose.

Common on roadsides and open forests. Nagarjunanagar. Fl. & Fr.: Oct.–Jan. *Vern.*: Chebura. The leaf juice is applied externally and internally and used for setting of fractured bones. Plant maceration in an infustion of rice is said to be antidote to snake poison. Koppanur, *VRK* 3257, 3260; Javuku, *VRK* 3923.

Rhinacanthus Nees

Rhinacanthus nasutus (L.) Kurz. in J. As. Soc. Bengal 39, pt. 2, Nat. Hist. 79. 1870. *Justicia nasuta* L. Sp. Pl. 16. 1763. *Rhinacanthus communis* Nees in Wall. Pl. As. Rar. 3: 109. 1832; FBI 4: 541; Gamble 1083 (758).

Erect herbs, up to 40 cm tall, stem subterete or obscurely angled; leaves elliptic, 4–10 × 2–5 cm, pubescent, acute, entire, base tapering; flowers white in terminal panicles; capsules narrow, velvety pubescent; seeds tuberculate, glabrous, black.

Occasional in forests. Fl. & Fr.: Sept.–Mar. *Vern.*: Nagamalle. Plant possesses antiseptic propesties. Leaves are chewed and reported to be used against cancer. Bruised leaves and roots with lime juice are used for eczema and ring-worm. Roots are used as an aphrodisiac and seeds for scenting clothes. Guntur, (*Uhl*, 1929).

Rostellularia Reichenbach

The determination of the various taxa belonging to the genus *Justicia* L. has been a problem, as the limits between them are very obscure and are based on intergrading characters. Santapau and other earlier workers also expressed similar opinion (Univ. Bomb. Bot. Mem. 2: 83. 1952). Bremekamp (in Verh. Nederl. Akad. Wet. 1948) has separated the species belonging to the section *Rostellularia* from the genus *Justicia* and treated them under a distinct genus *Rostellularia* Reichenb.

1. Spike stout, cylindric, bracteoles as long as or slightly longer than calyx lobes .... ***R. procumbens***
1. Spikes slender, linear, bracteoles shorter than calyx-lobes, margin scarious, mid rib green:
    2. Straggling herbs with stout root stock .... ***R. vahlii***
    2. Erect herbs without root stock .... ***R. prostrata***

Rostellularia procumbers (L.) Nees in DC. Prodr. 11: 371. 1847. *Justicia procumbens* L. Sp. Pl. 15. 1753; FBI 4: 539; Gamble 2: 1080 (756).

A diffuse, slender hispid herb, up to 30 cm tall, leaves elliptic, 1–2.5 × 0.7–1 cm, scabrous above, glabrescent to strigose below, apex acute, entire, base obtuse; flowers pale pink or purplish in terminal, dense, cylindric spikes; capsule oblong, sparsely, hairy, pointed, seeds 1 mm across.

Common in wastelands and open place in Nagarjuna University campus (Mohan Rao and Aasiya, 1995).

Rostellularia prostrata (Roxb. ex Clarke) Mujumdar in Bull. Bot. Soc. Bengal 25: 75. 1971. *Justicia prostrata* (Roxb. ex Clarke) Gamble Fl. Pres. Madras 2: 1081 (757). 1924. *J. diffusa* Willd. var. *prostrata* Roxb. ex Clarke in Hook. f. Fl. Brit. India 4: 538. 1885.

Erect or spreading much branched subquadrangular herbs, up to 40 cm tall; leaves ovate-elliptic or elliptic-lanceolate, 3–4 × 2–2.5 cm, glabrescent, obtuse, base acute; flowers pink in terminal spikes; bracts and bracteoles green, scarious; capsules oblong, 3 × 1.2 cm, shortly pointed, seeds minute, subconcentrically rugose.

Occasional weed of fields and roadsides, rare in forests. Fl. & Fr.: July–Dec. Paste of the plant is used for sprain and bone fracture and also for snake-bite. Bollapally, *VRK* 5752; Kondavedu Fort, *CAB* 4679 (MH); Nagarjunakonda, *KTH* 9644 (CAL).

Rostellularia vahlii (Roth) Nees var. **rupicola** Ellis in Bull. Bot. Surv. India 11: 435–436. 1969.

Straggling herbs with stout root stock; branches with long internodes, angular, scabrous; leaves oblong-ovate or ovate-elliptic, 1–6 × 0.5–1 cm, glabrous, acuminate; flowers pale pink with red streaks in axillary and terminal narrow spikes; capsules pubescent; seeds papillose.

Occasional in the forest undergrowth. Fl. & Fr: July–Jan. N. Varam, *VRK* 3223; Ethipothala, *VRK* 5799.

Ruellia Linnaeus

Ruellia tuberosa L. Sp. Pl. 635. 1752; Gamble 1017 (714).

Erect herbs up to 60 cm tall; stems 4-angled; leaves oblong-obovate, 6–12 × 3–4.5 cm, glabrescent, shining, obtuse, entire, base cuneate; flowers violet, axillary in pairs; capsules oblong, 2 × 0.3 cm, glabrous.

Occasional in forests and moist waste places. Native of tropical America. Fl. & Fr.: Sept.–May. V.P. South, *VRK* 6829.

Rungia Nees

1. Bracts dimorphic, fertile bracts villous, petiole ca. 2 cm long .... ***R. pectinata***
1. Bracts uniform, bracts with white membranous margin, petiole ca. 5 mm long .... ***R. repens***

Rungia pectinata (L.) Nees in DC. Prodr. 11: 469. 1847 *p.p. Justicia pectinata* L. Amoen. Acad. 4: 299. 1759. *Rungia parviflora* (Retz.) Nees var. *pectinata* (L.) Clarke in Hook. f. Fl. Brit. India 4: 550. 1885; Gamble 1071 (750).

A much-branched, annual ascending herbs, stem subterete, glabrous, subterete; leaves subsessile, 2–5 × 0.5–1 cm, ovate-elliptic, glabrous, entire, sub-obtuse, acute; flowers purple in short axillary spikes; bract ovate, hyaline, hairy; capsules ovoid; seeds small, orbicular, minutely echinulate, yellow.

Occasional weed on edges of forests. Fl. & Fr.: Dec.–Feb. Vinukonda RF, *VRK* 6904.

Rungia repens (L.) Nees in Wall. Pl. As. Rar. 3: 110. 1832; FBI 4: 549; Gamble 1070 (750). *Justicia repens* L., Sp. Pl. 15. 1753.

Procumbent herbs, up to 80 cm tall; often rooting at the nodes, branchlets villous, leaves elliptic-lanceolate, 4–6 × 0.5–1.5 cm, puberulous, acute at both ends; flowers blue in terminal spikes, up to 7 cm long; bracts 4-farious, with white margins, capsules elliptic-oblong, flat, pubescent; seeds orbicular, compressed.

Common in plains and rare in forests, outskirts of Chinakakani, Dholspur, Takikonda. Fl. & Fr.: Aug.–Feb. *Vern.*: Manga. The paste of whole plant is applied externally and internally on local swellings and snake and scorpion stings. Jayanthipuram, *VRK* 6929.

VERBENACEAE

1. Climbing shrubs, cymes capitate, bracts of the heads forming an involucre .... ***Sphaenodesme***
1. Other than climbers, no involucre:
   2. Leaves digitately 3–7-foliolate .... ***Vitex***
   2. Leaves simple:
      3. Inflorescence a spike, herbs or subshrubs:
         4. Fruit with 2,2-seeded pyrenes, branchlets sticky pubescent.—flowers white .... ***Priva***
         4. Fruit with 2 or 1, 1-seeded pyrenes, branchlets glabresent:
            5. Spikes long, slender, flowers blue, stamens 2 .... ***Stachytarpheta***
            5. Spikes capitate, ovoid or cylindric, flowers pinkish or purplish, stamens 4:
               6. Armed, drupes fleshy .... ***Lantana***
               6. Unarmed, drupes dry:
                  7. Prostrate herbs, leaves obovate-spathulate, corolla purplish-white, tube 1.5 mm, drupe subglobse .... ***Phyla***
                  7. Erect shrubs, leaves broadly elliptic-oblong, corolla pinkish, tube 4 cm long, drupe pyriform .... ***Lippia***
      3. Inflorescence a raceme or cymes, shrubs or trees:
         8. Corolla regular, stamens as many as corolla lobes, equal .... ***Tectona***
         8. Corolla bilabiate, lobes 5, stamens 4, didynamous:
            9. Stamens inserted below the tube, vegetative parts aromatic .... ***Clerodendrum***
            9. Stamens inserted near the middle of the tube, vegetative parts not aromatic:
               10. Flowers yellow, 2.5–3.5 cm across, fruits ellipsoid-globose, 2.5–4 cm long .... ***Gmelina***
               10. Flowers cream or white, 4 mm across, fruits sub-globose, 4–5 mm long .... ***Premna***

Clerodendrum Linnaeus

1. Leaves obovate to suborbicular, margins entire, apex obtuse, flowers in 3-flowered cymes, stamens inserted in the upper half of the tube .... ***C. inerme***
1. Leaves deltoid-ovate, margins serrate, apex acute to acuminate, flowers in many-flowered panicles, stamens inserted below the corolla tube .... ***C. phlomidis***

Clerodendrum inerme (L.) Gaertn. Fruct. Sem. Pl. 1: 271. 1788; FBI 4: 589; Gamble 1090 (769). *Volkameria inermis* L. Sp. Pl. 637. 1753.

Unarmed straggling, much-branched, aromatic shrubs, up to 2 m tall; leaves obovate to orbicular, 1.5–3 × 1.5–2 cm, coriaceous, glabrous or nearly so, base cuneate, apex obtuse; flowers white with pink lines in terminal cymes; drupe globose, 0.6 cm across, seated in the veined persistent enlarged calyx, separating into 4 woody pyrenes.

Common shrubs along the navigating canals in coast; often grown as an ornamental in gardens. Fl. & Fr.: Jan.–Apr. *Vern.*: Nilla vuppi. Kothapalem, *VRK* 5765; Nizampatnam, *DLR* 141 (NUH).

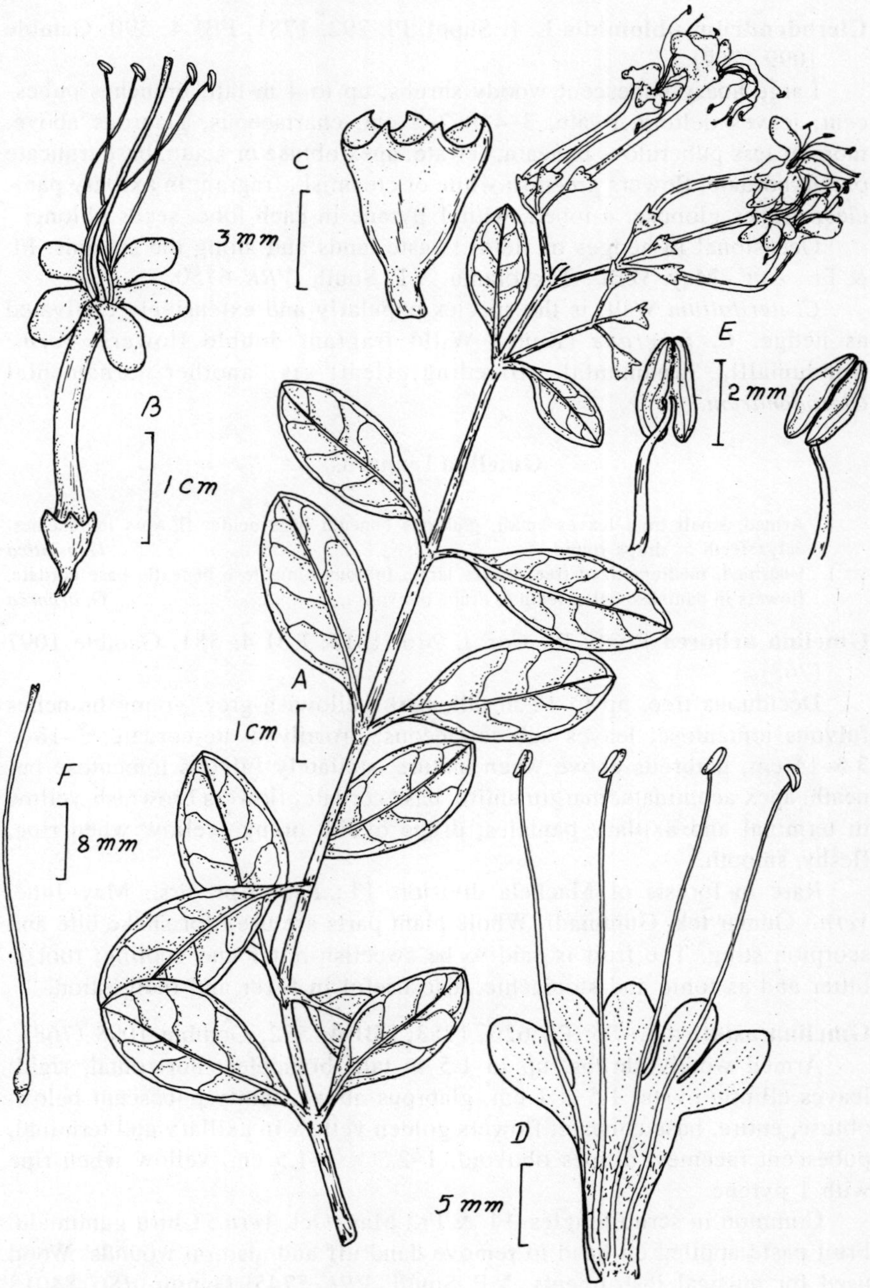

Fig. 18. *Clerodendrum inerme* (L.) Gaertn.
A. Twig, B. Flower, C. Calyx, D. Corolla open, E. Stamen, F. Pistil

Clerodendrum phlomidis L. f. Suppl. Pl. 292. 1781; FBI 4: 590; Gamble 1099 (769).

Large hoary pubescent woody shrubs, up to 4 m tall; branches pubescent; leaves deltoid, ovate, 3–4 × 2–3 cm, chartaceous, glabrous above, more or less puberulous beneath, serrate; apex obtuse or acute, base truncate or sub-cordate, flowers greenish white or creamish, fragrant in axillary panicles; drupes globose, 4-lobed, with 1 pyrene in each lobe, seeds oblong.

Occasional in hedges of fields, waste lands and along the streams. Fl. & Fr.: Aug.–May. *Vern.*: Takkolamu. V.P. South, *VRK* 6750.

C. nerifolium Wall. is the species popularly and extensively cultivated as hedge. *C. fragrans* (Kent.) Willd fragrant double flowers. *Vern.*: Bondumalli, ornamental. Bleeding Heart—is another ornamental *Clerodendrum.*

Gmelina Linnaeus

1. Armed, small tree; leaves small, glabrous beneath base acute; flowers in racemes, calyx teeth 5; drupe ovoid .... ***G. asiatica***
1. Unarmed, medium-sized tree, leaves large, fulvous tomentose beneath, base cordate, flowers in panicles, calyx teeth 4, drupe obovoid .... ***G. arborea***

Gmelina arborea Roxb. Pl. Cor. t. 246. 1815; FBI 4: 581, Gamble 1097 (768).

Deciduous tree, up to 12 m tall; bark yellowish-grey, young branches fulvous tomentose; leaves sub-coriaceous, broadly ovate-cordate, 3–18 × 3.5–14 cm, glabrous above when mature, stellately fulvous tomentose beneath, apex acuminate, margin entire, base cordate; flowers brownish-yellow in terminal and axillary panicles; drupe ovoid, orange-yellow when ripe, fleshy, smooth.

Rare in forests of Machela division. Fl.: Feb.–Apr. Frs.: May–June. *Vern.*: Gumar tek, Gummadi. Whole plant parts are used for snake bite and scorpion sting. The fruit is said to be sweetish-bitter and cooling; root is bitter and as tonic and stomachic, also useful in fever and indigestion.

Gmelina asiatica L. Sp. Pl. 626. 1753; FBI 4: 582; Gamble 1098 (768).

Armed woody shrubs, up to 1.5 m tall; branchlets horizontal, rigid; leaves elliptic-ovate, 1.5 × 1 cm, glabrous above, thinly pubescent below, obtuse, entire, base cuneate; flowers golden yellow in axillary and terminal, pubescent racemes; drupes obovoid, 1–2.5 × 1–1.5 cm, yellow when ripe with 1 pyrene.

Common in scrub jungles. Fl. & Fr.: May–Oct. *Vern.*: Chiru gummudu. Fruit paste applied on head to remove dandruff and also on wounds. Wood used for musical instruments. V.P. South, *VRK* 5745; Guntur, *JSG* 84018 (MH); Near Nambur AIR Station, Opp. NU campus.

Lantana Linnaeus

1. Plants unarmed; flowers white in elongate spikes .... ***L. wightiana***
1. Plants armed; flowers pink or red in condensed capitate spikes .... ***L. camara***

Lantana camara L. var. **aculeata** (L.) Moldenke in Torreya 34: 9. 1934. *L. aculeata* L. Sp. Pl. 627. 1753; FBI 4: 562; Gamble 1087 (761).

Armed bushy shrubs, up to 2 m tall; leaves ovate, 5–7 × 2–5 cm, scabrid, acute, serrate, base truncate; flowers variable in colour, orange, scarlet red or pink or in combination of these shades in terminal short capitate spikes; drupes globose, 0.2 cm across, black, shining.

Common weed on roadsides and forest edges. Native of tropical America. Fl. & Fr.: Throughout the year. *Vern.*: Akshintala pulu. Leaf juice squeezed into the eyes to treat eye diseases and used externally to treat cuts and skin diseases. Javuku, *VRK* 3236.

Lantana wightiana Wall. ex Gamble Fl. Madras 2: 1087 (761). 1924. *L. indica* Roxb. var. *albiflora* Wight ex Clarke in Hook. f. Fl. Brit. India 4: 562. 1885.

Unarmed subshrubs, up to 3 m tall; branches quadrangular, tender parts scabrous above, sub-villous beneath, base rounded to acute, apex subacute; flowers white in elongated spikes; berries brown-black when ripe.

Occasional in scrubs. Fl. & Fr.: May–Sept. N. Varam, *VRK* 3966.

Lippia Linnaeus

Lippia javanica (Burm. f.) Spreng. in L. Syst. Veg. (ed. 16) 2: 752. 1825. *Verbena javanica* Burm. f. Fl. Ind. 12. t. 6. f. 2. 1768. *Lippia alba* (Mill.) N.E. Br. ex Britton & Wilson. Bot. Porto Rico 6: 141. 1925. *Lantana alba* Miller, Gard. Dict. (ed. 8) 8. 1768.

Erect shrubs, up to 2 m tall; leaves broadly elliptic-oblong, 4–9 × 2–4 cm, thick-chartaceous, sparsely hairy-above, softly pubescent below, serrate-crenate, acute, base rounded to acute; flowers pinkish, in axillary spikes; drupes pyriform, enclosing 2, 1-seeded pyrenes.

Occasional along river banks. Fl. & Fr.: July–Dec. Guntur, (*Uhl,* 1929).

Phyla Loureiro

Phyla nodiflora (L.) Greene in Pittonia 4: 46. 1899. *Verbena nodiflora* L. Sp. Pl. 20. 1753. *Lippia nodiflora* (L.) Rich. in Michaux Fl. Bor. Amer. 2: 15. 1803; FBI 4: 563; Gamble 1088 (762).

Perennial, prostrate, much branched woody herbs, rooting at the nodes; leaves obovate, 1–3 × 0.8–1.5 cm, pubescent, base cuneate, apex rounded, deeply and sharply serrate in the upper part, flowers pale pink in axillary

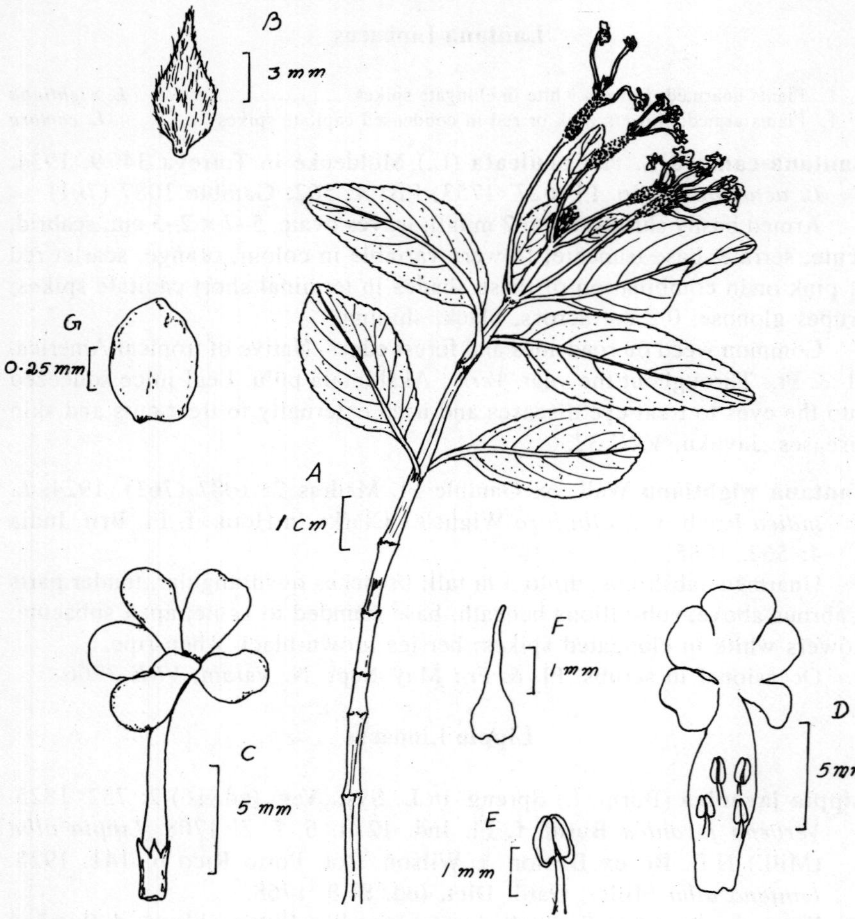

Fig. 19. *Lantana wightiana* Wall.
A. Twig, B. Bract, C. Flower, D. Corolla open, E. Stamen, F. Pistil, G. Capsule

dense, globose, long peduncled heads; drupes globose, splitting into two 1-seeded glabrous pyrenes; seeds minute.

Common in moist localities including interior forests. Fl. & Fr.: Throughout the year. *Vern.:* Bokkena, Mosalipappu. N. Varam, *VRK* 3588; Krishna river bed, *VRK* 6729; Krishna river, *KMS* 9854 (MH).

Premna Linnaeus *nom. cons.*

1. Flowers in cymose panicles, calyx shortly 5-lobed; leaves with tawny, yellow stellate tomentum .... *P. tomentosa*
1. Flowers in cymose corymbs, calyx 2-lipped; leaves with softly, greyish pubescence .. .. *P. latifolia*

Premna latifolia Roxb. var. **mollissima** Clarke in Hook. f. Fl. Brit. India 4: 578. 1885; Gamble 1096 (767).

Medium sized trees, up to 3 m tall; leaves broadly ovate, to (sub) orbicular, 4–7 × 3–5.5 cm, chartaceous, pubescent, obtuse, entire, base rounded; flowers greenish-white, scented in terminal corymbs; drupes globose, 6 mm across, black.

Occasional in forests. Fl. & Fr.: Mar.–Oct. *Vern*.: Pedda Nellikura. N. Varam, *VRK* 3959; Loyapalle, *VRK* 3859; Bollapalle, *VRK* 5750; Nagarjunakonda, *VRK* 6717; V.P. South, *VRK* 6830; Nakkarikal, *CAB* 17470 (MH).

Premna tomentosa Willd. Sp. Pl. 3: 314. 1800; FBI 4: 576; Gamble 2: 1095 (767). *Cornutia corymbosa* Lam. Encycl. 1: 54. 1783, auct non Burm. f. 1768.

Deciduous tree, up to 10 m tall; bark grey or light brown; branchlets with densely, soft, tawny tomentum of stellate hairs; leaves thick, chartaceous, 2.5–12.5 × 2–8 cm, ovate, tomentose among nerves above, densely tawny yellow, stellate tomentose, acuminate, entire, cordate; flowers greenish-yellow, scented in terminal, corymbose cymes; drupe sub-globose, 4 mm across, yellowish, puberulous.

Occasional, in deciduous forests. Fl.: Mar.–Apr. Fr.: May. *Vern*.: Naguru, Kokkiti, Narva. Fruit edible. An extract of the inner bark is stated to be used to arrest diarrhoea and a decoction of the roots is given in stomachache. The roots yield an aromatic oil which is also used in stomach disorders. Davupalli, *SSR* & *KSM* 21411.

Priva Adanson

Priva cordifolia L. f. Druce in Bot. Soc. Exch. Club Brit. Isles 1916.641. 1917. *Buchnera cordifolia* L.f. Suppl. Pl. 287. 1781. *Priva leptostachya* Juss. in Ann. Mus. Hist. Nat. 7. 70. 1806; FBI 4: 565; Gamble 2: 1091 (764).

Herb to 80 cm long, branchlets 4-angular, sticky-pubescent; leaves softly chartaceous, ovate, 4–7 × 3–5.2 cm, appressed-pubescent, acute, serrate, truncate to subcordate; flowers white, subsessile, in terminal elongate spikes; bracts persistent; drupe of 2,2-seeded pyrenes, enclosed with persistent calyx-base, echinate; seeds smooth.

Occasional, in plains, on old walls, rocks etc. Fl. & Fr.: Nov.–Feb. An infusion of leaves is used for inflammation of eye-ball. Paste of seeds is applied to sores. Loyapalli, *SSR* & *KSM* 21634.

Sphenodesme Jack

Sphenodesme involucrata (Presl.) Roxb. var. **paniculata** (Clarke) Munir in Gard. Bull. Singapore 21: 338. 1966. *S. paniculata* Clarke in Hook. f. Fl. Brit. India 4: 600. 1885; Gamble 1104 (773).

Climbing shrubs; leaves opposite, entire, shortly petioled, ovate to elliptic, acute, 4–6 × 2–3 cm, glabrescent, margin broadly serrate; flowers green in 7-flowered cymes in an involucre of 6 foliaceous bracts; fruit a globose drupe, included in the calyx; seeds globose.

Rare in forests. Fl. & Fr.: Feb.–Aug. Vinukonda RF, *VRK* 6897.

Stachytarpheta Vahl *nom. cons.*

Stachytarpheta jamaicensis (L.) Vahl Enum. Pl. 1: 206. 1804. *Verbena jamaicensis* L. Sp. Pl. 19. 1753. *Stachytarpheta indica* auct. non (L.) Vahl: FBI 4: 564; Gamble 1090 (763).

Erect herbs, up to 75 cm tall; branchlets glabrescent, leaves opposite, elliptic-obovate, 5–7 × 2–3.5 cm, glabrous, acute, serrate, base decurrent; flowers blue in terminal spikes; pyrenes 2, oblong, 6 mm; seeds 1 in each pyrene.

Occasional weed of wastelands and roadsides and rare in forests. Fl. & Fr.: July–Dec. V.P. South, *VRK* 6754.

Tectona Linnaeus f. *nom. cons.*

Tectona grandis L. f. Suppl. Pl. 151. 1781; FBI 4: 570; Gamble 1082 (765).

Deciduous trees, up to 12 m tall; bark grey or light greyish-brown; wood golden yellow, scented and oily to touch; branchlets tetragonous, stellately-tomentose; leaves broad-elliptic, 12–35 × 10–25 cm, glabrescent and rough above, stellate pubescent below, acute or obtuse, entire, rounded or cuneate; flowers white in terminal, large dichotomously branched panicles; drupes brown, globose, spongy epicarp, endocarp bony, thick; seeds oblong, numerous.

The teak wood. Occasional, in Davupalli forest. Planted for timber in hill slopes. Fl. & Fr.: Apr.–Oct. *Vern.*: Teku chettu. Timber is hard, and used for building, furniture, veneers for interior decoration. Strength and durability are hall marks of the wood. Near Kunchanapalli, Teak wood plantations are raised and designated as Green Gold Programme.

Vitex Linnaeus

Vitex negundo L. Sp. Pl. 638. 1753; FBI 4: 583; Gamble 1102 (771).

Grey-pubescent shrubs to small trees, up to 4 m tall; branchlets quadrangular, leaves 3–5-foliolate, leaflets oblanceolate, 8–10 × 2–2.5 cm, glabrous above, grey pubescent below, acute, entire, base narrowed; flowers blue or purple in terminal panicles; drupes globose, black.

Rare in the forests. Cultivated in several places. Fl. & Fr.: May–Feb. *Vern.*: Vavili. Warmed leaves applied for rheumatic swellings and stomach ache; agreebly aromatic when bruised, employed in local medicine, in inflammatory swellings of the joints. N. Varam, *VRK* 3965.

AVICENNIACEAE

Avicennia Linnaeus

1. Leaves obtuse at apex, elliptic-oblong or obovate, anthers exserted .... ***A. officinalis***
1. Leaves acute at apex, anthers included:
   2. Leaves ovate-lanceolate, flowers yellow, pneumatophores absent .... ***A. marina***
   2. Leaves lanceolate, flowers brown to yellow, pneumatophores present .... ***A. alba***

Avicennia alba Blume, Bijdr. 821. 1826. *A. officinalis* var. *alba* Clarke in Hook. f. Fl. Brit. India 4: 604. 1883; Gamble 1106 (775).

Small sized trees, up to 5 m tall; branches pale, pneumatophores conical; leaves opposite, coriaceous, entire, acute, flowers yellow, sessile in capitate cymes; capsule compressed; conical with 1.5–2 cm long beak, bright yellow and smooth when ripe.

Common mangrove tree along the sea çoast in back waters. Fl. & Fr.: Aug.–Nov. *V*ern.: Vilava mada. Nizampatnam, *VRK* 6910; Nizampatnam, *DLR* 125 (NUH).

Avicennia marina (Forssk.) Vierh. Akad. Wiss. Wien, Math. Naturwiss. Fl. Denkschr. 71: 435. 1907; Gamble 1106 (774). *Sceura marina* Forssk. Fl. Aeg.–Arab. 37. 1775.

Arboreous shrubs, up to 5 m tall; pneumatophores numerous; leaves obovate to oblanceolate, 6–10 × 2.5–4 cm, glabrous above, pubescent below, entire, acute, base acute to cuneate; flowers yellow in axillary and terminal cymes; drupes 1.5 × 1 cm.

Common in the mangrove forests of Repalle. Fl. & Fr.: May–Oct. Nizampatnam, *VRK* 6915; Kothapalem, *VRK* 5768, 5760, Nizampatnam, *DLR* 130 (NUH).

Avicennia officinalis L. Sp. Pl. 626. 1753; FBI 4: 604; Gamble 1105 (774).

Evergreen, small mangrove tree, up to 5 m tall; crown hemispherical, pneumatophores narrow; branchlets white-tomentose; leaves opposite,

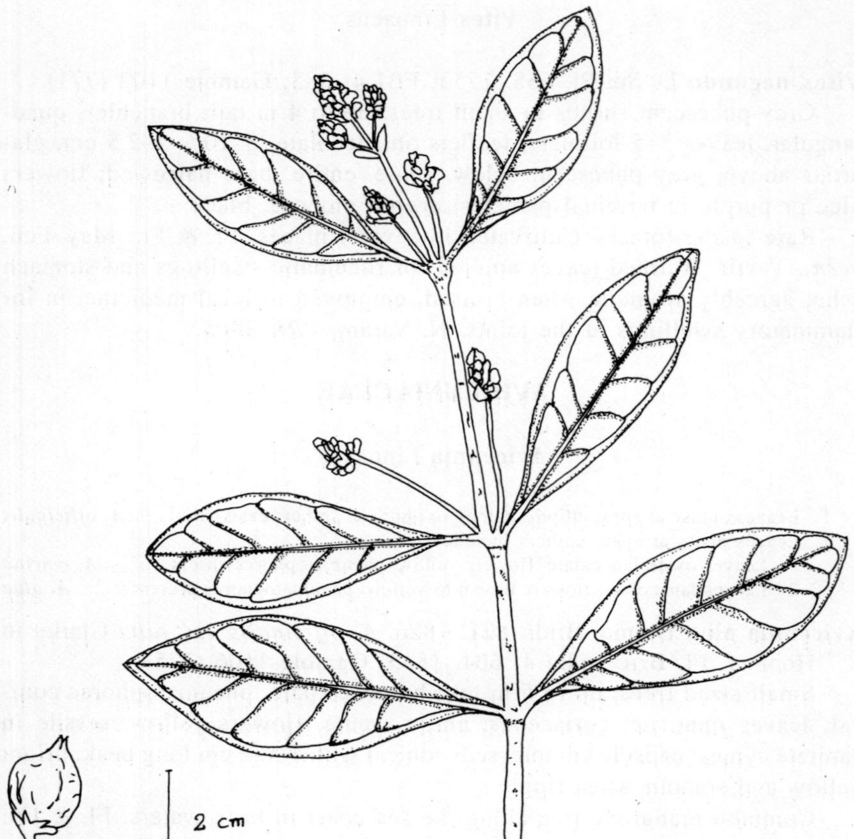

Fig. 20. *Avicennia marina* (Forssk.) Vierh.
A. Twig, B. Capsule

coriaceous, obovate or elliptic-oblong, 7–10.5 × 4–5 cm, glabrous, shining above, minutely and closely pubescent beneath, entire, obtuse; base attenuate, flowers yellow; sessile in capitate cymes; capsules ovoid, 2.5–3 cm long, beaked.

Common mangrove tree along sea coast in back waters. Fl. & Fr.: Aug.–Nov. *Vern.*: Nalla mada. Wood hard, used only for fuel and also used for cheap beams and door frames etc. The bark is astringent. A green, bitter and somewhat aromatic resin which oozes from the bark is used in medicine. Nizampatnam, *VRK* 6911, *DLR* 131 (NUH).

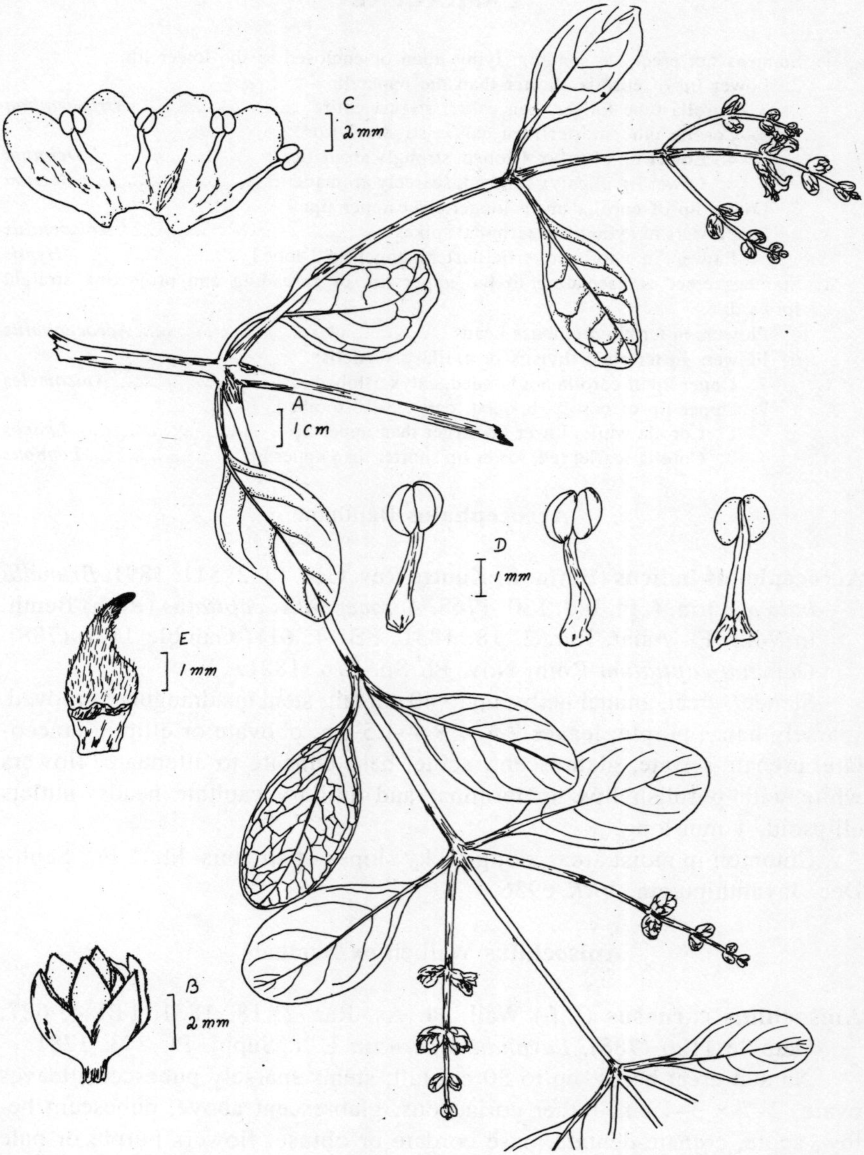

Fig. 21. *Avicennia officinalis* L.
A. Twig, B. Calyx, C. Corolla split open, D. Stamen, E. Pistil.

Immediate steps to be taken for reclamation and conservation of the mangrove species by '*in situ*' conservation technique.

LAMIACEAE

1. Stamens not erect, descending, lying upon or enclosed in the lower lip:
   2. Lower lip of corolla shorter than the upper lip:
      3. Corolla tube longer than calyx, stigma entire .... ***Orthosiphon***
      3. Corolla tube shorter than calyx, stigma 2-fid:
         4. Lower lip of calyx 4-lobed, strongly aromatic .... ***Ocimum***
         4. Lower lip of calyx 2-lobed, scarcely aromatic .... ***Becium***
   2. Lower lip of corolla much longer than upper lip:
      5. Flowers in cylindrical terminal spikes .... ***Anisochilus***
      5. Flowers in whorls or verticillasters, upper lip 2-lobed .... ***Hyptis***
1. Stamens erect or ascending to the upper lip, or spreading and projecting straight forward:
   6. Flowers in terminal globose heads .... ***Acrocephalus***
   6. Flowers in terminal thyrsus or axillary verticils:
      7. Upper lip of corolla not hooded, calyx 5-lobed .... ***Anisomeles***
      7. Upper lip of corolla hooded, calyx 5–10-toothed:
         8. Corolla white, lower lip larger than upper lip .... ***Leucas***
         8. Corolla scarlet red, lower lip shorter than upper lip .... ***Leonotis***

Acrocephalus Bentham

Acrocephalus indicus (Burm. f.) Kuntze Rev. Gen. Pl. 2: 511. 1891. *Brunella indica* Burm. f. Fl. Ind. 130. 1768. *Acrocephalus capitatus* (Roth) Benth. in Wall. Pl. Asiat. Rar. 2: 18. 1831; FBI 4: 611; Gamble 1126 (788). *Ocimum capitatum* Roth, Nov. Pl. Sp. 276. 1821.

Slender, erect, annual herbs, up to 40 cm tall; stem quadrangular, grooved, sparsely hairy, purple; leaves 2.5–4 × 1–1.5 cm, obovate or elliptic-lanceolate, crenate-serrate, glabrescent, acute, base cuneate to attenuate; flowers white with purplish lips, in terminal and axillary capitate heads; nutlets ellipsoid, 1 mm long.

Common in moist areas along rocky slopes of streams. Fl. & Fr.: Sept.–Dec. Jayanthipuram, *VRK* 6936.

Anisochilus Wallich ex Bentham

Anisochilus carnosus (L.f.) Wall. Pl. As. Rar. 2: 18. 1831; FBI 4: 627; Gamble 1126 (788). *Lavandula carnosa* L. f. Suppl. Pl. 273. 1781.

Annual erect herbs, up to 50 cm tall; stems sparsely pubescent; leaves ovate, 2–7 × 3–4 cm, rather coriaceous, glabrescent above, pubescent below, acute, crenate-dentate, base cordate or obtuse; flowers purple or pale violet in terminal long condensed spikes; nutlets suborbicular, compressed, brown.

Occasional in forests. Fl. & Fr.: Oct.–Mar. *Vern.:* Karpura valli. Plant is used as stimulant, expectorant and for cough and cold. Javuku, *VRK* 3293; Nagarjuna Sagar, *KMS* 9753 (MH).

Anisomeles R. Brown

1. Stems and branches acutely quadrangular, leaves ovate to obovate, corolla pink violet .... *A. indica*
1. Stems and branches obtusely quadrangular, leaves oblanceolate, corolla purple .... *A. malabarica*

Anisomeles indica (L.) Kuntze Rev. Gen. Pl. 2: 512. 1891; Gamble 1140 (797). *Nepeta indica* L. Sp. Pl. 571. 1753. *Anisomeles ovata* Ait. f. Hort. Kew. 3: 364. 1811; FBI 4: 672.

Perennial, erect, pubescent, woody herbs, up to 1.5 m tall; stem 4-angled, woolly; leaves ovate, 4–8 × 3–6 cm, chartaceous, pubescent, acuminate, crenate-serrate, base unequal or truncate; flowers pink in axillary clusters or in terminal dense spikes; nutlets black, polished, flattened.

Occasional in open forests. Fl. & Fr.: Oct.–Mar. *Vern.*: Adabera, Ranaberi. The dried leaves are rubbed and smoked to treat cold. The plant is used as carminative, astringent and for flavouring purposes. Javuku, *VRK* 3286; Koppanur, *VRK* 3897, 3857; Papaiah palem, *VRK* 5836; Vinukonda RF, *VRK* 6863.

Anisomels malabarica (L.) R. Br. ex Sims. in Curtis, Bot. Mag. t. 2071, 1819; FBI 4: 673; Gamble 1140 (797). *Nepeta malabarica* L. Mant. Pl. 566. 1771.

An undershrub to 2 m tall, silvery tomentose, stem tetragonous; leaves oblanceolate, 4–8 × 1–2.5 cm, thick, densely tomentose, acute at both ends, margin serrate; flowers purple, in dense whorls, forming a more or less interrupted spicate inflorescence; nutlets ellipsoid, compressed, smooth, brown.

Occasional, in waste lands, along roadsides or in the hills. Fl. & Fr.: Throughout the year. *Vern.*: Magabeera. Leaves are used in dyspepsia, fever, toothache and in rheumatism, recent demand as ingredient of an aphrodiriac. Davupalli, *SSR & KSM* 2693.

Becium J. Lindley

Note: While examining the Indian *Ocimum* material in the Kew Herbarium, R.M. Harley realised that *Ocimum adscendens* Willd. is a typical member of the related genus *Becium* (Kew Bull. 38: 96. 1983). The species appears to be wide spread in South India and in Sri Lanka.

Becium Lindl. has been segregated from *Ocimum* largely on account of its characteristic calyx. While in *Ocimum* the lateral calyx teeth are short but distinct, in *Becium*, these are obsolete and modified as by a row of many minute teeth or a row of dense cilia.

Becium filamentosum (Forssk.) Chiov. in Nuovo Giorn. Bot. Ital. (n.s.) 26: 162. 1919; R. Harley in Kew Bull. 38: 56. 1983. *Ocimum filamentosum* Forssk. Fl. Aeg.—Arab. 108. 1775. *O. adscendens* Willd. Sp. Pl. 3: 166. 1800; FBI 4: 609; Gamble 1112 (778).

Ascending sub-shrubs, up to 50 cm tall; branchlets puberulous, subtetragonous; leaves elliptic-oblanceolate, 2–4 × 1.5–2.5 cm, chartaceous, entire or serrulate, acute, base attenuate; flowers purple in terminal verticils; nutlets erect, basilar; seeds mucilaginous when wet.

Occasional in wastelands. Fl. & Fr.: Oct.–Jan. Kondavedu Fort, *CAB* 4644 (MH).

Hyptis N.J. Jacquin

Hyptis suaveolens (L.) Poit. Ann. Mus. Natl. Hist. Nat. 7: 472. t. 29. f. 2. 1806; FBI 4: 630; Gamble 1129 (789). *Ballota suaveolens* L. Syst. Nat. (ed. 10) 1100. 1759.

Perennial aromatic undershrubs, up to 2 m tall; branches tetragonous, sparsely glandular-hirsute; leaves ovate, 3–5 × 1.5–3 cm, glabrescent above, sparsely hirsute along nerves and glandular below; acute, crenate-serrate, base rounded or truncate; flowers violet or blue, scented, in axillary and terminal stalked cymes; nutlets 2, oblong.

Occasional in forests and roadsides. Fl. & Fr.: Aug.–Feb. *Vern.:* Sirna Tulasi. N. Varam, *VRK* 3227; V.P. South, *VRK* 3536; Krishna river bank, *VRK* 6730; Nagarjunasagar, *KMS* 9722 (MH).

Native of tropical America, a weed from Africa, India, Sri Lanka, Indonesia and Malaysia.

Leonotis (Persoon) R. Brown

Leonotis nepetifolia (L.) R. Br. Prodr. 504. 1810; FBI 4: 691; Gamble 1155 (808). (*nepetaefolia*). *Phlomis nepetiifolia* L. Sp. Pl. 586. 1753. (*nepetaefolia*).

Annual erect herbs, up to 2 m tall; branches 4-angled, grooved, pubescent; leaves ovate, 3–8 × 2–5 cm, chartaceous, glabrescent, acute, deeply crenate, base abruptly cuneate; flowers orange red in axillary, globose, 6 cm across verticillasters; nutlets erect, trigonous; seeds oblong.

Common in moist wastelands, roadsides and hedges. Rare in forests. Fl. & Fr.: Sept.–Mar. *Vern.*: Ranabheri, Hanumanta-beera, Magabira (also assigned to *Anisomeles*). Leaves in minor skin problem. The plant is used for curing paralysis. Ash of the plant with mustard oil is applied locally on the paralytic part. It is used for burns, breast swellings, ring worms, scalds, skin affections, malaria and rheumatic pain. Guntur (*Uhl*, 1929).

Leucas R. Brown

1. Plants diffuse herbs:
   2. Calyx-tube strigose, teeth triangular, coastal .... *L. diffusa*
   2. Calyx-tube villous, teeth linear, inland .... *L. stricta*
1. Plants erect herbs:
   3. Calyx mouth hairy within .... *L. cephalotes*
   3. Calyx mouth glabrous within or minutely pubescent .... *L. aspera*

Leucas aspera (Willd.) Link Enum. Hort. Berol. 2: 113. 1822; Spreng. Syst. Veg. 2: 743. 1825; FBI 4: 690; Gamble 1150 (803). *Phlomis aspera* Willd. Enum. Hort. Berol. 621. 1809.

Annual erect, scabrid herbs, up to 50 cm tall; leaves elliptic-linear, 2–6 × 0.2–0.6 cm, puberulous, acute, entire to minutely serrate, base cuneate; flowers white in terminal and axillary verticils; nutlets oblong, smooth, brown.

Common in wastelands, roadsides, cultivated fields and open forests. Fl. & Fr.: Oct.–June. *Vern.*: Tella Tummi. Extract of the leaves, (two or three drops) is put into the eye for treating fevers. Juice of the leaves, (two or three drops) is put into the nose and ears for treating headache, also useful in nonspecific coughs, and also skin irritation. Koppanur, *VRK* 1295; Anupu, *DMR* 18124; Bollapalle, *CAB* 17407 (MH); Nagarjunasagar, *KMS* 9718 (MH).

Leucas cephalotes Spreng. Syst. Veg. 2: 743. 1825; FBI 4: 689; Gamble 1151 (804). *Phlomis cephalotes* Roth Nov. Pl. Sp. 262. 1821.

Annual erect, pubescent herbs, up to 60 cm tall; leaves ovate to lanceolate, 3–5 × 1–3 cm, acute, crenate-serrate, base acute or cuneate flowers white in terminal large globose dense whorls, covered with foliar bracts; nutlets dark-brown.

Occasional in cultivated fields, roadsides and wastelands. Fl. & Fr.: Sept.–Feb. Leaf is used for curing old boils, particularly on head. It is an anthelmintic, antiseptic, diaphoretic, laxative, and for cough and cold, and scabies. Mangalagiri, *SSR & KSM* 21620; Bollapalle. *CAB* 17436 (MH); Nagarjunasagar, *KMS* 9725 (MH).

Leucas diffusa Benth. Labiat. Gen. Sp. 615. 1834. in DC. Prodr. 12: 531. 1848; FBI 4: 689; Gamble 1150 (804).

Diffuse hirsute herbs, up to 40 cm tall; stem and branches ribbed; hairs reflexed; leaves oblong, 1.3 × 0.4–0.8 cm, chartaceous, hispid below, sericeous, base cuneate, enitre, subacute to obtuse, subsessile; flowers white in verticils; calyx tube turbinate, strigose with teeth.

Occasional weed in sandy places near sea. Fl. & Fr.: Aug.–Dec. *Vern.*: Pedda Tummi. Nizampatnam, *VRK* 6815.

Leucas stricta Benth. in Wall. Pl. As. Rar. 1: 61. 1824; FBI 4: 688; Gamble 1150 (804).

Annual erect, sparsely-hairy herbs, up to 30 cm tall; leaves linear-lanceolate, 1.5–3 × 0.5–1 cm, pubescent-hairy, obtuse, crenate, base narrowed, flowers white in terminal verticils; nutlets oblong, brown.

Occasional in open forests. Fl. & Fr.: Aug.–Dec. N. Varam, *VRK* 3234.

Ocimum Linnaeus

1. Shurbs; lower calyx teeth shorter than upper lip .... ***O. gratissimum***
1 Herbs or with slightly woody below; lower calyx teeth larger than the upper lip:
   2. Peduncle and calyx green, peduncle branched to the third order .... ***O. americanum***
   2. Peduncle and calyx purplish, peduncle once or twice branched:
      3. Pedicels shorter than calyx, calyx hairy within, enlarged in fruit ...... ***O. basilicum***
      3. Pedicel shortly longer than calyx, calyx glabrous within, not enlarged in fruit .... ***O. tenuiflorum***

Ocimum americanum L. Cent. Pl. 1: 15. 1755; H. Keng in Steenis, Fl. Males. Ser. 1. 8. 476. 1928. *Ocimum canum* Sims in Curtis Bot. Mag. t. 2452. 1823; FBI 4: 607; Gamble 1111 (777).

Annual, scented, erect herbs, up to 50 cm tall; much branched, stems and branches subquadrangular, leaves elliptic-ovate, 2–3.5 × 0.5–1 cm, pubescent, acute at both ends, sparingly serrate; flowers white or creamish in terminal long racemes; nutlets black, pitted; seeds mucilaginous when wet.

Common in open forests and occasional in forests. Fl. & Fr.: Sept.–Feb. *Vern.*: Kukka thulasi. A medicinal herb in cough, mouth wash and flavour. Koppanur, *VRK* 3250; Mellavagu, *VRK* 5724; Nizampatnam, *VRK* 6816.

Ocimum basilicum L. Sp. Pl. 597. 1753; FBI 4: 608; Gamble 1111 (777).

Erect herbs, hispidly pubescent, stems and branches purplish; leaves elliptic ovate or lanceolate in whorls, entire, to serrulate, base and apex acute; flowers in terminal, loose, racemose whorls, white, pink or purplish; nutlets ellipsoidal, black and pitted, fruiting pedicel recurved in fruit.

Common on waste lands. Fl. & Fr.: Aug.–Dec. *Vern.*: Vibhudi patri. The plant has a strong aromatic flavour, and is employed as a seasoning for culinary purposes. The seeds are used medicinally. Koppanur, *VRK* 3249.

Ocimum gratissimum L. Sp. Pl. 1197. 1753; FBI 4: 608; Gamble 1111 (777).

A much branched sub-shrub to 2 m; branchlets glabrescent, 4-angular; leaves ovate, 3–10 × 2–5 cm, base acute, margin deeply serrate, apex acute to shortly acuminate; flowers white or pale greenish yellow in close whorls in racemes; nutlets 4, sub-globose, brown, calyx enlarged in fruit.

Occasional, in plains in NU campus. Fl. & Fr.: Oct.–Mar. Leaves lemon scented, mosquito repellant, a plant commercial mosquito repellant along with *Cymbopogon citratus* and has been suggested as measure of biological control of mosquitoes.

Ocimum tenuiflorum L. Sp. Pl. 579. 1753. *Ocimum sanctum* L. Mant. 1: 85. 1767; FBI 4: 609; Gamble 1111 (778).

Woody much branched aromatic herbs, up to 50 cm tall; branchlets hispid; leaves ovate to elliptic-oblanceolate, 1–2.5 × 0.5–1 cm, hairy, minutely dotted, acutely apiculate, serrate, undulate, base truncate, or cuneate; flowers purplish in terminal branched racemes; nutlets ellipsoid, smooth.

'Sacred Basil' common in roadsides and occasional in forests. Often cultivated in pots. Fl. & Fr.: Throughout the year. *Vern.*: Thulasi. Planted in temples for religious offerings. Plant parts used as medicinal. Koppanur, *VRK* 3278; Kondavedu Fort, *CAB* 4683 (MH).

Note: It is recommended that the specific epithet '*sanctum*' should be a *nom. cons.* The sacred Tulasi, cultivated and worshipped. Several medicinal uses are ascribed. It is widely used to relieve restlessness in skin disorders. Considered as a mosquito repellant. Considered as life saving drug.

Chemical analysis (1998 paper). Essential oils: Eugenol 73.2%; Carvacrol 3.2%, Citronellal 15.7%, Citral-aldehyde 68%, + Phenol and acetic acid and also two esters 4.8% as Methyl hepterone and citonellal.

The most recent merit makes the Tulasi a natural antipollulant in indoors as it adds to ozone content of the atmosphere. One plant can take care of 100 Sq. ft. area and keeps air clean for 24 hours.

Acc. to Dr. S.K. Padolen of Nagpur, the black form is essentially medicinal and the green form is antipollutant.

There is a nationwide relevance to the recommendation that this species should be cultivated extensively and also to harness wastelands.

Orthosiphon Bentham

Orthosiphon rubicundus (D. Don) Benth. in Wall. Pl. As. Rar. 2: 14. 1831; FBI 4: 614; Gamble 1114 (779). *Plectranthus rubicundus* D. Don Prodr. Fl. Nepal 116. 1825. (*Plectranthes rubicunda*).

Erect, much branched undershrubs, up to 1 m tall; lower leaves petiolate, upper leaves sessile, ovate-elliptic, 4–7 × 2–3.5 cm, glabrous, acute to shortly acuminate, crenate-serrate, base acute to obtuse, sweet-scented; flowers white in terminal racemes; nutlets subglobose, smooth.

Rare in forests as an undershrub. Fl. & Fr.: Sept.–Apr. *Vern.*: Kammagajjalaku. Juice of stem bark is applied to cure cauterize bleeding wounds. The tubers are eaten , used as remedy against colic. Vinukonda RF, *VRK* 6901.

Majorana hortensis Moench, *Mentha arvensis* L., *Plectranthus scutellariodes* (L.) R. Br.; *Salvia coccinia* Buc'hoz ex Etlinger; *S. officinalis* L. and *S. splendens* Sell ex Nees are cultivated as ornamentals.

NYCTAGINACEAE

1. Plants armed, straggling shrubs, stamens 4–7, far exerted, stigma 2-fid, ... ***Pisonia***
1. Plants unarmed, erect or diffuse herbs, stamens 1–3, as long as perianth tube, stigma peltate:
   2. Anthocarp 5-ribbed, < 3 mm long .... ***Boerhavia***
   2. Anthocarp 10-ribbed, > 7 mm long .... ***Commicarpus***

Boerhavia Linnaeus

1. Flowers sessile, Anthocarp club-shaped, glandular hairy, top round .... ***B. diffusa***
1. Flowers pedicelled, Anthocarp obconic, glabrous, top truncate .... ***B. erecta***

Boerhavia diffusa L. Sp. Pl. 3: 1753; Gamble 1163 (814). *B. repens* L. Sp. Pl. 3. 1753; FBI 4: 709.

Diffuse, ascending, much-branched, sparsely pubescent herb; leaves ovate to oblong-lanceolate, 2–4 × 0.5–2 cm, glabrescent above, thinly-pubescent below, acute, entire to sinuate, base truncate to sub-cordate; flowers pink in axillary and terminal umbellate panicles; anthocarps club-shaped, glandular-hairy, top rounded.

Very common on roadsides, waste lands, hedges and bushes. Fl. & Fr.: throughout the year. *Vern.:* Atikamamidi. Eng.: Horse-pursalne, Hog weed. Whole plant constitutes the drug 'Punarnava' which contains an active principle punarvine. Drug possess diuretic properties, and is used in dropsy. Also recommended for asthma. Koppanur, *VRK* 3269; V.P. South, *VRK* 6838; Kondavedu Fort, *CAB* 4652 (MH).

Boerhavia erecta L. Sp. Pl. 3: 1753. *B. punarnava* Saha & Krishnamurthy in J. Sci. Industr. Res. 210: 254. 1961.

Annual erect herbs, up to 50 cm tall; leaves ovate-elliptic, 1–2.5 × 0.8–1.5 cm, chartaceous, glabrescent, acute to apiculate, entire, base rounded; flowers with pink stripes in axillary and terminal slender panicles; anthocarps obconical, glabrous.

Common in open forests and wastelands. Fl. & Fr.: Sept.–Jan. Loyapalle, *VRK* 3850; Nagarjunakonda, *VRK* 6702.

Commicarpus Standley

Commicarpus chinensis (L.) Heimerl in Engl. & Prantl Nat. Pflanzenfam ed. 2. Ibe. 117. 1934. *Valeriana chinensis* L. Sp. Pl. 33. 1753. *Boerhavia rependa* Willd. Sp. Pl. 1: 22. 1797; FBI 4: 709; Gamble 1163 (814) *Boerhavia chinensis* (L.) Aschers. & Schweinf. Beitr. Fl. Aeth. 1: 167. 1867. (*Boerhaavia*).

Diffuse, subscandent, glabrescent herbs, up to 25 cm tall; leaves ovate, 5–10 × 3–8 cm, glabrous, sinuate, acute, base cordate; flowers pink or purple in axillary and terminal slender, long-peduncled umbels; anthocarps club-shaped, viscid-glandular.

Common in hedges and occasional in forests. Fl. & Fr.: Sept.–Feb. *Vern.*: Teega mamidi, Kanne komali. Sattenapally, *VRK* 5840; Kondavedu Fort, *CAB* 4604 (MH).

Pisonia Linnaeus

Pisonia aculeata L. Sp. Pl. 1026. 1753; FBI 4: 711; Gamble 1163 (815).

Armed scandent shrubs; prickles axillary, recurved; leaves elliptic-ovate, 4–7 × 3–4 cm, thick-chartaceous, base rounded, apex acute; flowers greenish to cream in axillary corymbose cymes; fruits oblong, 5-angled.

Occasional in Koppanur along the stream. Fl. & Fr.: Jan.–May. *Vern.*: Pisangi. Bark and leaves are used for swellings and rheumatic pains. Juice mixed with pepper and other ingredients given in pulmonary complaints of children.. Koppanur, *VRK* 3893 & 3891.

Bougainvillea spectabilis L. (Kagitapu puvvulu), *Mirabilis jalapa* L., *Pisonia grandis* R. Br. are commonly cultivated as ornamentals.

AMARANTHACEAE

1. Leaves filiform, decussate below, alternate above in the same plant .... ***Trichurus***
1. Leaves not filiform, either decussate or alternate (occasionally clustered):
    2. Flowers unisexual:
        3. Plants dioecious, inflorescence tomentose, white .... ***Aerva (javanica)***
        3. Plants monoecious, inflorescence glabrous, green .... ***Amaranthus***
    2. Flowers bisexual:
        4. Prostrate herbs, anthers 1-celled .... ***Alternanthera***
        4. Erect or straggling herbs or subshrubs, anthers 2-celled:
            5. Leaves decussate:
                6. Fertile flowers subtended by 2 sterile flowers, filaments free, bracts and bracteoles of staminodes hooked in fruits .... ***Pupalia***
                6. Fertile flowers single, filaments connate into tubular sheath, bracts and bracteoles smooth in fruits:
                    7. Spikes globose to elongate, axillary, more than 1.5 cm across .... ***Gomphrena***
                    7. Spikes elongate, terminal, less than 0.7 cm across ...... ***Achyranthes***
            5. Leaves alternate:
                8. Thyrses lax, less than 0.8 cm across, rose .... ***Digera***
                8. Spikes dense, more than 1.5 cm across, white or brownish:
                    9. Staminodes 5... ***Aerva (lanata)***
                    9. Staminodes 0:
                        10. Flowers single, mature spikes whitish pink, seeds several .... ***Celosia***
                        10. Flowers in groups of 3, mature spikes brownish, seeds solitary ... . ***Allmania***

Achyranthes Linnaeus

Achyranthes aspera L. Sp. Pl. 204. 1753; FBI 4: 730; Gamble 1176 (823).

Annual, erect, stiff herbs, branching from base, up to 30 cm tall; leaves decussate, variable, elliptic-obovate to orbicular, 2–7 × 1.5–4 cm, pubescent, obtuse or shortly acute, entire, base cuneate; flowers pink or greenish-white in terminal elongate spikes, bracts persistent, spiny; capsule 1-3 mm, seed cylindrical, smooth.

Abundant weed in open forests and roadsides. Fl. & Fr.: July–Jan. *Vern.*: Uttareni. Plant is used in cough and it's decoction is given in renal dropsy and bronchial infection. Plant extract used for piles, boils, skin eruptions etc. N. Varam, *VRK* 3222; Davupalli, *DMR* 21132.

Aerva Forsskal

1. Herb, leaves ovate-orbicular, spikes axillary, clustered, less than 3 cm long, leaves gradually smaller towards apices .... *A. lanata*
1. Subshrub, leaves oblanceolate to spathulate, spikes terminal, paniculate, more than 5 cm long, leaves well developed throughout .... *A. javanica*

Aerva javanica (Burm. f.) Juss. ex Schult. in L. Syst. Veg. 5: 565. 189; FBI 4: 727. *Iresine javanica* Burm. f. Fl. Ind. 212. t. 65. f. 2. 1768. *Aerva tomentosa* Forssk. Fl. Aeg.—Arab. 122. 170. 1775; Gamble 1178 (824).

Erect, woolly-tomentose undershrubs, up to 1.5 m tall; leaves alternate, oblanceolate, or linear-oblong, 3–7 × 1–2 cm, greenish-white, obtuse, acute, or apiculate, entire, base cuneate to alternate, sessile; flowers unisexual, white, in terminal dense tomentose panicles; seeds dark-brown, polished.

Occasional weed in wastelands, roadsides, hedges and occasional in forests. Fl. & Fr.: Dec.–Aug. *Vern.*: Peddapindi. Satrasala, *VRK* 6780.

Aerva lanata (L.) Juss. in Ann. Mus. Nat. Hist. paris 2: 131. 1803; FBI 4: 728; Gamble 1178 (825). *Achyranthes lanata* L. Sp. Pl. 204. 1753.

Prostrate or diffuse herbs, stems and branches pubescent when young; leaves elliptic-ovate, 1.5–4.5 × 0.5–3 cm, pubescent above, white woolly below, obtuse or apiculate, entire, base tapering; flowers white with pink tinge, in axillary and terminal group of spikes.

Common in open forests and roadsides and occasional in forests. Fl. & Fr.: Sept.–Mar. *Vern.:* Kondapinde; Sanskrit; Pashana bedhi. Root is medicinal for headache. Infusion in dissolving kidney stones. Koppanur, *VRK* 3259; Kondavedu Fort, *CAB* 4758 (MH).

Allmania R. Brown ex R. Wight

Allmania longepedunculata (Trimen) Gamble Fl. Pres. Madras 2: 1168 (818). 1925. *A. nodiflora* R. Br. var. *longepedunculata* Trimen Handb.

Fl. Ceylon 3: 394. 1895. *A. albida* (Willd.) R. Br. in Hook. f. Fl. Brit. India 4: 717. 1885. *Celosia albida* Willd. Sp. Pl. 1: 1197. 1798.

Slender herbs, with erect or spreading branches; leaves oblong-spathulate or elliptic, 2–4 × 0.5–1 cm, glabrescent, apiculate, entire, base narrowed; flowers orange yellow in axillary and terminal globose spikes; utricle circumcissile; seeds compressed.

Occasional in waste places, fields and in open forests. Fl. & Fr.: Aug.–Jan. *Vern.*: Errabadhi aku. Mellavagu, *VRK* 5718; Jayanthipuram, *VRK* 6971.

Alternanthera Forsskal

1. Bracts and bracteoles glabrous:
   2. Tepals dissimilar with a tuft of glochidiate bristles on the keel, leaves ovate to obovate .... *A. pungens*
   2. Tepals similar, without glochidiate bristles, leaves lanceolate to spathulate .... *A. sessilis*
1. Bracts and bracteoles barbellate hairy .... *A. paronychioides*

Alternanthera paronychioides St. Hil. Voy. Distr. Diamans Bresil 2: 439. 1833; Vedlk. in Blumea 19. 167. 1971.

Prostrate creeping herb, stem with long crisped hairs on the younger parts; leaves oblanceolate-elliptic, 1–1.5 × 0.5–0.7 cm, sessile, hairy, flowers numerous in heads; utricle orbicular, seed discoid, 1 mm across, faintly reticulate.

Rare, near powerhouse, Nagarjuna University Campus. Fl. & Fr.: Feb.–July.

Alternanthera pungens Kunth in H.B. & K. Nov. Gen. Sp. 2: 206. 1817. *Achyranthes repens* L. Sp. Pl. 205. 1253. *Alternanthera repens* (L.) Link Enum. Pl. Hort. Berol. 1. 154. 1821; (non Gmelin, 1791).

Spiny prostrate spreading herb with woody base to 50 cm, branches villous, leaves opposite with unequal pairs, obliquely elliptic to sub-orbicular, 1.5–3.5 × 0.4–1.4 cm, glabrous, obtusely apiculate, entire, cuneate to attenuate; flowers in axillary clusters, outer 2 tepals stronger, jointed above into a pungent awn; seed discoid, lenticular.

Common weed in wastelands, along roadsides and fields. (Mohan Rao and Aasiya, 1995). Fl. & Fr.: Throughout the year. *Vern.*: Mullu Ponnaganti.

Alternanthera sessilis (L.) R. Br. ex DC. Cat. Pl. Hort. Monop. 4: 77. 1813; FBI 4: 731. *Gomphrena sessilis* L. Sp. Pl. 225. 1753. *Alternanthera triandra* Lam. Encycl. 1: 95. 1783; Gamble 1179 (825).

Aquatic prostrate, spreading, much-branched herbs; often rooting at the nodes; leaves linear-oblong, 2–6 × 0.5–1 cm, glabrous, obtuse, entire, base cuneate; flowers white, shining, in axillary sessile globose heads; utricles compressed; seeds orbicular.

Common weed of moist places, hedges and occasional along streams and irrigation canals. Fl. & Fr.: Feb.–May. *Vern.*: Ponnaganti Kura. Plant extract with coconut oil promotes hairy growth. Ethipothala, *VRK* 5802; Sattenapalle, *VRK* 5835; Macherla RF, *VRK* 5878; Nallapadu, *VRK* 6754; Vinukonda RF, *VRK* 6875, Kondavedu Fort, *CAB* 4659 (MH).

Amaranthus Linnaeus

1. Plants armed, tepals and stamens 5 each .... ***A. spinosus***
1. Plants unarmed:
   2. Fruits circumscissile, leaf-base cuneate to attenuate:
      3. Bracteoles exceeding or equalling the perianth lobes, awned .... ***A. tricolor***
      3. Bracteoles shorter than the perianth lobes, acute .... ***A. graecizans***
   2. Fruits indehiscent, leafbase truncate to acute .... ***A. viridis***

Amaranthus graecizans L. Sp. Pl. 990. 1953. ssp. **thellungianus** (Nevski) Gusev in Bot. Zhurn. 57: 462. 1972. *A. thellungianus* Nevski in Act. Acad. Sci. U.R.S.S. 1(4): 311. 1937. *A. polygonoides* auct. non L.; Roxb. Fl. Ind. 3: 602. 1832; Nair in Bull. Bot. Surv. India 8.88. 1966. *A. polygamous* sensu Hook. f. in Fl. Brit. India 4: 721. 1885. *p.p.* Gamble 1176 (820).

Diffuse prostrate herb, stem red; leaves obovate, 1–2 × 0.5–1.5 cm, obtuse, rarely retuse, usually mucronate, entire, base cuneate, gland dotted, flowers in axillary clusters; utricle ovoid, slightly rugose or waved, usually circumscissile.

Common weed of cultivated fields and waste places. Fl. & Fr.: July–Dec. *Vern.*: Chirikura. Managalagiri, *SSR & KSM* 21617.

Amaranthus spinosus L. Sp. Pl. 991. 1753; FBI 4: 718; Gamble 1170 (819).

Armed, profusely branched undershrubs, up to 1 m tall; spines axillary, paired or clustered; leaves elliptic or ovate-lanceolate, 3–6 × 1.5–3 cm, glabrescent, emarginate or obtuse, apiculate, entire, base acute; flowers green with yellow anthers in axillary dense globose clusters and terminal aggregated spikes in terminal panicles, utricles circumcissile; seeds dark-brown, shining.

Abundant weed of wastelands and cultivated fields and hedges near villages. Fl. & Fr.: Throughout the year. *Vern.*: Mulla Thotakura. It is used as vegetable and fodder. Nallapadu, *VRK* 6770.

Amaranthus tricolor L. Sp. Pl. 989. 1763. *A. mangostanus* L. Cent. Pl. 1: 32. 1755; FBI 4: 720; Gamble 1171 (820). *A. gangeticus* L. Syst. Nat. (ed. 10) 1268. 1759; FBI 4: 719; Gamble 1171 (819).

Erect, diffuse, much branched, stout herbs, up to 30 cm tall; leaves obovate or rhomboid, 1–2 × 0.5–1 cm, glabrous, acute to retuse, entire, base shortly cuneate to attenuate; flowers light green in axillary fascicles; utricles ovoid; seeds black.

Common in open forest and cultivated fields. Fl. & Fr.: Throughout the year. *Vern.:* Chirraku. Eng.: Tampala, Chinese spinach, Amaranth (tender). Leaves and young stems are used as vegetable. Plant is astringent, and used in diarrhoea, dysentery and haemorrhage. Externally used in ulcerated conditions of throat and mouth, and wash for ulcers. Koppanur, *VRK* 3264.

Amaranthus viridis L. Sp. Pl. ed. 2. 1405; 1763; FBI 4: 720; Gamble 1171 (820). *Amaranthus gracilis* Desf. Tabl. Hort. Par. 43. 1804.

Erect, branched herbs, up to 75 cm tall; leaves deltoid-ovate, 2–5 × 1–3.5 cm, glabrous, acute, retuse, entire, base truncate to acute; flowers pale green in axillary and terminal panicled, spike-like racemes; utricles subcompressed, brownish; seeds black, shining.

Common weed in open forests. Fl. & Fr.: Throughout the year. *Vern.*: Chilaka thotakura. N. Varam, *VRK* 3225.

Celosia Linnaeus

Cleosia argentea L. Sp. Pl. 205. 1753; FBI 4: 714; Gamble 1166 (816).

Erect, glabrous herbs, up to 2 m tall; leaves elliptic-ovate below, linear-lanceolate above, 2–9.5 × 0.5–2 cm, acute to acuminate, entire, base cuneate; flowers pale white with pink tinge in terminal, dense cylindric or conical to oblong spikes; utricles ellipsoidal; seeds black, shining.

Occasional in open forests and as a weed of cultivation. Prominent in cultivated fields. (Mohan Rao and Aasiya, 1995). *Vern.*: Gurugaku. Seeds used for curing mouth sores and eye diseases.

Digera Forsskal

Digera muricata (L.) Mart. Beitr. Amarant. 2: 77. 1825. *Achyranthes muricata* L. Sp. Pl. ed. 2. 295. 1762. *Digera arvensis* Forssk. Fl. Aeg.—Arab. 65. 1775; FBI 4; 717; Gamble 1169 (818).

Annual slender herbs with spreading or erect branches, up to 50 cm tall; leaves deltoid to elliptic-ovate, 3–5 × 2–3.5 cm, glabrescent, obtuse to minutely apiculate, entire, base truncate or oblique; flowers pink in lax axillary or terminal thyrsiform spikes; nutlets indehiscent; seeds solitary.

Common in open forests. Fl. & Fr.: Throughout the year. *Vern.*: Chenchalikura. Tender twigs and inflorescences are used as vegetable. Koppanur, *VRK* 3241; V.P. South, *VRK* 5831; Kondavedu, *SSR* & *KSM* 21464; Nagarjunasagar, *KMS* 9846 (MH).

Gomphrena Linnaeus

1. Spikes whitish, terminal and axillary, leaves hairy, serrate, flowers 1.5 mm across .... ***G. serrata***
1. Spikes purple, terminal, leaves glabrous, entire, flowers 4.5 mm across .... ***G. globosa***

Gomphrena globosa L. Sp. Pl. 224. 1753; FBI 4: 732; Gamble 1179 (825).

Erect or ascending herbs, up to 50 cm tall; branchlets appressed white hairy; leaves obovate-lanceolate, 5–10 × 1.5–4 cm, glabrous, entire, acutely apiculate; flowers purple, in terminal globose spikes; utricles 2 mm; seeds shiny.

Occasional in plains as an escape. Fl. & Fr.: Aug.–Dec. *Vern.*: Bendu malli. Nagarjunakonda, *KTH* 9841 (CAL).

Gomphrena serrata L. Sp. Pl. 224. 1753; Mears in Taxon 29; 86. 1980. *G. decumbens* Jacq. Hort. Schoenbr. t. 482. 1805; Gamble 1179 (825), *G. celosioides* auct. non Mart. Bennet Fl. Howrah 142. 1979.

Annual, ascending or erect herbs, up to 40 cm tall; branches clothed with white, shaggy hairs; leaves obovate-lanceolate, 2–4 × 1–1.5 cm, glabrescent above, long white shaggy hairs below, obtusely apiculate, base cuneate; flowers white with yellow tinge in axillary and terminal compressed, cylindrical spikes; utricles enclosed by the hardened perianth; seeds brown, shiny.

Common weed in waste places, roadsides, hedges and fields. Native of South America. Fl. & Fr.: Throughout the year. *Vern.:* Neeru-Bogada Banthi. Nallapadu, *VRK* 6785; Vijayapuri, *KMS* 9845 (MH).

Note: The species, to which the name *G. serrata* L. was given, and renamed by Jacquin (1805) as *G. decumbens* without reference to Linnaeus species name. *G. disperma* Standl. a name applied to some phenotypes of the same variable species *G. serrata* L. is the most common and widely distributed non cultivated species of *Gomphrena*. It is frequently misidentified as *G. celosiodes* Mart., a closely related species of South America except in Cultivation.

Pupalia A.L. Jussieu *nom. cons.*

1. Leaves glabrescent above, sparsely pubescent below .... ***P. lappacea*** var. ***lappacea***
1. Leaves with white tomentum more on lower surface .... ***P. lappacea*** var. ***velutina***

Pupalia lappacea (L.) Juss. in Ann. Mus. Par. 2: 132. 1803. var. **lappacea**; FBI 4: 742; Gamble 1173 (821). *Achyranthus lappacea* L. Sp. Pl. 204. 1753. *A. atropurpurea* Lam. Encycl. 1: 346. 1785. *Pupalia atropurpurea* (Lam.) Moq. in DC. Prodr. 13: 371. 1849; FBI 4: 723; Gamble 1173 (821).

Erect or straggling herbs, up to 75 cm tall; branchlets pubescent; leaves decussate, ovate or elliptic, 3–5 × 1.5–2.5 cm, glabrescent above, pubescent below, acute to acuminate with apicule, entire, base obtuse; flowers greenish-white in terminal unbranched spikes, unibracteolate sterile flowers modified into hooked bristles in fruits, capsule depressed globose, circumscissle, surrounded by persistent hooked bristles.

Common in open forests and roadsides. Fl. & Fr.: Nov.–Jan. *Vern.*: Antreetha. Javuku, *VRK* 3563; *VRK* 3578; Loyapalli, *SSR* & *KSM* 21640; Kondavedu Fort, *CAB* 4636 (MH).

Pupalia lappacea (L.) Juss. var. **velutina** Hook. f. in Fl. Brit. India 4: 724; 1885; Gamble 1173 (821).

Leaves of this variety are with white tomentum more on the lower surface and less on the upper surface.

Koppanur, *VRK* 3253; N. Varam, *VRK* 3515.

The local populations in NU campus and its neighbourhood often show spikes with purple fruits referred as *P. atropurpurea.*

Trichurus Townsend

Trichurus monsoniae (L.f.) Towns. in Kew Bull. 29: 466. 1974. *Illecebrum monsoniae* L. f. Suppl. Pl. 161. 1781. *Aerva monsoniae* (L.f.) Mart. Nov. Act. Acad. Caes. Leop-Parol. 13. 281. 1826; FBI 4: 728; Gamble 1178 (825).

Annual, prostrate, tomentose herbs; branches pilose, rooting at the nodes; leaves filiform, up to 2 cm long, whorled at nodes, pubescent, acute, entire, base cuneate; subsessile; flowers white with pink tinge in axillary and terminal globose spikes; utricle circumscissile, longitudinally split, enclosed in persistent bracteoles and perianth-lobes; seed ovoid, shiny.

Common in open forests and forest undergrowth. Fl. & Fr.: June–Jan. *Vern.*: Erra pindi. N. Varam, *VRK* 3238; V.P. South, *VRK* 5805; Nagarjunasagar, *KMS* 9744 (MH).

Celosia argentea L. var. *cristata* (L.) Kuntze is cultivated as ornamental.

CHENOPODIACEAE

1. Stem jointed, leaves obscure, flowers irregular, embedded in depressions:
    2. Stem ascending, purplish, internodes narrowed at both ends, swollen in the middle, flowers in concealed cavities .... ***Salicornia***
    2. Stem erect, yellowish, internodes narrowed below, gradually broadened above, flowers in exposed cavities .... ***Arthrocnemum***
1. Stem not jointed, leaves well developed, flowers regular, flowers clustered .... ***Suaeda***

Arthrocnemum Moquin-Tandon

Arthrocnemum indicum (Willd.) Moq. Chenop. Monogr. Enum. 113. 1840; FBI 5: 12; Gamble 1183 (828). *Salicornia indica* Willd. Nova Acta Acad. Sci. Imp. Petrop. Hist. Nat. 5: 111. t. 4. f. 1799.

Succulent ascending herbs, up to 50 cm; branchlets succulent, terete, jointed, constricted at the nodes; leaves decurrent, obscure, those of each pair connate into a cup, 0.5–0.8 × 0.4–0.6 cm, thin towards the margin; flowers bisexual, sessile, embedded in cavities; utricles 1 mm; seeds minute.

Common, dense, reddish to purple succulent masses along the salty marshes of Nizampatnam coast. Fl. & Fr.: Jan.–Oct. *Vern.*: Koyya pippali, Sitamma-vari Dubbu. Nizampatnam Port, *VRK* 6912.

Salicornia Linnaeus

Salicornia brachiata Roxb. Fl. Ind. 1: 84. 1832; FBI 5: 12; Gamble 1183 (828).

Erect leafless much branched annual herbs, up to 30 cm long, branches terete, jointed, greenish-red, succulent; flowers in cylindric spikes; utricles membranous, ovoid.

Common in marshy coast of Nizampatnam and Kothapalem. Fl. & Fr.: Nov.–Jan. *Vern.*: Saakati-pusulu. Kothapalem, *VRK* 5757; Nizampatnam, *VRK* 6794; *DLR* 132 (NUH).

Suaeda Forsskal ex Scopoli

1. Tender leaves sickle-shaped, drying pale green, flowers bisexual .... ***S. maritima***
1. All leaves straight, drying black:
   2. Flowers bisexual, bracts and bracteole margin toothed .... ***S. nudifolra***
   2. Flowers unisexual, bracts and bracteole margin entire .... ***S. monoica***

Suaeda maritima (L.) Dumort. Fl. Beng. 22. 1827; FBI 5: 14; Gamble 1184 (829). *Chenopodium maritimum* L. Sp. Pl. 221. 1753.

Slender erect or nodding herbs; branchlets reddish; leaves linear-oblong, sickle shaped when young, 1.5–2.6 × 0.2–0.6 cm, fleshy, green or often tinged with pink; flowers minute, in long lax spikes in panicles.

Comon herbs in the saline marshy areas of the coast. Fl. & Fr.: Jan.–June. *Vern.*: Ilakura. Green leaves are reported to be eaten as vegetable by the poor in times of scarcity. It has also been reported to fix sand along sea-shores. Kothapalem, *VRK* 5759; Nizampatnam Port, *VRK* 6793, 6811.

Suaeda monoica Forssk. ex Gmel. Syst. Nat. (ed. 13) 2(1): 503. 1791; FBI 5: 13; Gamble 1184 (829).

Erect bushy herbs, up to 30 cm long; branches with leafscars; leaves linear or narrowly oblong, sessile, succulent, 1.5–2.2 × 0.3 cm, obtuse at both ends, flowers green in lax slender spikes; seeds ovoid.

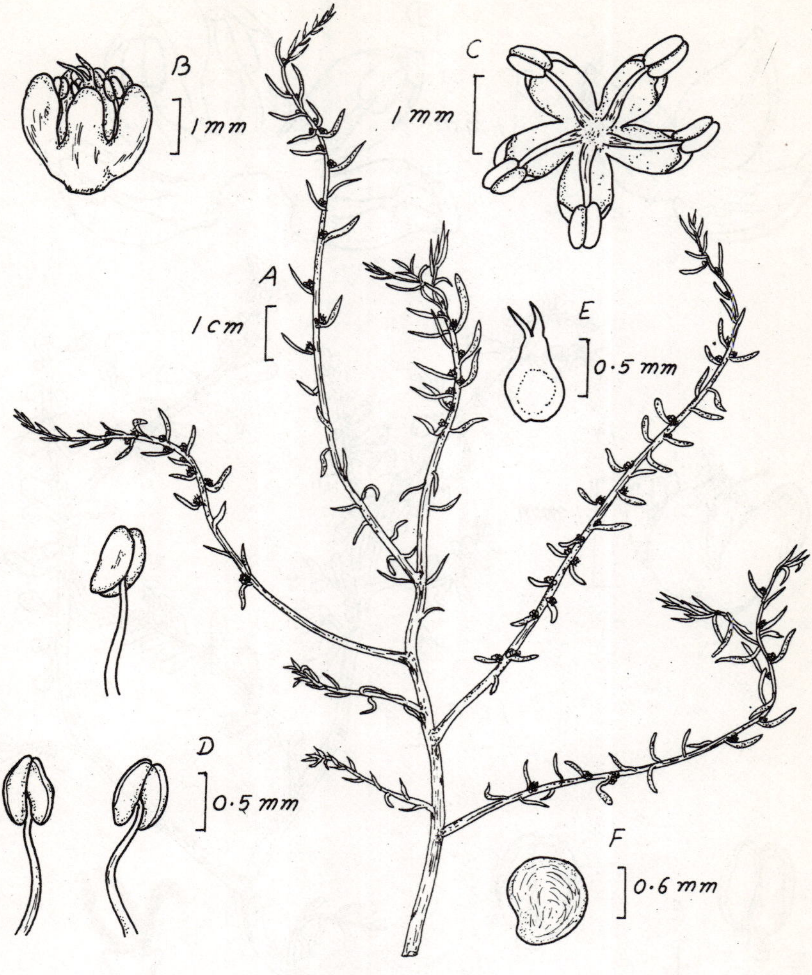

Fig. 22. *Suaeda maritima* (Forssk.) J. Gmel.
A. Twig, B. Flower, C. Perianth split open, D. Stamens, E. Pistil, F. Seed

Common marine herb in marshy back waters. Fl. & Fr.: May–Dec. *Vern.*: Koyyala kura. Nizampatnam, *VRK* 6804; *DLR* 144 (NUH).

Suaeda nudiflora Moq. Ann. Sci. Nat. (Paris) 23: 316. 1831 & in DC. Prodr. 13(2) 155. 1849; FBI 5: 14; Gamble 1184 (829). *S. indica* Moq. Ann. Sci. Nat. (Paris) 23: 316. 1831.

Erect subshrubs, up to 75 cm tall; leaves narrowly oblong or obovate, 1–2 × 0.1–0.3 cm, thick, base acute, apex subacute; flowers bisexual, in dense, globose clusters in the axils of reduced leaves; utricles thin-walled, concealed by the tepals.

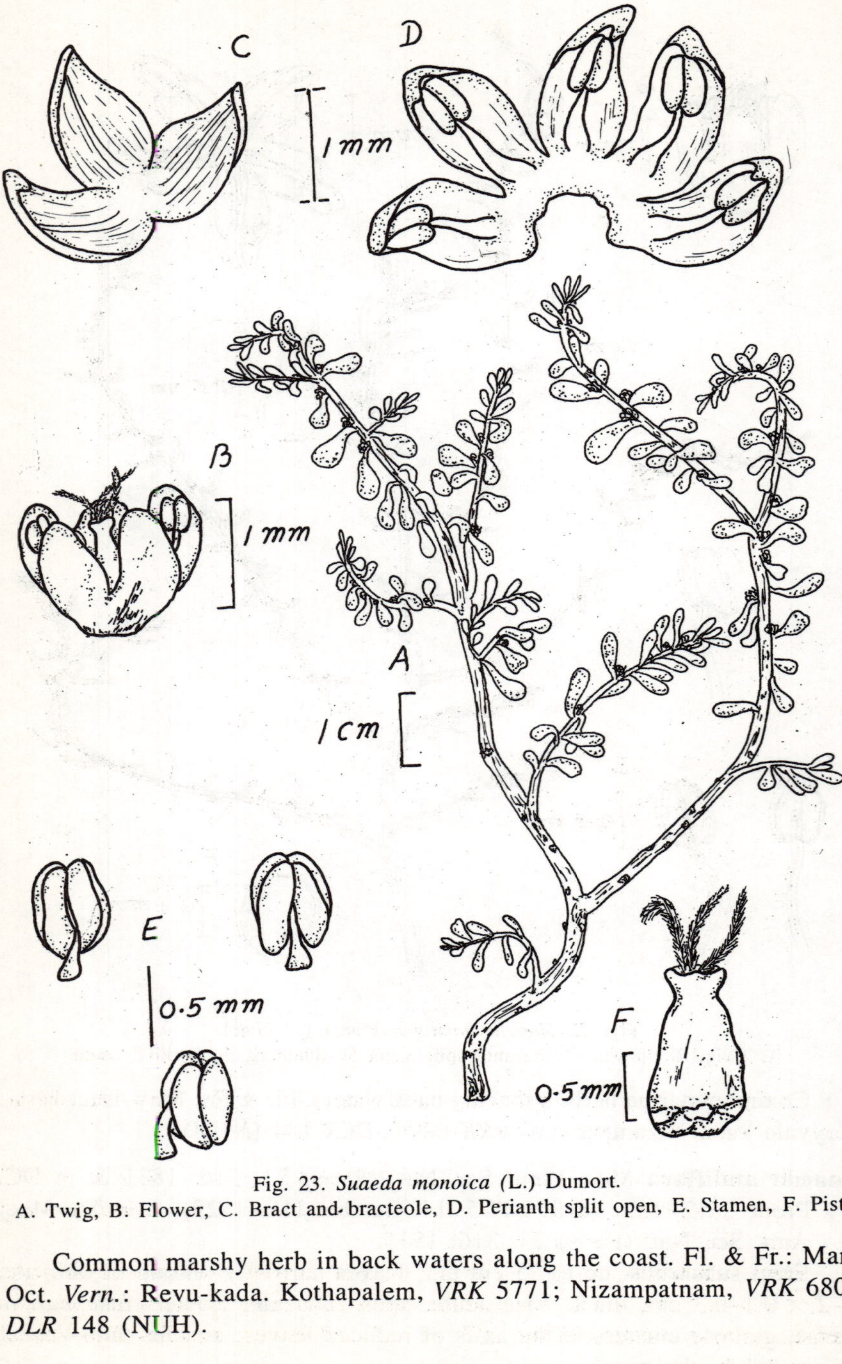

Fig. 23. *Suaeda monoica (*L.) Dumort.
A. Twig, B. Flower, C. Bract and bracteole, D. Perianth split open, E. Stamen, F. Pistil

Common marshy herb in back waters along the coast. Fl. & Fr.: Mar.–Oct. *Vern.*: Revu-kada. Kothapalem, *VRK* 5771; Nizampatnam, *VRK* 6808, *DLR* 148 (NUH).

BASELLACEAE

Basella Linnaeus

Basella alba L. Sp. Pl. 272. 1753. *B. rubra* L. Sp. Pl. 273. 1753; FBI 5: 20; Gamble 1185 (830).

Succulent, rhizomatous, twining herbs; branchlets terete; leaves simple, alternate, succulent, 3–8 × 2–6 cm, ovate, or lanceolate, entire, acute to acuminate, glabrous; flowers red or white, sessile in pedunculate cymes; berries globose, black when ripe.

The Indian spinach, cultivated herb and often as an escape in hedges. (Mohan Rao and Aasiya, 1995). Fl. & Fr.: Aug.–Jan. *Vern.*: Bachalikura. Leaves used as vegetable. Fleshy, berries are reported to yield dye which may be used for dyeing jellies and sweets.

POLYGONACEAE

1 Climbing herbs, inflorescence tip tendrilar .... *Antigonon*
1. Erect herbs, inflorescence not tendrilar .... *Polygonum*

Antigonon Endlicher

Antigonon leptopus Hook. & Arn. Bot. Beech. Voy. 308. t. 69. 1841; Gamble 1192 (835).

Large handsome climbing herbs with stout rootstock; leaves simple, alternate, hastate or cordate, 3–6 × 2–3.5 cm, acuminate with apicule, glabrous, entire or slightly 3-lobed; flowers pinkish-red or white with age; peduncles tendrillar.

Common climber in wastelands and often cultivated. Fl. & Fr.: July–Dec. Native of South America, introduced into India. *Vern.*: Teega batani. V.P. South, *VRK* 3934.

Polygonum Linnaeus

1. Erect glabrous undershrubs, leaves large, lanceolate, flowers in terminal panicles .... *P. glabrum*
1. Prostrate herbs, leaves small, oblong, flowers in axillary fascicles .... *P. plebeium*

Polygonum glabrum Willd. Sp. Pl. 2: 447. 1799; FBI 5: 35. *p.p.*; Gamble 1189 (832).

Erect, glabrous undershrubs, reddish, up to 1.5 m tall; leaves narrowly lanceolate, 10–20 × 1–2.5 cm, glabrous, gland dotted, acuminate, entire, base acute; flowers pink in terminal racemose panicles; nutlets biconvex; dark-brown.

Abundant in Ethipothala stream. Fl. & Fr. Throughout the year. *Vern.*: Burudagogu. Javuku, *VRK* 3581.

Polygonum plebeium R. Br. Prodr. 420. 1810 ('*plebejum*'); FBI 5: 27; Gamble 1188 (832).

Annual prostrate, diffusely branched herbs; leaves oblong or obovate, 0.5–0.8 × 0.15–0.3 cm, thick, glabrous, obtuse or acute, entire, base obtuse, stipules hyaline, flowers rose or pink in axillary fascicles; nutlets 3-gonous, with persistent style, black, shiny.

Dense weed of moist cultivated fields, on canal banks and moist floor of drying tanks. Fl. & Fr.: Throughout the year. V.P. South, *VRK* 6756; Vinukonda, *VRK* 6873.

ARISTOLOCHIACEAE

Aristolochia Linnaeus

1. Climbing shrubs, leaves oblong, 3-nerved from base .... *A. indica*
1. Prostrate herbs, leaves cordate, 5–7-nerved from base .... *A. bracteolata*

Aristolochia bracteolata Lam. Encycl. 1: 258. 1783. *A. bracteata* Retz. Obs. Bot. 5: 29. 1789; FBI 5: 76; Gamble 1201 (841).

Perennial decumbent or trailing herbs, leaves deeply cordate or broadly ovate, 2–6 × 3–6 cm, glabrous above, glabrescent or glaucous below, obtuse or rounded, entire, base cordate; flowers dark-purple, axillary, solitary; capsules oblong-ellipsoid; seeds cordate, partly covered by the remains of the placenta and dotted glands.

Common weed in cultivated dry crops and waste places. Fl. & Fr.: June–Dec. *Vern.*: Gadida-gadapaku. It is used as an anthelmintic and purgative. Decoction of roots is used for expelling round worms. The bruised leaf is mixed with castor-oil and applied to eczema on children's legs. Sattenapally, *VRK* 5864; Nizampatnam, *VRK* 6906. Phirangipuram, *CAB* 4609 (MH).

Aristolochia indica L. Sp. Pl. 960. 1753; FBI 5: 75; Gamble 1202 (841).

Slender, unarmed twiners; leaves oblong, 2–6 × 1.5–2.5 cm, 3-nerved from base, glabrous, entire, obtuse, base truncate to cordate; flowers dark purple, in axillary racemes; capsules 4 x 2 cm; seeds oblong, laterally winged.

Occasional on thickets in scrub jungles. Fl. & Fr.: Oct.–Mar. Plant to be conserved. *Vern.:* Eswari veru. Eng.: Indian birthwort. Dried stems and roots of plant constitute the drug, which is used in small doses. Drug promotes digestion and regulates menstruation. If used in higher doses, it may prove harmful. Also used as tonic, stimulant and for fevers. In moderate doses it is used as gastric stimulant and in dyspepsia. Root is much valued as stimulant, tonic and emmenagogue, and also used in intermittent fevers and bowel troubles of children. Active principle aristolochin identified; also local tradition shows as anti-snake bite. Jayanthipuram, *VRK* 6933.

LAURACEAE

Cassytha Linnaeus

Cassytha filiformis L. Sp. Pl. 35. 1753; FBI 5: 188; Gamble 868.

Leafless twining parasites; flowers white or yellow in axillary and terminal spikes; drupes globose, smooth, white; seeds-1, globose.

Common partial, leafless, yellowish, stem parasite all over the forest on *Mimosa, Prosopis* and *Acacia* and various other species. Fl. & Fr.: Aug.–Mar. *Vern.*: Sithammavari pegulu. Tender shoots contain tannin. Ethipothala Falls, *VRK* 3288; Loyapalli, *SSR & KSM* 21630.

HERNANDIACEAE

Gyrocarpus Jacquin

Gyrocarpus americanus Jacq. Strip. Amer. Hist. 282. t. 178. f. 80. 1763; Gamble 1242 (869). *G. jacquini* Roxb. Pl. Cor. t. 1. 1795; FBI 2: 461.

Medium sized trees, up to 5 m tall; tender parts sparsely pubescent; leaves broadly ovate, 8–14 × 8–12 cm, glabrous above, thinly tomentose below, acute, entire, base truncate; flowers pale yellow or light green in terminal panicles; drupes obovoid-globose, crowned by spathulate wings.

Common in Macherla RF. Fl. & Fr.: Jan.–Apr. *Vern.*: Pavurapukaya, Tella poniki. Wood useful for toys, boxes and in match industry. Javuku, *VRK* 3903.

Plants to be protected and conserved.

LORANTHACEAE

1. Bracts cupular, oblique, partly enclosing calyx lobes .... ***Dendrophthoe***
1. Bracts ovate or oblanceolate, free from calyx-base .... ***Taxillus***

Dendrophthoe Martius

Dendrophthoe falcata (L.f.) Etting in Denkschv. Kaiserl. Akad. Wiss. Math. Naturwiss. Kl. 32: 52–53, 58. t. 13. f. 14. 1872. *Loranthus falcatus* L. f. Suppl. 211. 1781. *L. longiflorus* Desv. var. *falcatus* (L. f.) Kurz. in For. Fl. 2: 321. 1877; FBI 5: 215; Gamble 1253 (877). *L. longiflorus* Desv. var. *amplexifolia* (DC.) Thw. Enum. 134. 1859; Gamble 1253 (877).

Branched, stout, partial stem parasites; leaves variable, elliptic-ovate or lanceolate, 5–14 × 0.5–3 cm, sub-coriaceous, puberulous, apex obtuse, entire, base acute-cuneate, lateral nerves faint; flowers yellow, scented in axillary racemes.

Common in forests and occasional in plains. Partial stem parasite on *Hardwickia binata*. Fl. & Fr.: May–Dec. *Vern.*: Bhajanaka, Badanika. Bollapalle, *VRK* 5749; Macherla RF, *VRK* 3892.

Taxillus Van Tieghem

Taxillus bracteatus (Wall.) Tieghem in Bull. Soc. Bot. Frace 42: 256. 1895.
Loranthus bracteatus Heyne ex Wall. in Roxb. Fl. Ind. 2: 220. 1824; FBI 5: 213; Gamble 1252 (876).

Partial stem parasite; branchlets tawny-tomentose, leves orbicular; 2–4 × 1.5–2 cm, coriaceous, pilose-pubescent, base rounded to cuneate, entire, apex obtuse; flowers greenish-yellow in axillary fascicles.

Occasional in forests. Fl. & Fr.: Apr.–Oct. Devupalli, *SSR & KSM* 21435.

VISCACEAE

Viscum Linnaeus

S.N. Dixit's (in Bull. Bot. Surv. Ind. 4: 49–55, 1962) opinion in treating *Viscum* Linn. under the family, *Viscaceae*, is followed in raising the subfamilial status to a family as propounded earlier by Miers in 1851; Dixit (op. cit.) takes into cognisance floral structure, development of embryo sac, endosperm, embryo, etc.

Viscum articulatum Burm. f. var. **dichotomum** Kurz. in For. Fl. Burma 2: 325. 1877; FBI 5: 226; Gamble 1259 (881).

Deciduous parasitic, leafless, jointed subshrubs, branchlets 2–3-chotomous; lower internodes terete, upper ones flattened, narrowed below, widened above, 3–5 × 0.5–1 cm, yellowish green; flowers green in axillary fascicles; berries ovoid-globose, glossy-white, apex truncate.

Parasite on trees. Fl. & Fr.: Oct.–Feb. *vern.:* Chettu Bajiniki. reported in Guntur Forest Division Work Plan.

SANTALACEAE

Santalum Linnaeus

Santalum album L. Sp. Pl. 349. 1753; FBI 5: 231; Gamble 2: 1261 (883).

Evergreen tree, to 8 m tall; branches slender, drooping; leaves elliptic-ovate, lanceolate, 4–7 × 2.5–4 cm, subcoriaceous, glabrous, apex sub-acute, margin entire, base acute; flowers brownish purple, in paniculate cymes; drupe globose, dark purplish, globose, 8 mm in diam., endocarp hard, ribbed; seed 1.

Occasional, in open dry deciduous forests and scrubs, grows mostly on red, ferruginous loam, overlying metamorpic rocks. It reproduces plentiful from seeds dispersed by birds also to certain extent from root-suckers and also coppices well. Fl.: June–Dec. Fr.: Jan.–Mar. *Vern.*: Tella chandanamu, Srigandham. Wood contains essential oil and is for mainly distillation. Wood powder used as cosmetic and paste applied externally for relieving head-ache. Both oil and wood credited with cooling, diaphoretic, diuretic and expectorant properties. Sandal wood finds several applications in house remedies. Fresh leaves yield a pale yellow wax. Wood is moderately hard to heavy used for burning as incense in temples. The wood is used for a variety of small carved articles, such as small boxes, cabinet panels, walking sticks, combs, picture frames, fan handles, pen holders, card cases, trays and similar articles, Davupalli, *SSR & KSM* 21416.

EUPHORBIACEAE

(Studies from NU, Guntur embraced embryology (Cf. Rao, 1978); leaf architecture 275 species (V.S. Raju, 1981), separation of *Chamaesyce* from *Euphoriba* (Raju & Rao) and ultrastructure of seed surface of *Chamaesyce* (B.V.R. Swamy, 1993, Ph.D.).

1. Plants mangroves, leaves opposite or alternate .... ***Excoecaria***
1. Plants inland, leaves alternate:
    2. Flowers reduced to cyathia with involucre:
        3. Branchlets zig zag, cyathia irregular, scarlet red .... ***Pedilanthus***
        3. Branchlets not zig zag, cyathia regular, greenish .... ***Euphorbia***
    2. Flowers distinct:
        4. Cells of ovary 1-ovuled:
            5. Trees:
                6. Deciduous tree, leaf margin irregularly dentate, base cordate flowers in racemose or paniculate cymes, fruit drupe, seed one .... ***Givotia***
                6. Evergreen tree, leaf margin entire, base cuneate, flowers in spikes, fruit capsular, 3–4 seeded .... ***Mallotus***
            5. Herbs or shrubs:
                7. Filaments inflexed in bud .... ***Croton***
                7. Filaments erect:
                    8. Calyx valvate in male flowers:
                        9. Petals present in male flowers .... ***Chrozophora***
                        9. Petals absent in male flowers:
                            10. Climbing herbs .... ***Tragia***
                            10. Erect herbs or trees:
                                11. Dioecious plants, filament much branched .... ***Homonia***
                                11. Monoecious plants, filament not branched:
                                    12. Stamens 8 or more, anthers verniculiform or linear, ovules on axile placentae, disc absent .... ***Acalypha***

12. Stamens 3–5, anthers not verniculiform, discrete, divaricate, ovules on lateral pendulous placentae, discof linear glands .... *Micrococca*
8. Calyx open or slightly imbricate in male flowers .... *Jatropha*
4. Cells of ovary 2-ovuled:
13. Fruit drupe or fleshy .... *Bridelia*
13. Fruit a capsule or berry:
14. Stamens numerous .... *Dimorphocalyx*
14. Stamens 7–8:
15. Plants dioecious .... *Securinega*
15. Plants monoecious:
16. Tepals biseriate, filaments free .... *Cleistanthus*
16. Tepals 1-seriate, filaments connate:
17. Stamens 5, ovules pendulous .... *Meineckia*
17. Stamens 3, ovules axile, collateral .... *Phyllanthus*

Acalypha Linnaeus

1. Male and female flowers on separate spikes:
2. Herbs, leaves large to 11 cm long, bracts smaller than flowers, stylar branches shorter than ovary .... *A. paniculata*
2. Shrubs, leaves small to 2.5 cm long, bracts larger than flowers, stylar branches longer than ovary .... *A. alnifolia*
1. Male and female flowers on same spike:
3. Bracts larger than capsules........ *A. indica*
3. Bracts smaller than capsules:
4. Bracts glabrous, bordered with gland-tipped hairs; capsules glabrous or nearly so .... *A. lanceolata*
4. Bracts hairy, not bordered with gland-tipped hairs; capsules hispid with bulbous based hairs .... *A. malabarica*

Acalypha alnifolia Klein ex Willd. Sp. Pl. 4: 524. 1805; FBI 5: 415; Gamble 1330 (930).

Erect woody undershrubs, up to 2 m tall; branchlets hispid; leaves elliptic-ovate, 1.5–2 × 1–1.5 cm, hairy, obtuse, crenate-serrate, base cordate; flowers pinkish-brown with yellow tinge, female in terminal short spikes, male in axillary slender elongated spikes; capsule 5 mm across, globose; seeds globose, smooth.

Occasional in forests and on roadsides and cultivated fields. Common along slopes near Mangalagiri Hill Temple, Sagar dam area—GNT Dt side. Fl. & Fr.: Apr.–Sept. *Vern.*: Mirapa buppinta. N. Varam, *VRK* 3518 & 3597; Loyapalli, *SSR & KSM* 21661; Kondavedu, *CAB* 4689 (MH).

Acalypha indica L. Sp. Pl. 1003. 1753; FBI 5: 416; Gamble 1330 (930).

Annual erect herbs, up to 50 cm tall; branches numerous, ascending; leaves broadly ovate, 3–6 × 2–3 cm, membranous, glabrous, petiole longer than leaf-blade, acute or subobtuse, serrate, base cuneate, somewhat 3-nerved, mosaic; flowers green in lax axillary catkins; male terminal, female basal, bracts leafy; capsules hairy; seeds ovoid, brown, shiny.

Common weed in open forests. Fl. & Fr.: May–Dec. *Vern.*: Koppinta, Mura pindaku. Whole plant constitutes drug which is used in asthma, pneumonia, bronchitis and rheumatism. Microbicide and for minor skin ailements. Davupalli, *SSR & KSM* 21612.

Acalypha lanceolata Willd. Sp. Pl. 4: 524. 1805; Gamble 1331 (931). *A. fallax* Muell.-Arg. Linnaea 34: 43. 1865; FBI 5; 416.

Erect herbs, up to 1 m tall; branchlets striate, thinly pubescent; leaves ovate-lanceolate, 3–11 × 2–8 cm, glabrous, crenate to serrate, subacute, base acute to cuneate, 3–5-nerved; flowers unisexual, male in slender, elongate, axillary spikes; female in terminal unbranched spikes; capsule longer than the bracts; seed ovoid, acute, pale yellowish-brown.

Occasional in moist and shady places. Fl. & Fr.: Throughout the year. Guntur, *Venkateswara Rao* 87105 (MH).

Acalypha malabarica Muell.—Arg. in Linnaea 34. 42. 1865; FBI 5: 416; Gamble 1330 (930).

Herb to 40 cm tall; branchlets striate; stem pubescent; leaves rather rigid, elliptic or elliptic-lanceolate, 2.5–4 × 1.5–2.5 cm, subacute, sub-serrate, base usually acute, 5-nerved; petiole slender; flowers in axillary spikes, the males few, terminal, minute, the females numerous; capsule globose, glabrous or thinly hispid at top, seeds smooth, globosely ovoid, pale-brown.

Occasional in moist shady places as forest undergrowth near the foot hills of Macherla forest division.

Acalypha paniculata Miq. Fl. Ned. Ind. 1(2): 406. 1857; FBI 5; 415; Gamble 1330 (930).

Annual erect herbs, up to 1 m tall; branchlets striate, thinly pubescent; leaves broadly ovate, 3–11 × 2–8 cm; glabrous above, hairy beneath, crenate to serrate, acuminate, base truncate; flowers unisexual, male flowers in slender, elongate, axillary spikes, or in panciles; female flowers in terminal fascicles; capsule globose, glandular ciliate.

Occasional in moist, shady places in plains. Fl. & Fr.: Throughout the year. Nagarjunakonda, *KTH* 9842 (CAL).

Bridelia Willdenow

Bridelia montana Willd. Sp. Pl. 4: 978. 1806, FBI 5: 269 *p.p.*; Gamble 1280 (896).

Much branched woody shrubs, up to 4 m tall; branchlets glabrous; leaves obovate, 2.5–7 × 1.5–3 cm, glabrous, obtuse to acute, serrulate to entire, base cuneate; flowers greenish yellow or white in axillary clusters; drupes globose, black.

Occasional in forests. Fl. & Fr.: Aug.–Apr. *Vern.*: Ulinda, Panchavoni. Fruits edible. Jayanthipuram, *VRK* 6944; Kondavedu Fort, *CAB* 4629 (MH).

Chrozophora A.H.L. Jussieu

Chrozophora rottleri (Geis) Juss. ex Spreng. Syst. 3: 850. 1826; FBI 5: 410; Gamble 1317 (921). *Croton rottleri* Geis Crot. Monogr. 54. 1807. *Chrozophora plicata* Hook. f. Fl. Brit. India 5: 409. 1887.

Annual, erect herbs, up to 50 cm tall; branchlets densely hispid or stellate-tomentose; leaves broadly ovate-orbicular to oblong, 3–4 × 6–8 cm, stellate-tomentose, crenulate-crenate, sub-acute, base truncate or cuneate; flowers unisexual, in axillary or terminal bracteate racemes; male crowded at upper part, female at lower part of the raceme; capsules subglobose, reddish-purple when mature.

Abundant weed of cultivated fields in wet and dry lands. Fl. & Fr.: Feb.–June. *Vern.*: Linga mirapa. Guntur (*Uhl*, 1929).

Cleistanthus Hooker f. ex Planchon

Cleistanthus collinus (Roxb.) Benth. ex Hook. f. in Fl. Brit. India 5: 274. 1887; Gamble 1282 (897). *Cluytia collina* Roxb. Pl. Cor. t. 169. 1802.

Small trees, up to 10 m tall; branchlets terete; leaves elliptic-ovate, 3–10 × 1.5–5.8 cm, glabrous, rounded, often retuse at apex, entire, base obtuse; flowers pale green in axillary silky clusters; capsules globose, 3-valved, dark brown, shining; seeds 3, globose, shiny.

Common in forests. Fl. & Fr.: Mar.–Oct. *Vern.:* Vodisa, Kodisa. The leaf, the root and the fruits are irritants, and they are used as fish poison. Tender leaves are deadly poisonous. Wood is durable, used for house posts, fences etc. Vinukonda RF, *VRK* 6845, 6902, 6947.

Croton Linnaeus

Croton bonplandianum Baill. in Adansonia 4: 339. 1864. *Croton sparsiflorus* Morong. Ann. N.Y. Acad. Sci. 7: 221. 1893; Gamble 1316 (920).

Annual, erect, diffuse, much-branched herbs, up to 50 cm tall; stem with viscous greyish latex; leaves ovate-lanceolate, 3–5.5 × 1–2 cm, glandular, acute, serrulate, base acute, alternate below and whorled above; flowers yellowish-white in terminal androgynous spikes; male above and female below; capsules 3-angled, stellate-hairy; seeds with caruncle.

Abundant weed of open forests and plains. Native of South America, introduced. Fl. & Fr.: Throughout the year. *Vern.*: Kukka mirapa. Koppanur, *VRK* 3256; Vijayapuri, *K.M. Sabastine*, 9830 (MH).

Dimorphocalyx Thwaites

Dimorphocalyx glabellus Thw. Enum. Pl. Zeyl. 278. 1861; FBI 5: 403; Gamble 2: 1337 (934).

Evergreen tree, 5–10 m tall; bark reddish-brown, deeply furrowed, spongy; wood red; branchlets striate, glabrous, whitish; leaves elliptic or elliptic-oblong, 3–15 × 2–6 cm, glabrous, apex obtuse or acute, margin entire, base rounded; flowers white, solitary or few, male ones axillary, 1-few in branched cymes; female flowers axillary or terminal, 1-few in umbellate; capsule 1–1.5 cm long, subglobose, slightly hairy, scarcely lobed; seeds ovoid, 3.

Occaisonal, in semievergreen forests and scrubs near the sea. Fl.: Feb.–Aug. Fr.: Aug.–Dec. Davupalli, *SSR & KSM* 21429.

Euphorbia Linnaeus

1. Plants armed, deciduous:
   2. Flowering twigs leafless, branchlets angled with leafscars .... *E. antiquorum*
   2. Flowering twigs leafy, branchlets terete with leafscars:
      3. Subshrubs, less than 1 m tall:
         4. Cyathia scarlet-yellow, showy in cymes, leaves spathulate .... *E. milli*
         4. Cyathia pinkish-green or purplish brown in axillary clusters, leaves oblong .... *E. caducifolia*
      3. Trees, up to 8 m tall .... *E. nivulia*
1. Plant unarmed, leafy:
   5. Shrubs, branchlets articulate, subsucculent .... *E. tirucalli*
   5. Herbs, branchlets not so:
      6. Leaves alternate below, opposite above, estipulate:
         7. Upper leaves entire, uniformly green, glands cupular .... *E. heterophylla*
         7. Upper leaves sinuate, yellow to red at base, glands bilabiate .... *E. cyathophora*
      6. Leaves opposite, decussate or distichous, stipulate:
         8. Leaves broadly oblong-elliptic, erect or ascending herbs:
            9. Appendage of glands aciculate, floral leaves distichously imbricate .... *E. elegans*
            9. Appendage of glands entire, floral leaves distantly decussate:
               10. Erect herbs, cyathia stalked .... *E. indica*
               10. Ascending herbs, cyathia subsessile .... *E. hirta*
         8. Leaves elliptic-oblong, spreading herbs:
            11. Cyathia axillary in subterminal clusters, appendages of glands conspicuous, male florets 4–10 .... *E. thymifolia*
            11. Cyathia axillary, solitary or a few, appendages of glands inconspicuous, male florets 1–4 .... *E. heyneana*

Euphorbia antiquorum L. Sp. Pl. 450. 1753; FBI 5: 255; Gamble 1277 (894).

Armed succulent leafless shrubs with 3–5 angular branches, trunk stout, cylindric or fluted, latex milky; spines up to 1 cm long, divariate, sharp;

leaves small, obovate-oblong, fleshy, entire, cuneate, deciduous early; cyathia reddish brown in axillary lax cymes; capsules trigonous.

Abundant in forests. Often cultivated for hedges. Fl. & Fr.: Dec.–Mar. *Vern.*: Bomma jemudu. The plant is considered to be purgative, digestive, pungent and bitter. The juice of the plant is used as an irritant in rheumatism. It is popular application for warts and other cutaneous affections. The latex is used as fish poison. Javuku, *VRK* 3916; Loyapalle, *VRK* 3816; Mellavagu, *VRK* 5847.

Euphorbia caducifolia Haines in Ind. For. 11: 54. 1922; Gamble 1275 (893).

Succulent much-branched, thorny shrubs, up to 2 m tall; latex milky; branches cylindrical with stipular, divaricate persistent sharp spines; leaves very small, oblong, deciduous; cyathia pinkish-green or purplish brown in axillary clusters; capsule 1 cm across, trigonous, keeled; seeds 2 mm, smooth.

Abundant in scrub jungles and occasional in plains. Fl. & Fr.: Dec.–Apr. V.P. South, *VRK* 3954.

Euphorbia cyathophora Murr. in Commentat. Soc. Regiae. Scio. Gott. 7: 81. 1786. *Poinsettia cyathophora* (Murr.) Klotzch & Garcke Monastber. Deutsch. Akad. Wiss. Berlin 1859; 253. 1859. *Euphorbia heterophylla* auct. non L., Gamble 1278 (894).

Shrub to 2 m; dichotomously branched; branchlets terete; striate, leaves alternate below, decussate to whorled above; obovate-oblanceolate or panduriform, 5–8 × 1–4 cm, base acute to attenute, entire to sinuate, apex acute; floral leaves yellow to red at the basal half; cyathia 12–15 in terminal clusters; capsule 4 mm across, seed oblong-globose, 3 mm long.

Occasional weed in moist places. (Mohan Rao and Aasiya, 1995). Fl. & Fr.: June–Jan.

Euphorbia elegans Spreng. in L. Syst. Veg. (ed. 16) 3: 794. 1826. FBI 5: 246; Gamble 1274 (891).

Ascending herbs up to 30 cm tall, dichotomously branched; leaves decussate, broadly elliptic-oblong, or 1–25 × 0.5–1.5 cm; 3-nerved from base, softly hairy inside, entire to serrulate, obtuse, base truncate to cordate; cyathia 1–3, axillary, within the distichously imbricate leaves; capsules 3 mm across, pubescent; seeds yellowish-brown, 4-angled, prominently furrowed, glabrous.

Occasional in forests. Fl. & Fr.: Aug.–Mar. Nagarjunasagar, *KMS* 9758 (MH).

Euphorbia heterophylla L. Sp. Pl. 453. 1753. *Euphorbia geniculata* Ort. Nov. Rar. Pl. Hort. Matrit. Dec. 18. 1797; FBI 5: 239; Gamble 1275 (893).

Annual erect stout herbs; up to 1 m tall; latex milky; leaves elliptic-ovate, or obovate, 3–8 × 2–4 cm, glabrescent, acute, serrate, base acute or narrowed; cyathia greenish-white or greenish-yellow in terminal clusters.

Occasional weed of wastelands, hedges and cultivated fields. Fl. & Fr.: Jan.–July. Mellavagu, *VRK* 5730.

Euphorbia heyneana Spreng. in L. Syst. Veg. (ed. 16) 3: 791. 1826. *E. microphylla* Heyne ex Roth Nov. Pl. Sp. 229. 1821. non Lam. 1788; FBI 5: 252 *p,p,* Gamble 1275 (893) *p.p.*

Glabrous spreading herbs, hispid with long often yellowish hispid hairs, leaves opposite, elliptic or oblong or obovate to lanceolate, 0.5–1 × 0.3–0.5 cm; glabrous, dark green above, pale beneath, acute to obtuse, entire, base obtuse; cyathia pink in axillary clusters; capsules 3-lobed, glabrous; seeds 4-angled.

Abundant weed in cultivated fields and wet places in open forests. Fl. & Fr.: Jan.–Sept. V.P. South, *VRK* 5786, 6841; Davupalli, *SSR* & *KSM* 21434.

Euphorbia hirta L. Sp. Pl. 454. 1753; Gamble 1275 (892). *Euphorbia pilulifera* Hook. f. Fl. Brit. India 5: 250. 1889.

Annual erect or ascending hispid herbs, up to 60 cm tall; latex milky; leaves elliptic-ovate, 2.5–4 × 1.5–2 cm, reddish, shining above, white villous below, acute to obtuse, serrate, base unequal; cyathia greenish red in axillary and terminal crowded cymes.

Common weed in open forests and roadsides. Fl. & Fr.: Throughout the year. *Vern.*: Alumu. Reddivarinanubalu. In respiratory disorders often called the Australian Asthma plant. The leaves and roots are used in medicine for curing dysentry. N. Varam, *VRK* 3224; Nagarjunasagar, *KMS* 9726 (MH).

Euphorbia indica Lam. Encycl. Method. Bot. 2: 423. 1786; Raju & Rao in Ind. J. Bot. 2: 202–208. 1979. *E. hypericifolia* L. Sp. Pl. 454. 1753; FBI 5: 249; Gamble 1275 (892).

Annual erect or spreading glabrescent herbs with milky latex, up to 60 cm tall; leaves broadly oblong, 1.5–2.5 × 0.8–1.5 cm, obtuse, serrulate, base unequal; cyathia pink or greenish white with red tinge in axillary and terminal lax cymes; capsule sub-globose, hairy, seeds smooth, 1.5 cm long.

Common weed of cultivated fields. Fl. & Fr.: Throughout the year. Mellavagu, *VRK* 5733; Sattenapally, *VRK* 5862.

Euphorbia milli Des. Moul., Bull. Hist. Nat. Soc. Linn. Bordeaux 1: 27. t. 1. 1826. *E. splendens* Boj. ex Hook., Bot. Mag. t. 2902 1829; Gamble 1278 (894).

Shrub, fleshy, branches radiating;stem greenish, spinous; leaves alternate, small, spathulate, glabrous, mucronate, entire, flowers scarlet-yellow, showy in cymes; fruit capsule.

Native of Madagascar, running wild and cultivated in gardens (Mohan Rao and Aasiya, 1995). Fl. & Fr.: Jan.–Apr.

Euphorbia nivulia Buch.-Ham. in Trans. Linn. Soc. London 14: 286. 1824; FBI 5: 255; Gamble 1277 (893).

Armed deciduous, succulent trees, up to 6 m tall; stems cylindrical; with stipular, well developed, divaricate, persistent, sharp spines, latex milky; leaves linear-oblong or oblanceolate, 10–15 × 2.5–5 cm, fleshy, obtuse, base cuneate; cyathia greenish-red in sub-terminal lax cymes; capsules trigonous, 6 x 8 mm, glabrous; seed 4-angular, smooth.

Occasional in forests. Fl. & Fr.: Dec.–Mar. *Vern.*: Akujamudu, Erra Kalli. The leaf juice is purgative and diuretic. It is said to be used for earache when mixed with neem oil. It is applied externally in rheumatism. The bark is suitable for making composite cork boards with *Hevea* rubber latex as binder. It is also used for plastering the handles of sickles. Loyapalle, *VRK* 3819.

Euphorhia thymifolia L. Sp. Pl. 454. 1753; FBI 5: 252; Gamble 1276 (893). *Chamaesyce thymifolia* (L.) Millsp. in Pub. Field Columb. Bot. Mus. Ser. 2: 412. 1916.

Annual prostrate herb, dichotomously branched; leaves opposite, very small, numerous, oblong, 3–7 × 1.5–4 mm, glabrous above, glaucous and usually slightly pubescent beneath, apex rounded, entire, base unequal sided; petiole short; cyathia solitary or 2 to 3 in axils; capsule 1.5 mm, obtusely keeled, pubescent; seeds quadrangular, bluntly pointed, 5–6-furrowed.

Common in plains and lower hills (Mohan Rao and Aasiya, 1995) Fl. & Fr.: Aug.–Mar. *Vern.*: Erra-chitti payalaku. It is an astringent, anthelmintic, stimulant, laxative and antidote to snake-bite, and for bowel complaints and ringworm.

Euphorbia tirucalli L. Sp. Pl. 452. 1753; FBI 5: 254; Gamble 1276 (893).

Succulent shrub or small tree, to 5 m tall, branchlets terete, glabrous, whorled or fascicled, articulate, leaves linear-oblong, deciduous, 3–12 × 1–2 mm, apex obtuse to subacute, entire, cuneate; cyathia few, terminal; capsule subglobose, glabrous, smooth; seed ovoid, 3–4 mm across, smooth, carunculate.

Introduced in field hedges and roadsides (Mohan Rao and Aasiya, 1995). Fl.: Jan.–Mar. Fr.: Apr.–May. *Vern.*: Manchi Jemudu. Wood is said to yield good charcoal powder.

Native of Africa and usually seen as hedge plant.

Excoecaria Linnaeus

Excoecaria agallocha L. Syst. Nat. (ed. 10) 2: 288. 1759. et Sp. Pl. (ed. 2) 1451. 1763; FBI 5: 472; Gamble 1344 (941).

Small, poisonous evergreen, trees or large shrubs, with white highly acrid juice, up to 5 m tall; latex milky; bark with prominent lenticels; leaves alternate or opposite, ovate-elliptic, 4–8.5 × 2.5–4 cm, glabrous, shortly acuminate at apex; entire or shallowly sinuate-crenate, male flowers in catkin like spikes; female flowers in spikes or racemes; capsules smooth, deeply 3-lobed; seeds subglobose, smooth.

Common in tidal saline sandy soils in coasts. Fl. & Fr.: May–Aug. *Vern.*:Tilla. The juice of the plant boiled in oil and applied in rheumatism, leprosy and paralysis. The bark is emetic and purgative. The latex is used as fish poison. Kothapalem, *VRK* 5763, 5775.

The tree is sometimes called 'The Blinding tree'. The juice is extremely acrid, producing blisters on the skin. It hardens into a kind of a black caoutchouc.

Givotia W. Griffith

Givotia moluccana (L.) Sreem. in Taxon 24: 696. 1975. *Croton moluccanum* L. Sp. Pl. 1005. 1753 excl. ref. Burman. *Givotia rottleriformis* Griff. Calcutta J. Nat. Hist. 4: 388. 1844; FBI 5: 395; Gamble 1342 (938); Smith in Kew Bull. 22: 504. 1968.

Tomentose, unarmed, deciduous trees, up to 15 m tall; bark smooth, peeling off in circular thick bosses; latex red; leaves alternate, broadly ovate, 6–18 × 5–20 cm, coriaceous, glabrous above, yellowish tomentose below, 5-nerved from base, coarsely-dentate, shortly acuminate, base subcordate, 5-nerved; flowers unisexual, yellow in lax panicles; drupes subglobose or ellipsoid, to 3 × 2.3 cm, brown; seeds globose or ellipsoid, hairy.

Common in dry deciduous forests. Fl. & Fr.: Apr.–Aug. *Vern.*: Tella poliki. Wood light and used for toys. Seeds are source of lubricating oil. Kondadavedu fort, *SSR & KSM* 21479; Kondaveedu Fort, *CAB* 17499 (MH).

Note: Sreemadhavan (l.c.) has argued for reinstating the Linnean epithet, *moluccaunum* for this species which Smith (l.c.) earlier has considered as *nomen. confusum*.

Homonoia Loureiro

Homonoia retusa (J. Graham) Muell.-Arg. Linnaea 34: 200. 1865; FBI 5: 456; Gamble 1333 (933). *Adelia retusa* J. Graham, Cat. Pl. Bombay 186. 1839.

Dioecious shrubs, up to 3 m tall; branchlets stout, woody; leaves obovate to oblanceolate, 2–4 × 1–2 cm, coriaceous, margin serrate to denate, apex obtuse to retuse, acute-cuneate at base, flowers in axillary spikes; capsules 3.5 mm across, globose; seed rounded on the back, smooth, yellow.

Common shrub in Krishna river bed. Fl. & Fr.: Jan.–Mar. *Vern.*: Nilla Jeni. Wood soft, very light used for carved images, toys, lacqured articles and catamarans. The seeds give an oil. Krishna river, *VRK* 6748.

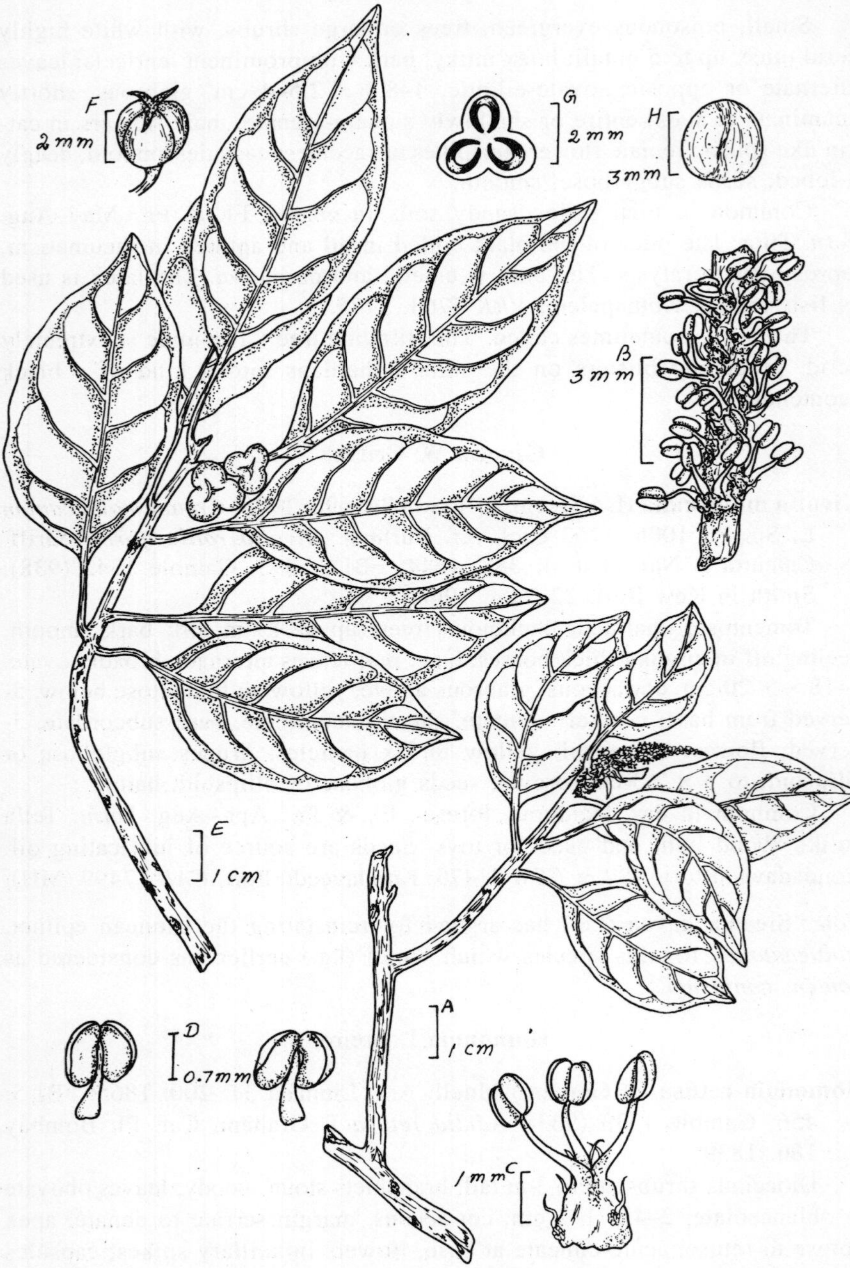

Fig. 24. *Micrococca mercurialis* (L.) Benth.
A. Male twig, B. Spike (part), C. Flower, D. Stamens, E. Female twig, F. Flower, G. Ovary T.S., H. Seed

Jatropha Linnaeus

1. Leaves entire or divided from above the middle, shrubs:
   2. Leaves obovate-oblanceolate, entire, acuminate, flowers pink in terminal racemes .... ***J. integerrima***
   2. Leaves cordiform, shallowly 5-lobed, acute, flowers green in axillary cymes . . . . ***J. curcas***
1. Leaves divided to below the middle; sub shrubs:
   3. Leaf margin entire, glandular-hairy .... ***J. gossypifolia***
   3. Leaf margin serrate, serrations gland tipped .... ***J. glandulifera***

(E.S. Rao 1996. Ph.D. Thesis, Kakativa University on Systematics of *Jatropha* genus provided update seed data, SEM data, chemical data, nomenclature and identity of species).

Jatropha curcas L. Sp. Pl. 1006. 1753; FBI 5: 383; Gamble 1340 (937).

Shrubs, up to 1.5 m tall; stem greyish; bark greenish-white, smooth, peeling off in thin flakes; leaves 10–12 cm across, entire or 3–5-lobed; cordate, apex acute, base truncate to cordate; flowers greenish-yellow in axillary long peduncled panicled cymes; capsules 2–2.5 cm long, subglobose.

Common weed on roadsides and in hedges around cultivated fields. Fl. & Fr.: May–July. Native of tropical America. Eng.: Physic nut. *Vern.*: Nepalamu. Seed oil is used for manufacturing candles, soap and as lubricant and for illumination; also used as purgative. Tender twigs are reported to be used for cleaning teeth. Native of tropical America and naturalised in India. Nizampatnam, *VRK* 6800.

Jatropha glandulifera Roxb. Fl. Ind. 3: 688; 1832; FBI 5: 382; Gamble 1340 (937).

Shrubs up to 2 m tall; latex watery; stems grey-white; leaves deeply 3–5-lobed, 4–9 × 6–10 cm, lobes obovate, 3–7 × 1.5 cm, acuminate, serrate, gland-tipped; stipules branched, filiform; flowers yellowish green in terminal cymes.

Common in waste places, roadsides. Fl. & Fr.: Throughout the year. *Vern.:* Dundagachettu. Latex for dental diseases. Seed oil as purgative, externally in rheumatism and paralytic affection; root pounded with water is recommended to children suffering from abdominal enlargement. Kothapalem, *VRK* 5774. AIR Relay Station, Nambur.

Jatropha gossypifolia L. Sp. Pl. 1006. 1753; FBI 5: 383; Gamble 1340 (937).

Erect subshrubs, up to 3 m tall; branchlets glandular-hairy; leaves deeply 3–5-lobed, lobes obovate, 7.5–13 × 6–15 cm, base cordate, margin glandular hairy, acute; flowers unisexual in axillary cymes; capsules 3-lobed; seeds oblong, 3-gonous.

Common in plains and open forests. Fl. & Fr.: Throughout the year. *Vern.:* Seema-Nepalamu. Native of Brazil and naturalised in India. Leafy extract is used for eczema and itches. Vinukonda, *VRK* 6997; NU campus, *SSR* & *KSM* 21492.

Jatropha integerrima Jac. Select. Am. 256. 1763. *J. pandurifolia* Andr. Bot. Repos. 4: t. 267. 1802.

Erect unarmed shrubs, up to 3 m tall; leaves obovate-oblanceolate, 5–7 × 2–3 cm, base cuneate, entire, acuminate, glabrous above, glabrescent below; flowers pink in terminal racemes; capsules trigonous; seeds oblong with arils.

Occasional ornamental shrub and rarely as an escape in forests. Fl. & Fr.: Throughout the year. Ethipothala, *VRK* 3579.

Mallotus Loureiro

Mallotus philippensis (Lam.) Muell.—Arg. in Linnaea 34: 196. 1865; FBI 5: 5: 442; Gamble 1322 (924). *Croton philippensis* Lam. Encycl. 2: 206. 1786.

Evergreen tree, up to 12 m tall; bark grey, nearly smooth, exfoliating in small woody flakes, branchlets striate, densely grey-tomentose; leaves elliptic-ovate or ovate-lanceolate, 4–19 × 2–10 cm, glabrous above, pubescent and numerous red orbicular glands beneath, apex acuminate, margin entire, base cuneate; flowers pale yellow in axillary or terminal spikes; capsule 3-lobed, red glandular, 1 cm across, smooth; seeds globose, black, 4 mm across.

Common, on hills of deciduous forests and open scrub jungles. It is good soil improver, sensitive to drought, coppices well and produces root suckers. Fl. & Fr.: Oct.–Mar. *Vern.*: Kumkumachettu, Sindhuramu, Nagarakta. The red dye known as "Kamala dye" obtained from the red glands on the fruit extensively used for dyeing wool and silk. Kamala dye has been used for colouring food stuffs and beverages. The fruit is bitter, anthelmiatic and styptic, used in external applications for parasitic affections of the skin. The wood of no value as timber but forms a good fuel, which is suitable for rafters, tool handles, match boxes etc. Guntakaramma bavi, *DMR* 18198.

Meineckia (Baillon)

Meineckia parviflora (Wight) Webster in Acta Bot. Neerl. 14: 342. 1965. *Peltandra parvifolia* Wight Ic. t. 1892. 1852. *Neopeltandra suberosa* (Muell.-Arg.) Gamble Fl. Pres. Madras 2: 1286 (900) 1925. *Phyllanthus suberosus* Muell.-Arg. in Linnaea 32: 10. 1863; FBI 5: 287.

Herb to 40 cm tall, monoecious, branchlets zig-zag, quadrangular, glabrous, leaves 1.5–4.8 × 1–3 cm, ovate, acute-rounded at base, entire, acute; flowers axillary, solitary, long pedicellate; capsule 5 mm, depressed globose; seeds pitted.

Occasional, in the hills and scrub jungles as forest undergrowth. Fl. & Fr.: Jan.-Mar. Kondavedu, *SSR & KSM* 21462; Mellavagu, *SKW* 3695; 3696 (BLAT).

Micrococca Bentham

Micrococca mercurialis (L.) Benth. in Hooker Niger. Flora 503. 1849; Gamble 1328 (928). *Tragia mercurialis* L. Sp. Pl. (ed. 3) 1391. 1764. *Claoxylon mercuriale* Thw. Enum. Pl. Zeyl. 271. 1861; FBI 5: 412.

Erect annual herbs, up to 40 cm tall; branchlets pubescent, leaves simple, ovate-lanceolate, 3–5 × 1.5–3 cm; crenate-serrate, acute to acuminate, mosaic; flowers green in thin axillary spikes or racemes, up to 8 cm; capsules hairy, 3-lobed.

Common along roadsides and wastelands. Fl. & Fr.: Throughout the year. Mellavagu, *VRK* 5736.

Pedilanthus Necker ex Poitean

Pedilanthus tithymaloides (L.) Poit. in Ann. Mus. Natl. Hist. Nat. 19. 390 t. 19. 1812; FBI 5: 239; Gamble 1346 (942). *Euphorbia tithymaloides* L. Sp. Pl. 453. 1753.

Erect or scandent subshrubs, up to 1 m tall; latex milky; branchlets zigzag, succulent; leaves alternate; distichous, elliptic-ovate, 5–7 × 3–5 cm, variegated, acute, subsessile, base attenuate; cyathia stalked, terminal, cymose, subtended by petaloid, lanceolate bracts.

Occasional in hedges. Fl. & Fr.: Throughout the year. *Vern.:* Kanchipala. Root is powerful emetic. J.M.J. College, Tenali, *SSR* 21649.

Phyllanthus Linnaeus

1. Shrubs or trees:
    2. Shrubs, berries dark blue .... ***P. reticulatus***
    2. Trees, capsules/drupes green:
        3. Capsules dry, dehiscent, less than 0.5 cm across .... ***P. polyphyllus***
        3. Drupes fleshy, indehiscent, more than 1.5 cm across .... ***P. emblica***
1. Herbs or subshrubs:
    4. Leaves elliptic-ovate, > (0.8) 1 cm across, disc (in male flowers) of star shaped glands .... ***P. deblis***
    4. Leaves oblong-lanceolate (elliptic-orbicular, *P. pinnatus*), < 5–7 mm across, disc (in male flowers) annular or of lobed glands:
        5. Tepals 5, disc-glands 5, anthers dehiscence transverse .... ***P. amarus***
        5. Tepals 6, disc-glands 6, anthers dehiscence vertical:
            6. Anthers didymous or reniform, slit very short, stipules peltate, sub sagilttate, leaf base slightly cordate .... ***P. gardnerianus***
            6. Other than above:
                7. Leaf base cuneate, capsules smooth:

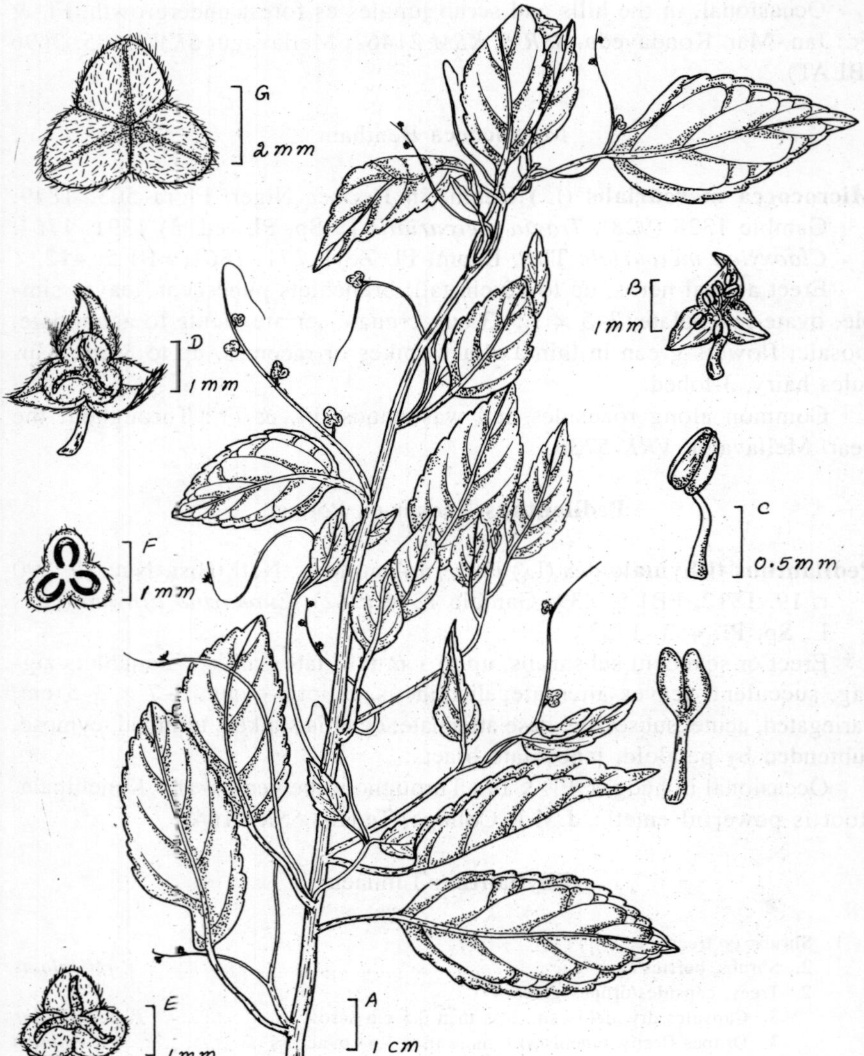

Fig. 25. *Micrococca mercurialis* (L.) Benth.
A. Twig, B. Male flower, C. Stamens, D. Female flower, E. Pistil, F. Ovary T.S., G. Capsule

8. Subshrubby, leaf-apex acute, stamens 6, free .... ***P. pinnatus***
8. Herbs, leaf-apex retuse or obtuse, stamens 3, connate .... ***P. maderaspatensis***
7. Leaf base truncate or obtuse, capsules verrucose .... ***P. virgatus***

Phyllanthus amarus Schum. & Thonn. in Kongl. Danske Vidensk. Selskr. 4: 195. 1829. *Phyllanthus niruri* auct. non L. 1753; FBI 5: 298; Gamble 1290 (903).

Annual, erect, glabrous herbs, up to 40 cm tall; branches resembling pinnate leaves; leaves elliptic-oblong, 0.4–1 × 0.3–0.6 cm, glaucous below, obtuse, entire, base obtuse; flowers unisexual, greenish, axillary, petals usually 5, free; capsules smooth, dry.

Common weed in open forests. Fl. & Fr.: July–Feb. *Vern.*: Nela usiri. Koppanur, *VRK* 3266.

P. Kantikumari (1986) M.Phil., NU, studied phenoplasticity and ecophysiology of this ruderal medicinal weed from Guntur populations.

Infusion of plant known since decades as antihepatitis (jaundice). Chemical analysis was thoroughly done in India between 1930–1970. The active principle as specific remedy for Hepatiitis was registered as patent in US which is challenged. The history of this emphasises rethinking on Indian Patenting Act of Biological products.

Phyllanthus deblis Klein ex Willd. Sp. Pl. 4: 582–83. 1804; FBI 3: 299; Gamble 1290 (903).

An erect very slender herb, to 60 cm tall; stem terete; leaves elliptic-ovate, 1–2 × 0.8–1.2 cm, glaucous below, base acute to rounded, apex broadly acute; flowers solitary, or in axillary clusters, greenish; capsule 3-valved, 4 mm across, smooth; seeds ribbed.

Common in cultivated fields, on lower hill slopes and in waste lands. Fl. & Fr.: July–Dec. Jayanthipuram, *VRK* 6967.

Phyllanthus emblica L. Sp. Pl. 982. 1753; FBI 5: 289. *Emblica officinalis* Gaertn. Fruct. 2: 122. t. 108. 1790; Gamble 1295 (906).

Deciduous, monoecious trees, up to 10 m tall; branches spreading; leaves distichous, pinnate, linear-oblong, 1–2 × 0.2–0.4 cm, glabrous, obtuse, entire, base unequal; flowers yellow in axillary fasicles, on leaf bearing branchlets; drupes indehiscent, depressed globose, yellowish-green, juicy, 3-celled; seeds reniform, 3-gonous, shining, reddish-brown.

Rare in V.P. South section and uncommon in rest. Fl. & Fr.: July–Mar. *Vern.*: Racha usiri.It is a light demander and sensitive to drought. Ingredients of Triphala churna. Enormous biodiversity occurs at enzyme levels. Fruits acidic, cooling, refrigerent, diuretic and laxative contains much gallic acid and are made into pickles; wood hard, heavy, durable. Under water, useful for poles, agricultural implements and furniture; it yields excellent charcoal and fire wood. N. Varam, *VRK* 3964.

Phyllanthus gardnerianus (Wight) Bail. Etudi. Gen. Euphorb. 628. 1853; Gamble 1289 (902). *Macraea gardneriana* Wt. Ic. t. 1902. f. 3. 1852. *Phyllanthus simplex* Retz. var *gardneriana* (Wt.) Muell.–Arg. in Linnaer 32: 32. 1863; FBI 5: 295.

Under shrub, monoecious; branches slender, long; leaves elliptic-oblong, slightly cordate at base, revolute, obtuse, subsessile, flowers axillary, males in fascicles with short pedicel; female solitary with long filiform pedicels, capsule depressed globose, minutely verrucose; seeds minutely tubercled.

In crevices of shaded hard rocks of Nagarjuna University campus (Mohan Rao and Aasiya, 1995). Fl. & Fr.: June–Nov.

Phyllanthus maderaspatensis L. Sp. Pl. 982. 1753; FBI 5: 292: Gamble 1289 (902).

Erect glabrous herbs, up to 75 cm tall; leaves linear-obovate, 1–1.8 × 0.5–1 cm, base cuneate; flowers greenish-yellow, axillary, male flowers above, clustered; female flowers below, solitary; capsules depressed-globose, smooth.

Common weed of open moist forests and plains. Fl. & Fr.: Throughout the year. *Vern.*: Nela usiri. An infusion of leaves is used for headache. Seeds possess laxative, carminative and diuretic properties. Koppanur, *VRK* 3279; V.P. South, *VRK* 5832; Phirangipuram, *CAB* 4611 (MH).

Phyllanthus pinnatus (Wight) Webster in J. Arnold Arbor. 38: 52. 1957. *Chorisandra pinnata* Wight, Icon. Pl. Ind. Orient. t. 1994; Gamble 1285 (899).

Subshrubs up to 5 m tall; branchlets glaucous; leaves elliptic to orbicular, 1–1.5 × 0.5–0.7 cm, acute, entire, base rounded; flowers unisexual, male more in number than female, in axillary or terminal clusters, female fewer long stalked; capsules depressed-globose of 3, bivalved cocci.

Occasional in forests. Fl. & Fr.: June–Feb. Kondaveedu Fort, *CAB* 1902 (MH).

The authors opine that this plant deserves a separate generic status due to free stamens and peculiar seed structure.

Phyllanthus polyphyllus Willd. Sp. Pl. 4: 586. 1805; FBI 5: 290; Gamble 1288 (901).

Deciduous armed shrubs to small trees, up to 5 m tall; leaves pinnate, 0.5–2 × 0.2–0.5 cm, glabrescent, acute, entire, base truncate; flowers green in axillary clusters; capsules 3-valved.

Occasional in scrub forests. Fl. & Fr.: July–Mar. *Vern.:* Pureedu usiri. It can be used for tanning. Satrasala, *VRK* 6777.

Phyllanthus reticulatus Poir. in Lam. Encycl. 5: 298. 1804; FBI 5: 288. *Kirganelia reticulata* (Poir.) Baill. Etudes Gen. Euph. 613. 1874; Gamble 1294 (905).

Large bush like glabrous shrubs, up to 4 m tall; bark thin, greyish-brown; leaves elliptic-oblong, 2–3 × 1–1.5 cm, acute, entire; flowers male purple or green with pink tinge in axillary fascicles; female solitary; berry globose, dark purple when ripe.

Occasional in hedges and forests. Fl. & Fr.: May–Mar. *Vern.*: Poolichettu, Nallapulugudu. Roots are source of red dye, and leaves are used in medicines as diuretic and cooling. Juice of leaves mixed with cubebs and camphor, is used for bleeding gums, also for diarrhoea in infants. Decoction of root is given to children for cough and catarrh. N. Varam, *VRK* 3600; Koppanur, *VRK* 3265; Nizampatnam, *VRK* 6788; Loyapalli, *SSR & KSM* 21637; Kondavedu, *CAB* 4650 (MH).

Phyllanthus virgatus Forst. f. Fl. Ins. Prodr. 65. 1786. *P. simplex* Retz. Obs. Bot. 5: 29. 1788; Gamble 1289 (902).

Erect woody herbs, with reddish stout root stock, up to 50 cm tall; leaves oblong-lanceolate, 1.5–2.5 × 0.3–0.6 cm, often reddish, obtuse, apiculate, entire, base truncate; flowers pale white or yellow, axillary; capsules globular.

Common weed in forests and plains. Fl. & Fr.: Throughout the year. *Vern.*: Nela usiri. Fresh leaves bruised and mixed with butter milk are used as wash for itch in children. Roots used as external application of mammary abscesses. Leafy juice is remedy for bleeding gums. Javuku, *VRK* 3571; Achempeta, *VRK* 5869; Kondavedu, *CAB* 4641 (MH).

Securinega Commerson ex A.L. Jussieu

1. Armed shrubs, capsules white .... ***S. leucopyrus***
1. Unarmed shrubs, capsules grey, ... ***S. virosa***

Securinega leucopyrus (Willd.) Muell.-Arg. in DC. Prodr. 15: 451. 1866. *Fluggea leucopyrus* Willd. Sp. Pl. 4: 757. 1806; FBI 5: 328; Gamble 1292 (907).

Erect shrubs with straggling greyish branches, branchlets stiff, ending with spines, bearing leaves and flowers; leaves obovate, 1–2.5 × 0.5–1.5 cm, pubescent and glaucous below, obtuse, retuse or apiculate; cuneate, flowers greenish in axillary clusters; capsules globose, white.

Occasional in scrubs and roadsides. Fl. & Fr.: June–Oct. *Vern.*: Tella pulugudu. Fruits are edible, and paste of leaves combined with tobacco is used to kill worms in sores. It is said to make an excellent hedge. N. Varam, *VRK* 3955; Bollapalle, *VRK* 5743; Ethipothala, *VRK* 5798; Satrasala, *VRK* 6784; Kondavedu Fort, *CAB* 17489 (MH).

Securinega virosa (Roxb. ex Willd.) Baill. in Adansonia 6: 334. 1866. emend. Pax & Hoffm. in Engler Nat. Pfl. Fam. (ed. 2) 190: 60. 1931. *Phyllanthus virosus* Roxb. ex Willd. Sp. Pl. 4: 578. 1805. *Fluggea virosa* (Roxb. ex Willd.) Baill. Etude Euphorb. 593. 1858; Gamble 1296 (907). *F. microcarpa* Blume Bijdr. 580. 1820; FBI 5: 328.

Erect, much branched shrubs to small trees, up to 3 m tall; leaves variable in shape, elliptic-ovate to obovate, 1–5 × 1–4 cm, thinly pubescent, obtuse or acute, entire, base cuneate-acute; flowers pale yellow, scented in axillary clusters; capsules globose.

Occasional in scrub forests. Fl. & Fr.: June–Oct. *Vern.*: Tella-pulcheri. Alakolids such as securinine and virosecurinine have been reported to occur in different parts of plant. Leaf juice in combination with tobacco is used to kill worms in sores. Ripe fruits are edible, and species is said to make an excellent hedge. Mellavagu, *VRK* 5735.

Tragia Linnaeus

Tragia plukenetii R. Sm. in Kew Bull. 37: 688. 1983. *T. cannabina* L. f. Suppl. Pl. 415. 1781. *nom. illeg.;* Gamble 1333 (932). *T. involucrata* L. var. *cannabina* Muell.—Arg. in DC. Prodr. 15(2): 944. 1862; FBI 5: 465.

Bushy herbs, with stinging hairs; leaves oblong-ovate to palmately-lobed, 3–6 × 1–2 cm, obtuse to acute, base truncate; flowers yellowish-green in terminal racemes.

Common in roadside hedges and open forests. Fl. & Fr.: June–Dec. *Vern.:* Chinna-Dulagondi. Javuku, *VRK* 3547.

Note: Linnaeus f. when uniting *Croton hastatus* L. and *C. urens* L. and transferring them to *Tragia* L. did not choose one of these epithets. Hence, his *T. cannabina* becomes superfluous and illegitimate. As these epithets viz. *hastatus* and *urens* have since been preoccupied, A. Radcliffe-Smith has selected a new epithet *plukenetii* in memory of Leonard Plukenett upon whose plants Linnaeus based his concept of *hastatus* and *urens.*

Acalypha tricolor L., *Codiaeum variegatum* L., are commonly grown as garden ornamentals. The Tapioca or Cassava, *Manihot esculenta* Crantz is a cultivated crop for its tubers. The Castorbean *Ricinus communis* L. is cultivated for the castor oil from seeds in upland agriculture. The Tapioca tubers are edible as vegetable, in making chips and sago.

URITICACEAE

Pouzolzia Gaudichaud—Beaupre

Pouzolzia auriculata Wight, Icon. Pl. Ind. Orient. t. 1980. f. 2. 1853 & t. 2099. f. 37. 1853; FBI 5: 582; Gamble 1383 (967).

Gregarious scabrid shrubs, up to 2 m tall; tender parts white hairy; leaves alternate, elliptic-lanceolate, 4–10 × 2–4 cm, appressed hairy, entire, acute to acuminate, base acute to rounded; flowers in sessile clusters, unisexual; achenes not winged.

Occasional on banks of streams in the forests. Fl. & Fr.: June–Oct. Nagarjunakonda, *KTH* 9862 (CAL).

ULMACEAE

1. Fruit, dry, winged, compressed samara, stamens 7–9, flowers in fascicles ...... *Holoptelia*
1. Fruit a drupe, stamens 5, flowers in racemes .... *Celtis*

Celtis Linnaeus

Celtis timorensis Spanoghe Linnaea 15: 343. 1841. *C. cinnamomea* Lindley ex Planchon Ann. Sci. Nat. Bot. (Ser. 3) 10: 303. 1848; FBI 5: 482; Gamble 1349 (944). *C. trinervia* Beddome, Fl. Sylv. S. India t. 302. 1873, non Roxb.

Tree, up to 20 m tall; branchlets pubescent; leaves chartaceous, ovate-elliptic, 5.5–11 × 2.2–5.2 cm, glabrous, blackish when dry, entire or serrate, acute to acuminate, base rounded; flowers small, greenish in terminal racemes; drupes ovoid-globose, 6 mm across, orange when ripe.

Common in plains. Fl. & Fr.: Mar.–Apr. *Vern.:* Koti-Bira, Kaki-Mushti. Nagarjunakonda, *KTH* 9871 (CAL).

Holoptelea J.E. Planchon

Holoptelea integrifolia (Roxb.) Planchon in Ann. sci. Nat. Bot. (Ser. 3) 10: 266. 1843; FBI 3: 431; Gamble 1349 (943). *Ulmus integrifolia* Roxb. Pl. Cor. t. 78. 1798.

Deciduous tree, up to 10 m tall; bark whitish-grey, pustular, smooth, exfoliating in quadrilateral scales, branchlets terete, glabrous, lenticellate; leaves chartaceous, elliptic-ovate, 6–14 × 3–8 cm, puberulous along nerves when young, apex abruptly acuminate, margin entire, base rounded; flowers yellow in axillary racemes or in fascicles; samara orbicular, 3 cm in dia. flattened, yellow.

Common, in deciduous forests and in villages along roadsides. It is moderate light-demander, coppices well. Wood light to moderately hard, used for cheap grades of furniture, match boarding, panelling, door frames, wall plates, carts and carriages, combs, shoe heels, mathematical instruments, suitable for plywood, packing cases, slate frames, match boxes, splints, paper pulp, pulp and also for fuel. Twigs used as tooth brushes. Davupalli, *SSR* & *KSM* 21413.

MORACEAE

1. Flowers enclosed in recaptacle; fruit achene .... *Ficus*
1. Flowers exposed; fruit berry .... *Streblus*

Ficus Linnaeus

1. Figs on special shoots, leaves decussate:
   2. Leaves scabrid above, margin toothed, figs dioecious .... *F. hispida*
   2. Leaves glabrous on both surfaces, margin entire, figs monoecious .... *F. racemosa*
1. Figs on branchlets, leaves alternate:
   3. Branchlets and figs fulvous-tomentose .... *F. mollis*
   3. Branchlets and figs glabrous:
      4. Leaf tip (cusp) more than 4 cm long .... *F. religiosa*
      4. Leaf tip (cusp) less than 2 cm long:
         5. Leaf nerves without prominent intercostals or secondary lateral nerves, ovary white .... *F. amplissima*
         5. Leaf nerves with prominent intercostals or secondary lateral nerves, ovary red at the base or on the stylar side:
            6. Figs < 1 cm across, leaf with secondary lateral nerves .... *F. microcarpa*
            6. Figs. > 1.5 cm across, leaf with intercostals .... *F. benghalensis*

Ficus amplissima J.E. Sm. in Rees. Cyclop. 14. n.t. 1810. *F. tsiela* Roxb. ex Buch.-Ham. in Trans. Linn. Soc. London 15: 149. 1826; FBI 5: 515; Gamble 1362 (953).

Glabrous, spreadng trees, up to 10 m tall; aerial roots absent; leaves ovate-oblong, 8–12 × 5–7 cm, coriaceous, acute, entire, base rounded or truncate; figs axillary, globose, achenes smooth.

Occasional in forests. Fl. & Fr.: Nov.–June. *Vern.*: Juvvi, Pitta Juvvi. Leaves and twigs are lopped for fodder. The wood is used for furniture and fuel. Loyapalle, *VRK* 3822.

Ficus benghalensis L. Sp. Pl. 1059. 1753; FBI 5: 499; Gamble 1361 (952).

Large evergreen trees, up to 30 m tall; sending down many aerial roots from the spreading branches and form support, thus extending the growth of the tree indefinitely; leaves elliptic-ovate, 10–20 × 7–10 cm, coriaceous, obtuse or blunt, entire, base truncate; figs globose, axillary, paired; red when ripe.

The Banyan tree. Common in plains and rare in forests. It is sacred tree of Hindus and most popular shade providing tree. Fl. & Fr.: Aug.–Mar. *Vern.*: Marri. Fruits are edible and lac is grown on this species. Milky juice, leaves, bark, infusion of young buds, seeds and tips of young roots are used medcinally. V.P. South, *VRK* 3941; Satrasala, *VRK* 6779.

Ficus hispida L.f. Suppl. Pl. 442. 1781; FBI 5: 522; Gamble 1367 (956).

Shrubs to small trees, up to 5 m tall; branches hispid; leaves opposite, broadly oblong; 15–22 × 8–12 cm, coriaceous, scabrid above, hispid below, acute, entire, base truncate; figs globose, grey-tomentose.

Occasional near streams. Fl. & Fr.: Jan.–July. *Vern.*: Bramha medi. Leaves in leucoderma, Munching figs induces vomitings for expelling poisonous food consumed unawares. Fruits are edible. Bark yields a rough fibre. The leaves are used for poulticing boils. Javuku, *VRK* 3865; Bollapalle, *CAB* 17426 (MH).

Ficus microcarpa L.f. Suppl. Pl. 442. 1781. *F. retusa* auct. non. L.; King in Ann. Roy. Bot. Gard. Calcutta 1: 50: t. 61. 1887; FBI 5: 511; Gamble 1362 (952). *F. retusa* auct. non. L. var. *nitida* auct. non. Thunb. King l.c. 1: 50. t. 62. 1887; FBI 5: 511; Gamble 1362 (952).

Large evergreen tree with few aerial roots, all parts glabrous; trunk cylindric; leaves elliptic-obovate, 5–8 × 1.5–5 cm, glabrous, shining, apex rounded to acute or retuse, margin entire, base narrowed or cuneate, 3-nerved, lateral nerves 10–13, closely and reticulately veined; receptacles monoecious, 8 mm diam, sessile in axillary pairs, rarely solitary, globose, yellowish-red when ripe.

Occasional, in Nambur. Syconia ripening by Oct. *Vern.*: Juvvi, Konda Juvvi, Yerra Juvvi. Leaves and twigs are lopped for fodder. The wood is used for furniture or fuel.

Ficus mollis Vahl Symb. Bot. 1: 82. 1790. *F. tomentosa* Roxb. ex Willd. Sp. Pl. 4: 1136. 1806; FBI 5: .501; Gamble 1361 (952).

Evergreen tree, up to 12 m tall; bark greenish-white, smooth, scaly; branchlets fulvous-tomentose; aerial roots present; leaves elliptic-ovate, 4–18 × 3–11 cm, glabrous or nearly so and dotted above, tomentose below, acute, entire, base truncate or cordate with a gland at the base of the mid rib beneath, subcoriaceous; figs axillary, paired or clustered, subglobose.

Occasional in forests and plains. Fl. & Fr.: Apr.–Aug. *Vern.:* Juvvi, Kal-Juvvi. Nagarjunakonda, *VRK* 6707.

Ficus racemosa L. Sp. Pl. 1060. 1753. *F. glomerata* Roxb. Pl. Cor. 2: 13. t. 123. 1799; FBI 5: 535; Gamble 1364 (954).

Large deciduous trees, up to 10 m tall; aerial roots few or absent; branchlets minutely pubescent; leaves oblanceolate, 8–12 × 4–5 cm, glabrous, acute, entire, base obtuse; or oblique; figs globose, white-pilose; purplish-red when ripe; on short leafless separate branchlets, pedunculate.

Common in forests and along river banks, roadsides, village surrounds. Fl. & Fr: Apr.–Dec. *Vern.*: Medi. Athi. Bark ingredient of cooling tonics, Dermatitis relief, preparations. Leaves are frequently used as fodder and fruits are edible. V.P. South, *VRK* 3944.

Ficuss religiosa L. Sp. Pl. 1059. 1753; FBI 5: 513; Gamble 1363 (953). *Urostigma religiosum* (L.) Gasp. Ric. Caprifico 82. t. 7. ff. 1–5. 1845.

Large trees up to 20 m tall; leaves alternate, broadly ovate, 5–13 × 4–12 cm, coriaceous, 3-nerved from base, glabrous, sinuate, acuminate, base truncate; figs monoecious, axillary, raised, sessile, obovoid or globose, pink on ripe; achenes smooth.

The Sacred Peepul Tree. Common in plains, often planted in temples. Fl. & Fr.: Feb.–Aug. *Vern.*: Ravi. Aswatha. Avenue Tree, roadsides in plains. Bark insecticide, roots, figs-nervine tonic ingredients. Mangalagiri, *SSR & KSM* 21610.

STREBLUS Loureiro

Streblus asper Lour. Fl. Cochinch. 615. 1790; FBI 5: 489; Gamble 1353 (940).

Evergreen scraggy or bushy tree to 7 m tall; bark smooth, grey, wood yellowish; branchlets hairy; leaves thick-coriaceous, elliptic-obovate or rhomboid, 1.7–6.8 × 1–3 cm, rough on both surfaces, apex acute, margin irregularly dentate, base acute or cuneate; flowers dioecious, male flowers in globose heads, cream; female ones solitary axillary, fruit globose, 6 mm across, yellow, enclosed in enlarged fleshy sepals, style persistent; seed 1, globose, smooth, 3 mm across.

Common, in edges, out skirts of forests in dry deciduous forests of Davupalli. Fl. & Fr.: Jan.–Apr. *Vern.*: Stianike, Barranke, Bamigi. Wood hard and tough used for making small articles and cart wheels. The rough leaves are used to polish wood. Fruits are edible. Bark yields fibre which is used for making paper. Wood is said to be an excellent fuel. Twigs as substitutes for tooth brush.

CASUARINACEAE

Casuarina Adanson

Casuarina littorea L. Diss. Stickman 12: 1754. *C. equisetifolia* Forst. & Forst. f. Char. Gen. Pl. 104. f. 53. 1776; FBI 5: 598; Gamble 1389 (672).

Tall, leafless trees, up to 15 m tall; branchlets arising from the axils of scales; leaves scaly, alternating with ribs of the upper node; male flowers in terminal catkins, female in globose-ovoid cone like heads; nutlets woody, globose, compressed, winged; seeds minute.

The Casuarina Tree, Beafwood tree, *C. equisetifolia* is better conserved as *nom. cons.* Widely planted in gardens. Fl. & Fr.: Feb.–Aug. *Vern.:* Sarugudu. Javuku, *VRK* 3566.

Widely cultivated as wind shield, reclamation of wastelands, coastal as well as inland, cyclone shelter plant. Energy species encouraged by BESI (Bio energy Society of India) & DNES (Department of Nonconventional Energy Sources).

Initiation and development of root nodules in *Casuarina*. John G. Torrey, Cabot Foundation, Harvard Univ., Peterson, Mass 01366,45 Sp. A non leguminous genus forming root nodules, response to invasion by an actinomycete like organisms. They are nitrogen fixing. This entitles its use in wasteland development.

CERATOPHYLLACEAE

Ceratophyllum Linnaeus

Ceratophyllum demersum L. Sp. Pl. 992. 1753; FBI 5: 639; Gamble 1391 (973).

Submerged rootless herbs; branchlets brittle, green, glabrescent, leaves 7–12 in close verticels; 2–3 cm, swollen near middle, margin denticulate, apex acute; flowers solitary in upper axils; nutlets ellipsoid, black.

Koppanur stream, in tanks at Sangamjagarlamudi, Amnitaher Vasuchenuni.. Fl. & Fr.: Jan.–Mar. Koppanur, *VRK* 3894.

HYDROCHARITACEAE

1. Leaves radical, stems 0, spathes peduncled:
   2. Leaves broadly ovate-orbicular, distinctly petioled, spathes winged .... ***Ottelia***
   2. Leaves linear-lanceolate or oblong, without a distinct petiole, spathes not winged .... ***Vallisneria***
1. Leaves cauline, stems present, spathes sessile:
   3. Leaves whorled, male flowers solitary .... ***Hydrilla***
   3. Leaves alternate, male flowers several .... ***Nechamandra***

Hydrilla Richard

Hydrilla verticillata (L.f.) Royle Ill. Bot. Himal. t. 376. 1839; FBI 5: 659; Gamble 1396 (977). *Serpicula verticillata* L. f. Suppl. Pl. 416. 1781.

Slender, leafy, submerged, fresh water herbs; branches with a short sheathing leaf at the base, leaves whorled, oblong-linear, 0.8–1.5 × 0.2–0.4 cm, glabrous, acute to obtuse, apiculate; flowers white; axillary, solitary; fruit sub-terete; seeds ellipsoid.

Common in lakes and streams in the forests. Fl. & Fr.: Throughout the year. Sattenapally, *VRK* 5873.

Nechamandra Planchon

Nechamandra alternifolia (Roxb.) Thw. Enum. Pl. Zeyl. 332. 1864; Subram. & Balakr. in Bull. Bot. Surv. India 3: 23. 1962. *Vallisneria alternifolia* Roxb. Fl. Ind. 3: 750. 1832. *Lagarosiphon roxburghii* (Planchon) Benth. in Benth. & Hook. f. Gen. Pl. 3: 451. 1880; FBI 5: 659. *L. alternifolia* (Roxb.) Druce. Bot. Exch. Club. Soc. Brit. Isles 1916: 630. 1917; Gamble 1396 (977).

Submerged slender aquatic herbs; branches elongate, slender, 10–60 cm, glabrous; leaves alternate, apically sub-verticillate; oblong, linear or elliptic-lanceolate, 1.5–7.5 × 0.2–0.5 cm, membranous, base amplexicaul, margin minutely serrulate-subentire, obtuse-acute; flowers dioecious, in sessile spathe, axillary, solitary; fruit ovoid; seeds ellipsoid.

Occasional in canals and streams in flowing waters. Fl. & Fr.: Throughout the year. Macherla RF, *VRK* 5894; Repalle, *JSG* 81929.

Ottelia Persoon

Ottelia alismoides (L.) Pers. Syn. Pl. 1: 400. 1805; FBI 5: 662; Gamble 1398 (978). *Stratiotes alismoides* L. Sp. Pl. 535. 1753.

Aquatic herbs in shallow waters, up to 50 cm tall; leaves all radical, submerged ones oblong, 4–7 × 0.6–2.5 cm, floating ones broadly ovate-suborbicular, 6–15 × 5–12 cm, glabrous; flower white with yellow tinge, axillary, solitary; fruit oblong-ellipsoid.

Common in perennial waters in streams, ponds and tanks. Fl. & Fr.: Aug.–Apr. Krishna river, *VRK* 6738; Macherla, *VRK* 6823.

Vallisneria Linnaeus

Vallisneria natans (Lour.) H. Hara in J. Jap. Bot. 49: 136. 1974. *Physkium natans* Lour. Fl. Cochinch. 663. 1790. *Vallisneria gigantea* Graeb. Bot. Jahrb. Syst. 49: 68. 1912. *V. spiralis* auct. non L.; 1753. FBI 660 *p.p.*; Gamble 1368 (977).

Submerged, tufted herbs at the muddy bottoms of water; leaves radical, 10–35 × 0.3–0.8 cm, green, translucent, entire or serrulate at the apex, flowers unisexual, light greenish white in spathes; peduncle of female spathe coiled; fruit linear, included in the spathe, many-seeded.

Abundant in Ethipothala falls area, banks of Krishna river at shallow zones—and various tanks and ponds. Fl. & Fr.: Aug.–Mar. Eng.: Eel grass, wild celery. Leaves, rich source of phosphorus, calcium, and iron are edible. Javuku, *VRK* 3885; Macherla, *VRK* 5896.

ORCHIDACEAE

1. Epiphytic herbs, flowers yellow, in racemes .... ***Vanda***
1. Terrestrial herbs, flowers white .... ***Habenaria***

Habenaria Willdenow

Habenaria roxburghii (Pers.) R. Br. Prodr. 1: 312. 1810. *Orchis roxburghii* Pers. Syn. Pl. 2: 503. 1807. *Habenaria platyphylla* (Willd.) Spreng. in L. Syst. Veg. (ed. 16) 690. 1826; FBI 6: 140; Gamble 1470 (1027).

Annual erect, terrestrial herbs; stems up to 40 cm, tall; leaves in a rosette, orbicular, flat on the ground, fleshy, 5–10 × 3–7 cm, entire, apex acute; flowers white, fragrant in long peduncled racemes; capsules elliptic.

Occasional in forests under shade of trees. Fl. & Fr.: July–Dec. Jayanthipuram, *VRK* 6931.

Vanda P. Miller

Vanda tessellata (Roxb.) Hook. ex G. Don in Loud. Hort. Brit. 372. 1830; Gamble 1445 (1010). *Epidendrum tessellatum* Roxb. Pl. Cor. t. 42. 1795. *Vanda roxburghii* R. Br. Bot. Reg. 61 t. 506. 1820; FBI 6: 52.

Epiphytic herbs; stem scandent; simple or branching roots; leaves up to 20 x 2 cm, thickly coriaceous, recurved, conduplicate, keeled, linear-oblong; flowers brown in 6–10-flowered racemes; sepals and petals yellow tessellate with brown lines and white margins; petals shorter than sepals, capsules clavate—oblong with acute ribs.

Common on trees of the forests. Fl. & Fr.: Aug.–Nov. Roots are used for curing rheumatism and bronchitis. Leaves are used for fever. Juice of leaves is also used for ear troubles. Gunture (*Uhl*, 1929).

MUSACEAE

The Banana, *Musa paradisiaca* L. of Musaceae. Several cultivars are extensively cultivated in the district.

CANNACEAE

The Canna, *Canna indica* L. of Cannaceae is a common ornamental herb grown in gardens and parks.

ZINGIBERACEAE

Alpinia Roxburgh *nom. cons.*

Alpinia calcarata Roscoe in Trans. Linn. Soc. Lond 8: 347. 1807; FBI 6: 254; Gamble 1493 (1043).

Perennial herbs with horizontal root stock; leaves oblong-lanceolate, up to 20 cm, glabrous, acuminate, entire, sheath covering the stem; flowers white with variegated lip; numerous, large, in dense panicles; fruit globose with membranous aril, red.

Cultivated aromatic herbs and often as an escape in plains. Fl. & Fr.: Mar.–Aug. V.P. South, *VRK* 3947.

Curcuma domestica Valeton and *Zingiber officinale* L. (Eng.: Ginger; Tel.: Allumu) are cultivated for their rhizomes.

AMARYLLIDACEAE

Crinum Linnaeus

Crinum latifolium L. Sp. Pl. 291. 1753; FBI 6: 283; Gamble 1504 (1051).

Herbs, with underground large, subglobose bulbs; neck short, stout; leaves numerous, strap-shaped, bright green, glabrous, acuminate, margin slightly scabrous; scape inserted in the neck of the bulb, stout, purple; flowers fragrant, white, tinged with purple in 10–20-flowered umbels; fruit subglobose.

Occasional in the fields and wet shady forest areas. Fl. & Fr.: Aug.–Jan. Guntur (*Uhl*, 1929).

AGAVACEAE

1. Robust stout shrub, leaf margin and apex spinous, scape exceeding the leaves by a few meters .... *Agave*
1. Fleshy herb, leaves unarmed, scape as long as leaves .... *Sansevieria*

Agave Linnaeus

Agave americana L. Sp. Pl. 323. 1753; FBI 6: 277; Gamble 1505 (1052).

Armed shrubs; leaves radical, succulent, rigid, glaucous, margins dentate spiny, apex cuspidate; flowers light yellow in long stalked inflorescence, 2–5 m long, in cymose bunches; bulbils present; capsule oblong-clavate, beaked, ca. 4 cm long; seeds shining.

Most common hedge plant along the fields all over the district and checking soil erosion. Also useful for seaside landscaping. Fl. & Fr.: Mar.–July. Eng.: Centuary plant. *Vern.*: Kalabanda, Kitanara. Fibre from leaves used in preparation of ropes. Reported in Guntur Forest Division Work Plan.

Sansevieria Thunberg *nom. cons.*

Sansevieria roxburghiana Schult. & Schult. f. in Roem. & Schult. in L. Syst. Veg. 7: 357. 1829; FBI 6: 271; Gamble 1520 (1061).

Perennial shrubs; root stock creeping; leaves basal, 8 or 9 in a tuft, 50–75 × 1–3 cm, glossy, glabrous, base sheathing; flowers pale green tinged with violet, in fascicles of 3–6 on 30–60 cm long racemes; berries globose, membranous, indehiscent; seeds 1–3, large-globose, fleshy, ripening outside the pericarp.

Occasional in forests, open places in plains at bases of Palms near Nambur. Fl. & Fr.: June–Dec. Eng.: Indian bowstring hemp. *Vern.*: Sagamatta;

Fibre, obtained from succulent leaves, is used for mats and cordage. Plant is slightly warmed, squeeze the juice, about three drops applied in the ear to cure ear-ache. This process is continued early in the morning and at bed time for about 2 days. A paste is prepared from the fresh rhizome, is applied topically to the spot bitten by poisonous snake. It is also applied externally to releive pain due to poisonous sting of scorpion. Loyapalle hill slopes, *VRK* 3820.

LILIACEAE

1. Leaves reticulately veined, stipules tendrillar, flowers umbellate .... ***Smilax***
1. Leaves parallel veined, stipules not tendrillar, flowers racemose or solitary:
   2. Plants scandent or climbers, stem branched:
      3. Under shrubs, cladodes present, leaves scaly, armed, leaftip not tendrillar, flowers minute in racemes, fruit a berry .... ***Asparagus***
      3. Herb, Cladodes absent, leaves well developed, unarmed, leaf-tip tendrillar, flowers large, showy, solitary, fruit a capsule .... ***Gloriosa***
   2. Not as above:
      4. Root stock with fleshy or tuberous roots, capsules winged .... ***Chlorophytum***
      4. Root stock a bulb or corm, capsules not winged:
         5. Leaves fleshy, margins spinous .... ***Aloe***
         5. Leaves thin, margins entire:
            6. Flowers close appearing with leaves, seeds globose .... ***Scilla***
            6. Flowers on scape, not appearing with leaves, seeds winged .... ***Urginea***

Aloe Linnaeus

Aloe vera (L.) Burm. f. Fl. Ind. 83. 1768; Gamble 1520 (1602). *A. perfoliata* L. var. *vera* L. Sp. Pl. 320. 1753.

Stoloniferous herbs, up to 50 cm tall; leaves dense, all radical, ensiform, 30–40 × 2–6 cm, succulent, margin spiny; narrowed from base to apex; flowers reddish-yellow in terminal racemes; capsules ellipsoid-oblong.

Common in open forests and hedges. Introduced from Canary Islands. Naturalized. Also grown as an ornamental. Fl. & Fr.: Sept.–Dec. Eng.: Barbados aloe. *Vern.:* Manchi Kalabanda, Musambaram. Leaves are source of resinous drug which is used mainly as purgative. Dried paste for swellings, leaves in eye troubles. Javuku, *VRK* 3572.

Asparagus Linnaeus

Asparagus racemosus Willd. Sp. Pl. 2: 152. 1799; FBI 6: 316; Gamble 1517 (1060).

Straggling armed shrubs; spines slightly recurved; branchlets angular, glabrous; leaves scaly, triangular, 4–6 mm, stiff-acuminate, cladodes 2–9, filiform, 1.5 × 2 cm long; flowers white, fragrant, in simple (rarely branched)

5–15 cm long racemes; berries globose, red when ripe; seeds 3–6, globose or angled.

Common in the forests. Fl. & Fr.: June–Dec. *Vern.*: Pillithegalu. Consumed as vegetable after cooking in water or roasting in embar. Root is used as demulcent, aphrodisiac, diuretic, antidysenteric, and as demulcent in veterinary medicine. In excessive menstrual flow, about 2 teaspoons of juice of roots, mixed with butter milk given before bed time for 5 consecutive days from first day of mensus. N. Varam, *VRK* 3963; Loyapalle, *VRK* 3817; Bollapalle, *CAB* 17406 (MH).

Chlorophytum Ker-Gawler

Chlorophytum tuberosum (Roxb.) Baker in J. Linn. Soc. Bot. 15: 332. 1876; FBI 6: 334; Gamble 1526 (1066). *Anthericum tuberosum* Roxb. Fl. Ind. 1: 149. 1832.

Herbs, with conspicuous cylindric, tuber like roots; leaves 6–12, linear, membranous, sessile, 15–30 × 1.5–2 cm, flat; flowers white in dense, simple or branched racemes; capsules orbicular, triquetrous, strongly nerved, glabrous, seeds 4–6, subquadrate, irregularly pitted.

Occasional weed of cultivated fields and in forests. Fl. & Fr.: June–Oct. Roots are edible and leaves are used as vegetable. The fresh leaves and tuberous root are made into a paste, plastering the paste over the spot bitten to relieve pain due to poisonous sting of scorpions. V.P. South, *VRK* 6833.

Gloriosa Linnaeus

Gloriosa superba L. Sp. Pl. 305. 1753; FBI 6: 358; Gamble 1519 (1061).

Glabrous tendril climbers; root stock tuberous, solid, fleshy, white, cylindric, leaves alternate, oblong-lanceolate, 6–12 × 1.5–2.5 cm, flat, glabrous, acuminate, tendril like, entire, base cordate, margin entire; flowers red and reddish yellow, axillary, solitary; capsule ellipsoid-oblong, 4.5 × 1.5 cm, green.

Occasional in hedges. Fl. & Fr.: Sept.–Nov. Eng.: Glory lily. Tiger's claws. *Vern.*: Nabhi. Roots poisonous used in medicines as anthelmintic, and leaf juice is reported to kill lice in hair. Loyapalle, *VRK* 3998; Davupalli, *DMR* & *SSR* 21136.

Recommended for micropropagation to maintain large population.

Scilla Linnaeus

Scilla hyacinthina (Roth) Macbride Contr. Gray. Herb (n.s) 56: 14. 1918. *Ledebouria hyacinthina* Roth, Nov. Pl. Sp. 195. 1821. *Scilla indica*

Baker. Sand. Refug. Bot. 3: App. 12. 1870; FBI 6: 348; Gamble 1527 (1067).

Perenial herbs, milky tunicated bulbs; leaves 2–5, usually mottled with red-purple spots above, radical, 3.5–7.5 × 1–2.5 cm, oblanceolate, glabrous; flowers pale to bright pinkish-purple, in long racemes of 6–15 cm long; capsules violet-purple, globose or ovoid.

Rare in open areas in the moist localities. Fl. & Fr.: June–Sept. Eng.: White squill, Indian drug squill, sea-onion. *Vern.*: Adavi tellagadda. Bulbs a source of raticide, and drug which is used as an expectorant, stimulant and cardio-tonic in small doses. It is used chiefly in chronic bronchitis and asthma. Alcoholic extracts of the bulbs possess anticancer activity against human epidermoid cardinoma. Also employed in dropsy, rheumatism and skin troubles. Externally used for removing corns and warts. Guntur (*Uhl*, 1929).

Smilax Linnaeus

Smilax zeylanica L. Sp. Pl. 1029. 1752; FBI 6: 309; Gamble 1518 (1061). *S. macrophylla* Roxb. Fl. Ind. 3: 793. 1832; FBI 6: 310.

Shrubby, tendrillar climbers; stems smooth, angled, prickly; leaves 10–12 × 6–7 cm, orbicular, oblong-broadly elliptic or ovate-lanceolate, nerves 3–5 from base, margin entire, apex obtuse, somewhat notched, cuspidate; leafbase modified into tendrils; flowers pale green in pedunculate umbels; fruits globose, smooth, remaining green for a long time, becoming red when ripe.

Occasional climber in the forests. Fl. & Fr.: July–Jan. *Vern.:* Kondaguriva teega. Young branches and leaves are used as vegetable. Roots are used as substitute for sarsaparilla in treatment of venereal diseases; applied for rheumatism and also used in bloodless dysentery. Reported in Guntur Forest Division Work Plan.

Urginea Steinheil

Urginea coromandeliana Hook. f. Fl. Brit. India 6: 347. 1892; Gamble 1527 (1066). *U. wightiana* Hook. f. Fl. Brit. India 5: 347. 1892.

Scapigerous bulbous herbs; leaves radical, simple, entire, narrow, linear, up to 20 cm; flowers white in scapaceous racemes appearing before the leaves; capsules triquetrous, globose; seeds numerous, black.

Common in dry places in open forests and plains especially in rocky areas. Fl. & Fr.: Apr.–Oct. *Vern.*: Verri vulli. V.P. South, *VRK* 6825.

Plant recommended for micropropagation and for study as a source of the alkaloid colchicene.

Allium cepa L. and *Allium sativum* L. are cultivated crops.

PONTEDERIACEAE

Eichhornia Kunth

Eichhornia crassipes (C. Martius) Solms-Laub, in A. DC. Monogr. Phan. 4: 527. 1883; Gamble 1530 (1069). *Pontederia crassipes* C. Martius, Nov. Gen. Sp. Pl. 9. t. 4. 1823.

Free-floating rooted herbs, stolon slender; roots elongate, fibrous; leaves broadly ovate-rhomboid, 5–10.5 cm long, as long as broad, thin coriaceous, obscurely curvi-nerved, glabrous, base cuneate to rotund, margin entire, apex obtuse; petiole spongy, bulbous; flowers bluish-lilac in terminal spikes.

Abundant aquatic weed covering the entire water surfaces of canals, ponds and lakes all over Guntur district. Fl. & Fr.: Jan.–May. Eng.: Water hyacinth. *Vern.*: Budagatamara. Nizampatnam, *VRK* 6814.

The Blue Angel that became a blue Devil, World's worst weed.

COMMELINACEAE

1. Fertile stamens 3, staminodes present:
   2. Inflorescence terminal in paniculate cymes and not arising from spathaceous bract; fruit naked .... ***Murdannia***
   2. Inflorescence axillary in 1–2-cymes, arising from spathaceous bracts; fruits concealed .... ***Commelina***
1. Fertile stamens 6, staminodes absent:
   3. Inflorescence enclosed in leaf-sheath .... ***Amischophacelus***
   3. Inflorescence not enclosed in leaf-sheath .... ***Cyanotis***

Amischophacelus R.S. Rao et R.V. Kammathy

1. Creeping herbs, leaves succulent, flowers pink-purple, capsules ellipsoid, seeds pitted .... ***A. axillaris***
1. Prostrate herbs, rooting at nodes, leaves not succulent, flowers blue-violet, capsules oblong, seeds glabrous .... ***A. cucullata***

Amischophacelus axillaris (L.) Rao & Kammathy in J. Linn. Soc. Bot. 59: 306. 1966. *Commelina axillaris* L. Sp. Pl. 42. 1753. *Cyanotis axillaris* (L.) D. Don Prodr. Fl. Nepal 46. 1825; FBI 6: 388; Gamble 1550 (1082).

Annual slender prostrate herbs, rooting at nodes; leaves linear-lanceolate, 5–10 × 0.2–0.5 cm, succulent, acute to acuminate, entire; flowers purple or violet in axillary, sessile clusters; capsules ellipsoid.

Common weed of cultivated fields and moist places in plains and forests. Fl. & Fr.: Sept.–Mar. Mellavagu, *VRK* 5720; Jayanthipuram, *VRK* 6945.

Amischophacelus cucullata (Roth) R. Rao & Kammathy in J. Linn. Soc. 59: 306. 1966. *Tradescantia cucullata* Roth, Nov. Pl. Sp. 189. 1821. *Cyanotis cucullata* (Roth) Kunth Enum. Pl. 4: 107. 1843; FBI 6: 389; Gamble 1550 (1082).

Prostrate or suberect herbs, rooting at nodes; leaves linear-lanceolate, 1.5–4.5 × 0.3–0.5 cm, glabrous; flowers blue-violet in axillary fascicles; capsules glabrous, ellipsoid-oblong; seeds minute.

Occasional weed of moist places. Fl. & Fr.: July–Dec. Nagarjunasagar, *KMS* 9731 (MH).

Commelina Linnaeus

1. Spathes funnel-shaped:
   2. Leaves less than 8 mm broad, capsules 2–celled, spathes solitary, pedunculate .... *C. ensifolia*
   2. Leaves more than 1.5 cm broad, capsules 3-celled, spathes clustered, sessile:
      3. Erect or decumbent herbs, rooting at lower nodes, capsules ovoid ..... *C. paludosa*
      3. Diffuse or sub-erect herbs, not rooting at lower nodes, capsules ellipsoid:
         4. Cleistogamous flowers present, seeds 5, pitted .... *C. benghalensis*
         4. Cleistogamous flowers absent, seeds 3, smooth .... *C. erecta*
1. Spathes ovate-lanceolate:
   5. Seeds appendaged at one or both ends:
      6. Spathes sagittate-auriculate at base, capsules 2-celled, seeds 4 .... *C. attenuata*
      6. Spathes rounded-cordate at base, capsules 3-celled, seeds 5 .... *C. longifolia*
   5. Seeds without appendages:
      7. Capsules 2-celled, constricted in the middle, sheaths more than 2.5 cm long .... *C. clavata*
      7. Capsules 3-celled, not constricted, sheaths less than 1.5 cm long .... *C. diffusa*

Commelina attenuata Koen. ex Vahl Enum. 2: 168. 1806; FBI 6: 372; Gamble 1538 (1074). *C. rajamahalensis* Clarke in J. Linn. Soc. 11: 444. 1870.

Annual, slender, much-branched herbs; stem and branches angled; stem 30–60 cm long, slender, much branched, leaves linear-lanceolate, 3–10 × 0.5–1 cm, chartaceous, sparsely pubescent, acuminate, entire, base obtuse; flowers blue or rarely purple; capsules oblong; seeds whitish-appendaged at ends.

Occasional in forest undergrowth. Fl. & Fr.: July–Dec. Koppanur, *VRK* 3246.

Commelina benghalensis L. Sp. Pl. 41. 1753; FBI 6: 370; Gamble 1539 (1075).

Branched, diffuse herbs, with dimorphic flowers; leaves ovate to ovate-oblong, 3–7 × 1.5–4 cm, glabrescent; base subtruncate, margin ciliate, apex acute-obtuse; aerial flowers blue, axillary, underground flowers white, cleistogamous, solitary in pear-shaped spathe; capsule pyriform, membranous, 5-seeded; seed oblong, closely pitted.

Common weed of cultivated fields, waste places and roadsides. Fl. & Fr.: Aug.–Dec. *Vern.*: Yanadra aku, Venna veduru. It is used as fodder and its young leaves are eaten as vegetable. Macherla RF, *VRK* 3914.

Commelina clavata Clarke in Comm. & Cyrt. Beng. 5: 15. 1874; FBI 6: 371; Gamble 1539 (1074).

Annual branched herbs; branches diffuse, rooting at the nodes; leaves elliptic-lanceolate to linear, 3.5–5 × 0.6–1 cm, puberulous or pubescent on one or both surfaces, acute-acuminate, entire, base narrowed; sheath narrowed into the blade; flowers pale violet with brown streaks; capsules oblong, club-like; seeds 4, oblong, truncate at one end, rounded at the other, reticulately pitted.

Occasional along the irrigation canals. Fl. & Fr.: Aug.–Dec. Macherla, *VRK* 5887; Vinukonda RF, *VRK* 6858.

Commelina diffusa Burm. f. Fl. Ind. 18. t. 7. f. 2. 1768. *C. nudiflora* auct. non L.; FBI 6: 369; Gamble 1538 (1074).

Diffuse succulent herbs; branches prostrate or subscandent; leaves oblong-lanceolate, 3.5–6 × 0.5–1 cm; cymes usually two, 1–10, 3-flowered; flowers blue, spathes ovate-lanceolate, peduncled; capsules 3-celled, oblong, acuminate, coriaceous, 5-seeded; seeds oblong-cylindric, tuberculate and reticulate, brown.

Occasional along the canals and streams and rare in forests. Fl. & Fr.: Dec.–May. Krishna river, *VRK* 6741.

Commelina ensifolia R. Br. Prodr. 269; FBI 6: 373; Gamble 1539 (1075). *C. undulata* R. Br. var. *setosa* Clarke in DC. Monogr. Phan. 3: 179. 1881; FBI 6; 373; Gamble 1539 (1075).

Diffuse, slender herbs, with rhizomatic roots; stem slender, leaves linear-lanceolate, 4–6 × 0.8–1 cm, glabrous above, puberulous below, acute-obtuse, entire, base narrowed; flowers blue, axillary, solitary; capsules 2-celled; seeds 1–3, ovoid, smooth.

Occasional along the canal banks. Fl. & Fr.: Sept.–Mar. Javuku, *VRK* 3564; Macherla. *VRK* 5886; Bollapalle, *CAB* 17404 (MH).

Commelina erecta L. Sp. Pl. 41. 1753. *C. undulata* R. Br. Prodr. 270. 1810. *C. kurzii* Clarke in J. Linn. Soc. Bot. 11: 144 et in DC. Monogr. Phan. 3: 185. 1891; FBI 6: 373; Gamble 1580 (1075).

Straggling slender herbs; leaves up to 14 x 3.5 cm, elliptic-lanceolate, scaberulous or hirsute above, glabrous below, acuminate; sheaths ciliate; spathes sessile; capsules 7-valved; seeds 3, ashy, broadly ellipsoid.

Common weed of cultivated fields and wastelands. Fl. & Fr.: June–Dec. Karempudi, *CAB* 17449 (MH).

Commelina longifolia Lam. Tabl. Encycl. 1: 129: 1791; R. Rao in Notes Roy. Bot. Gard. Edinburg 25: 179. 1965. *C. salicifolia* Roxb. Fl. Ind. 1: 176. 1820; FBI 6: 370; Gamble 1538 (1074).

Annual prostrate herbs, stem slender, decumbent, sometimes rooting, leaves linear or linear-lanceolate, 3–8 × 0.5- cm, chartaceous, glabrous or nearly so, base obtuse, entire, acute-acuminate; sheath to 2 cm, margin ciliate; flowers blue, small, polygamous, in ovate-narrowly lanceolate spathes;

branches of the cymes equal, usually 1–2-flowered, capsules oblong, 3-celled; seeds black, smooth, ovoid or subglobose, appendaged at one end with a whitish membranous appendage.

Occasional in cultivated field. Fl. & Fr.: July–Dec. V.P. Ṣouth, *VRK* 6840.

Commelina paludosa Blume Enum. 1: 2. 1827. *Commelina obliqua* Ham. ex. D. Don Prodr. Fl. Nepal 45: 1825; non Vahl 1805; FBI 6: 372; Gamble 1539 (1075).

Stout subscandent, branched herbs, up to 75 cm tall; leaves elliptic-ovate or elliptic-lanceolate, 6–8 × 1.5–3 cm, acute or caudate-acuminate, membranous, flowers blue in simple racemes; spathes sessile, solitary or crowned in terminal heads, funnel-shaped; capsules trigonous, obovoid.

Common along the streams. Fl. & Fr.: July–Dec. Its young shoots are eaten as vegeable. Javuku, *VRK* 3295, 3873; Karempudi, *CAB* 17449 (MH).

Cyanotis D. Don *nom. cons.*

1. Branchlets cottony, cymes lax, roots not tuberous .... *C. fasciculata*
1. Branchlets glabrous, cymes corymbose, roots tuberous .... *C. tuberosa*

Cyanotis fasciculata (Heyne ex Roth) Schult. f. Mant. Syst. Veg. 7: 1150. 1830; FBI 6: 387; Gamble 1550 (1082). *Tradescantia fasciculata* Heyne ex Roth Nov. Pl. Sp. 189. 1821.

Erect annual herbs, up to 20 cm tall; floccosely silky; stems reddish purple; leaves lanceolate to ovate-lanceolate, 1.5–9 × 0.2–2.2 cm, hairy, subfleshy, acute or acuminate, straight or recurved; flowers pink in compact axillary and terminal cymes; capsules 0.2–0.4 cm, oblong, pilose at the apex; seed oblong, faintly rugose.

Occasional weed of cultivated fields, wastelands and forests. Fl. & Fr.: July–Jan. Kondaveedu Fort, *CAB* 4642 (MH).

Cyanotis tuberosa (Roxb.) Schult. & Schult. f. in Roem. & Schult. in L. Syst. Veg. 7(2): 1153. 1830; FBI 6: 386; Gamble 1549 (1081). *Tradescantia tuberosa* Roxb. Pl. Cor. t. 108. 1799.

Annual erect villous herbs, up to 20 cm tall; roots tuberous; basal leaves oblong, 10–30 × 1–1.5 cm, upper ones linear-ensiform, up to 3 × 1 cm, chartaceous, glabrous, villous above, downy-sericeous below, and often purple, ciliate, subacute, base obtuse; sheaths of radical leaves glabrous, those of cauline silky; flowers purple in terminal or axillary cymes; capsules ellipsoid, up tc 2.5 mm; seeds pyramidal, rugose, brown.

Occasional, in plains and forests. Fl. & Fr.: June–Dec. Tubers made into powder given for fever. Nagarjunakonda, *KTH* 9867 (CAL).

Murdannia Royle

Murdannia spirata (L.) Brueckner in Engler Nat. Pfl. Fam. (ed. 2) 15; 173. 1930. *Commelina spirata* L. Mant. Pl. 176. 1767. *Aneilema spiratum* R. Br. Prodr. 271. 1810; FBI 6: 377; Gamble 1545 (1078).

Ascending herbs, up to 20 cm tall; roots fibrous; branches decumbent and rooting below, leafy; leaves ovate-lanceolate, 1.5–3.5 x 0.5–0.7 cm, chartaceous, glabrous, entire, acute, base amplexicaul; flowers pink in dichotomously branched, few-flowered, terminal panicles; capsules oblong, mucronate, smooth, shining, 3-gonous; seeds cuboid, minute.

Common weed on wet grounds and fields. Fl. & Fr.: Aug.–Dec. Jayanthipuram, *VRK* 6951.

ARECACEAE

1. Leaves simple, palmate, unarmed .... *Borassus*
1. Leaves compound, feather like, armed:
   2. Stragglers, leaf tip acuminate .... *Calamus*
   2. Trees, leaf tip spinous .... *Phoenix*

Borassus Linnaeus

Borassus flabellifer L. Sp. Pl. 1187. 1783; FBI 6: 482; Gamble 1562 (1090).

Tall unbranched palms, up to 20 m tall; stems with semicircular scars of the fallen leaves; leaves terminal, simple, palmately fan-shaped, plicately multifid, segments 60–80 cm, margins spinulose, petiole spinous; spadices interfoliar, peduncles sheathed with open spathes; male flowers small, clustered, female larger, solitary, globose; perianth accrescent; drupes subglobose, brown, fibrous.

The Palmyrah Palm. Common in plains especially along coasts. Fl. & Fr.: Mar.–Aug. *Vern.*: Thadi. Fruits edible (Endosperm). Stems used for house building. "The 'Kalpa Vriksha' of villagers for the multiple uses. Leaves for rooves, fruits edible, floral shoots for alcohol (Neera); seedlings sold as food in market. Nagarjuna University campus, *VRK* 6982.

Calamus Linnaeus

Calamus rotang L. Sp. Pl. 325. 1753; FBI 6: 447; Gamble 1568 (1094).

Stout tufted shrubs, often climbing in favourable conditions; stem prickly, young parts covered with powdery tomentum; leaves pinnatisect; leaflets 30 x 15 cm, linear-lanceolate, bristly along margins; petiole spinous; flowers small, unisexual in spikes; fruits sub-globose.

Sometimes grown in gardens. Fl. & Fr.: Sept.–Feb. *Vern.:* Pemu Bettamu. Stems are used for making baskets and chairs. Young tender shoots are used as a vegetable. Guntur (*Uhl*, 1929).

Phoenix Linnaeus

Phoenix sylvestris (L.) Roxb. Fl. Ind. 3: 787. 1832; FBI 6: 425; Gamble 1559 (1088). *Elate sylvestris* L. Sp. Pl. 1189. 1753.

Trees, up to 10 m tall; trunk covered with persistent leaf bases; crown large, hemispherical; leaves 2–3 m long, leaflets numerous, 10–25 x 1–2 cm, glabrous with spinous apex, greyish green, flowers yellowish-green numerous, angular, in axillary pistillate spadix; drupes oblong, brownish, red when ripe; seed solitary, rounded at the ends, pale brown.

Common near rivulets and occasional in wastelands and canal banks. Fl. & Fr.: Nov.–Aug. Eng.: Wild date. *Vern.*: Eetha chettu. The leaves are used for making mats, baskets, brooms, and fans. The petioles are beaten and the fibre twisted into ropes which are employed in drawing water from wells. Fruits are edible. From the juice obtained by tapping the tree, a coarse sugar is manufactured by boiling. Nallapadu, *VRK* 6775.

The Coconut—*Cocos nucifera* L. is a commonly cultivated palm throughout the district along the coast and interior. *Caryota urens* L. is a common garden ornamental.

Other cultivated palms: *Areca catechu* L. (Betal nut palm), Krishnanagar, Guntur; Royal palm (*Roystonea regia*), *Pritchardias*.

PANDANACEAE

Pandanus Parkinson

Pandanus odoratissimus L. f. Suppl. 64. 1781. *P. fascicularis* Lam. Encyl. 1: 372. 1785; FBI 6: 485. *P. tectorius* Soland. ex Balf. f. in J. Linn. Soc. 17: 63. 56. 1879; Gamble 1570 (1095).

Large shrub-like trees, up to 10 m tall; stem supported by stilt roots; leaves glaucous-green, coriaceous, ensiform, caudate-acuminate, margins spinous; flowers unisexual, in long spikes, enclosed by long yellow fragrant spathes; fruits woody, red when ripe, oblong or globose, syncarpium; drupes numerous.

The Screw Pine. Common in sandy places near the sea-coast, water shed hill slopes and near perennial streams. Fl. & Fr.: June–Dec. *Vern.*: Mogali, Guntur (*Uhl*, 1929).

Roots prevent abortion, floral bracts in perfumery and incense sticks and soaps. Leaves are used for making baskets, cordage, umbrellas and paper-making. Also used in leprosy, small-pox, scabies and diseases of heart and brain. Sweet smelling spathes are used by women as hair ornaments.

Note: Stone (1983) in the "Flora of Tamil Nadu Carnatic" discussed the nomenclature of the Indian *Pandanus*.

TYPHACEAE

Typha Linnaeus

Typha angustata Bory & Chaub. Expect. Sci. Morea. Bot. 1: 338; 1833; FBI 6: 489; Gamble 1571 (1096).

A robust aquatic plant, up to 2 m tall; leaves up to 1 m long, semi-terete above the sheath, flattened at apex, leaf sheath auricled, scarious along the margin; male and female spikes are separated, brown in colour.

The Giant Water Seed. Common in stagnant waters, marshy places and in streams. Fl. & Fr.: Aug.–Dec. *Vern*.: Jammu, Jambu. Thatching of huts with leaves, stuffing the pillows with cotton material from spikes. Javuku, *VRK* 3299; Nagarjunasagar, *KMS* 9731 (MH). Common in and around Nagarjunanagar in stagnant water channels and puddles.

ARACEAE

1. Plants aquatic, free-floating herbs, leaves pubescent, obdeltoid .... ***Pistia***
1. Plants semi-epiphytic climbers, leaves glabrous, oblong .... ***Rhaphidophora***

Pistia Linnaeus

Pistia stratiotes L. Sp. Pl. 963. 1753; FBI 6: 497; Gamble 1573 (1097).

Aquatic, stoloniferous floating herbs; with a peculiar muriatic odour; leave obovate, bright silvery green, white pubescent, obovate, rounded or retuse at the apex, male and female flowers separated by a disc like structure.

The water lettuce. Abundant in the ponds and lakes and occasional in streams. Fl. & Fr.: Dec.–June. *Vern.*: Antaratamara. The Guntur populations were studied by Rao and Reddy, 1983; Hydrobiologia 119: 15–19, also Reddy, A.S. 1985. Biology of *P. stratiotes,* Ph.D. Thesis, Nagarjuna University, Guntur; also Rao and Reddy (1986, 87, 92); Rao (1991).

Rhaphidophora Hasskarl

Rhaphidophora pertusa (Roxb.) Schott. Bonplandia 5: 45. 1857; FBI 6: 546; Gamble 1590 (1109). *Pothos pertusus* Roxb. Fl. Ind. 1: 455. 1820. '*pertusa*'. *Monstera pertusa* (Roxb.) Schott. Wiener Z. Kunst 1830. 781. 1830. *Scindapsus pertusus* (Roxb.) Schott in Schott & Endl. Melet. Bot. 21. 1832.

Evergreen, semi-epiphytic, unarmed, root climbers; stem cylindric; leaves pinnately lobed; ovate to suborbicular, acute, unequal-sided, 22–45 × 15–28 cm; petiole 15–40 cm long with a withering sheath; flowers bisexual, sessile, in spathes covered by spathaceous bracts; berries many-seeded; seeds oblong or reniform.

Occasional root climber on trees as ornamental. Rare in forests. Fl. & Fr.: Jan.–Aug. V.P. South, *VRK* 6727.

Amorphophallus campanulatus Blume and *Colocasia antiquorum* schott. are cultivated for vegetables.

LEMNACEAE

Spirodela Schleiden

Spirodela polyrhiza (L.) Schleiden in Linnaea 13: 392. 1839. *Lemna polyrhiza* L. Sp. Pl. 970. 1753; FBI 6; 557.

Small floating herbs; fronds often in groups of 3–5, suborbicular-rotund, 5–7 × 4–5 mm; dorsal side flat, smooth; ventral side turgid, pigmented; stipe fugacious, base obtuse, apex rounded; ventral scale covering the base of roots, primary root 1; secondary roots 2; root cap acute; spathe 2-lipped; stamens 2.

Common in ponds and lakes. Fl. & Fr.: July–Dec. Ethipothala, *VRK* 3887.

NAJADACEAE

Najas Linnaeus

Najas graminea Del. Descr. Egypte Hist. Nat. 282. t. 50. f. 3. 1813; FBI 6: 569; Gamble 1604 (1118).

Submerged aquatic herbs, stems smooth, terete, up to 50 cm long; rooting at lower nodes; leaves in pseudo whorls, filiform; obtuse or acute with a spinously dentate margin, 1–2.5 × 0.05–0.1 cm, sheaths spinously denticulate; flowers white, closed in spathes; fruits ellipsoid-oblong, attenuate at the apex.

Common aquatic herbs in streams and canals. Fl. & Fr.: July–Mar. Krishna river, *VRK* 6744.

APONOGETONACEAE

Aponogeton Linnaeus f. *nom. cons.*

Aponogeton echinatus Roxb. Fl. Ind. 2: 210. 1832; FBI 6: 564; Bruggen in Blumea 18: 483 t. 7. 1970.

Submerged, tuberous perennial herbs; leaves sub-dimorphic, submerged ones short-petioled, membranous, floating ones long petioled, oblong-elliptic to elliptic-lanceolate, 11–20 × 2–3 cm, subacute, entire, base subcordate, margin entire, apex subacute; flowers white in dense spikes; fruit ovoid, echinate; seed oblong, flat on back, turgid on the upper surface.

Infrequent, in stagnant waters. Fl. & Fr.: June–Dec. Vinukonda, *CAB* 17401 (MH).

ERIOCAULACEAE

Eriocaulon Linnaeus

Eriocaulon quinquangulare L. Sp. Pl. 87. 1753; FBI 6: 582; Gamble 1620 (1128).

An aquatic herb, up to 25 cm tall; leaves radical, linear, acuminate, flat, glabrous, membranous, purplish or red when old; flowers in globose heads, greyish white; bracts obovate, hairy at apex; capsules, sub-globose; seed ellipsoid or subglobose, pale yellow, quite smooth.

Common in marshy lands. Fl. & Fr.: Nov.–Apr. Javuku, *VRK* 3291.

CYPERACEAE

Note: Kern (1971) is followed for the generic delimitation. The genera *Kyllinga* Rottb. and *Pycreus* P. Beauv. are retained which are included under *Cyperus* L. by Kern. This has also been followed by T. Koyama (1977).

1. Leaves reduced to sheaths:
    2. Style base not dilated, continuous with the nut .... ***Schoenoplectus***
    2. Style base dilated and constricted or articulated above the nut:
        3. Hypogynous bristles present, syle-base persistent with nuts .... ***Eleocharis***
        3. Hypogynous bristles absent, style-base deciduous .... ***Fimbristylis p.p.***
1. Leaves with well developed blades:
    4. Hypogynous bristles or scales present, leaves distinct.............................. ***Fuirena***
    4. Hypogynous bristles or scales absent:
        5. All, or upper glumes spiral:
            6. Style breaking off from base, leaving persistent bulbiform button on nut .... ***Bulbostylis***
            6. Style breaking off as a whole .... ***Fimbristylis p.p.***
        5. All glumes distichous:
            7. Rachilla of spikelets articulate at base, spikelets falling entire .... ***Kyllinga***
            7. Rachilla of spikelets not articulate, glumes falling separately from rachilla:
                8. Stigmas 2, nuts flattened, with one edge against rachilla .... ***Pycreus***
                8. Stigmas 3, nuts trigonous or if flattened one edge not against rachilla ... . ***Cyperus***

Bulbostylis Kunth *nom. cons.*

1. Bracts as long as or shorter than the heads, glumes 2–2.2 mm with a short straight mucro at apex .... ***sub sp. barbata***

1. Bracts much longer than the heads, glumes 3 mm with a recurved mucro at apex .... ***sub sp. pulchella***

Bulbostylis barbata (Rottb.) Clarke in Hook. f. Fl. Brit. India 6: 651. 1893; Gamble 1662 (1153). ssp. **barbata.** *Scirpus barbatus* Rottb. Descr. Pl. Rar. 27. 1772 & Descr. Icon. Rar. 52. t. 17. f. 4. 1773.

Annual, tufted herbs, with fibrous roots; leaves reduced, filiform, glabrous except at the mouth of the sheath; spikelets brown, linear-oblong, in terminal heads; nut obovoid, trigonous.

Common weed of open forests. Fl. & Fr.: July–Mar. *Vern.:* Sama chamonti gadda. Mellavagu, *VRK* 5721; V.P. South, *VRK* 5820; Madinepalle, *JSG* 17464 (MH).

sub sp. **pulchella** (Thw.) T. Koyama in Bot. Mag. (Tokyo) 93: 341. 1980. *Isolepis pulchella* Thw., Enum. Pl. Zeyl. 350. 1864. *Bulbostylis barbata* Kunth ex Clarke var. *pulchella* (Thw.) Clarke in Hook. f Fl. Brit. India 6: 652. 1893.

Plants more rigid than sub. sp. *barbata*. Head 10–15 mm across, more densely, bearing numerous spikelets, glumes 3 mm long, woolly subdensely pilose with often tubercle-based brownish hairs, the mid rib thick, not clearly nerved, projecting beyond the glumes apex forming a recurved awn-like cusp.

Rare, in coastal areas and sandy places. Fl. & Fr.: Aug.–Nov. Macherla, *VRK* 3524.

Cyperus Linnaeus

1. Spikelets digitately or stellately arranged or capitate:
    2. Inflorescence compact in a head or cluster without rays:
        3. Glumes keeled, glabrous .... ***C. niveus***
        3. Glumes not keeled, pubescent:
            4. Nuts ellipsoid, spikelets indefinite .... ***C. dubius***
            4. Nuts obovoid, spikelets definite:
                5. Inflorescence pseudolateral, glumes muticous, spikelets 10–20 in a cluster, stigmas 2 .... ***C. laevigatus***
                5. Inflorescence terminal, glumes mucronate, spikelets 5–7 in a cluster, stigmas 1 .... ***C. teneriffae***
    2. Inflorescence open, with 1 or more rays:
        6. Spikelets indefinite, glumes muticous, nuts as long as glumes .... ***C. difformis***
        6. Spikelets 10–20, in spikes, glumes mucronate, nuts less than half as long as glumes:
            7. Involucre bracts 3–5, spikelets 1.5 cm long, inflorescence terminal .... ***C. compressus (p.p.)***
            7. Involucre bracts 1 or 2, spikelets 1 cm long, inflorescence pseudolateral .... ***C. arenarius***
1. Spikelets spicately arranged:
    8 Rachilla of spikelets wingless or obscurely winged:
        9. Glumes cuspidate .... ***C. compressus (p.p.)***

9. Glumes obtuse .... *C. iria*
8. Rachilla of spikelets winged:
10. Leaves reduced to sheaths .... *C. corymbosus*
10. Leaves well developed:
11. Spikes oblong, longer than broad, stems more than 1 m long .... *C. nutans*
11. Spikes narrowly or broadly ovoid, stems less than 1 m long:
12. Leaves more than 1 cm wide, spikelets purplish, ... *C. pangorei*
12. Leaves less than 2.5 mm wide, spikelets reddish brown .... *C. rotundus*

Cyperus arenarius Retz. Obs. Bot. 4: 9. 1786; FBI 6: 602; Gamble 1640 (1140).

Perennial rhizomatous runners; plant pale or glaucous green; rhizome elongate, creeping, striate, rigid, dichotomously branched, clothed with lanceolate brown sheaths; stems sparsely tufted, rigid, grooved; leaves up to 10 cm long, terete, coriaceous, acute; sheaths brown; spikes terminal, up to 3 cm; spikelets digitate, 15–20 in a cluster, oblong; rachilla winged; nuts obovoid, smooth.

Common runner on the beaches along the coast. Sand binder. Fl. & Fr.: Sept.–Feb. Nizampatnam port, *VRK* 6925; Bapatla, *KJ* 17504 (MH).

Note: The species can be easily identified by its single dense spikelets and long rhizomes. Usually this species forms large communities along sandy sea shores.

Cyperus compressus L. Sp. Pl. 46. 1753 emend Dandy in Exell. Cat. S. Tome 3357. 1944; FBI 5: 605; Gamble 1640 (1140).

Annual erect herbs, up to 45 cm tall; stems tufted, erect, slender, triquetrous; leaves linear, shorter or longer than the stem, finely acuminate, 1-nerved; sheaths purplish-grey; spikelets pale green in condensed umbellate spikes, glumes broadly ovate, 3–4 mm; nuts obovate, triquetrous, dark-brown.

Occasional in marshy places in plains. Fl. & Fr.: June–Dec. Mellavagu, *VRK* 5722.

Cyperus corymbosus Rottb. Descr. Icon. Rar. 42. & 7. f. 4. 1773; FBI 6: 612; Gamble 1641 (1140).

Erect herbs, up to 1 m tall; rhizome creeping, clothed with dark brown scales; stems rigid, triquetrous; leaves reduced to sheaths; spikes in compound inflorescence; spikelets 5–10, purplish, linear, sub-terete, rusty-brown, 15–30-flowered; nuts obovoid, triquetrous, tapering towards the base, apiculate.

Occasional sedge in marshy places and near canals; stems are used for mats. Fl. & Fr.: July–Dec. Nagarjunakonda, *KTH* 9882 (CAL).

Cyperus difformis L. in Torner. Cent. Pl. 2: 6. 1756; FBI 6: 599; Gamble 1640 (1139).

Annual, tufted erect herbs, up to 30 cm tall; stems tufted, weak, triquetrous; leaves flat, obscurely nerved, sheaths brown; spikelets linear-lanceolate, many in globose heads; glumes ovate-orbicular, membranous; nuts obovoid, brown; style much shorter than the nut; stigmas 3, linear, short.

Occasional in marshy lands and streams in plains. Fl. & Fr.: June–Feb. Javuku, *VRK* 3548.

Cyperus dubius Rottb. Descr. & Ic. 20 t. 4 to 5. 1773; Kern in Reinwardtia 6: 65. 1961. *Mariscus dregeanus* Kunth Enum 2: 120. 1837; FBI 6: 620; Gamble 1642 (1141).

Erect perennial herbs; base of stem covered by turgid, brown sheathy leaves, glabrous, broad, weak; sheath dilated, reddish; spikelets ovoid, bearing 2–6 flowers; nuts oblong or sub-ovoid, black.

Occasional in moist localities. Fl. & Fr.: Throughout the year. Kondaveedu Fort, *CAB* 17501 (MH).

Cyperus iria L. Sp. Pl. 45. 1773; FBI 6: 606; Gamble 3. 1640 (1140). *C. parviflorus* Nees in Wight, Contr. Bot. India 87. 1834 excl syn. non Vahl, 1806. *C. iria* var. *parviflorus* (Nees) Mov. in Fl. Ind. Bot. 3: 270. 1859. *C. iria* var. *paniciformis* (Franch. & Sav.) Clarke in Hook. f. Fl. Brit. India 6. 607. 1893.

Annual, erect herb, up to 50 cm tall; culms trigonous, striate, rigid, leaves flat or channeled, 30 cm long, 6 mm wide, multistriate, acuminate, flaccid; sheath purplish; spikelets 0.7 cm long, alternate, in terminal umbellate panicles; peduncles up to 13 cm long, glumes ovate-orbicular, apiculate, obtuse at base, hyaline along margin towards top, nuts brown, obovate, trigonous.

Occasional, weed in wet and floody paddy fields (Mohan Rao & Aasiya, 1995). Fl. & Fr.: July–Dec.

Cyperus laevigatus L. Mant. Alt. 179. 1771; Kueken in Engl. Pfl. Fam. 101. 321. 1935. *Juncellus laevigatus* (L.) Clarke in Hook. f. Fl. Brit. India 6: 596. 1895; Gamble 1629 (1133).

Parennial erect herbs, up to 30 cm tall; stems densely leafy near the base, crowded on the rootstock; rhizome creeping, horizontal; leaves spreading and recurved, narrowly linear, obtuse or acute; spikelets in simple or compound umbels, crowded on a globose head; spikelets ellipsoid or ovoid, pale brown, 12–30-flowered; glumes ovate or oblong, acute, mucronulate; anthers linear, yellow with a reddish-brown acute tip.

Occasional weed of cultivated fields and moist localities. Fl. & Fr.: Throughout the year. Nagarjunasagar, *K.M. Sebastine* 9761 (MH).

Cyperus niveus Retz. Obs. Bot. 5: 12. 1788; FBI 6: 601; Gamble 1640 (1139).

Perennial, tufted, glaucescent, rhizomatous herbs, 20–70 cm tall, leaves linear, 20–50 cm, narrowly linear, glabrous; spikelets pale yellow,

1–3 × 2–3 cm long, elliptic-oblong, much compressed, 20–30-flowered, in terminal capitate heads; nuts triquetrous, dark brown, obovoid.

Occasional, seen in sandy localities of river beds, along the rice fields and on open hill slopes. Fl. & Fr.: Jan.–Apr. Nakrekal, *CAB* 17474 (MH).

Cyperus nutans Vahl ssp. ***eleusinoides*** (Kunth) T. Koyama in Gard. Bull. Straits Settlem: 30: 136. 1977. *C. eleusinoides* Kunth Enum. Pl. 2: 39. 1837; FBI 6: 608; Gamble 1640 (1140).

Erect perennial herbs; stems trigonous, up to 30 cm tall; leaves as long as the stem; involucral bracts leaflike, 7, longest to 30 cm long, spreading; spikelets in compound umbels, rayed, unequal, brown; nuts trigonous.

Occasional in damp localities. Fl. & Fr.: July–Dec. Nagarjunasagar, *KMS* 9757 (MH).

Cyperus pangorei Rottb. Descr. Pl. Rar. Programme 18: 1772; Gamble 1640 (1140). *Cyperus tegetum* Roxb. Fl. Ind. 1: 211. 1828; FBI 6: 613.

Glaucous, perennial rhizomatous herbs, up to 1.5 m tall; culms to 130 cm tall, 7 mm wide, rigid, trigonous, ribbed, sulcate; leaves few, shorter than stems with long, brown sheaths; inflorescence loose panicle, 5–10 rayed, secondary bracts conspicuous; bracts long and leafy exceeding rays; spikes broadly ovate, spikelets linear, reddish-brown, glumes elliptic.

Common in marshy aquatic places. Fl. & Fr.: Aug.–Dec. Javuku, *VRK* 3283.

Cyperus rotundus L. Sp. Pl. 45. 1753; FBI 6: 614; Gamble 1641 (1140).

Perennial aquatic rhizomatous herbs, up to 75 cm tall; stem triquetrous; leaves linear, flat, sheath reddish-brown; spikelets narrow-oblong in simple or compound spikes, involucre bracts leafy, 10 cm long, glumes ovate; membranous; nuts oblong.

The Purplenut Sedge. World's worst weed No. 1 (Holm. from Hawaii).

Abundant weed in streams and moist places and cultivated fields. Fl. & Fr.: July–Apr. Eng.: Nut grass. Rhizome yields an essential oil which is used in perfumery, soap making and as insect repellent. Rhizome with stem of *Tinospora cordifolia* made into decoction, taken in internally to cure fever. Sattenapalli, *VRK* 5865; Nakarekal, *CAB* 17478 (MH).

Cyperus rubicundus Vahl. Enum. Pl. 2: 308. 1805. *Cyperus teneriffae* Poir. in Lam. Encycl. 7: 245. 1806; FBI 6: 601; Gamble 1639 (1139).

Annual, tufted erect herbs, up to 30 cm tall; stems trigonous; leaves usually shorter than the stem or half (or less) as, narrow-linear, folded; spikelets sessile, oblong, pink or purple, 5–11 in a cluster; glumes ovate or oblong, red-purple sides; nuts broadly elliptic-obovoid.

Common in forests along streams. Fl. & Fr.: Sept.–Mar. Jayanthipuram, *VRK* 6969.

Eleocharis R. Brown

Eleocharis atropurpurea (Retz.) K.B. Presl. Rel. Haenk. 1; 196. 1828; FBI 6: 627; Gamble 1648 (1145). *Scirpus atropurpureus* Retz. Obs. Bot. 5: 14. 1788.

Annual tufted, filiform herbs; up to 25 cm tall; culms terete, capillary, smooth, sulcate; leaves absent, sheaths purple; spikelets solitary, ovoid, 5 × 3 mm, dark brown, glume ovate-oblong, purplish; nuts obovoid, biconvex, to 1 mm long, glossy.

Occasional weed in streams in the forests. Fl. & Fr.: Sept.–Feb. Macherla, *VRK* 5881; Jayanthipuram, *VRK* 6949.

Fimbristylis Vahl

1. All glumes distichous or sub-distichous, spikelet solitary .... *F. ovata*
1. All glumes spirally imbricating:
   2. Stigmas 3, nuts trigonous:
      3. All or upper leaves surrounding culm bases, bladeless .... *F. miliacea*
      3. Leaves surrounding culm bases, bladed .... *F. falcata*
   2. Stigmas 2, nuts biconvex:
      4. Spikelets solitary, leaves 0 .... *F. acuminata*
      4. Spikelets many, umbellate:
         5. Nuts prominently striate, with 6–8 vertical trabeculate ribs:
            6. Annuals, glumes 1-nerved .... *F. bisumbellata*
            6. Perennials, glumes 3-nerved .... *F. dichotoma*
         5. Nuts smooth or faintly ribbed:
            7. Stems leafless, or if present very short .... *F. ferruginea*
            7. Stems leafy, leaves well developed .... *F. cymosa*

Fimbristylis acuminata Vahl Enum. Pl. 2: 285. 1805; Gamble 1658 (1150).

Small erect herbs; stems terete, ridged; to 35 cm tall, obtusely trigonous, leaves 2-rowed; sheath inflated at apex; blades triangular; flowers in terminal spikelets; spikelets solitary, ovate, acute, 10 × 3.5 mm; glumes ovate-elliptic, brown, rounded; nuts ovate-orbicular, compressed, with wavy ridges.

Common on sandy sea coasts. Fl. & Fr.: Throughout the year. Kothapalem, *VRK* 5772.

Fimbristylis bisumbellata (Forssk.) Bub. Dodec. 30. 1850; Gamble 1658 (1151). *Scirpus bisumbellatus* Forssk. Fl. Aeg.—Arab. 1: 15. 1775. *Fimbristylis dichotoma* auct. non Vahl; FBI 6: 635.

Annual caespitose herbs; up to 15 cm tall; leaves flat, pubescent, sheaths grey, membranous; spikelets oblong, 3–7 × 0.1–0.2 mm, in terminal clusters; glumes ovate-oblong, reddish-brown; nuts obovoid.

Occasional in streams and canals. Fl. & Fr.: Oct.–June. Krishna river, *VRK* 6743; Nagarjunasagar, *KMS* 9807 (MH).

Fimbristylis cymosa R. Br. Prodr. 228. 1810. *F. spathacea* Roth Nov. Pl. Sp. 24. 1821; FBI 6: 640; Gamble 1659 (1151).

Perennial aquatic herbs, up to 50 cm tall; stems trigonous to terete; leaves linear, 15 cm long, grouped at the base, sheaths ash-brown; spikelets brownish, globose-ovoid, 3–15 × 0.1–0.2 cm, in terminal decompound umbellate spikes; involucral bracts 3, linear, glumes orbicular to ovate, 1.5 mm, membranous; nuts ovoid, trigonous.

Occasional along the streams. Fl. & Fr.: July–Mar. Javuku, *VRK* 3545; Nizampatnam, *VRK* 6914.

Fimbristylis dichotoma (L.) Vahl Enum. Pl. 2: 287. 1806; FBI 6: 635; Gamble 168 (1151). ssp. **podocarpa** (Nees & Mayen) T. Koyama, Micronesica 1: 87. 1964. *Scirpus dichotomous* L. Sp. Pl. 50. 1753. *Fimbristylis podocarpa* Nees & Maeyen in Wight Contrib. Bot. India 98. 1834.

Annual erect herbs; stems rigid, up to 70 cm tall, leaves canaliculate, rarely flat, scabrid, apex acute; sheaths rusty or pinkish; spikes simple, 5–11 cm; spikelets solitary, ovoid or oblong-ovoid; rachis winged; nuts obovoid, glossy, stipitate.

Occasional in open forests. Fl. & Fr.: July–Dec. Nagarjunakonda, *KTH* 9877 (CAL).

Fimbristylis falcata (Vahl) Kunth, Enum. Pl. 2: 239. 1837. *Scirpus falcatus* Vahl, Enum. Pl. 2: 275. 1806. *Trichelostylis junciformis* Nees in Wight, Contrib. Bot. India 106. 1834. *Fimbristylis junciformis* (Nees) Kunth, Enum. Pl. 2: 239. 1837; FBI 6: 647; Gamble 1660 (1152).

Perennial, erect, rhizomatous herb; rhizome oblique to 7 cm long, thickly clothed with brown leaf sheaths; culms solitary or sparsely tufted, rigid, 4-angled, deeply striate; leaves crowded, shorter than the stem, spreading and recurved with triangular acute tip; sheaths short, ligule 0, umbel compound; solitary, purplish, rhachilla winged, achenes obovoid, trigonous, stramineous.

Grows on rocky surfaces and open grassy places. Fl. & Fr.: June–Nov.

Fimbristylis ferruginea (L.) Vahl, Enum. Pl. 2: 291. 1506; FBI 6: 638; Gamble 1658 (1151). *Scirpus ferrugineus*. L. Sp. Pl. 50. 1753.

Perennial erect herb, rhizome absent or scarcely any; root fibrous stout and slender; culm subtrigonous, rigid; leaves absent or few, short, < 10 cm long, scabrid, apex acuminate; ligule a dense fringe of short hairs; sheath pubescent, thin, margin hyaline, umbels subcompound of few spikelets, to 5 cm long; involucral bracts 2, lowest involucral bract shorter than inflorescense; spikelets solitary, ovoid or oblong, pale reddish brown, pubescent; to 2 cm long; lowest spikelet bractiform; rhachilla angular and winged, achenes obovoid, biconvex, stipitate, strameneous.

Rare, found in wet places, open swamps, marshy areas and in brackish waters. Fl. & Fr.: Aug.–Dec. Nizampatnam estuary, *DLR* 106.

Fimbristylis miliacea (L.) Vahl Enum. Pl. 2: 287. 1806; FBI 6: 644; Gamble 1660 (1152). *Scirpus miliaceus* L. Syst. Nat. (ed. 10) 2: 868. 1759. *Fimbristylis littoralis* Gaudich. in Gaudich. Voy. Bonite, Bot. 413. 1826. *F. miliacea* auct. non (L.) Vahl; FBI 6: 544.

Annual caespitose herbs, up to 60 cm tall; culms compressed, 4-angled below, triquetrous above, leaves linear, 20–35 × 0.3–0.6 cm, apex acuminate, exceeding stem, sheaths stramineous; spikelets subglobose, purplish-grey in decompound umbels; glumes ovate; nuts obovoid.

Common along streams and canals. Fl. & Fr.: May–Nov. Sattenapally, *VRK* 5874; Jayanthipuram, *VRK* 6940.

Fimbristylis ovata (Burm. f.) Kern in Blumea 15: 126. 1967. *Carex ovata* Burm. f. Fl. Ind. 194. 1768. *Cyperus monostachyos* L. Mant. Pl. 180. 1771. *Fimbristylis monostachyos* (L.) Hassk. Pl. Jav. Rar. 61. 1846; Gamble 1660 (1152).

Densely tufted erect herbs, up to 35 cm tall; stem slender; trigonous; leaves flat, up to 15 cm, margins inrolled, scabrid; spikelets solitary, broadly ovate, much compressed; nuts obovoid, trigonous, tubercled.

Occasional in marshy places. Fl. & Fr.: Aug.–Dec. Kondaveedu Fort, *CAB* 17492 (MH); Nakrekal, *CAB* 17477 (MH).

Fuirena Rottboll

Fuirena ciliaris (L.) Roxb. Fl. Ind. 1: 184. 1820. *Scirpus ciliaris* L. Mant. Pl. 182. 1771. *Fuirena glomerata* Lam. Tabl. Encycl. 1: 150. 1791; FBI 6: 666. 1893. *F. rottboelii* Nees in Wight, Contrib. Bot. Ind. 94. 1834.

Annual, slender erect herbs, up to 1 m tall; stem angular, rigid, scaberulous; leaves flaccid, 2–8 × 0.3–0.5 cm; sparsely hairy above, glabrous below; flowers greenish in umbellate spikes; glumes oblong-obovate, to 2 mm, thin, 3-nerved, short-puberulous; nuts to 0.8 mm, fuscous, smooth.

Occasional weed in moist grounds and rice fields. *Vern.*: Meesala tunga. Fl. & Fr.: Throughout the year. Vinukonda RF, *VRK* 6867.

Kyllinga Rottboll *nom. cons.*

1. Rhizome short or absent, stem tufted .... ***K. bulbosa***
1. Rhizome long, stems solitary in linear rows along the length of the rhizome .... ***K. nemoralis***

Kyllinga bulbosa Beauv., Fl. d'Oware & Benin 1: 11. t. 8. f. 1. 1804. *Kyllinga triceps* Rottb. Descr. Icon. Rar. 14. t. 4. f. 6. 1773; *nom. illegit.*, FBI 6: 587; Gamble 1623 (1130). *Cyperus triceps* Endl. Cat. Horti Vindob. 1: 94. 1842.

Perennial tufted herbs, up to 20 cm tall; stems triquetrous; leaves few, flat, blades flattish-plicate, sheaths yellow; spikelets oblong, 1-flowered, in capitate spike; glumes narrowly linear, nuts oblong.

Occasional in moist places in forests and plains. Fl. & Fr.: May–Nov. Nagarjunakonda, *VRK* 6714; Vinukonda RF, *VRK* 6835.

Kyllinga nemoralis (Forst & Forst. f.) Dandy ex Hutchins & Dalziel Fl. W. Trop. Africa 2; 486, 487. 1936. *Thryocephalon nemoralis* Forst. & Forst. f. Char. Gen. Pl. 130. 1776; FBI 6: 588; Gamble 1624 (1134). *Cyperus kyllinga* Endl. Cat. Hort. Acad. Vinaob. 1: 94. 1842.

Perennial, erect marshy rhizomatous herb, about 30 cm tall; culms triquetrous, glabrous; leaves many, longer than the stem, linear, 15 × 0.4 cm; sheaths 5 cm long, glabrous; flowers green, several in terminal, ovoid heads; foliar bracts ca. 15 × 0.4 cm, linear; glumes ovate-acuminate, 2-toothed at apices; nuts brown, biangular.

Occasonal, found in roadsides and waste places (Mohan Rao & Aasiya, 1995). Fl. & Fr.: Throughout the year.

Pycreus P. Beauvois

1. Rhachilla not winged, nuts biconvex, spikes fuscous .... *P. flavidus*
1. Rhachilla winged:
    2. Glumes less than 1.5 mm long, spikelets 2 mm wide .... *P. pumilus*
    2. Glumes more than 2 mm long, spikelets 1.5 mm wide .... *P. polystachyos*

Pycreus flavidus (Retz.) T. Koyama in J. Jap. Bot. 51: 313. 1976 & in Gard. Bull. Straits Settlem. 30: 150. 1977. *Cyperus flavidus* Retz. Obs. Bot. 5: 13. 1788. *Pycreus globosus* (All.) Reich. Fl. Germ. Excuss. 1: 140 (10). 1830; Gamble 1627 (1132). *Cyperus globosus* All. Fl. Pedem. Auct. 49. 1789. *Pycreus capillaris* Nees ex Clarke in Hook. f. Fl. Brit. India 6: 591. 1893.

Annual tufted aquatic herbs, up to 50 cm tall; stems trigonous, leaves few, setaceous, flat, 10–16 × 0.1–0.3 cm, blades slenderly linear, plicate, sheaths purple; spikelets oblong, in compound spikes, involucral bracts 3, up to 15 cm long, glumes ovate, purplish; nuts ellipsoidal.

Common along the canal bank. Fl. & Fr.: July–Mar. Javuku, *VRK* 3548.

Pycreus polystachyos (Rottb.) Beauv. Pl. Oware 2: 48. t. 86. f. 2. 1807; FBI 6: 592. *Cyperus polystachyos* Rottb. Des. Pl. rar. 21. 1772. *Pycreus odoratus* Urban·Symb. Antill. 2: 164. 1900; Gamble 1627 (1132).

Annual tufted herbs, up to 45 cm tall; stem triquetrous, leaves flat, shorter than stem, sheaths purplish-brown; spikelets oblong in sub-compound spikes; glumes ovate-oblong, greenish yellow; nut oblong.

Occasional in streams and canals. Fl. & Fr.: June–Nov. Macherla, *VRK* 5880.

Pycreus pumilus (L.) Nees ex Clarke in Hook. f. Fl. Brit. India 6: 591. 1893; Gamble 627 (1132). *Cyperus pumilus* L. Cent. Pl. II. 6. 1756. *C. nitens* Retz. Obs. Bot. 5: 13. 1787. *Pycreus nitens* (Retz.) Nees Nov.

Act. Acad. Caes. Leop.—Carol. Nat. Cur. 19 (Suppl. 1): 53. 1843; FBI 6: 591.

Annual tufted herbs; up to 20 cm tall; roots fibrous; leaves many, flat, 5–8 × 0.2–0.6 cm, acuminate, involucral bracts 3–5, 7–12 × 0.3–0.6 cm; spikelets linear, 1 × 0.2 cm, in compound spike; glumes obovate, greenish; nuts ellipsoid, compressed.

Occasional in moist localities and rare in cultivated fields. Fl. & Fr.: Sept.–Dec. Nagarjunakonda, *VRK* 6715.

Schoenoplectus (Reich.) Palla

Schoenoplectus articularis (L.) Palla in Bot. Jahrb. Syst. 10: 299. 1888. *Scirpus articularis* L. Sp. Pl. 47. 1753; FBI 6: 656; Gamble 3: 1666 (1156).

Erect herbs, about 50 cm tall; culm tufted, terete, spongy; leaves reduced to stramineous-brown sheaths; spikelets ovate-oblong in stellately-arranged globose clusters; glumes ovate, pale green; nuts obovate, black.

Occasional, in marshy localities of Nagarjuna University campus area (Mohan Rao & Aasiya 1995). Nizampatnam coast. Fl. & Fr.: Aug.–Dec. *Vern.*: Arala tunga.

POACEAE

1. Spikelets 1- or many-flowered, usually laterally compressed or terete:
 2. Tall woody arborescent or shrubby bamboos, leaf-blades with petioled base:
  3. Spikelets in globose heads, reed-like shrubs .... ***Dendrocalamus***
  3. Spikelets not in globose heads, trees, more than 20 m tall .... ***Bambusa***
 2. Perennial or annual herbs, with herbaceous culms, leaf-blades sessile:
  4. Inflorescence of solitary or digitate spikes or rarely spiciform racemes:
   5. Spikelets breaking up at maturity above the persistent glumes:
    6. Spikes solitary:
     7. Spikelets clustered in heads .... ***Melanocenchris***
     7 Spikelets not clustered in heads.................................... ***Tetrapogon***
    6. Spikes 2 or more:
     8. Lemmas awned, rhachilla produced above, bearing 1–3-sterile lemmas .... ***Chloris***
     8. Lemmas awnless, rhachilla not produced above, empty lemmas 0 or rarely 1 .... ***Cynodon***
   5. Spikelets falling entire at maturity, either singly or in clusters:
    9. Glumes awned, lower glume present, lemmas 1-nerved .... ***Perotis***
    9. Glumes awnless, lower glumes 0 or reduced .... ***Tragus***
  4. Inflorescence usually a panicle, either open or contracted, rarely spiciform:
   10. Glumes very minute or suppressed, spikelet 1- or 3-flowered, palea 3-nerved, stamens 6 .... ***Leersia***
   10. Glumes usually well developed, palea 2-nerved, stamens 3 or fewer:
    11. Spikelets 1-flowered:
     12. Lemmas with 3-branched awns, glumes exceeding lemmas ..... ***Aristida***
     12. Lemmas awnless, glumes shorter than lemmas .... ***Sporobolus***

11. Spikelets 2- or 3-flowered:
    13. Tall reed-like grasses with plumose panicles .... ***Phragmites***
    13. Slender grasses without large plumose panicles:
        14. Lemmas 1–3-nerved:
            15. Inflorescence of digitate, whorled of capitate spikes:
                16. Spikes terminating in a sharp point, spikes digitate .... ***Dactyloctenium***
                16. Spikes terminating in a spikelet:
                    17. Lemmas hairy along nerves, glumes rounded on the back ..... ***Acrachne***
                    17. Lemmas glabrous along nerves, glumes strongly keeled ...... ***Eleusine***
            15. Inflorescence of open or contracted panicles or spiciform racemes:
                18. Lemmas 2–4-toothed or lobed .... ***Diplachne***
                18. Lemmas entire:
                    19. Glumes long-aristate, spikelets sessile, 2 or 3-flowered ..... ***Dinebra***
                    19. Glumes acute or obtuse, muticous or shortly mucronate:
                        20. Spikelets prominently secund, sessile .... ***Eragrostiella***
                        20. Spikelets not secund, pedicelled .... ***Eragrostis***
        14. Lemmas 9–11-nerved, spikelets clustered in heads .... ***Aeluropus***

1. Spikelets 2-flowered, usually dorsally compressed:
    21. Spikelets solitary, rarely paired, glumes usually membranous, upper lemmas papery, usually awnless:
    22. Spikelets subtended or surrounded by 1 or more bristles:
        23. Inflorescence of spiciform racemes, involucre of bristles thickened into a hard cup .... ***Cenchrus***
        23. Inflorescence of effuse or spiciform panicles, involucre of bristles not thickened as above:
            24. Upper lemmas rugose, bristles rigid, involucre persistent .... ***Setaria***
            24. Upper lemmas smooth, bristles soft, involucre deciduous .... ***Pennisetum***
    22. Spikelets neither subtended nor surrounded by bristles:
        25. Inflorescence globose, dioecious, bract pungent, leaves thick-coriaceous, rigid, coastal .... ***Spinifex***
        25. Inflorescence not globose, not dioecious, inland:
            26. Spikelets awned:
                27. Upper lemma aristate, lower lemma muticous; margins of upper glume densely ciliate or winged .... ***Alloteropsis***
                27. Upper lemma muticous, lower lemma awned:
                    28. Racemes 1- or 2-rowed, culms slender, creeping, lower glumes awned .... ***Oplismenus***
                    28. Racemes in 4- or more rows, culms tough, erect or suberect, lower glumes muticous .... ***Echinochloa***
            26. Spikelets unawned:
                29. Inflorescence of spiciform racemes:
                    30. Spikelets 1–6 in a cluster, mixed with scaly empty glumes .... ***Trachys***
                    30. Spikelets solitary, 2- or 3-nate:
                        31. Lower glumes present:
                            32. Lower glumes adaxial .... ***Urochloa***
                            32. Lower glumes abaxial .... ***Brachiaria***
                        31. Lower glume 0, or minute and scaly:
                            33. Spikelets orbicular, oblong-ovate .... ***Paspalum***
                            33. Spikelets elliptic, oblong or lanceolate .... ***Digitaria***
                29. Inflorescence of effuse or spiciform panicles:
                    34. Lower glumes and lowest rachilla-internode swollen into a callus .... ***Eriochloa***

34. Lower glumes and lowest rachilla-internode not swollen into callus .... ***Panicum***

21. Spikelets paired with 1 sessile and the other pedicelled, glumes as long as the spikelet and enclosing the florets, upper lemma often with a geniculate awn:

35. Inflorescence supported by spathes:

36. Racemes solitary .... ***Eremopogon***

36. Racemes in digitate or corymbose panicles:

37. Involucral spikelets present... ***Iseilema***

37. Involucral spikelets 0:

38. Basal joint of racemes bulbous, spikelets in threes .... ***Apluda***

38. Basal joint of racemes not bulbous, spikelets 2-nate .... ***Cymbopogon***

35. Inflorescence not supported by spathes:

39. Spikelets unawned:

40. Inflorescence of silky panicles, spikelets of each pair well developed .... ***Saccharum***

40. Inflorescence of solitary, fascicled or corymbose racemes, spikelet of each pair different .... ***Ophiuros***

39. Spikelets awned:

41. Racemes solitary:

42. All pairs of spikelets heterogamous, upper glumes of sessile and pedicelled or sessile spikelet awned, awns not hygroscopic .... ***Sehima***

42. Lower pairs of spikelets homogamous, upper heterogamous, all glumes awnless, awns hygroscopic .... ***Heteropogon***

41. Racemes more than one:

43. Spikelets in three, one sessile, other two pedicelled, raceme branches capillary .... ***Chrysopogon***

43. Spikelets binate:

44. Inflorescence a panicle .... ***Vetiveria***

44. Inflorescence not a panicle:

45. Lower glumes of sessile spikelets 2- or 3-toothed at apex .... ***Lophopogon***

45. Lower glumes of sessile spikelets not as above:

46. Awn from the sinus of the 2-toothed, upper lemma well developed .... ***Ischaemum***

46. Awn from the tip of the upper lemma reduced to the hyaline base .... ***Dichanthium***

Acrachne Wight & Arnott

Acrachne racemosa (Heyne ex Roem. & Schult.) Ohwi in Bull. Tokyo Sci. Mus. n. 18: 1. 1947; Bor 487. *Eleusine racemosa* Heyne ex Roem. & Schult. in L. Syst. Veg. 2: 583. 1817. *E. verticillata* Roxb. Fl. Ind. 1: 346. 1820; FBI 7: 295.

Annual herbs, culms erect, 30–80 cm tall; stout or slender, glabrous; leaf sheaths 3–12 cm long, appressed, shorter than the internodes; ligule hairy; blades 6–25 × 0.6–1.4 cm, lanceolate, apex acuminate; inflorescence of whorled or scattered spiciform racemes, up to 20 cm long; spikelets 4–5 mm, 8–20-flowered; glumes small, ovate, keeled, apex finely acuminate; lemma 3–4 mm long, broadly ovate, 3-nerved, keel excurrent, lateral nerves ending in small teeth; palea ciliate; anthers 3, 0.5 mm long; caryopsis oblong, rugose, grooved, 1 mm long, pericarp caducous.

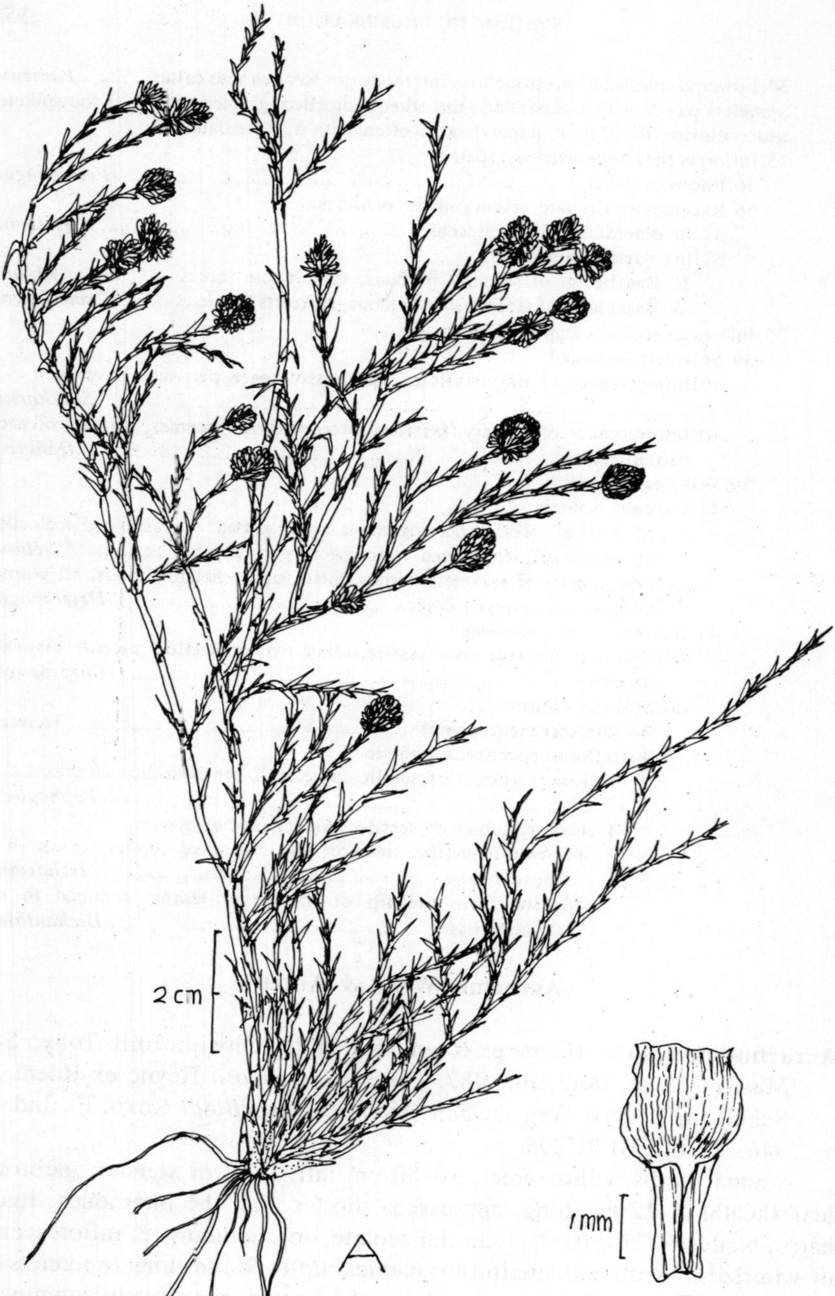
2 cm
1 mm
A
B

Occasional in moist places and in cultivated crops. Fl. & Fr.: Aug.–Nov. Ethipothala falls, *VRK* 5792.

Aeluropus Trinius

Aeluropus lagopoides (L.) Trin. ex Thw. Enum. Pl. Zeyl. 374. 1864; Gamble 1843 (1276); Bor 380. *Dactylis lagopoides* L. Mat. Pl. 33. 1767. *Aeluropus villosus* Trin. ex C. Meyer, Verz. Pfl. Casp. Meer. 18. 1831; FBI 7: 334.

Perennial creeping glabrous herbs; leaf sheaths compressed, 3–5 mm; ligule ciliate; blades lanceolate, 1–1.5 cm long, narrow, flat, folded, pungent, scaberulous on margin; heads subglobose, pedunculate, terminal, spikelets up to 7 mm; 4–8-flowered, crowded, elliptic-oblong, 2.5–3.5 mm, 4–8-flowered; caryopsis ovoid.

Common grass of coastal area in mangroves. Fl. & Fr.: Feb.–Aug. Nizampatnam, *VRK* 6809; Nizampatnam, *DLR* 149 (NUH).

Alloteropsis K.B. Presl

Alloteropsis cimicina (L.) Stapf in Oliv. Fl. Trop. Africa 9: 487. 1919; Gamble 1766 (1223); Bor 276. *Milium cimicinum* L. Mant. Pl. 184. 1771. *Axonopus cimicinus* (L.) Beauv. Ess. Agrost. 12. 1812; FBI 7: 64.

Annual tufted hairy erect or ascending herbs, up to 60 cm tall; leaf blades ovate-lanceolate, 3–7 × 0.4–1.5 cm, acute, ciliate, base amplexicaul, sheaths up to 3 cm long, ligule hairy, panicles terminal, spikes 3–9, whorled, 6–15 cm long, spikelets solitary, purplish green, pedicelled, up to 3 mm long, two-flowered; caryopsis ellipsoid.

Occasional weed of the forests. Fl. & Fr.: Aug.–Dec. Macherla RF, *G.H. Madhuram* 136 (CAL); Nakrekal, *CAB* 17476 (MH).

Apluda Linnaeus

Apluda mutica L. Sp. Pl. 82. 1753; Gamble 750 (1212); Bor 93. *A. aristata* L. Amoen.Acad. 4: 303. 1756. *A. varia* Hackel ssp. *aristata* (L.) Hackel Monogr. Andropog. in DC. Monogr. Phan. 6: 95. 1889; FBI 7: 150.

Perennial, tufted erect herbs, up to 2 m tall; culms cylindrical; leaves linear or lanceolate, 5–18 × 0.5–1 cm, glabrous above, scaberulous below and on margins; panicles terminal with many spike like racemes; spikelets 3, one sessile, the rest pedicelled, awns up to 1.5 cm long.

Common in scrub jungles of Loyapalle section. Fl. & Fr.: Aug.–Feb. It

---

Fig. 26. *Aeluropus lagopoides* (L.) Trin.
A. Twig, B. Ligule

is used as fodder. Loyapalle, *VRK* 3844; B.K. Palem, Macherla RF, *VRK* 3274; Kondavedu Fort, *CAB* 4736 (MH).

Aristida Linnaeus

1. Glumes muticous, panicles contracted, narrow, awns sessile .... ***A. adscensionis***
1. Glumes aristate, panicles not contracted:
   2. Panicles effuse, pyramidal with spreading branches .... ***A. hystrix***
   2. Panicles contracted, narrow-oblong, with erect branches:
      3. Awns with a twisted column, lower glume longer than upper one .... ***A. funiculata*** var. ***funiculata***
      3. Awns sessile, leaf-blades cauline, lower glume shorter than upper one .... ***A. setacea***

Aristida adscensionis L. Sp. Pl. 82. 1753; FBI 7: 224. *p.p.;* var. **adscensionis**; Bor 407. *A. depressa* Retz. Obs. Bot. 4: 22. 1786; Gamble 1809 (1252).

Perennial, tufted, ascending herbs; leaves narrow-linear, 5–20 × 0.1 cm, glabrous, sheath smooth, ligule hairy; panicles interruped, terminal, spikelets light violet, green, awns 3, middle one larger than laterals.

Common weed of forests. Fl. & Fr.: July–Feb. *Vern.*: Oobagaddi. Panicles are used for making brooms. Javuku, *VRK* 3552; Macherla RF, *P.C. Nanda* 135 (CAL, DD); Kondavedu, *CAB* 4710 (MH).

Aristida funiculata Trin. & Rupr. var. **funiculata** Sp. Gram. 159. 1842; FBI 7: 226; Gamble 1809 (1253); Bor 410.

Annual, tufted, erect herbs, up to 40 cm tall; branched from the base; leaves narrow-linear, 2.5–6 × 0.1 cm, scabrous, sheath pilose, ligule ciliate; panicles terminal, lax, branches erect, spikelets variable in length, glumes awned, central awn longer than laterals.

Occasional in forests. Fl. & Fr.: July–Feb. Macherla RF, *S.C. Agarwal* 134 (CAL); Kondavedu, *CAB,* 4699 (MH).

Aristida hystrix L.f. Suppl. Pl. 113. 1781; FBI 7: 225; Gamble 1809 (1252); Bor 410.

Perennial xerophytic tufted herbs, culms up to 50 cm, ascending with spreading woody root stock; leafsheaths 3–5 cm, densely pilose; blades 5–15 × 0.1–0.2 cm, narrow-linear, sparsely pilose along the margins; spikes in terminal lax panicles; spikelets linear, yellowish-green; caryopsis linear, 8 mm, prickly.

Occasional in plains and foot hills. Fl. & Fr.: July–Dec. Nagarjunakonda, *KTH* 9877 (CAL).

Aristida setacea Retz. Obs. Bot. 4: 22. 1786; FBI 7: 225; Gamble 1809 (1152); Bor 412.

Perennial, tufted, ascending herbs, with stout woody root stocks; leaf blades narrow, linear, 10–25 × 0.1–0.3 cm; coriaceous, ligule hairy; panicles lax, branches distant or compact, spikelets brownish green or greenish, fascicled; awns long, middle one longer than lateral ones.

Common in hedges, waste places, roadsides. Fl. & Fr.: Sept.–Apr. *Vern.*: Cheepuru gaddi. Panicles are used for making brooms and brushes. V.P. South, *VRK* 5821; Macherla RF, *S.C. Agarwal* 132 (CAL); Nagarjuna University campus, *VRK* 6993.

Bambusa Schreber

Bambusa arundinacea (Retz.) Willd. Sp. Pl. 2: 245. 1799; Beddome, Fl. Sylv. S. India tt. 321. 1872; Gamble 1859 (1286); FBI 7: 395. *Bambos arundinacea* Retz. Obs. Bot. 5: 24. 1788. *Bambusa bambos* (L.) Voss in P.L. Vilm. Blumengartherei t: 1189. 1896. *Arundo bambos* L. Sp. Pl. 81. 1753.

Large thorny shrubs; culms thick, erect 9–30 m tall; internodes hollow, sheaths hairy, triangular, 15–20 × 2.5–3.0 cm, glabrous, leafsheaths linear, 15 × 2.5 cm, blade trigonous, broad, petiole short; inflorescence a pseudo panicle of spikelets, clustered in heads, in spicate branchlets; spikelets oblong, 1–2 × 4 mm, 3–6-flowered, supported by 3 empty glumes; glumes broadly ovate, mucronate, paleas to 7 mm, keeled, stamens 6; styles 1; stigmas 3.

A rare shrub in the Vinukonda range. Not found in other places of Guntur. Fl. & Fr.: Very infrequent. Eng.: Thorny bamboo. *Vern.*: Veduru. Culms variously used for domestic purposes for mats, scaffolds and as raw material for paper pulp. Young buds and grains are consumed as food. Green bamboo is scratched with a knife and the powder thus obtained is applied externally to heal fresh wounds. Fractured bones are properly set with bamboo strips, plant paste applied for healing and tied with cloth for one month. Mellavagu, *VRK* 5738.

Brachiaria Griseback

1. Rachis of racemes ribbon like, more or less flat, wings narrow:
    2. Spikelets ellipsoid, glabrous .... ***B. distachya***
    2. Spikelets ovate-lanceolate, base with silky hairs .... ***B. mutica***
1. Rachis of racemes triquetrous, wings very prominent:
    3. Panicles linear, racemes closely grouped .... ***B. eruciformis***
    3. Panicles oblong, lanceolate, racemes apart:
        4. Spikelets up to 2.5 mm long .... ***B. reptans***
        4. Spikelets more than 3 mm long .... ***B. ramosa***

Brachiaria distachya (L.) Stapf in Oliver, Fl. Trop. Africa 9: 565. 1919; Gamble 1769 (1224); Bor 281. *Panicum distachyum* L. Mant. Pl. 183. 1771; FBI 7: 37.

Erect herbs, up to 30 cm, leafsheaths 1–3 cm, glabrescent; ligule to 0.5 mm, blades linear, 1–2 × 0.3–0.5 cm, cartilaginous on margins; racemes with 2–3 branches to 3 cm, spikelets ellipsoid, 3 x 1 mm; caryopsis ovoid.

Common grass in the forests. Fl. & Fr.: July–Dec. Macherla RF, *M.P. Guha* 128 (CAL).

Brachiaria eruciformis (J.E. Smith) Griseb. in Ledeb. Fl. Ross. 4: 469. 1853; Gamble 1769 (1226); Bor 283. *Panicum eruciforme* J.E. Smith in Sibth. & Sm. Fl. Graeca 1: 44. t. 59. 1806. *P. isachne* Roth ex Roem. & Schult. in L. Syst. Veg. 2: 458. 1817; FBI 7: 28.

Annual tufted herbs, up to 50 cm tall; suberect or ascending; leaves 2.5–7 × 0.2–0.4 cm, flat, lanceolate, appressed-hairy; panicles 6–7.5 cm long; spikes racemosely arranged, 1–3 cm long, sessile or shortly pedicellate.

Occasional in plains and forests. Fl. & Fr.: Throughout the year. Narasaraopet, *CAB* 4543 (MH).

Brachiaria mutica (Forssk.) Stapf in Prain. Fl. Trop. Afr. 9. 526. 1919; FBI 7: 34; Gamble 1770 (1226); Bor 284. *Panicum muticum* Forssk. Fl. Aeg. -Arab. 20. 1775.

Perennial aquatic tufted herbs; culms decumbent, erect, 60–150 cm tall; nodes bearded (villous); leafsheaths hairy, 6–17 cm; ligule pubescent; blades 15–20 × 0.4–1.7 cm, pilose, midrib prominent; acuminate; inflorescence of spiciform branched racemes, 12–18 cm; rachis flat, ribbon like with wavy margins; spikelets solitary, ovate-lanceolate, base with silky hairs; lower glume ovate; upper glume ovate-mucronate; lemma obovate; palea elliptic.

Cultivated as fodder grass in farm and occasionally as a weed in rice fields. Fl. & Fr.: Oct.–Dec. An excellent fodder grass for livestock. *Vern.*: Perai Gaddi. Sattenapally, *VRK* 5875.

Brachiaria ramosa (L.) Stapf in Oliver Fl. Trop. Africa 9: 542. 1919; Gamble 1770 (1226); Bor 284. *Panicum ramosum* L. Mant. Pl. 29. 1767; FBI 7: 36. *p.p.*

Annual, tufted, erect or ascending hairy herbs, up to 60 cm tall; rooting at nodes, leaf blades lanceolate, 5–10 × 0.5–1 cm, sparsely hairy; acuminate, margins ciliate, sheaths up to 6 cm long, hairy, ligule membranous; panicles terminal, spikes distant, rachis hairy; spikelets ellipsoid, greenish, densely hairy.

Common weed of open forests. Fl. & Fr.: Aug.–Feb. It serves as good fodder. N. Varam, *VRK* 3514.

Brachiaria reptans (L.) Gardn. & Hubb. in Hook. Icon. Pl. Sub. t. 3363. 1938; Bor 285. *Panicum reptans* L. Syst. Nat. (ed. 10) 870. 1759. *P. prostratum* Lam. Tabl. Encycl. 1: 171. 1791; FBI 7: 33. *Urochloa reptans* (L.) Stapf in Oliver Fl. Trop. Africa 9: 601. 1920; Gamble 1775 (1230).

Annual, creeping herbs, rooting at nodes; leaf blades lanceolate, 3–5 × 0.3–0.7 cm, acuminate to acute, glabrous; sheaths up to 4 cm long, margins densely hairy, ligule woolly; panicles terminal, rachis tomentose; spikes up to 3 cm long; spikelets 1–2 mm long, greenish; caryopsis ellipsoid.

Occasional weed in the forests. Fl. & Fr.: Sept.–Jan. Javuku, *VRK* 3551; Macherla RF, *B.D. Patil* 127 (CAL); Guntur, *Supt. Lam Farm* 94393 (MH).

Cenchrus Linnaeus

1. Involucral bristles retrorsely barbellate, rigid, spinose, lower glume absent .... ***C. biflorus***
1. Involucral bristles antrorsely barbellate, flexuous, not spinose, lower glume present ... . ***C. ciliaris***

Cenchrus biflorus Roxb. Fl. Ind. 1: 238. 1820; FBI 7: 89; Bor 287. *C. barbatus* Schum. Beskr. Buin. Pl. 43. 1827; Gamble 1794 (1242). *C. cathariticus* Del. Cat. Hort. Monsp. 1838. 42. 1839; FBI 7: 90.

Annual ascending herbs; leaves lanceolate, 6–12 × 0.2–0.5 cm, glabrous or sparsely hairy, sheaths glabrous; ligule pubescent; spike solitary, dense, greenish yellow; spikelets bristly; caryopsis orbicular-oblong, compressed, pale brown.

Occasional weed of cultivated fields and wastelands, rare in forests. Fl. & Fr.: Aug.–Dec. Its seeds are eaten in times of scarcity. Young shoots are acceptable to stock. V.P. South, *VRK* 5810.

Cenchrus ciliaris L. Mant. Pl. 302. 1771; Gamble 1749 (1242); Bor 287. *Pennisetum cenchroides* Rich. in Pers. Syn. Pl. 1: 72. 1805; nom. Superfl; FBI 7: 88.

A perennial, tufted grass; culms up to 50 cm tall, erect or decumbent, much branched; leaf sheaths 4–6 cm long, keeled, glabrous, ciliate near the mouth, ligule a pubescent ring, blade 7–15 × 0.2–0.5 cm, linear-lanceolate, base rounded, apex acuminate, glabrous-sparsely tuberculate pilose; racemes 6–10 cm long, solitary; involucral bristles antrorsely barbellate; outer filiform, not ciliate, inner flattened, thickened base, densely ciliate, connate below into a ring; spikelets 6 mm long; glumes ovate, hyaline, caryopsis oblong, apex beaked.

Common. Eng.: Bunch grass, African foxtail, Buffel grass. It is used as fodder and for soil conservation. Seeds are edible.

Chloris O. Swartz

1. Perfect florets 2-Spikelets 4-awned, culms puberulous below the inflorescence .... ***C. bournei***
1. Perfect floret only one:
   2. Empty lemmas above the fertile florets rudimentary .... ***C. dolichostachya***
   2. Empty lemmas above fertile florets well developed:

3. Sterile lemmas 2, keel of the fertile lemmas densely ciliate on the back .... C. **barbata**
3. Sterile lemmas 3, keel of the fertile lemmas glabrous on the back .... *C. montana*

Chloris barbata Sw. Fl. Ind. Occid. 1: 200. 1797; FBI 7: 292; Gamble 1838 (1272); Bor 465.

Annual tufted ascending herbs, up to 60 cm tall; leaf blades linear, 5–15 x 0.2–0.4 cm, flat, acuminate, sheaths 3–7 cm long; caryopsis oblong.

Occasional in forests. Fl. & Fr.: Aug.–Feb. *Vern*.: Uppu gaddi. Macherla RF, *M.P. Guha* 166 (CAL); Nagarjuna University campus, *VRK* 6983; Krishna river, *KMS* 9859 (MH); Macherla, *P.C. Nanda* 166 (DD).

Chloris bournei Rang. et Tad. in J. Indian Bot. 2: 189. 1921; Gamble 1838 (1272); Bor 466.

Perennial (rarely annual) tufted herbs; leaf blades 5–15 × 0.1–0.3 cm, glabrous, acute, sheaths up to 4 cm long; spikes umbelled, 3 to many, purplish brown, spikelets 2–3-flowered, perfect florets 2; caryopsis oblong.

Occasional in forests. Fl. & Fr.: Aug.–Dec. Macherla RF, *S.C. Agarwal* 130 (CAL).

Chloris dolichostachya Lagasca Gen. Sp. Pl. 5. 1816; Bor 466. *C. incompleta* Roth Nov. Pl. Sp. 60. 1821; FBI 7: 290; Gamble 1838 (1272).

Perennial, tufted, erect herbs, up to 1.5 m tall; base geniculate; leaf blades linear, 15–25 × 0.1–0.5 cm, glabrous, sheaths up to 20 cm long, ligule ciliate; spikes umbelled, 10–20 cm long, greenish-brown; spikelets 2-flowered, perfect flower 1, awns brown, up to 1.5 cm long; caryopsis oblong.

Common on hill slopes. Fl. & Fr.: Sept.–Feb. It is considered good fodder grass. Vinukonda RF, *VRK* 6852.

Chloris montana Roxb. Fl. Ind. 1: 331. 1820; FBI 7: 292; Gamble 1838 (1272); Bor 466. *C. montana* var. *glauca* Hook. f. in Trim. Handb. Fl. Ceyl. 5: 276. 1900.

Perennial, tufted, erect herbs; culms up to 40 cm long; leaves flat, linear, 4–11 cm long, glabrous; spikes 3–10 cm long, digitately arranged; rachis glabrous; spikelets 0.2 cm long, purplish; caryopsis linear, oblong.

Occasional in plains and forests. Fl. & Fr.: July–Oct. Nagarjunasagar, *KMS* 9756 (MH).

Chrysopogon Trinius *nom. cons.*

1. Culms geniculately ascending; pedicels half or as long as sessile spikelets, awn of sessile spikelet 3–5 mm long .... ***C. aciculatus***
1. Culms erect; pedicels shorter than half the length of sessile spikelet; awn of the sessile spikelet 20–25 mm long .... ***C. fulvus***

Chrysopogon aciculatus (Retz.) Trin. Fund. Agrost. 188. 1820; Gamble 1738 (1205); Bor 115. *Andropogen aciculatus* Retz. Obs. Bot. 5: 22. 1789; FBI 7: 188.

A perennial, tufted grass; culms up to 60 cm tall, geniculately ascending below; leaf sheaths 3–5 cm, glabrous; ligule a minute fimbriate rim, blades 2–7 × 0.4–0.6 cm, oblong, flat, base rounded, margin scabrid, apex acute; panicle narrow, up to 10 cm long; spikelets 2-nate; caryopsis laterally compressed.

In Nagarjuna University campus area (Mohan Rao & Aasiya 1995). Fl. & Fr.: July–Oct.

Chrysopogon fulvus (Spreng.) Choiv. Fl. Somal. 327. 1929; Bor 116. *Pollinia fulva* Spreng. Pl. Minus Cognit. Pug. 2: 10. 1815. *Chrysopogon montanus* Trin. ex Spreng. Neue Entdeck. 2: 93. 1821; Gamble 1739 (1205). *Andropogon monticola* Roem. & Schult. in L. Syst. Veg. 2: 665. 1827; FBI 7: 192.

Perennial, densely tufted, erect herbs, up to 1.5 m tall; leaves narrow, linear, 8–15 × 0.2–0.6 cm, acuminate, margins scabrid; panicles open spreading, spikelets 3-nate, one sessile, 2 pedicelled, yellowish-green, compressed.

Common weed in forests. Fl. & Fr.: July–Dec. It is used as fodder. N. Varam, *VRK* 3535; Krishna river, *KMS* 9806 (MH).

Cymbopogon K.P.J. Sprengel

1. Panicle narrow, of short, dense fascicles of racemes pairs, joints and pedicels villous all over, hairs long, concealing the sessile spikelets .... *C. jwarancusa*
1. Panicle large, decompound or compound, sometimes narrow, but the hairs not covering the joints and pedicels:
    2. Lower glumes of the fertile spikelets deeply channeled .... *C. caesius*
    2. Lower glumes of the fertile spikelets not channeled, slightly depressed:
        3. Pedicel of the lowest pedicelled spikelet in the sessile raceme not swollen .... *C. flexuosus*
        3. Pedicel of the lowest pedicelled spikelet in the sessile raceme swollen .... *C. coloratus*

Cymbopogon caesius (Hook. & Arn.) Stapf in Bull. Misc. Inform. 1906; 341 & 360. 1906; Gamble 1757 (1217); Bor 125. *Andropogon caesius* Nees ex Hook. & Arn. in Hook & Arn. in Hook. Bot. Beechey Voy. 244. 1833. *A. schoenanthus* L. ssp. *schoenanthus* var. *caesius* (Nees) Hackel in A. & C. DC. Mongr. Phan 6: 610. 1889; FBI 7: 205.

Perennial, tufted erect aromatic herbs, up to 1 m tall; nodes woolly; leaves narrow-linear, 10–20 × 0.2–0.4 cm, glaucous, ligule glabrous; panicles terminal, racemes unequal, paired, one sessile, other one pedicelled; spikelets greenish, awns up to 1.5 cm long.

Occasional weed of wastelands and forests. Fl. & Fr.: Sept.–Mar. *Vern.*: Kokchi grass. Leaves yield an essential oil which is used in soap manufacture. Nallapadu, *VRK* 6763; Nagarjuna University campus, *VRK* 6981.

Cymbopogon coloratus (Hook. f.) Stapf in Kew Bull. 1906: 321. 1906. Gamble 1756 (1217); Bor 126. 1960. *Andropogon nardus* L. var. *coloratus* Hook. f. FBI 7: 206. 1896.

A tufted perennial grass; culms up to 2 m tall, nodes bearded; leaf sheaths 5–15 cm, basal tomentose, upper glabrous; ligule fimbriate, membrane; blade 10–40 × 0.4–0.8 cm, flat, linear-lanceolate, base rounded, apex acuminate, scabrid; panicles dense, up to 8 cm, racemes 4 cm, 2-nate, glumes keeled, lemmas hyaline, epaleated, stamens 3.

Common in forests and plains of Nagarjuna University area (Mohan Rao & Aasiya, 1995). Fl. & Fr.: Aug.–Dec. *Vern.*: Bodagaddi, Doddanchi grass. Good raw material for paper manufacture. Dried grass used for thatching of hut.

Cymbopogon flexuosus (Nees ex Steud.) Wats. in Atkins. Gaz. N.W. Pro. Ind. 392. 1882; var. **flexuosus**; Gamble 1756 (1216); Bor 127. *Andropogon flexuosus* Nees ex Steud. Syn. Pl. Glum. 1. 388. 1854. *A. nardus* L. var. *flexuosus* Hack. in DC. Monogr. Phan. 6. 603. 1889; FBI 7: 207.

Perennial erect aromatic herbs; culms up to 3 m tall; nodes glabrous; leafsheaths 10–30 cm long, tomentose; blades lanceolate, glucous, 15–35 × 1–1.5 cm, lax, slender, flexuous, greyish green; spikelets 2-nate, sessile and pedicellate, rachis and pedicels softly pilose, slender; caryopsis ovoid.

Occasional in hills. Fl. & Fr.: July–Dec. *Vern.* East Indian lemon grass. Culms are used for thatching. The lemon oil, which is obtained from the leaves, is used in perfumery and cosmetics. Lemon grass oil is basic raw material for manufacture of many aromatic chemicals which find application in soap, perfumery and cosmetic industries. Oil is also used in many pharmaceutical preparations, such as pain balm and disinfectants. Writing and printing papers are reported to have been made from spent grass. Nagrjunakonda, *KTH* 9878 (CAL).

Cymbopogon jwarancusa (Jones) Schult. Syst. Veg. 2. Mant. 458. 1824; Bor 128. *Andropogon jwarancusa* Jones in Asiat. Res. 4: 109. 1875; FBI 7: 203.

A perennial, tufted grass; culms up to 70 cm tall; creeping at base; nodes glabrous; leafsheaths glabrous; ligule ciliate; blades 10–25 × 0.1–0.4 cm, linear, flat or convolute, glabrous, apex acuminate; panicle very variable; racemes unequal; joint and pedicels densely villous with long hairs; spikelets 2-nate, sessile spikelet 4 mm, lanceolate, bisexual; pedicelled spikelet longer than sessile spikelet, lanceolate.

Vern.: Kovi grass. Roots are used in treatment of fevers.

Cynodon Richard ex Persoon *nom. cons.*

Cynodon dactylon (L.) Pers. Syn. Pl. 1: 85. 1805; FBI 7: 288; Gamble 1835 (1270); Bor 469. *Panicum dactylon* L. Sp. Pl. 58. 1753.

Perennial stoloniferous procumbent herbs; leaf blades linear, 2–8 × 0.7–0.3 cm, flat, glabrous, obtuse, sheaths up to 2.5 cm long, ligule membranous; ciliolate; panicle digitate; spikes 3–5, up to 4.5 cm long, filiform; spikelets greenish, secund, dense, up to 3 mm long, sessile; caryopsis ellipsoid.

Occasional weed of wastelands and forests. Fl. & Fr.: Sept.–Mar. Eng.: Bermuda grass, Bahama grass. *Vern.*: Garika. Krishna river, *VRK* 6745; Guntur, *Suptd. Lam Farm* 94394 (MH).

Dactyloctenium Willdenow

1. Stoloniferous, spikes 2–5 cm long, tip of the rachis shortly produced up to 2 mm long .... *D. aegyptium*
1. Non-stoloniferous, spikes 0.5–2 cm long, tip of the rachis up to 4 mm long .... *D. aristatum*

Dactyloctenium aegyptium (L.) P. Beauv. Essai Agrost. Expl. t. 15. 1812; Gamble 1840 (1273); Bor 489. *Cynosurus aegyptius* L. Sp. Pl. 72. 1753. *Eleusine aegyptia* (L.) Desf. Fl. Atlant. 1: 85. 1798; FBI 295.

Annual, geniculately ascending herbs, rooting at nodes; leaf blades linear, 5–10 × 0.4–1 cm, sparsely hairy on nerves, acute, margin with bulbous based hairs, sheaths up to 3 cm long, ligule ciliate; panicle digitate, spikes 3–5, 1.5–4 × 0.4–0.6 cm, spikelets sessile, brownish-green, 3–5-flowered, glumes minutely hairy, mucronate; caryopsis reddish, subglobose.

Occasional weed of open forests. Fl. & Fr.: June–Dec. Eng.: Crow-foot. seeds are consumed as food, and straw as fodder. Koppanur, *VRK* 3242.

Dactyloctenium aristatum Link Hort. Berol. 1. 59. 1827; Bor 489. *Eleusina aristata* Hook. f. Fl. Brit. India 7: 296. 1896.

Perennial, slender, prostrate, proliferously branched grass; culm smooth, stiff, widely spreading, rooting at the distant thickened woody leafy nodes; blade linear-lanceolate, 2–4.8 × 0.1–0.8 cm, acuminate, glaucous, sparsely hairy, margin ciliate; leaf sheaths 0.5–1.75 cm long, glabrous or sparsely hairy; spike solitary, terminal, 1.5–2 cm long; rachis slender, bearded at base; spikelets 4 mm long, laterally compressed, crowded, sessile; caryopsis globose with rugose margins.

Rare in plains of Guntur district (Mary *et al.*, 1996).

Dendrocalamus Nees

Dendrocalamus strictus (Roxb.) Nees in Linnaea 9: 476. 1834; FBI 7: 405; Gamble 1858 (1286). *Bambusa stricta* Roxb. Pl. Cor. t. 18. 1798.

Perennial bamboos in erect clumps, up to 6 m tall; culms 10 cm in diameter; leaves 12 × 1.2 cm, ovate-acuminate; spikelets 0.8 cm long in clustered heads, along branched panicles; caryopsis beaked.

Occasional in degraded forests. Fl. & Fr.: Apr.–May. It is used in manufacture of paper, baskets, brushes and cellulose gum. Cellulose gum has many industrial applications such as in textile, paper, cosmetics, drugs, paints, leather and lacqures. Also found suitable for rayon-grade pulp. Reported in Guntur Forest Division Work Plan.

Dichanthium Willemet

1. Lower glumes of the sessile spikelets not hairy below the middle .... ***D. annulatum***
1. Lower glumes of the sessile spikelets hairy below the middle:
   2. Lower glume of the sessile spikelet pitted .... ***D. pertusum***
   2. Lower glume of the sessile spikelet without pit .... ***D. pseudoischaemum***

Dichanthium annulatum (Forssk.) Stapf in Oliv. Fl. Trop. Africa 9: 178. 1917; Gamble 1741 (1206); Bor 133. *Andropogon annulatus* Forssk. Fl. Aeg .-Arab. 173. 1775; FBI 7: 196.

Perennial, densely tufted, erect or ascending herbs, up to 1.5 m tall; leaf blades linear, 10–20 × 0.2–0.7 cm, sparsely hairy, acuminate, sheaths up to 10 cm long, ligule ciliate, panicles cluster like, spikes 6–9 cm long, spikelets greenish yellow, margins of lemma ciliate, awns brown, up to 2 cm long.

Common in forest undergrowth, Fl. & Fr.: Throughout the year. It is used as fodder. Guntur, *Suptd. Lam Farm* 96103 (MH).

Dichanthium pertusum (L.) Clayton in Kew Bull. 32: 4. 1977; Deshpande in Fasc. Fl. India 15: 22. 1984. *Holcus pertusus* L. Mant. Alt 301. 1771. *Andropogon pertusus* (L.) Willd. Sp. Pl. 4: 922. 1806; FBI 7: 173. *Amphilophis pertusus* (L.) Nash ex Stapf in Agric. News. Ind. 15: 179. 1916; Gamble 1731 (1200). *Bothriochloa pertusa* (L.) A. Camus in Ann. Soc. Linn. Lyon 1930; Bor 109.

Perennial, erect, tufted herbs; culms up to 50 cm tall; leafsheaths lax, 2.5–6 cm, glabrous; blades aggregated at the base, oblong, 8–15 × 0.2–0.4 cm, flat, acuminate; spikelets in digitate racemes; both sessile and pedicellate; caryopsis ellipsoid.

Occasional in plains in forests. Fl. & Fr.: July–Jan. Nagarjunakonda, *KTH* 9882 (CAL).

Dichanthium pseudoischaemum C. Nees ex. Steud. Jain et Deshpande in Bull. Bot. Surv. Ind. 20: 134. 1978 (1979); Despande in Fasc. Fl. Ind.

15: 23. 1984. *Andropogen pseudoischemum* Nees ex Steud. Sy. Pl. Glum 1: 380. 1854; FBI 7: 172. *A. ischaemum* sensu Roxb. Fl. Ind. 1: 264. 1820. non L. (1753). *Amphilophis pseudoischaemum* (Nees ex Steud.) Fischer Fl. Madras 1731 (1200). *Bothrichloa pseudoischaemum* (Nees ex Steud.) Henr. in Blumea 3: 457. 1940; Bor 109.

Perennial grass, leaf blade oblong, 15– 30 × 0.2–0.5 cm, scaberulous, rounded at the base, apex setaceous; 5–15 cm, glabrous, ligule 3–5 mm, membranous, fimbriate, spikes fasciculate, 4–10, 8 cm long; joints and pedicels pilose along the margins. caryopsis ellipsoid, 1.5 mm.

Common on hills. Fl. & Fr.: July–Sept. Loyapalle, *VRK* 3863.

Digitaria Heister ex Fabricius

1. Inflorescence of large effuse panicle; spikelets solitary .... ***D. tomentosa***
1. Inflorescence of spike like racemes, digitate; spikelets binate or ternate:
    2. Spikelets ternate .... ***D. abludens***
    2. Spikelets binate:
        3. Spikelets homomorphic .... ***D. ciliaris***
        3. Spikelets heteromorphic .... ***D. bicornis***

Digitaria abludens (Roem. & Schult.) Veldkemp. in Blumea 21: 53. 1973. *Panicum abludens* Roem. & Schult. in L. Syst. Veg. 2: 457. 1817. *Digitaria granularis* (Spreng.) Henrard Monogr. Digitaria 302. 1950; Bor 301. *Paspalum pedicellare* Trin. ex Hook. f. Fl. Brit. India 7: 19. 1896. *Digitaria pedicellaris* (Hook. f.) Prain Bengal Pl. 1182. 1903; Gamble 1765 (1223).

Annual tufted slender herbs; culms up to 30 cm long; leafsheaths 2–7 cm, glabrous; ligule short, scarious; blades 5–15 × 0.2–0.4 cm long, subterminal on slender peduncles; spikelets elliptic-oblong, 2–5-nate; caryopsis oblong, lightly enclosed in lemma and palea.

Occasional in plains. Fl. & Fr.: July–Oct. Macherla, *S.C. Agarwal* 170 (DD).

Digitaria bicornis (Lam.) Roem. & Schult. in L. Syst. Veg. 470. 1817; Bor 299. *Paspalum bicorne* Lam. Tabl. Encycl. 1: 176. 1791. *P. sanguinalis* Lam. var. *communatum* Hook. f. Fl. Brit. India 7: 15. 1896. *Digitaria marginata* Link. Enum. Hort. Berol. 1: 102. 1821; Gamble 1764 (1222).

Annual herbs; culms 20–40 cm tall; leaves 20 cm long, linear; spikes slender, 3–10, digitate, up to 8 cm long; spikelets 2-nate on abbreviated peduncles; caryopsis oblong, whitish.

Common grass in wastelands and fields. Fl. & Fr.: June–Dec., Vijayapuri, *KMS* 9839 (MH).

Digitaria ciliaris (Retz.) Koel. Descr. Gram. 27. 1802. *Panicum ciliare* Retz. Obs. Bot. 4: 16. 1786. *Digitaria marginata* Link var. *fimbriata*

(Link) Stapf in Oliver Fl. Trop. Africa 9: 440. 1919; Gamble 1764 (1222). *D. adscendens* (H.B.&K.) Henrard in Blumea 1: 92. 1934; Bor 298.

Annual, erect herbs; culms up to 50 cm tall; leaves oblong-linear or lanceolate, 2–8 × 0.3–0.8 cm, flat, sparsely pubescent, sheaths lax; inflorescence digitate, spikes 3, greenish brown, glumes linear, lanceolate or triangular.

Occasional weed of forests. Fl. & Fr.: Throughout the year. Eng.: Crab grass. Koppanur, *VRK* 3248.

Note: A very variable grass in which the digitate spikes range from 2-10 in number. This species is distinguished by glabrous homomorphous spikelets during early stages and sparsely pubescent spikelets at maturity.

Digitaria tomentosa (Willd.) Henrard in Blumea 1: 100. 1934; Bor 306. *Milium tomentosum* Koen. ex Willd. Ges. Naturf. Freunde Berlin Neue Schriften 4: 220. 1803. *Panicum subeglume* Trin. Mem. Acad. Imp. Sci. St. Petersb. Ser. 6, Sci. Math. 3: 292. 1834; FBI 7: 51; Gamble 1783 (1235).

Perennial, tufted, erect herbs, up to 1 m tall; culms yellowish; leaf blades linear, 5–15 × 0.4–1 cm, sparsely hairy, acuminate, sheaths hairy, ligule ciliate; panicles effuse, 15–30 × 5–10 cm; racemes lax, much branched; spikelets yellowish, distantly arranged, solitary, pubescent, elliptic, up to 2.5 × 0.75 cm; caryopsis ellipsoid.

Occasional weed of the moist waste places and rare in forests. Fl. & Fr.: Aug.–Feb. Ethipothala, *VRK* 5791.

Dinebra Jacquin

Dinebra retroflexa (Vahl) Panz. Denksch. Kongl. Acad. Wiss. Munchen 270. t. 12–1814; Gamble 1841 (1274); Bor 491. *Cynosurus retroflexus* Vahl, Symb. Bot. 2: 20. 1791. *Dinebra arabica* Jacq. Fragm. Bot. 77. t. 121. f. 1. 1807; FBI 7: 297.

Annual, tufted, ascending herbs, leaf blades lanceolate, 5–10 × 0.2–0.5 cm, acuminate, sheaths up to 6 cm long, ligule membranous; panicle pyramidal, spicate; spikes racemosly arranged along the axis, alternate; spikelets sessile, brownish-green; caryopsis ellipsoid-oblong.

Occasional weed of cultivated fields and open forests. Fl. & Fr.: Aug.–Dec. Macherla RF, *VRK* 5897; Krishna river, *KMS* 9877 (MH).

Diplachne P. Beauvois

Diplachne fusca (L.) P. Beauv. Ess. Agrost. 80: 163. 1812; FBI 7: 329; Gamble 1829 (1266); Bor 492. *Festuca fusca* L. Sp. Pl. ed. 2. 109. 1762.

Annual tufted, erect herbs; culms up to 10 m tall; leaves linear, flat, 20–30 cm long, rigid, convolute, glabrous; panicles 7–30 cm long, lax; spikelets 4–13-flowered, distant on slender rachis.

Occasional in plains. Fl. & Fr.: Aug.–Jan. Nagarjunasagar, *KMS* 9738 (MH).

Echinochloa P. Beauvois

1. Upper glume and lower lemma equally acute or cuspidate, unawned .... *E. colona*
1. Upper glume and lower lemma acuminate-cuspidate, awned .... *E. crusgalli*

Echinochloa colona (L.) Link Hort. Berol. 2: 209. 1833; Gamble 1776 (1230); Bor 308. *Panicum colonum* L. Syst. Nat. (ed. 10) 870. 1759; FBI 7: 32.

Annual tufted slender herbs, up to 75 cm tall; leaf blades linear, 5–20 × 0.2–0.7 cm, glabrous, acuminate, margins scaberulous; panicle terminal, spikes 5–12, racemosely arranged on rachis, 1–2 × 0.2–0.5 cm, spikelets globosely ovoid, 2–3 × 1–1.5 mm, greenish, glumes ciliate.

Common weed of rice fields and occasional in other fields, waste lands and along the streams. Fl. & Fr.: Throughout the year. Eng.: Shama millet. *Vern.*: Udarlu. It is used as fodder. Grains are eaten in times of scarcity. Macherla, *VRK* 5899; Krishna river, *VRK* 6737.

Echinochloa crusgalli (L.) P. Beauv. Ess. Agrost. 53: 161. 1812; Gamble 1777 (1231); Bor 310. *Panicum crusgalli* L. Sp. Pl. 56. 1753; FBI 7: 31.

Annual erect herbs, up to 1 m tall; leaf blades linear, 10–20 × 0.4–0.8 cm, glabrous, acuminate; sheaths up to 15 cm long, imbricate, ligule membranous; panicles spiciform, spikes 5–12, terminal much longer than the others, up to 6 cm long; spikelets globosely ovoid, greenish, 2.5–3.5 × 1–2 mm, margins of glumes hairy; awns up to 2.5 cm long.

Occasional weed of cultivated fields and rare along the streams. Fl. & Fr.: Throughout the year. It is used as forage. Grains are used for food in scarcity. Vinukonda RF, *VRK* 6860.

Eleusine Linnaeus

Eleusine indica (L.) Gaertn. Fruct. Sem. Pl. 1: 8. 1788; FBI 7: 293; Gamble 1839 (1273); Bor 493. *Cynosurus indicus* L. Sp. Pl. 72. 1753.

Annual erect tufted, herbs; culms up to 40 cm tall; compressed, slender, nodes glabrous; leafsheaths 2–6 cm, sparsely pilose, compressed, ligule hairy; blades linear, flat or plicate, crowded at base, 8–25 × 0.3–0.6 cm, glabrous; spikes 2–7, digitate, slightly recurved; spikelets ovate or oblong, 3–6-flowered, glabrous; caryopsis oblong, trigonous, rugose.

Occaional in plains, usually on the bunds of paddy fields. Fl. & Fr.: June–Sept. Used as good fodder grass and also serve as useful soil-binder. Nagarjunakonda, *KTH* 9897 (CAL).

Eragrostiella Bor

1. Leaves narrow, filiform, apex setaceous, keel of the palea broadly winged ... *E. bifaria*
1. Leaves broader, much reduced, apex obtuse, keel of the palea not broadly winged .... *E. brachyphylla*

Eragrostiella bifaria (Vahl) Bor in Ind. For. 66: 270. 1940. & Grass. Ind. 494. 1960. *Poa bifaria* Vahl Symb. Bot. 2: 19. 1791. *Eragrostis bifaria* (Vahl) Wight ex Steud. (Nom. Bot. ed. 21: 562. 1840) Syn. Pl. Glum. 1: 264. 1854; FBI 7: 325; Gamble 1828 (1265).

Perennial, tufted, ascending herbs, up to 70 cm tall; leaf blades narrow-linear, 8–20 × 0.1 cm, glabrous, acuminate; sheaths up to 4 cm long, old ones hairy, ligule ciliate; spike long, whip like; spikelets secund, distichously spreading on rachis, violet green, 8–20-flowered, laterally compressed; caryopsis globose.

Common weed of waste lands and occasional in cultivated fields and forests. Fl. & Fr.: June–Dec. Kondavedu Fort, *CAB* 17493 (MH).

Eragrostiella brachyphylla (Stapf) Bor in Ind. For. 66: 270. 1940 & Grass. Ind. 494. 1960. *Eragrostis brachyphylla* Stapf in Hook. f. Fl. Brit. India. 7: 327. 1896; Gamble 1828 (1265).

Perennial, tufted, erect herbs; culms up to 40 cm tall; leaves all basal, linear, 5–10 × 0.2–0.4 cm, glabrous; sheaths up to 5 cm, glabrous; spikes 5–10 cm long with distichously arranged spikelets; spikelets lanceolate, 1–2 cm, 2-seriate, 5–15-flowered; glumes subequal, lemmas with rounded tip; caryopsis ellipsoid.

Common in wastelands. Fl. & Fr.: July–Dec. Kondavedu Fort, *CAB* 17493 (MH).

Eragrostis P. Beauvois

1. Spikelets breaking up from downwards to apex, rachis tough:
   2. Pedicels of the spikelets with grandular bands; glumes less than 1 mm long .... *E. tenuifolia*
   2. Pedicels of the spikelets without glandular bands; glumes more than 1 mm long ... . *E. atrovirens*
1. Spikelets breaking up from apex to downwards, rachis fragile:
   3. Lemmas ciliate on the margin .... *E. coarctata*
   3. Lemmas not ciliate on the margin:
      4. Stamens 2, grain obovoid .... *E. diarrhena*
      4. Stamens 3, grain ovoid:
         5. Plant viscous, sweet scented, rachis glabrous at nodes .... *E. viscosa*
         5. Plant not viscous, not sweet scented, rachis bearded at the nodes .... *E. tenella*

Eragrostis atrovirens (Desf.) Trin. ex Steud. Nom. Bot. (ed. 2) 1: 562. 1840; Bor 503. *Poa atrovirens* Desf. Fl. Atlant. 1: 73. t. 14. 1798.

Perennial grass. Leaf blade linear, 10–30 × 0.3–0.4 cm, involute, sheaths 4–12 cm, glabrous; ligule 2–4 mm. Panicle more or less effuse, 10–25 cm long, pedicels 2–6 cm. Spikelets oblong, acute, 4–20-flowered, glaucous, caryopsis oblong, fusiform, 0.6 mm long.

Common grass occurring, in waste places in plains. Fl. & Fr.: Aug.–May. Macherla, *VRK* 5883.

Eragrostis coarctata Stapf in Hook. f. Fl. Brit. India 7: 313. 1896; Bor 507; Gamble 1825 (1263).

Perennial grass; culms up to 70 cm tall; wiry, tufted, nodes glabrous, leafsheaths 4–10 cm, glabrous; ligule 2–3 mm, hairy, blades linear, 3–10 × 0.2–0.3 cm, glabrous; panicle 6–12 × 0.5–1.5 cm, narrowly pyramidal, dense, branches hispid, spikelets 3–4 mm long, laterally compressed.

Occasional in forests. Fl. & Fr.: Nov.–June. Macherla RF, *B.D. Patil* 129 (CAL, DD).

Eragrostis diarrhena (Schult.) Steud. Syn. Pl. Glum. 1: 266. 1854; Gamble 1826 (1264); Bor 507. *Poa diarrhena* Schult. Veg. 2. Mant. 616. 1827. *Eragrostis interrupta* P. Beauv. var. *diarrhena* Stapf in Hook. f. Fl. Brit. India 7: 316. 1896.

Perennial tufted erect herbs, up to 75 cm tall; leaf blades linear, 10–20 × 0.1–0.3 cm; sheaths up to 10 cm long, ligule membranous; panicles terminal, contracted, 20–30 × 1.5–3 cm, racemes solitary; spikelets 6–7-flowered, brownish-green; carypsis obovoid.

Occasional weed of cultivated fields. Fl. & Fr.: Sept.–May. Culms are reported to be used for broom sticks. Nagarjunakonda, *VRK* 6713; Krishna river, *KMS* 9799 (MH).

Note: This species is sometimes called *E. interrupta* (Lamk.) P. Beauv. Ess. Agrost. 71. 1812. i.e. based on *Poa interrupta* of Lamarck. But Beauvois did not base his name on Lamarck's species, but on Robert Brown's *Poa interrupta* R.Br. *Eragrostis interrupta* (R. Br.) Beauv. an Australian grass is quite distinct from *E. diarrhena.*

Eragrostis tenella (L.) P. Beauv. ex Roem. & Schult. in L. Syst. Veg. 2: 576. 1817. var. *tenella*; FBI 7: 315; Bor 513. *Poa tenella* L. Sp. Pl. 69. 1753.

Annual tufted, erect herbs; culms up to 50 cm tall; leaves lanceolate, 6–12 × 0.2–0.4 cm, glabrous; sheaths up to 4 cm, ciliate near the mouth; panicles 6–15 cm long, lax, with loosely spreading spikelets; spikelets up to 4 mm, 3–9-flowered; caryopsis ovoid.

Occasional in forests. Fl. & Fr.: July–Dec. Phirangipuram, *CAB* 4605 (MH); Kondavedu Fort, *CAB* 4712 (MH).

Eragrostis tenuifolia Hochst. ex Steud. Syn. Pl. Glum. 1: 268. 1854; Gamble 1826 (1264); Bor 515. *Poa tenuifolia* A. Rich, Tent. Fl. Abyss. 2: 425. 1851.

Perennial, tufted, prostrate or geniculately ascending viscid herbs, branches filiform; leaf blades narrow linear, 5–7 × 0.1 cm, sticky, sheaths up to 2 cm long, ligule long, hairy; panicle oblong, 5–15 × 1–3 cm, rachis sticky; spikelets purplish-brown or brownish-green, 6–12-flowered; caryopsis ovoid.

Common weed of the forests. Fl. & Fr.: June–Dec. Koppanur, *VRK* 3247.

Eragrostis viscosa (Retz.) Trin. Mem. Acad. Imp. St. Petersburg. Ser. 6. Sci. Math. 1: 397. 1830; Gamble 1826 (1264); Bor 515. *Poa viscosa* Retz. Obs. Bot. 4: 20. 1786. 187. *Eragrostis tenella* var. *viscosa* (Retz.) Stapf in Hook. f. Fl. Brit. India 7: 315. 1896.

Perennial viscid scented grass; culms erect, 10–50 cm tall, 1–2 mm wide, tufted, glandular patches scattered throughout, nodes glabrous, leafsheaths 2–6 cm, ciliated at mouth, viscid below the collar; ligule a fimbriate membrane; blades 1.5–8 × 0.1–0.3 cm, glabrous, sticky, apex setaceous; panicle dense, narrowly cylindrical, 5–13 × 0.5–2.5 cm, glandular, branches numerous, spreading; caryopsis ovoid.

Abundant grass in waste places in plains and forests and fields. Fl. & Fr.: July–Mar. *Vern.*: Bankagaddi. Relished by cattle both fresh and dry. Mellavagu, *VRK* 5714; Guntur, *S.C. Agarwal* 887 (DD).

Eremopogon Stapf

Eremopogon foveolatus (Del.) Stapf in Prain, Fl. Trop. Africa 9. 183. 1917; Gamble 1751 (1213); Bor 148. *Andropogon foveolatus* Del. Descri. Egypte. Hist. Nat. 160. t. & f. 2. 1812; FBI 7: 168.

Perennial, erect, tufted herbs; culms up to 60 cm, ascending, leaves linear, 5–18 x 0.2–0.4 cm, pubescent, acuminate; sheaths up to 5 cm, glabrous, racemes solitary, subtended by spatheoles; spikelets binate, one sessile, awned, other pedicelled, unawned; caryopsis ellipsoid.

Common in waste lands and cultivated fields. Fl. & Fr.: July–Dec. Namala chettu. It is considered as good fodder grass. Narasaraopet, *CAB*, 4546 (MH).

Eriochloa Humboldt, Bonpland et Kunth

Eriochloa procera (Retz.) C.E. Hubb. Bull. Misc. Inform. 1930. 256. 1930; Gamble 1767 (1224); Bor 312. *Agrostis procera* Retz. Obs. Bot. 4. 19. 1786. *Eriochloa polystachya* Hook. f. Fl. Brit. Ind. 7: 20. 1897.

Perennial, tufted herbs; culms up to 60 cm tall; glabrous with pubescent nodes; leaves linear, 10–30 × 0.2–0.6 cm, glabrous, sheaths to 15 cm, terete;

panicle 15–20 cm long; spikelets in 5–10 cm long racemes, paired, ovate-lanceolate to 4 mm, light-purple; glumes unequal, caryopsis ellipsoid.

Occasional in moist localities. Fl. & Fr.: Aug.–Dec. Tufted grass used as fodder. Kondavedu Fort, *CAB* 4708 (MH), *SSR* & *KSM* 21606.

Heteropogon Persoon

Heteropogon contortus (L.) P. Beauv. ex Roem. & Schult. in L. Syst. Veg. 2: 836. 1817; Gamble 1743 (1208); Bor 163. *Andropogon contortus* L. Sp. Pl. 1045. 1753; FBI 7: 199.

Perennial, tufted, erect or ascending herbs, up to 1.25 m tall; leaves linear, 7–15 × 0.2–0.6 cm, scaberulous, acuminate, sheaths up to 5 cm long, ligule fimbriate; racemes terminal, solitary; spikelets hairy, lower ones awnless, upper ones long awned, up to 7 cm long, hygroscopic, twist about each other.

Common weed of the forests. Fl. & Fr.: Aug.–Feb. Eng.: Spear grass. Used as fodder, and culms are made into mats. Also used for soil conservation and in manufacture of paper and boards. Koppanur, *VRK* 3243; V.P. South, *VRK* 5823; Nagarjunakonda, *VRK* 6709; Kondavedu Fort, *CAB* 4706 (MH).

Ischaemum Linnaeus

Ischaemum rugosum Salisb. Icon. Strip. Rar. 1. t. 1. 1781; FBI 7: 127; Gamble 1722 (1194); Bor 184.

Aquatic, annual, erect, tufted herbs, up to 60 cm tall; leaf blades lanceolate, 7–15 × 0.2–0.6 cm, glabrescent, nerves minutely ciliate below, acuminate; sheaths up to 8 cm long, glabrous, ligule hairy; panicle spiciform; spikes 2, jointed up to 1/3 of the panicle length; spikelets up to 6 mm long, rugose, ridges deep, shining yellow, glabrous; awns brown, up to 5 cm long.

Common weed of rice fields. Occurring in wet places. Fl. & Fr.: Aug.–Dec. Young plants are eaten by cattle and horses, and grain is used as food in times of scarcity. Vinukonda, *VRK* 6861.

Iseilema Anderson

1. Pedicelled spikelets shorter than and hidden by the involucral spikelets .... ***I. venkateswarlui***
1. Pedicelled spikelets projecting beyond the tips of the involucral spikelets:
    2. Glumes coriaceous, lower glumes hairy at base, ciliate on the margins .... ***I. anthephoroides***
    2. Glumes membranous, lower glumes glabrous at base, not ciliate on the margins:
        3. Spathes broad, glabrous, peduncles 3–3.5 mm long, hispid .... ***I. laxum***
        3. Spathes narrow, tubercled on the keel and margins .... ***I. prostratum***

Iseilema anthephoroides Hack. in A. & DC. Monogr. Phan. 6: 683. 1889; FBI 7: 219; Gamble 1748 (1211); Bor 187.

Perennial, tufted, decumbent herbs; nodes glabrous; leaves narrow, linear, 4–7.5 × 0.1–0.3 cm, glabrous, sheaths up to 3 cm long; panicles axillary and terminal, clustered in foliar bracts; spikelets linear, compressed, awns brown, up to1 cm long.

Common in the hedges and bushes. Fl. & Fr.: July–Apr. It is considered to be an excellent fodder grass. Loyapalle, *VRK* 3996, 3999; Phirangipuram, *CAB* 4620 (MH).

Iseilema laxum Hack. in A. & C. DC. Monogr. Phan. 6: 682. 1889; FBI 7: 218; Gamble 1748 (1211); Bor 188.

Perennial ascending herbs; culms up to 80 cm long; nodes glabrous; keeled; blades cauline, 2–8 × 0.2–0.4 cm, lanceolate, apex subobtuse, scaberulous at the apex; panicles 10–12 cm long, flexuous; spikelets both sessile and pedicellate; caryopsis obovoid.

Occasional in moist places. Fl. & Fr.: Aug.–Mar. *Vern.*: Nendra gaddi. A good fodder grass and also acts as an efficient soil binder. Nagarjunakonda, *KTH* 9891 (CAL).

Iseilema prostratum (L.) Anders. in Nov. Act. Soc. Sci. Upsal. Ser. 3. 2: 251. 1856; Gamble 1747 (1211); Bor 188. *Andropogon prostratus* L. Mant. Act. 204. 1771. *Iseilema wightii* Anders in Nov. Act. Soc. Sci. Upsal. Ser. 3. 2: 251. 1856.

Perennial, tufted, erect herbs; culms up to 1 m tall, with reddish tinge; leaves purple, 18–30 cm, glabrous; sheaths with long base; racemes compound, axillary; spikes with long peduncles; spikelets sessile; awns hairy, violet green; caryopsis ellipsoid.

Common in forests. Fl. & Fr.: July–Dec. It is an excellent fodder. Krishna river at Vijayapuri, *KMS* 9821 (MH).

Iseilema venkateswarlui Sathyavathi in J. Bombay Nat. Hist. Soc. 65. 667. 1969.

An annual prostrate grass, culms 30–40 cm tall; rooting at nodes, round and purple or white in colour; involucral spikelets 6.5–7.5 mm, oblong-acute, markedly umbonate at base; pedicels 1.25 mm, densely ciliate on the outer margins; spikelets 2-nate.

Rare. Reported by U.R.K. Murty and Satyavati from Guntur District near Lam Farm, Guntur. Fl. & Fr.: June–Feb.

Leersia O. Swartz

Leersia hexandra Sw. Nov. Gen. Pl. 21. 1788; FBI 7: 94; Gamble 1845 (1277); Bor 599.

Annual ascending herbs, up to 1 m; nodes bearded; leaf sheaths up to 9.5 cm; ligule up to 4.5 mm, membranous; blades lanceolate, 5–20 × 0.5–1 cm, scaberulous, setaceous; panicles up to 15 cm; branches up to 7 cm; spikelets solitary, laterally compressed, oblong, 1-flowered; glumes absent; caryopsis ovoid.

Occasional weed in rice fields and rare in other places. Eng.: Rice grass. Fl. & Fr.: July–Oct. Nallapadu, *VRK* 6771.

Lophophogon Hackel

Lophophogon tridentatus (Roxb.) Hack. in Engl. Nat. Pfl. 2(2): 26. 1887; FBI 7: 149; Gamble 1724 (1195); Bor 190. *Andropogon tridentatus* Roxb. Fl. Ind. 1: 261. 1820.

Perennial tufted, ascending herbs, up to 30 cm tall; leaf blades narrow linear, 5–15 × 0.1–0.4 cm, glabrous, sheaths up to 5 cm long, ligule ciliate, panicles axillary and terminal, enclosed by ovoid, flattened, caudate-acuminate sheath; spikes 2, appearing like one; spikelets brownish, imbricate, callus with a tuft of hairs at the tip; awns up to 2 cm long, brown; caryposis ovoid.

Common weed on rocky soil especially river bank. Fl. & Fr.: Aug.–Feb. It is a useful grass and cattle nibble it only before flowering. Krishna river, *VRK* 6753; Krishna river, *KMS* 9828 (MH).

Melanocenchris Nees

Melanocenchris jacquemontii Jaub. ex Spach. Ill. Pl. Or. 4: 36. 325. 1851; Bor 473. *M. royleana* Nees ex Steud. Syn. Pl. Glum. 1: 218. 1825; Gamble 1831 (1268). *Gracilea royleana* Hook. f. Fl. Brit. India 7: 284. 1896.

Annual tufted ascending herbs, up to 15 cm tall; leaves narrow-linear, 2–6 × 0.1–0.2 cm, acute, convolute; panicle terminal, 3–7 cm long, spikelets clustered in heads, silky, yellowish green, awns brown; caryopsis ovoid.

Common weed of dry wastelands and cultivated fields. Fl. & Fr.: July–Dec. Used as fodder. Mellavagu, *VRK* 5713; Ethipothala, *VRK* 5790.

Ophiuros Gaertner

Ophiuros exaltatus (L.) Kuntze. Rev. Gen. Pl. 2: 780. 1891; Gamble 1758 (1218). *Aegilops exaltata* L. Mant. Pl. 575. 1771. *Rottboellia corymbosa* L.f. Suppl. Pl. 114. 1781. *Ophiuros corymbosus* Gaertn. f. Suppl. Carp. 3:4 t. 181. f. 3. 1805; FBI 7: 160.

Perennial rhizomatous, tufted, erect herbs; culms up to 2 m tall; nodes glabrous; leafsheaths 2–6 cm long; ligule membranous, fimbriate; blades linear, lanceolate, 3–1.5 × 0.3–0.6 cm, glabrous, pilose, flat, acuminate; spikes 3–6 in a cluster in corymbose racemes; spikelets oblong, 2–3 mm, awnless; caryopsis oblong, slightly compressed.

Occasional in black cotton soils. Fl. & Fr.: July–Nov. The grass is used for thatching. Macherla, *S.C. Agarwal* 178 (DD).

Oplismenus P. Beauvois

Oplismenus burmanii (Retz.) P. Beauv. Agrost. 54. 168–169. 1812; FBI 7: 68; Gamble 1778 (1232); Bor 317. *Panicum burmannii* Retz. Obs. Bot. 3: 16. 1783.

Annual slender creeping herbs; culms diffusely branched, up to 50 cm long; nodes glabrous; leafsheaths 1–4 cm long, glabrous-sparsely pilose, blades ovate-lanceolate, 2–6 × 0.5–1.2 cm, sparsely pilose, with long slender white hairs; racemes 2–6 cms long; spikelets 2–2.5 cm long, 2-nate, pedicels short; caryopsis oblong, up to 1 mm long.

Occasional grass in the shade of the trees in forests. Fl. & Fr.: Aug.–Dec. It is good lawn grass in humid tropical regions. Jayanthipuram, *VRK* 6952.

Panicum Linnaeus

1. Perennials, culms usually creeping .... *P. repens*
1. Annuals, culms erect:
   2. Leaf blade base rounded or truncate, caryopsis ovoid:
      3. Spikelets gaping at maturity; spikelets persistent .... *P. trypheron*
      3. Spikelets not gaping at maturity; spikelets deciduous .... *P. miliaceum*
   2. Leaf blade base amplexicaul, caryopsis ellipsoid .... *P. notatum*

Panicum miliaceum L. Sp. Pl. 58. 1753; FBI 7: 45; Gamble 1782 (1234); Bor 327. 1960.

Annual, tufted grass; culms up to 1 m tall, sparingly branched; leaf sheath 7 cm long, loose, hirsute; ligule a ciliate rim; blade 5–20 × 0.4–1 cm, lanceolate, base rounded, margins involute, apex acuminate, sparsely hairy on both surfaces, panicle 20 cm long, narrowly oblong; spikelets ovate-oblong, apiculate, sparsely hairy on both surfaces, panicle 20 cm long, narrowly oblong; spikelets ovate-oblong, apiculate, glumes persistent, unequal; caryopsis ovoid.

Occasionally cultivated for its grains and often as an escape, found in Nagarjuna University campus area (Mohan Rao & Aasiya, 1995). Fl. & Fr.: July–Nov. *Vern.*: Varigalu.

Panicum psilopodium Trin. Gram. Panic. 217. 1826; FBI 7: 46; Gamble 1782 (1234); Bor 329.

Annual, tufted, erect herbs, culms up to 40 cm tall; glabrous, flattened, angled; leaves spathaceous, lanceolate, acuminate; ligules hairy; spikes in terminal panicles; spikelets, 0.2–0.25 cm long, on slender pedicel; caryopsis ovoid.

Occasional in moist localities. Fl. & Fr.: Sept.–Feb. Plants are used as fodder. Guntur, *Lam Farm Superintendent* 94392 (MH).

Panicum repens L. Sp. Pl. ed. 2. 87. 1762; FBI 7: 49; Gamble 1783 (1235); Bor 330.

Perennial; culms up to 1 m, rhizomatous, rooting from the nodes; leaf sheath margins ciliate; ligule membranous, fimbriate; blades distichous, lanceolate, 5–15 × 0.3–0.7 cm, involute, sparsely pubescent; panicle oblong, up to 12 cm; spikelets ovate-lanceolate, 3 mm long, 2-flowered, unawned; glumes unequal, lower glume 1 mm long, upper glume up to 3 mm; caryopsis ellipsoid, 1.5 mm.

Occasional, semiaquatic grass in Nagarjuna University campus (Mohan Rao & Aasiya, 1995). Fl. & Fr.: Mar.–Dec. *Vern.*: Karigaddi.

Panicum trypheron Schult. Mant. Syst. Veg. 2: 244. 1824; FBI 7: 47; Gamble 1783 (1235); Bor 331.

Annual slender herbs, up to 60 cm tall; leaf blades linear, 8–15 × 0.4–0.8 cm, glabrous, sheaths up to 7 cm long, ligule ciliate, panicles terminal, spikes branched; spikelets ellipsoid, gaping widely at anthesis.

Occasional in forests along the streams. Fl. & Fr.: Sept.–May. Seeds are said to be used as bread in times of scarcity. Macherla RF, *S.C. Agarwal,* 131 (CAL); Phirangipuram, *CAB* 4606 (MH).

Paspalum Linnaeus

1. Culms rhizomatous, perennial, spikelets elliptic .... *P. paspaloides*
1. Culms non-rhizomatous, annual, spikelets orbicular .... *P. scrobiculatum*

Paspalum paspalodes (Michx.) Scribn. in Mem. Torr. Bot Cl. 5. 29. 1894. *(paspaloides)*. *Digitaria paspaloides* Michx. Fl. Bor. Amer. 1: 46. 1803. *Paspalum distichum* auct. non L. 1759. nom. confus.; FBI 7: 12. 1896; Bor 338.

Perennial decumbent herbs; stolons long, forming mats; leaves linear-lanceolate, 4–8 × 0.2–0.5 cm, glabrous, acute, sheaths loose, ligule ciliate, spikes 2, rarely 3–4, spikelets obovate-oblong, 2-seriate; caryopsis elliptic, depressed.

Occasional weed of moist localities. Fl. & Fr.: Aug.–Dec. N. Varam, *VRK* 3507.

Paspalum scrobiculatum L. Mant. Pl. 1: 29. 1767; FBI 7: 10; Gamble 1772 (1227); Bor 340.

Annual, tufted, erect herbs; culms up to 60 cm tall; leaves linear-lanceolate, 5–30 × 0.2–0.6 cm, acuminate, glabrous; sheaths glabrous; spikes 2–5, up to 8 cm long, in shortly peduncled racemes; spikelets orbicular, up to 3.5 mm, 2-flowered, upper glume glabrous; caryopsis ovoid.

Cultivated, rarely as an escape. Fl. & Fr.: Throughout the year. *Vern.*: Arikelu. Grain used as food, straw as fodder. A decoction of roots is used as an alterative in child-birth and the stem juice is useful for corneal opacity. Krishna river, *K.M. Sebastine* 9865 (MH).

Pennisetum L.C. Richard ex Persoon

1. Spikelets subsessile, not gaping, longest involucral bristles 1–5 cm, wild .... ***P. purpureum***
1. Spikelets prominently pedicelled, gaping and exposing grain, longest involucral bristle 5–7 mm, cultivated .... ***P. americanum***

Pennisetum americanum (L.) Leeke. Z. Naturwiss. 79: 52. 1907. *Panicum americanum* L. Sp. Pl. 56. 1753. *Pennisetum typhoides* (Burm. f.) Stapf & Hubb. Bull. Misc. Inform. 1933: 271. 1933; Gamble 1792 (1241); Bor 350. *Alopecurus typhoides* Burm. f. Fl. Ind. 27. 1768. *Pennisetum typhoideum* Rich. in Pers. Syn. Pl. 1: 72. 1805; FBI 7: 82.

Annual erect herbs, up to 2 m tall; leaves linear, flat or convolute, inflorescence of cylindrical panicles; spikelets 2–5 in a cluster, subtended by scabrid, involucral bristles; glumes small, hyaline, lower glume often reduced or 0; upper glume sometimes as long as spikelet; lower lemma with a male floret; upper lemma with a bisexual floret; caryopsis oblong, compressed dorsally.

Common cultivated plant. Rarely as an escape. Fl. & Fr.: Aug.–Dec. *Vern.*: Sajjalu, V.P. South, *VRK* 5822.

Pennisetum purpureum Schum. Beskr.-Guin. Pl. 44. 1827; Bor 430.

Erect herbs; culms decumbent, to 1 m; nodes bearded; leafsheaths to 10 cm; ligule to 3 mm; blades linear, 10–30 × 0.7–1 cm, panicles 15 × 1.5 cm, spikelets subterete, to 6 mm.

The Napier Grass. Occasional in forests as an escape. Fl. & Fr.: July–Dec. Cultivated as a fodder grass. Vinukonda RF, *VRK* 6854.

Perotis W.T. Aiton

Perotis indica (L.) Kuntze Rev. Gen. Pl. 2: 787. 1891; Gamble 1814 (1256); Bor 611. *Anthoxanthum indicum* L. Sp. Pl. 28: 1753. *Perotis latifolia* Ait. Hort. Kew. 1. 96. 1789; FBI 7: 98.

Perennial ascending tufted herbs; culms up to 40 cm, tall; geniculate at base; nodes glabrous; leaves ovate-lanceolate, 1–3.5 × 0.2–0.6 cm, glaucous; sheaths 1.5–7 cm, spikes terminal, raceme like, 5–15 cm long, slender;

spikelets 2 mm long, linear, shortly pedicelled, 1-flowered; caryopsis linear, terete.

Common in sandy tracts, scrubs and in open fields. Fl. & Fr.: July–Nov. *Vern.* Nakka thoka. Used as fodder. Nagarjuna University campus, *VRK* 6983; Bapatla, *N.K. Diwakar* 96015 (MH).

Phragmites Adanson

Phragmites karka (Retz.) Trin. ex Steud. Nom. Bot. (ed. 2) 2: 324. 1841; FBI 7: 304; Gamble 1807 (1251); Bor 416. *Arundo karka* Retz. Obs. Bot. 4: 21. 1786–87.

Erect marshy shrubs, up to 4 m tall; leafsheath ciliate; ligule shortly hairy; blades linear to narrowly lanceolate, 60 × 2 cm, flat, glabrous above, scabrid below; panicles 60 × 10 cm; spikelets up to 1 cm long, 4–10-flowered; rachilla jointed; glumes persistent; lower glume 1-nerved, upper glume 3-nerved; stamens 3.

Abundant in marshy area in Ethipothala falls. Fl. & Fr.: Dec.–Apr. Its stems are made into pipes, and are also used for making wrighting and printing papers. Culms are also used for making chairs, baskets and mats. Ethipothala, *VRK* 3882.

Saccharum Linnaeus

Saccharum spontaneum L. Man. Pl. 183. 1771; FBI 7: 118; Gamble 1709 (1185); Bor 214.

Perennial, densely tufted herbs, up to 2 m tall; leaves narrow-linear, 20–40 × 0.2–0.5 cm; coriaceous, acuminate, ligule membranous; panicles conical or lanceolate, contracted, spikelets paired, one pedicellate, the other sessile, silky hairy or woolly.

Occasionally along streams. Fl. & Fr.: Sept.–Apr. *Vern.*: Kaki chettu. Used as sand-binder and also for paper pulp. Javuku, *VRK* 3573.

Sehima Forsskal

Sehima nervosum (Rottl. ex Willd.) Stapf in Oliver Fl. Trop. Africa 9: 36. 1917; Gamble 1724 (1195). *Andropogon nervosus* Rottl. in Willd. Ges. Naturf. Freunde Berlin Neue Schriten 4: 218. 1806. *Ischaemum laxum* R.Br. Prodr. 205. 1810; FBI 7: 136.

Perennial, densely tufted, erect herbs; culms up to 90 cm tall; nodes glabrous; leaves linear, 5–20 × 0.3–0.5 cm, apex setaceous; sheaths light, terete, up to 7 cm long, sparsely hirsute; racemes solitary, spike like, up to 12 cm long; rachis jointed, fragile; spikelets 2-flowered, compressed, glumes subequal; caryopsis ellipsoid.

Occasional in plains and open forests. Fl. & Fr.: July–Dec. Considered to be good fooder grass. Also used for thatching. Kondaveedu Fort, *CAB* 4702 (MH).

Setaria P. Beauvois

1. Involucral bristles retrorsely barbellate .... ***S. verticillata***
1. Involucral bristles antrorsely barbellate:
   2. Leaves more or less plicate, culms robust .... ***S. palmifolia***
   2. Leaves flat, not plicate, culms slender:
      3. Cultivated, upper lemma faintly transversely rugose, upper glume equal to the length of upper lemma .... ***S. italica***
      3. Wild, upper lemma distinctly trasversely rugose, upper glume smaller than the upper lemma:
         4. Panicles not continuous, interrupted, involucral bristles 3–4, spikelets up to 2 mm long .... ***S. intermedia***
         4. Panicle continuous, involucral bristles 6–11, spikelets more than 2 mm long .... ***S. pumila***

Setaria intermedia (Roth) Roem & Schult. in L. Syst. Veg. 2: 489. 1817; FBI 7: 79; Gamble 1789 (1239); Bor Append. II 15. *Panicum intermedium* Roth Nov. Pl. Sp. 47. 1821. *Setaria tomentosa* (Roxb.) Kunth Rev. Gramin. 1: 47. 1829; Bor 365. *Panicum tomentosum* Roxb. Fl. Ind. 1: 303. 1820.

Annual herbs; culms up to 1.5 m tall; decumbent at base, nodes glabrous; leafsheaths loose, 3–8 cm, keeled; sparsely villous; blades 20–50 × 0.4–1 cm; flaccid, linear-lanceolate, finely acuminate, flat; panicle up to 10 cm long, narrowly pyramidal; spikelets broadly ovoid, rachis glabrous, grooved; caryopsis ellipsoid, pale brown.

Occasional in moist localities. Fl. & Fr.; July–Dec. Nagarjunakonda, *KTH* 9904 (CAL).

Setaria italica (L.) Beauv. Ess. Agrost. 51. 170. 178. 1812; FBI 7: 78; Gamble 1789 (1239); Bor 362. *Panicum italicum* L. Sp. Pl. 56. 1753.

An annual, cultivated, tufted grass, usually branched, rooting towards the base; culms up to 1 m tall; nodes glabrous, leaf sheath 3–10 cm long, ciliate on the margins, ligule membranous, blade 10–35 × 0.5–2 cm, linear, flat, base cordate, apex acuminate, minutely scaberulous; panicle, dense, spike like, up to 10 cm long, contracted; spikelets 2 mm, ovoid, compressed; stamens 3, stigmas plumose; caryopsis ellipsoid.

The Italian millet extensively cultivated as a food grain. Fl. & Fr.: Sept.–Nov. *Vern.*: Korralu. N.S. Canal sight, *VRK* 3531.

Setaria palmifolia (Koen.) Stapf in J. Linn. Soc. Bot. 42. 186. 1914; Gamble 1789 (1239); Bor 363. *Panicum palmaefolium* Koen. Natur. 23: 203. 1733. *P. plicatum* Willd. Enum. Hort. Berol. 1033. 1809. non Lam. 1791; FBI 7: 55.

Annual erect herbs, up to 1 m tall; culms stout, nodes glabrous, leaf blades lanceolate, 20–25 × 2–4 cm, flat, shortly pubescent; panicles pyramidal; spikelets ellipsoid or ovoid; lower glume ovate; upper glume acute, awns short, up to 0.5 cm; caryopsis ovoid.

Occasional weed in the moist localities. Fl. & Fr.: July–Dec. Grown for fodder. Javuku, *VRK* 3557.

Setaria pumila (Poir.) Roem. & Schult. in L. Syst. Veg. 2: 891. 1817. *Panicum pumilum* Poir. Encycl. (Suppl.) 4: 273. 1816. *Setaria pallidefusca* (Schum.) Stapf & C.E. Hubb. Bull. Misc. Inform. 1930: 259. 1930; Gamble 1789 (1239); Bor 363. *Panicum pallide-fuscum* Schum. Beskv. Guin. Pl. 58. 1827. *Setaria glauca* Hack. Bot. Soc. Brot. 3; 135. 1884. non Beauv. 1812; FBI 7: 78. *p.p.*

Annual, tufted, ascending herbs, up to 1 m tall; leaf blades linear, 5–15 × 0.4–0.8 cm, scabrid, flat, sheaths up to 7 cm long, ligule ciliate; panicles spiciform, oblong, 1–8 × 0.5–1 cm, compact, rachis hairy, spikelets greenish, ovoid or slightly elliptic; caryopsis ellipsoid.

Occasional weed in forests. Fl. & Fr.: Aug.–Apr. Papaiahpalem, *VRK* 5844; Nagarjuna University campus, *VRK* 6965.

Setaria verticillata (L.) P. Beauv. Agrost. 51. 178. 1812; FBI 7: 80; Gamble 1789 (1239); Bor 365. *Panicum verticillatum* L. Sp. Pl. ed. 2: 82. 1762.

Annual, erect or ascending tufted herbs; culms up to 1 m tall; nodes glabrous; leaves lanceolate; 4–12 × 0.4–2 cm flat; glabrescent—sparsely hairy, acuminate; panicles spiciform, 3–10 cm long; spikelets ellipsoid-ovoid, 1.5–2 mm, glabrous; caryopsis ellipsoid.

Occasional in moist localities. Fl. & Fr.: July–Dec. *Vern.*: Chiklenta. Grains are used as food. B.K. Palem, Macherla RF, *VRK* 3255; Krishna river at V.P. South, *KMS* 9805 (MH).

Spinifex Linnaeus

Spinifex littoreus (Burm. f.) Merr. in Philipp. J. Sci. 7. 229. 1912; Gamble 1706 (1183); Bor 366. *Stipa littorea* Burm. f. Fl. Ind. 29. 1768. *Spinifex squarrosus* L. Mant. Pl. 300. 1771; FBI 7: 63.

Armed perennial herbs with woody culms, creeping at base; nodes glabrous; leaf sheaths up to 3.5 cm; ligule setose, tufted; blades oblong, 5–20 cm, convolute, thick, stiff-scabrid along margins; inflorescence globose, dioecious; male inflorescence 8–15 × 7–14 cm; bract lanceolate, 2–4.5 cm; spikelet ovoid; glumes coriaceous; bisexual inflorescence stiff-bristly, 30 × 15 cm; spikelets oblong, to 1 cm; caryopsis ellipsoidal.

Common coastal sand binder throughout the beach. Fl. & Fr.: Aug.–Mar. *Vern.*: Ravanasurudi Meesalu. Nizampatnam beach, *VRK* 6805.

Sporobolus R. Brown

1. Panicles effuse, branches spreading .... *S. coromandelianus*
1. Panicles contracted, narrow, branches appressed to the rachis .... *S. tremulus*

Sporobolus coromandelianus (Retz.) Kunth Rev. Gramin. 1: 68. 1829; FBI 7: 252; Gamble 1817 (1258); Bor 627. *Agrostis coromandeliana* Retz. Obs. Bot. 4: 19. 1786.

Annual, tufted, spreading herbs; culms up to 20 cm long, leafy at base; leaves linear, 1–4 × 0.2–0.4 cm, flat, base cordate, spinulously toothed; sheaths 0.8–2 cm long; panicles 3–5 cm long; ovate or pyramidal, effuse; spikelets 1–1.5 mm, pedicelled; caryopsis ellipsoid, compressed.

Rare on hard surfaced soil. Fl. & Fr.: Sept.–Dec. Phirangipuram, *CAB* 4619 (MH).

Sporobolus tremulus (Willd.) Kunth Rev. Gramin. 1: 67. 1829; FBI 7: 250; Gamble 1817 (1258); Bor 633. *Agrostis tremula* Willd. Sp. Pl. 1: 372. 1797.

Annual tufted prostrate herbs; culms up to 30 cm long; from hard stoloniferous stock; leaves filiform, 0.5–2 cm long, involute, narrowed from the base; leafsheaths 1–2 cm long, glabrous; panicles subspiciform, 1.5–4 cm long; spikelets 1–2.5 mm, ellipsoidal, articulate on short pedicels; caryopsis ellipsoid.

Occasional grass in fallow fields. Fl. & Fr.: July–Dec. Common pasture and fodder grass, and also serves as sand binder. Narsaraopet, *CAB* 4721 (MH).

Trachys Persoon

Trachys muricata (L.) Pers. ex Trin. in Mem. Acad. Imp. Sci. St. Petersb. Ser. 6. Sci. Math. 3.189. 1834; Gamble 1813 (1255); Bor 369. *Cenchrus muricatus* L. Mant. Pl. 302. 1771. *Trachys mucronata* Pers. Syn. Pl. 1: 85. 1805; FBI 7: 96.

Annual, ascending herbs; leaf blades lanceolate, 10–18 × 0.4–1 cm, villous, hairs up to 2 mm long, acuminate; sheaths up to 5 cm long; ligule thin, membranous; spikes paired, terminal, 3–6 × 0.3–1 cm, spikelets inserted in joints, greenish or brownish green; caryopsis ovoid.

Occasional weed of cultivated fields; periphery of coastal sands. Fl. & Fr.: Sept.–Feb. *Vern.*: Pappu gaddi. V.P. South, *VRK* 5794.

Ecological indicator of typical coast land and sand dunes.

Tragus Haller *nom. cons.*

Tragus roxburghii Panigr. in Kew Bull. 29: 495. 1974. *Lappago biflora* sensu Roxb. Fl. Ind. 1: 284. 1820. *Tragus biflorus* Sensu Schult. Mant.

2: 205. 1824; Gamble 1813 (1255); Bor 682. *T. racemosus* auct. non Scop. 1977; FBI 7: 97.

Annual, diffuse ascending tufted herbs; leaf blades lanceolate, 2–7 × 0.2–0.4 cm, glabrous, margins hairy, acute, sheaths up to 1.2 cm long, ligule ciliate; panicle cylindric, terminal, spiciform, spikelets lanceolate, paired, facing each other and appearing like a single spikelet, prickly, sticky; caryopsis ellipsoid.

Occasional as forest undergrowth in moist localities. Fl. & Fr.: June–Dec. N. Varam, *VRK* 3508.

Tetrapogon Desfontaines

1. Spikes solitary, spikelets large up to 6 mm long .... *T. tenellus*
1. Spikes 1–3, spikelets up to 3 mm long .... *T. villosus*

Tetrapogon tenellus (Roxb.) Chiov. in Ann. Ist. Bot. Roma 8: 352. 1908; Bor 475. *Chloris tenella* Roxb. Fl. Ind. 1: 230. 1820; FBI 7: 290; Gamble 1837 (1271).

Annual tufted erect or ascending herbs, up to 60 cm tall; stems weak; leaf blades linear, 8–15 × 0.2–0.4 cm, glabrous, acuminate, sheaths up to 7 cm long, ligule membranous, spike solitary, terminal; spikelets yellowish green, cuneate, bifarious, awns 3, up to 1 cm long; caryopsis obovoid.

Occasional in waste places. Fl. & Fr.; Sept.–Dec. *Vern.*:Antrinta gaddi. Mellavagu, *VRK* 5824; V.P. South, *VRK* 5778.

Tetrapogon villosus Desf. Fl. Atlant. 2: 389. t. 255. 1799; Bor 475. *Chloris villosa* Pers. Syn. Pl. 1: 87. 1805; FBI 7: 291.

Annual tufted erect herbs; culms up to 30 cm tall, geniculate; leaves 4.5–7 × 0.2–0.5 cm, ovate-lanceolate, flat, sparsely hairy; spikes 2–3.5 m long, solitary; spikelets 0.4–0.5 cm long, cuneate, bifarious, 3-awned; caryopsis ellipsoid, compressed.

Occasional in plains and rare in forests. Fl. & Fr.: July–Dec. Macherla, *P.C. Nanda* 137 (DD).

Urochloa P. Beauvois

Urochloa panicoides P. Beauv. Agrost. 53. t. 11. f. 1. 1812; Gamble 1775 (1230); Bor 372. *Panicum javanicum* Poir. in Lam. Encycl. Suppl. 4: 274. 1816; FBI 7: 35 *p.p.*

Annual tufted erect or ascending herbs, culms up to 60 cm long; nodes pubescent; leaves ovate-lanceolate, 2–12 × 0.6–1.2 cm; finely hirsute, wavy margin; sheaths loose, tuberculate hairy; racemes spiciform, flexuous, 2–5 cm long; spikelets solitary, ovate-elliptic, glabrous; caryopsis ellipsoid, compressed.

Occasional on the bunds of cultivated fields. Fl. & Fr.: Aug.–Oct. Used as fodder. Grains are also edible. Phirangipuram, *CAB* 4615 (MH).

Vetiveria Bory

Vetiveria zizanioides (L.) Nash in Small. Fl. S.E.U.S. 67. 1903; Gamble 1733 (1201); Bor 258. *Phalaris zizanioides* L. Mant. Pl. 2: 183. 1771. *Andropogon squarrosus* auct. non L. f. FBI 7: 186.

Perennial robust tufted herbs; culms up to 2 m tall; rootstock branching with spongy aromatic roots; nodes glabrous; leaves linear, 15–30 × 0.4–1 cm; rigid, flat, margins spinulously rough; sheaths compressed, 8–14 mm long, glabrous; panicles 15–30 cm long; spikelets 2-nate, sessile and pedicellate; on 6 cm long racemes; caryopsis ellipsoid, compressed.

Occasional in moist places and lower hills. Fl. & Fr.: Aug.–Oct. *Vern.*: Vattiveru. Dried roots are made into aromatic scented mats, fans and ornamental baskets, and are also woven into curtains which when moistened, cool and scent the air. An essential oil, obtained from roots, is used in perfumery, cosmetics, soaps and for flavouring sherbets. Oil is stimulant, diaphoretic and refrigerent. Dried roots are employed to scent linen clothes. Grass is also used in manufacture of paper and board. A paste of root is applied to swellings.Reported in Guntur Forest Division work Plan.

Oryza sativa L., *Pennisetum typhoides* Stapf. *Sorghum bicolor* (L.) Moench. and *Zea mays* L. are cultivated crops throughout the district.

FERNS AND FERN ALLIES

MARSILIACEAE

Marsilia Linnaeus

Marsilia quadrifolia L. Sp. Pl. 1753; Gupta Marsilia Monogr. 22. 1962.

Rhizomatous aquatic slender herbs, rooting at nodes; leaves 4-lobed; leaflets obliquely obovate, 1–2 × 0.5–1 cm, glabrous, entire, base narrowed; sporocarps brownish, on long stalks.

Occasional in marshy places. Stalks and leaves of plants are eaten as pot-herb. Javuku, *VRK* 3559.

PARKERIACEAE

Ceratopteris Brongniart

Ceratopteris thalictroides (L.) Brongn. in Bull. Sci. Soc. Philom. paris 1860. 1821; Bedd. Handb. Ferns. Brit. Ind. 123. 1883; Bedd. Fern.

South India 6. 75. 1875. *Acrostichum thalictroides* L. Sp. Pl. 2: 1070. 1753.

Erect acquatic herbs, up to 1 m tall; fronds dimorphic, succulent in texture, 2–3-pinnatifid, sterile ones narrowly oblong, pinnae usually alternate, rarely opposite, fertile ones much taller than sterile; sporangia borne in a single line along marginal veins.

Koppanur forest streams, rare. Koppanur, *VRK* 3896.

PTERIDACEAE

Pteris Linnaeus

Pteris vittata L. Sp. Pl. 2: 1074. 1753. *P. longifolia* sensu Bedd. Handb. Fern. Brit. Ind. 106. f. 55. 1883.

Rhizomatous ascending herbs, up to 1 m tall; rhizomes densely scaly; frond 1-pinnate, 25–75 × 10–20 cm, pinnae sessile, oblique, 4–8 × 0.3–0.8 cm, sori brownish, sub-marginal, continuous along the margins.

Occasional in shady moist areas; along the stream in Javuku. Javuku, *VRK* 3877.

POLYPODIACEAE

Actinopteris Link

Actinopteris radiata (Sw.) Link, Fl. Sp. Hort. Reg. Bot. Beroli Cult., 80. 1841; Bedd. Handb. Fern. Brit. Ind. 197. 1883; Bedd. Fern. South India t. 124. 1865. *Asplenium radiatum* Sw. in Schrad. Journ. Bot. 2: 50. 1801.

Erect rhizomatous herbs, up to 30 cm tall; rhizomes scaly; fronds tufted, desiccated, fan like, lobes 1–3 × 0.2 cm, segments of fertile fronds longer than sterile one; sori in sub-marginal lines, indusium membranous.

Occasional in hill slopes. *Vern.*: Mayurasikhi. Branded as one of the species with 'Sanjeevani' properties. The root powder is mixed with cow milk, about 5 teaspoons of powder suspension once a day, are given orally for about 42 days early in the morning to promote fertility. The leaves are pounded with black pepper, garlic, castor oil and cock blood made into a paste, applied on externally to treat rheumatism. Loyapalle, *VRK* 3843; Jayanthipuram, *VRK* 6968.

SALVINIACEAE

Salvinia Seguir

Salvinia molesta Mitch., Br. Fern. Gaz. 10: 251 (1973); Dixit, Cens. Ind. Pterid. 174. 1964.

Aquatic free floating plants; stem spongy, terete, brown, branched with nodes and internodes, submerged leaf modified into root-like organ covered by brown, septate hairs; normal leaves opposite, erect, floating, sessile, oblong, pale green, lower surface glabrous, upper surface with dense hairs borne on the internal areas; sporocarps borne in clusters on submerged leaves, about 10 in number.

Common in tanks, ponds, border of rivers and streams.

AZOLLACEAE

Azolla Lamarck

Azolla pinnata R. Br. Prodr. Fl. Nov. Holt. 167 (1810); Holttum, Fl. Mal. 2: 621 (1954); De Vol, Fl. Taiwan 1: 60 (ed. 2); Dixit, Cens. Ind. Pterid. 174 (1984).

Stem horizontal, profusely branched, zig-zag, bearing roots which are covered by dense hairs; leaves alternate, arise from dorsal lobes, aerial, more or less rectangular, sessile, grey green, thick; ventral lobe submerged, broadly ovate, base cuneate, margin entire, brownish, glabrous.

Common in canals of Krishna river, irrigation channels, water tanks, reservoirs. Biofertilizer, purposely introduced into rice fields as a biofertilizer.

BIBLIOGRAPHY

Aiton, W. 1789. *Hortus Kewensis*. (ed. 1). London.

Aiton, W.T. 1810–1813. *Hortus Kewensis* (ed. 2). London.

Ali, S.I. 1967. The genus *Goniogyna* DC. *Taxon* **16**: 463–464.

Ali, S.I. 1968. Revision of the genus *Indigofera* L. from West Pakistan and Nor West Himalayas. *Bot. Notiser*, **3**: 543–577.

Ananda Rao, T and A.R.K. Sastry. 1970. New distributional records for coastal plar from Andhra Pradesh, *J. Bombay Nat. Hist. Soc.* **67**: 614–615.

Ananda Rao, T and A.R.K. Sastry. 1974. An outline of the coastal vegetation of Indi *Bull. Bot. Surv. India*, **16**: 101–105.

Anderson, T. 1867. An enumeration of Indian species of Acanthaceae. *J. Linn. Soc Bot.* **9**: 425–526.

Asha Kumari, J. 1989. Preliminary observations on some ecological parameters in Energy plantations and their weeds in harnessing wastelands of Nagarjunanagar. M.Phil. Thesis, Nagarjuna University.

Ashwini Kumar. 1987. Petrocarp Resources of Rajasthan. In: Proc. Bio-Energy So . Fourth Convention and Symposium, Udaipur, 98–102.

Austin, D.F. 1977. *Ipomoea carnea* Jacq. Vs. *Ipomoea fistulosa* Mart. ex Choisy. *Taxon* **26**: 235–238.

Backeb, C.V. 1960. *Die Cactaceae*. Band IV. Jena.

Backer, C.A. and R.C. Bakhuizen van der Brink. 1963–68. *Flora of Java*. 3 Vols. Netherlands.

Banerjee, L.K., A.R.K. Sastry and M.P. Nayar. 1989. *Mangroves in India*, BSI, Calcutta.

Banerjee, S.P. 1962. Identity of *Trichodesma amplexicaule* DC. (non Roth) (Boraginaceae). *Bull. Bot. Surv. India*, **16**: 10.

Beauvois, A.M.F.J.P. de 1812. *Essai d'une nouvella Agrostographie*, Paris.

Beddome, R.H. 1868–74. *Icones plantarum Indiae Orientalis*. Madras.

Beddome, R.H. 1869–74. *The Flora Sylvatica for Southern India*. 2 Vols. Madras.

Beddome, R.H. 1883. *Handbook of the Ferns of British India, Ceylon and the Malay Peninsula*. Calcutta.

Benerji, R.G. Misra and S.K. Nigam, 1983. Non-edible oil-seeds—Potential Raw Material for Industry—*J. Scient. Ind. Res.* 142: 686–93.

Bennet., S.S.R. 1986. Name changes in flowering plants of India and adjoining regions. Dehra Dun.

Bentham, G. 1835. *Scrophularineae Indicae, A synopsis of the Eastern Indian Scrophulariaceae*. London.

Bentham, G. and J.D. Hooker. 1862–1883. *Genera Plantarum*. 3 Vols. London.

Bhandari, M.M. 1964. *Artabotrys hexapetalus*: Correct name for *A. odoratissimus* R. Br. *Baileya* **12**: 147–150.

Bhandari, M.M. 1978. *Flora of the Indian Desert*. Jodhpur.

Bhatia, V.K. 1984. Petrocarps availability and scope for utilization as a source of hydrocarbons. In: Proc. Bio-Energy Society. First Convention and Symposium, New Delhi, 118–121.

Bhirava Murthy, P.V. and R. Krishna Mohan. 1991. A contribution to the Flora of Prakasam district, Andhra Pradesh. *J. Econ. Tax. Bot.* **15**: 49–63.

Bibi Aasiya. 1987. *Flora of Guntur*. M.Phil. dissertation, Nagarjuna University.

Biswas, K. and C. Calder. 1937. *Handbook of common water and marshy plants of India and Burma.* Delhi (ed. 2. 1955, Calcutta).

Blatter, E. 1939, Revision of the Flora of Bombay Presidency. *J. Bombay Nat. Hist. Soc.* **37**: 15–35, 265–277, 764–779.

Blatter, E. and C. McCann. 1935. *The Bombay grasses*. Delhi.

Blatter, E. and W.S. Millard. 1954. *Some Beautiful Indian Trees.* Bombay.

Blume, C.L. Vonb. 1835–1849. *Rumphia sive Commentationes botanicae imprimis de plantis Indiae Orientalis*. Leiden-Batavorum.

Bogle, A.L. 1970. The genera of Molluginaceae and Aizoaceae in the South-eastern United States. *J. Arnold Arbor.* **51**: 431–462.

Bor, N.L. 1960. *The grasses of Burma, Ceylon, India and Pakistan (excluding Bambuseae)*. London (ed. 2. 1973).

Borssum Walkes, J.V. 1966. Malesian Malvaceae revised. *Blumea* **14**: 1–213.

Brandis, D. 1906. *Indian Trees*. New Delhi.

Bremekamp, C.E.B. 1944. Materials for a Monograph of the Strobilanthinae (Acanthaceae). *Verh. Nederl. Akad. Wet.* **41**: 1–306.

Buchanan-Hamilton, F. 1807. *A Journey from Madras through the countries of Mysore, Canara, Malbar.* 3 Vols. London.

Burman, N.L. 1768. *Flora indica*. Leiden. Batavorum.

Calvin, M. 1979. *Bioscience*, **29**: 5333–38.

Candolle, A.L.P.P. de and A.C.P. de Candolle. 1878–1896. *Monographiae phanerogamarum*, 9 Vols. Paris.

Candolle, A.P. de 1824–1874. *Prodromus systematics Naturalis Regni vegetabilis*. 17 Vols. and Index. Paris. (Vols. 8–17 by A.L.P. de Candolle).

Chakravarthy, H.C. 1959. Monograph of the Indian Cucurbitaceae. *Rec. Bot. surv. India* **17**: 1–234.

Chakravarthy, H.C. 1982. Cucurbitaceae; *Fascicles of Flora of India*. No. 11. Howrah.

Champion, H.G. and S.K. Seth. 1968. *A revised survey of the forest types of India*. New Delhi.

Chandrabose, M. and N.C. Nair. 1981. The genus *Polygala* L. (Polygalaceae) in Andhra Pradesh, Kerala and Tamil Nadu (South India). *Proc. Indian Acad. Sci.* (Plant Sci.) **90**: 107–127.

Clarke, C.B. 1876. *Compositae Indica descriptae Secus genera Benthamii ordinatae*. Calcutta.

Cook, T. 1901–1908. *The Flora of the Presidency of Bombay.* 3 Vols. London. (Reprinted ed. 1958. BSI Calcutta).

Cronquist, A. 1981. *An Integrated System of Classification of Flowering plants.* Boston.

Delile, A.R. 1813. *Description de l'Egypte Historoire naturelle.* 2 Vols. Paris.

Desfontaines, R.L. 1798–1799. *Flora Atlantica* etc. 2 Vols. and Atlas. Paris.

Ding Hou, L. 1962. Celastraceae-I; in *Fl. Males.* Ser. 1. **6**: 227–291.

Don, D. 1825. *Prodromus florae Nepalensis*. London.

Don, G. 1831–1838. *A General History of dichlamydeous plants.* London.

Elliot, W. 1859. *Flora Andhrica.* Madras.

Ellis, J.L. 1987. *Flora of Nallamalais, A.P. on the Eastern Ghats of India*, Vol. 1, Calcutta.

Endlicher, S.L. 1836–1840. *Genera plantarum.* Vienna.

Engler, H.G.A. 1900–1937. *Das Pflanzenreich Regni vegetabilis Conspectus*. Leipzig.

Engler, H.G.A. and K.A.E. Prantl. 1887–1915. *Die Naturalichen Pflanzen familien*. etc. 23 Vols. ed. 1. Leipzig.

Forsskal, P. 1775. *Flora Aegyptiaco-Arabica.* Copenhagen.

Forster, J.R. and J.G.A. Foster. 1775. *Characters Generum Plantarum* etc. London.

Fyson, P.F. 1931. *The Flora of South Indian Hill Stations.* Madras.

Gaertner, J. 1788–1807. *De fructibus et Siminibus Plantarum.* 3 Vols. Stuttgart.

Gamble, J.S. and C.E.C. Fischer. 1915–1935. *Flora of the Presidency of Madras* (repr. ed. 1957. Calcutta).

Gmelin, J.F. 1791–1792. *Linnaeus Systema Naturae.* ed. 13. Leipzig.

Gray, A. 1848. *A Manual of Botany of the northern United States.* Boston and Cambridge (ed. 2. New York).

Greuter, W. *et al.* 1988. *International code of Botanical Nomenclature.* Konigstein (Germany).

Hackel, E. 1889. *Monographia Andropogonearum.* Paris.

Haines, H.M. 1921–25. *The Botany of Bihar and Orissa.* 3 Vols. (repr. ed. 1961, Calcutta).

Harley, R.M. 1983. *Becium* Lindl., a new genus of Labiatae, new to India and Sri Lanka. *Kew Bull.* **38**: 56.

Haworth, A.H. 1812. *Synopsis Plantarum Succulentarum.* London.

Heine, H. 1962–63. Trophical African Plants, XXVI. Some West African Acanthaceae. *Kew Bull.* **16**: 173–178.

Heipko, P. 1984. Opiliaceae; in *Fl. Males.* Ser. 1. **10**: 31–52.

Herny, A.N., K. Vivekananthan and N.C. Nair, 1978. Rare and threatened flowering plants of South India. *J. Bombay Nat. Hist. Soc.* **75**: 684–697.

Hooker, J.D and T. Thomson 1855. *Flora Indica.* London.

Hooker, J.D. *et al.* 1872–1897. *Flora of British India.* 7 Vols. London.

Hooker, W.J. and G.A.W. Arnott. 1830–1841. *The Botany of Captain Beecheys Voyage.* London.

Huber, H. 1973. In Abeywickrama, B.A. *et al. A revised handbook of Flora of Ceylon.* Vol.1. Part I. University of Ceylon, Peradenia.

Humboldt, F.H.A., A.J.A. Bonpland and C.S. Kunth. 1816–1825. *Nova Genera et species Plantarum* etc. Paris.

Jacobs, M. 1960. Capparidaceae; in *Fl. Males. Ser.* 1.**6**: 61–226.

Jacquin, N.J. 1781–1795. *Icones plantarum rariorum.* Vindobonae.

Jacquin, N.J. 1787–1797. *Collectanea ad Botanicam, Chemiam et Historium naturalem Spectantia.* 5 Vols. Vindobonae.

Jain, S.K. 1978. Welcome address. *125th Anniversary Souvenir of Madras Herbarium.* Bot. Surv. India, Southern Circle, Coimbatore.pp. 1–2.

Jain, S.K. 1982. *Flora of India-Tasks before 2000 AD.* Panchanam Maheswari Memorial Lecture, Indian Botanical Society.

Jain, S.K. and R.R. Rao 1977. *Field and Herbarium methods.* New Delhi.

Jeffrey, C. 1980. Further notes on Cucurbitaceae of the Indian Subcontinent. *Kew Bull.* **34**: 789–810.

Jussieu, A.L. de. 1789. *Genera plantarum Secundum Ordines Naturales Disposita.* (ed.2. 1791), Paris.

Karthikeyan, S., M.P. Nayar, and R.S. Raghavan. 1981. *An annotated Bibliography of Taxonomic Botany of Peninsular India 1959–1978.* Botanical Survey of India. Howrah.

Kern, J.H. 1974. Cyperaceae; in *Fl. Males. Ser.1.* **7**: 510–742.

Koyama T, 1976. Cyperaceae. J. Jap. bot. **51**: 313.

Koyama T, 1977. Cyperaceae. *Gard. Bull. Staraits Settlem.* **30**: 150–163.

Krishnamohan, P. 1986. *Flora of Prakasam District.* Ph.D.Thesis, Andhra University, Visakapatnam.

Kunth, C.S. 1833–1850. *Eumeratio plantarum omnium huccusque congitarûm.* Stuttgart.

Kuntze, O. 1891–1898. *Revisio Generum Plantarum Vascularium Omnium.* Leipzig.

Lakshma Reddy, D. 1985. *Ecofloristic studies on Nizampatnam Mangals.* Ph.D. Thesis. Nagarjuna University, Guntur.

Lakshminarayana, K., P. Venkanna and T. Pullaiah. 1997. *Flora of Krishna District, Andhra Pradesh, India.* M.D. Publications, New Delhi.

Lamarck, J.B.A. P.M. de. 1783–1817. *Encyclopedie methodique; Botanique.* 8 vols. and 5 suppl., Paris. (Vol. 5–13 by J.L.M. Poiret).

Lamarck, J.B.A. P.M. de. 1791–1823. *Tableau Encyclopedique et Methodique.* 3 Vols. and Suppl. Paris (Vol. 2, Part 5, 2 and Vol. 3 by J.L.M. Poiret).

Lawrence, G.H.M. 1951. *Taxonomy of vascular plants.* New York.

Lewis, W.H. 1961. Merger of North American *Houstonia* and *Oldenlandia* under *Hedyotis. Rhodora* **63**: 216–223.

Linnaeus, C. 1753. *Species Plantarum.* 2 Vols. Stockholm.

Linnaeus, C. 1755–56. *Centuria Plantarum.* 2 Vols. Stockholm.

Linnaeus, C. 1758–59. *Systema Naturae.* ed. 10. Vol. 2, Botany, Holmie.

Linnaeus, C. 1767. *Mantissa Plantarum.* Pages 1–142. Holmie.

Linnaeus, C. 1771. *Mantissa Plantarum Altera.* Pages 143–587. Holmie.

Linnaeus, f. 1781. *Supplementum Plantarum Systematis Vegetabilium.* etc. Brunsvigae.

Lushington, A.W. 1902. Identification of Loranthaceae by their leaves. *Indian For.* **28**: 58–66.

Lushington, A.W. 1915. *Vernacular list of trees, shrubs and woody climbers in the Madras Presidency.* Madras.

Manilal, K.S. 1988. *Flora of Silent valley.* Calicut.

Mary, T.N., M.R.P. Satyakeerthi and N. Hanumantha Rao. 1996. Biosystematic Survey of *Dactyloctenium aegyptium* complex from Andhra Pradesh II. *Dactyloctenium aristatum* Link Geobios New Reports **15**: 151–152.

Mathew, K.M. 1981–1983. *The Flora of Tamilnadu Carnatic.* 3 vols. Tiruchirapalli.

Mathew, K.M. 1982. *Illustrations on the Flora of Tamilnadu Carnatic.* Tiruchirapalli.

Mathew, K.M. 1988. *Further illustrations on the Flora of the Tamilnadu Carnatic.* Tiruchirapalli.

Mears, J.A. 1980. The Linnaean species of *Gomphrena* L. (Amaranthaceae). *Taxon* **29**: 85–89.

Merril, E.D. 1917. *An interpretation of Rumphius Herbarium Amboinense. Bur. Sci. Publ. Manila* **9**: 1–595.

Michaux, A. 1803. *Flora Boreali-Americana.* 2 Vols. Paris.

Miller, P. 1768. *The Gardeners Dictionary* (ed. 8). London.

Mitra, R.L. and S.K. Jain. 1985. Concept of *Phyllanthus niruri* (Euphorbiaceae) in Indian Floras. *Bull. Bot. Surv. India* **27**: 161–176.

Mohana Rao, P.R. and Bibi Aasiya. 1995. Flora of Nagarjuna University Campus and its immediate surrounding environs. *J. Econ. Tax. Bot.* **19**: 141–155.

Naidu, B.A. 1953. A new species of *Sesamum* L. *J. Bombay Nat. Hist. Soc.* **51**: 697.

Nair, N.C. and A.N. Henry. 1983. *Flora of Tamil Nadu, India*, Vol. I and II. Coimbatore.

Nayar, M.P. and G.S. Giri. 1982. A synopsis of the genus *Tribulus* L. (Zygophyllaceae) in India. *Bull. Bot. Surv. India* **24**: 160–164.

Nayar, M.P., D.C.S. Raju and M. Ahmedullah. 1986. The dwindling flora of Andhra Pradesh. A call for conservation. *Indian J. For.* **9**: 283–286.

Nayar, M.P., M. Ahmad and D.C.S Raju. 1984. Endemic and Rare plants of Eastern Ghats. *Indian J. For.* **7**: 35–42.

Nees, C.G.D. 1841. *Florae Africae Australioris Illustrationes monographicae.* Glogaviae.

Nees, C.G.D. 1847. Acanthaceae in DC., Prodr. **11**: 46–519.

Nicolson, H. and M. Sivadasan. 1980. Identification of *Gomphrena hispida* L. with *Acrocephalus* (Lamiaceae). *Taxon* **29**: 324–325.

Ohwi, J. 1947. New or Noteworthy Grasses from Asia. *Bull. Tokyo Sci. Mus.* n. 18.

Panigrahi, G. 1977. Proposal to reject the name *Limonia acidissima* L. *nom. ambig. rejic.* in favour of *L. elephantum* (Corr.) *comb. nov. Taxon* **26**: 576–577.

Panigrahi, G. 1987. *Echites antidysenterica* (L.) Roxb. ex Fleming and *Holarrhena antidysenterica* (L.) Wall., Validly published synonyms of *Wallida antidysenterica. Taxon* **36**: 464–467.

Panigrahi, S. 1976. Nomenclatural notes on 2 species of *Rotala* L. (Lythraceae) from the Indian region. *Indian For.* **102**: 766–767.

Pennell, F.W. 1935. Scrophulariaceae of Eastern temperate North America. *Acad. Nat. Sci. Philad. Monogr.* **1**: 137–159.

Pennell, F.W. 1943. A second summary of the Scrophulariaceae of New Guinea. *J. Arnold Arbor.* **24**: 243–274.

Prain, D. 1963, *Bengal plants.* (repr. ed.). 2 vols. Calcutta.

Prain, L.C. 1909. *Botany in Imperial Gazetter of India.* Provinicial series, Hyderabad state. Calcutta.

Pullaiah, T. 1997. *Flora of Andhra Pradesh. Vol. III, Monocotyledons*, Scientific Publishers, Jodhpur.

Pullaiah, T. and D. Ali Moulali. 1997. *Flora of Andhra Pradesh, Vol. II. Rubiaciae—Urticaceae.* Scientific Publishers, Jodhpur.

Pullaiah, T. and E. Chennaiah. 1997. *Flora of Andhra Pradesh, Vol. I. Ranunculaceae—Alangiaceae.* Scientific Publishers, Jodhpur.

Pullaiah, T. and N. Yesoda. 1989. *Flora of Anantapur district.* Dehra Dun.

Pullaiah, T. and P. Surya Prakash Babu. 1998. *Flora of Andhra Pradesh, Vol. IV. Illustrations.* Scientific Publishers, Jodhpur.

Pullaiah, T. and P.V. Prasanna and G. Obulesu. 1992. *Flora of Adilabad district.* C.B.S. Publishers, New Delhi.

Pullaiah, T. and S. Sandhya Rani. 1999. *Trees of Andhra Pradesh, India.* Regency Publications, New Delhi.

Radford, A.E. *et al.* 1974. *Vascular Plant Systematics,* New York.

Raghava Rao, S.R. 1987. *Flora of Mahabubnagar district.* Ph.D. Thesis, Osmania University, Hyderabad.

Raghavaswamy, B.V. and P.N. Rao. 1987. Further additions to the Flora of Andhra Pradesh. *J. Econ. Tax. Bot.* **9**: 489–492.

Raina, A.K. 1984. *Jatropha curcas*—a fence against energy crisis. In: Proc. Bio-Energy Society First Convention and Symposium, New Delhi, 114–117.

Rajagopal, M.V. (ed.) 1973. *Guntur district;* in Andhra Pradesh District Gazetteers, Hyderabad.

Rajagopal, T. 1973. *Flora of Hyderabad including study of the foliar epidermal characters of species as an aid to taxonomy.* Ph.D. Thesis, Osmania University, Hyderabad.

Raju, D.C.S. 1966. Plants from Papi hill in Andhra Pradesh, *Indian For.* **92**: 482–492.

Raju, V.S. 1984. New angiospermous taxa described from the state of Andhra Pradesh after its formation. *J. Econ. Tax. Bot.* **5**: 469–472.

Rambabu, A.V.S., B.V. Prasad, D. Lakshmana Reddy, P.N. Rao and M. Balaparameswara Rao. 1984. The impact of a Fishing harbour under construction on the distribution and zonation of the Flora and the Molluscan Fauna of a Tropical Mangrove Ecosystem. *Wat. Sci. Tech.* **16**: 425–431.

Randeria, A.J. 1960. The compositae genus *Blumea*, a taxonomic revision. *Blumea* **10**: 173–317.

Rangarajan, T.V. 1983–84. *Guntur forest Division working Plan.* 2 Vols. Guntur.

Rao, P.N., V.S. Raju and R.S. Rao. 1983. Floristic Research work in Andhra Pradesh. *Bull. Bot. Surv. India* **23**: 90–95.

Rao, R.S. and S. Hara Sreeramulu. 1986. *Flora of Srikakulam district, Andhra Pradesh, India.* Meerut.

Rao, R.S. 1956. *Parthenium hysterophorus* L., A new record for India. *J. Bombay Nat. Hist. Soc.* **54**: 218–220.

Rao, R.S. and R.V. Kammathy. 1966. Notes on Indian Commelinaceae. V. *J. Linn. Soc.* (London) *Bot.* **59**: 305–308.

Rao, R.S. , P. Venkanna and T. Appi Reddy. 1986. *Flora of West Godavari district, Andhra Pradesh, India.* Meerut.

Reddy, A.S. 1985. Studies on biology of two Aquatic weeds. *Pistia stratiotes* L. an acropleustophyte and *Ammannia baccifera* L. a rhizophyte. Ph.D. Thesis, Nagarjuna University.

Retzius, A.J. 1779–1791. *Observations botanicae.* Leipzig.
Ridley, H.N. 1930. *The dispersal of plants throughout the world.* Kent.
Roemer, J.J. and J.A. Schultes. 1817–1830. *Linnaeus Systema Vegetabilium* ed. 15. 7 vols. Stuttgart.
Roth, A.W. 1821. *Novae plantarium species praesertum Indae orientalis ex collection Doct. B. Henu cum descriptionibum et observationibus.* Halberstadt.
Roth, A.W. 1827. *Enumeration Plantarum Phanerogamarum* etc. Leipzig.
Roxburgh, W. 1795–1819. *Plants of coast of Coromandel.* 3 vols. London.
Roxburgh, W. 1814. *Hortus benghalensis.* Serampore.
Roxburgh, W. 1832. *Flora indica.* 3 vols. Serampore.
Saldanha, C.J. 1963. The genus *Striga* Lour. in Western India. *Bull. Bot. Surv. India.* **5**: 67–70.
Saldanha, C.J. 1986. *Flora of Karnataka.* Vol. 1. New Delhi.
Saldanha, C.J. and D.H. Nicolson. 1976. *Flora of Hassan District.* New Delhi.
Santapau, H. 1954. Annual excursion of the Indian Botanical Society, Hyderabad, Deccan. *J. Ind. Bot. Soc.* **33**: 78–92.
Santapau, H. 1955. *Botanical collector's manual.* Calcutta.
Santapau, H. 1958. *Flora of Purandhar.* New Delhi.
Santapau, H. 1967. The Flora of Khandala on the Western ghats of India. (ed.3). *Rec. Bot. Surv. India.* **16**: 1–32.
Santapau, H. and S.K. Wagh. 1964. Nomenclatural changes in Indian plan's. *Bull. Bot. Surv. India* **5**:105–109.
Sastry, A.R.K. and T.A. Rao. 1973. Studies on the Flora and Vegetation of Coastal Andhra Pradesh. *Bull. Bot. Surv. India.* **15**: 92–107.
Saxena, H.O. and M. Brahman. 1994–95. *The Flora of Orissa.* Vol. 1, Orissa Forest Development Corporation, Bhubaneswar.
Sedgwick, L.J. 1919. On the *Alysicarpus rugosus* DC. and its allied species. *J. Ind. Bot. Soc* **1**: 14–18.
Seshagiri Rao, K. and M.N.V. Prasad. 1987. Petrocarp Resources of Andhra Pradesh—4 Preliminary Survey. In: Proc. Bio-Energy Society, Fourth Convention and Symposium, Udaipur, 103–105.
Shah, G.L. 1978. *Flora of Gujarat state.* 2 vols. Vallabh Vidhyanagar.
Sharma, B.D. and M. Sanjappa. 1993. *Flora of India*, Vol. 3 (Portulacaceae—Ixonanthaceae), Botanical Survey of India, Calcutta.
Sharma, B.D. and N.P. Balakrishnan. 1993. *Flora of India*, Vol. 2 (Papavaraceae—Caryophyllaceae), Botanical Survey of India, Calcutta.
Sharma, B.D., N.P. Balakrishnan, R.R. Rao and P.K. Hajra. 1993. *Flora of India*, Vol. 1 (Ranunculaceae—Barclayaceae), Botanical Survey of India, Calcutta.
Sivarajan, V.V. 1981. Taxonomic notes of the genus *Portuluca* L. in India. *J. Bombay Nat. Hist. Soc.* **78**: 256–260.
Sivarajan, V.V. and P. Mathew. 1983. The genus *Lindernia* All. (Scrophulariaceae) in India. *J. Bombay Nat. Hist. Soc.* **80**: 131–140.
Sprengel, C.P.J. 1824–1828. *Linnaeus Systema Vegetabilium.* ed. 16. Gottingae.
Sreemadhavan, C.P. 1968. *Indoneesiella* a substitute name in Acanthaceae. *Phytologia* **16**: 466.
Srivastava, G.S. and V.K. Bhatia, 1985. Potential petrocarps in India—A Critical Review. In: Proc. Bio-Energy Society, Second Convention and Symposium, Hyderabad, 119–124.
Srivastava, J.G. 1964. Some Tropical American Weeds that have invaded the state of Bihar. *J. Indian Bot. Soc.* **43**: 102–112.
Stapf, O. 1917–1930. Gramineae; in Prain, *Fl. Trop. Africa* 9:1–768.
Subba Rao, G.V. and T.R. Kumari. 1967. Contribution to the flora of Karimnagar district, A.P. *Bull. Bot. Surv. India* **9**: 95–113.
Subramanyam, K. 1962. *Aquatic Angiosperms.* New Delhi.

Tadulingam, M.C. and G. Venkata Narayana. 1932. *A handbook of some south Indian Weeds.* Madras.

Takhtajan, A. 1980. Outline of the classification of flowering plants (Mangnoliophyta). *Bot. Rev.* **46**: 225–259.

Thothathri, K. 1964. *Sida ovata* Forssk: A new Record for South India. Curr. Sci. **33**: 593–594.

Thothathri, K. 1964. Studies in Leguminosae-4. New variety of *Crotalaria paniculata* Willd. var. *nagarjunakondensis. Bull. Bort. Surv. India* **6**: 67–68.

Thothathri, K. 1964. The Flora of Nagarjunakonda valley and the surrounding hills. *Indian For.* **90**: 539–546.

Thunberg, 1794–1800. *Prodromus Plantarum capensium.* Upsala.

Thwaites, G.H.W. 1858–64. *Enumeratio Plantarum Zeylaniae.* London.

Tirvengadum, D.D. 1978 a. Reestablishment of the genus *Catunaregum* Wolf. (Rubiaceae). A contribution towards the plants described in Rheed's Hortus Malabaricus. *Taxon* **27**: 512–517.

Tirvengadum, D.D. 1978 b. A synopsis of the Rubiaceae-Gardenieae of Ceylon (Sri Lanka). *Bull. Mus. Hist. Nat.* (Paris) Ser. 3. **35**:3–33.

Tirvengadum, D.D.1983. New taxa and name changes in Tropical Asiatic Rubiaceae. *Nord. J. Bot.* **3**: 45–469.

Torrey, J. 1943. *A Flora of the state of New York.* 2 vols. Albany.

Trimen, H. 1894. *A handbook to the flora of Ceylon.* London.

Trimen, H. and J.D. Hooker, 1893–1900. *A handbook to the flora of Ceylon.* Vol. 6. (Suppl.) London (1931 ed by Alston).

Uhl, Lemon, L. 1929. *Guntur Flora.* Lutheran Press, Guntur.

Vahl, M. 1779–94. *Symbolae botanicae.* 9 vols. Copenhagen.

Van Steenis, C.G.G.J. (ed.) 1950–1984. *Flora Malesiana.* 10 vols.

Venkata Raju, R.R. and T. Pullaiah. 1995. *Flora of Kurnool.* Bishen singh Mahendra Pal Singh, Dehra Dun.

Verdcourt, B. 1976. Rubiaceae; in *Flora of Tropical East Africa:* Rubiaceae 1. p.p. 339–379 (ed. R.M. Polhill). London.

Voight, J.O. 1845. *Hortus Suburbanus Calcuttensis.* Calcutta.

Wagh, S.K. 1960. *Studies on the flora of Andhra State.* Ph.D. Thesis, Bombay University, Bombay.

Wallich, N. 1828–1849. *A Numerical list of Dried specimens of Plants in the East India Company's Museum.* London.

Wallich, N. 1829–32. *Plantae Asiaticae Rariores.* 3 vols. London.

Wang, S. and J.B. Huffmann, 1982. Botano-chemicals supplements to Petro-chemicals. *Econ. Bot.* **35**: 369–82.

Wight, R. 1834. *Contribution to the Botany of India.* London.

Wight, R. 1840–50. *Illustrations to Indian Botany.* London.

Wight, R. and G.A. Walker Arnott. 1834. *Prodromus Florae Peninsulae Indiae orientalis.* London.

Wight, R.1938–1853. *Icones Plantarum Orientalis* 6 vols. Madras.

Willdenow, C.L. 1797–1830. *Species Plantarum.* (ed.4 of L., Sp. Pl.) 4 Vols. Berlin.

Willdenow, C.L. 1809. *Enum. Hort. Berol. Berlin.*

Willis, J.C. 1973. *A dictionary of flowering plants and ferns.* (8th rev. ed. by H.K. Airy Shaw). Cambridge.

INDEX TO FAMILIES, GENERA AND SPECIES (ITALICS INDICATE SYNONYMY)

A

Q

T

U

V

Species marked with asterisk (*) are cultivated.